T0202103

OXFORD HANDBOOK OF
Medical Sciences

Published and forthcoming Oxford Medical Handbooks

Oxford Handbook for the Foundation Programme 5e

Oxford Handbook of Acute Medicine 4e

Oxford Handbook of Anaesthesia 4e

Oxford Handbook of Cardiology 2e

Oxford Handbook of Clinical and Healthcare Research

Oxford Handbook of Clinical and Laboratory Investigation 4e

Oxford Handbook of Clinical Dentistry 7e

Oxford Handbook of Clinical Diagnosis 3e

Oxford Handbook of Clinical Examination and Practical Skills 2e

Oxford Handbook of Clinical Haematology 4e

Oxford Handbook of Clinical Immunology and Allergy 4e

Oxford Handbook of Clinical Medicine—Mini Edition 10e

Oxford Handbook of Clinical Medicine 10e

Oxford Handbook of Clinical Pathology

Oxford Handbook of Clinical Pharmacy 3e

Oxford Handbook of Clinical Specialties 11e

Oxford Handbook of Clinical Surgery 4e

Oxford Handbook of Complementary Medicine

Oxford Handbook of Critical Care 3e

Oxford Handbook of Dental Patient Care

Oxford Handbook of Dialysis 4e

Oxford Handbook of Emergency Medicine 5e

Oxford Handbook of Endocrinology and Diabetes 3e

Oxford Handbook of ENT and Head and Neck Surgery 3e

Oxford Handbook of Epidemiology for Clinicians

Oxford Handbook of Expedition and Wilderness Medicine 2e

Oxford Handbook of Forensic Medicine

Oxford Handbook of Gastroenterology & Hepatology 2e

Oxford Handbook of General Practice 5e

Oxford Handbook of Genetics

Oxford Handbook of Genitourinary Medicine, HIV, and Sexual Health 3e

Oxford Handbook of Geriatric Medicine 3e

Oxford Handbook of Infectious Diseases and Microbiology 2e

Oxford Handbook of Integrated Dental Biosciences 2e

Oxford Handbook of Humanitarian Medicine

Oxford Handbook of Key Clinical Evidence 2e

Oxford Handbook of Medical Dermatology 2e

Oxford Handbook of Medical Imaging

Oxford Handbook of Medical Sciences 2e

Oxford Handbook for Medical School

Oxford Handbook of Medical Statistics

Oxford Handbook of Neonatology 2e

Oxford Handbook of Nephrology and Hypertension 2e

Oxford Handbook of Neurology 2e

Oxford Handbook of Nutrition and Dietetics 2e

Oxford Handbook of Obstetrics and Gynaecology 3e

Oxford Handbook of Occupational Health 2e

Oxford Handbook of Oncology 3e

Oxford Handbook of Operative Surgery 3e

Oxford Handbook of Ophthalmology 4e

Oxford Handbook of Oral and Maxillofacial Surgery 2e

Oxford Handbook of Orthopaedics and Trauma

Oxford Handbook of Paediatrics 2e

Oxford Handbook of Pain Management

Oxford Handbook of Palliative Care 3e

Oxford Handbook of Practical Drug Therapy 2e

Oxford Handbook of Pre-Hospital Care

Oxford Handbook of Psychiatry 4e

Oxford Handbook of Public Health Practice 3e

Oxford Handbook of Rehabilitation Medicine 3e

Oxford Handbook of Reproductive Medicine & Family Planning 2e

Oxford Handbook of Respiratory Medicine 3e

Oxford Handbook of Rheumatology 4e

Oxford Handbook of Sport and Exercise Medicine 2e

Handbook of Surgical Consent

Oxford Handbook of Tropical Medicine 4e

Oxford Handbook of Urology 4e

OXFORD HANDBOOK OF
Medical
Sciences

THIRD EDITION

EDITED BY

Prof Robert Wilkins

Associate Professor in Physiology, University of Oxford;
American Fellow in Physiology, St Edmund Hall, University of
Oxford, Oxford, UK

Prof Ian Megson

Head, Health Research & Innovation, University of the Highlands
and Islands, Inverness, UK

Dr David Meredith

Senior Lecturer in Biochemistry & Biomedical Science, Oxford
Brookes University, Oxford, UK

OXFORD
UNIVERSITY PRESS

OXFORD
UNIVERSITY PRESS

Great Clarendon Street, Oxford, OX2 6DP,
United Kingdom

Oxford University Press is a department of the University of Oxford.
It furthers the University's objective of excellence in research, scholarship,
and education by publishing worldwide. Oxford is a registered trade mark of
Oxford University Press in the UK and in certain other countries

© Oxford University Press 2021
except pp184–193 section of chapter 2 (© Keith Frayn)

The moral rights of the authors have been asserted

First Edition published in 2006
Second Edition published in 2011
Third Edition published in 2021
Impression: 1

Published in the United States of America by Oxford University Press
198 Madison Avenue, New York, NY 10016, United States of America

British Library Cataloguing in Publication Data
Data available

Library of Congress Control Number: 2020938135

ISBN 978–0–19–878989–5

DOI: 10.1093/med/9780198789895.001.0001

Printed and bound in China by
C&C Offset Printing Co., Ltd.

Preface to the first edition

This Handbook is written by Biomedical Scientists and Clinicians to disseminate the fundamental scientific principles that underpin clinical medicine. Frequent cross-referencing with the *Oxford Handbook of Clinical Medicine* helps to highlight the clinical relevance of specific issues. The challenge of this ambitious undertaking was to distil the contents of information found in several voluminous textbooks into a conveniently-sized handbook, without loss of important information. This has been achieved by sifting out the key facts that are crucial to understanding fundamental principles and presenting them as clear, brief statements or straightforward diagrams. The Handbook has been deliberately divided into sections that mirror modern medical teaching strategies; it builds from a clear, easily digestible account of basic cell physiology and biochemistry to an investigation of the traditional piers of medicine (anatomy, physiology, biochemistry, pathology, and pharmacology), intertwined in the context of each of the major systems relevant to the human body. Although primarily aimed at medical students, this Handbook will also provide a useful reference for science students studying any of the traditional biomedical sciences.

We hope you find this Handbook helpful. Feedback on errors and omissions would be much appreciated. Please post your comments via the OUP website: *www.oup.com/uk/medicine/handbooks*.

RW, SC, IM, and DM
2006

Preface to the second edition

The aim of this Handbook, written by Biomedical Scientists and Clinicians, remains unchanged from the first edition, that is to disseminate the fundamental scientific principles that underpin clinical medicine by distilling the content of several voluminous textbooks into a conveniently-sized handbook, without losing important information. While keeping the same overall format in this second edition, we have rewritten and updated sections to reflect recent advances, incorporated a chapter covering the place of medicine in society and reordered and reformatted material to provide more uniform and accessible coverage of topics.

In addition, updated and frequent cross-referencing to the *Oxford Handbook of Clinical Medicine*, 8e, the *Oxford Handbook of Clinical Specialties*, 8e, and the *Oxford Handbook of Practical Drug Therapy*, 2e helps to highlight the clinical significance of specific information. Although the Handbook is aimed at medical students, with its structure mirroring the modern medical teaching strategy, its content will also serve as a key reference for science students studying biomedical sciences courses.

We hope that you will find this second edition of the Handbook as useful as previous readers have found the first. Feedback on errors or omissions would be much appreciated. Please post your comments via the OUP website: *www.oup.com/uk/medicine/handbooks*.

RW, SC, IM and DM
2011

Preface to the third edition

As for its earlier editions, the aim of this Handbook remains the elucidation of the fundamental scientific principles that underpin modern medical science in a conveniently sized handbook format, covering core concepts without oversimplifying their treatment. While keeping the same overall format in this third edition, we have updated chapters to reflect recent advances, and included many new and full-colour figures to illustrate the text.

In addition, updated and frequent cross-referencing to the *Oxford Handbook of Clinical Medicine*, 10e, the *Oxford Handbook of Clinical Specialties*, 11e and the *Oxford Handbook of Practical Drug Therapy*, 2e helps to highlight the clinical significance of specific information. Although the Handbook will be of obvious utility to medical students, we hope that its content will also serve as a strong foundation for students studying the biomedical sciences. We hope that you will find this third edition of the Handbook at least as useful as previous readers have found the first two. Feedback on errors or omissions would be much appreciated. Please post your comments via the OUP website: *www.oup.com/uk/medicine/handbooks*.

RW, IM and DM
2020

Acknowledgements

The original edition of this book would not have been possible without the input of the additional contributors, who so readily provided material on subjects outside our fields. Their contributions remain an invaluable part of this third edition. In addition, we remain especially grateful for the contribution to the first two editions from Professor Simon Cross, who has stepped down from being an editor of *OHMS* to focus on other projects.

The assistance of Dr Judith Collier and Miriam Longmore with the cross-referencing of the contents of this book to the *Oxford Handbook of Clinical Medicine* is greatly appreciated. We again gratefully acknowledge that many figures in this book have been reproduced from the *Oxford Textbook of Functional Anatomy*, second edition (2005) by Pamela MacKinnon and John Morris (Oxford: Oxford University Press).

Finally, we thank the readers of the previous editions for their constructive feedback and colleagues at Oxford University Press for guiding us through the preparation of this edition of the Handbook.

Contents

Contributors to the first edition

Donald Bissett
Royal Infirmary,
Aberdeen

Garry Brown
Division of Life and Environmental
Sciences, University of Oxford

Jennifer Brown
Division of Medical Sciences,
University of Oxford

Joseph Browning
The John Radcliffe Hospital,
Oxford

Helen Christian
Division of Medical Sciences,
University of Oxford

David Dockerell
Department of Infection and
Immunity, University of Sheffield

Alexander Foulkes
King's College Hospital,
London

Keith Frayn
Division of Medical Sciences,
University of Oxford

Philip Larkman
Division of Neuroscience,
University of Edinburgh

Helen Marriott
Department of Infection and
Immunity, University of Sheffield

Rosie McTiernan
The John Radcliffe Hospital,
Oxford

Symbols and abbreviations

⮌	cross reference
→	leading to
∴	therefore
3D	three-dimensional
5-HT	5-hydroxytryptamine (serotonin)
ACE	angiotensin-converting enzyme
ACh	acetylcholine
AChE	acetylcholine esterase
ACTH	adrenocorticotropic hormone (corticotropin)
ADH	antidiuretic hormone (vasopressin)
ADP	adenosine diphosphate
AF	atrial fibrillation
AIDS	acquired immunodeficiency syndrome
ALH	ascending loop of Henle
AMP	adenosine monophosphate
AMPA	A-amino-3-hydroxy-5-methyl-4-isoxazole propionic acid
ANP	atrial natriuretic peptide
APC	antigen-presenting cell
AQP2	aquaporin II
ARDS	acute respiratory distress syndrome
AT	angiotensin
ATP	adenosine triphosphate
AV	atrioventricular
AVN	atrioventricular node
BAC	bacterial artificial chromosome
cAMP	cyclic adenosine monophosphate
CCK	cholecystokinin
CFTR	cystic fibrosis transmembrane conductance regulator
cGMP	cyclic guanosine monophosphate
CJD	Creutzfeldt–Jakob disease
CNS	central nervous system
CO	carbon monoxide
CoA	coenzyme A
COMA	Committee on Medical Aspects of Food Policy
COMT	catechol-O-methyl transferase
COPD	chronic obstructive pulmonary disorder
COX	cyclo-oxygenase
CPR	cardiopulmonary resuscitation
CRH	corticotropin-releasing hormone
CSF	cerebrospinal fluid
CTL	cytotoxic T-lymphocyte
CVS	cardiovascular system
DAG	diacylglycerol
DHAP	dihydroxyacetone phosphate
DLH	descending loop of Henle
DNA	deoxyribonucleic acid
DXA	dual-energy absorptiometry
2,3-DPG	2,3-diphosphoglycerate
ECF	extracellular fluid
ECG	electrocardiogram
ECV	effective circulating volume
EDHF	endothelium-derived hyperpolarizing factor
EDRF	endothelium-derived relaxing factor
EEG	electroencephalogram
E_m	membrane potential
ELISA	enzyme-linked immunosorbent assay
EMG	electromyography
ENaC	epithelial Na^+ channel
EOG	electro-oculography
EPMR	equipotent molar ratio
EPP	end plate potential
EPSP	excitatory postsynaptic potential
ER	endoplasmic reticulum
ET-1	endothelin 1
ETC	electron transport chain
FAD	flavine adenine dinucleotide
$FADH_2$	reduced flavine adenine dinucleotide
FSH	follicle-stimulating hormone
G protein	heterotrimeric GTP-binding protein
G-1-P	glucose-1-phosphate
G-6-P	glucose-6-phosphate

G6PD or G6PDH	glucose-6-phosphate dehydrogenase	LCAD	long-chain acyl-CoA dehydrogenase
GABA	γ-aminobutyric acid	LDH	lactate dehydrogenase
GAG	glycosaminoglycan	LDL	low-density lipoprotein
GALT	gut-associated lymphoid tissue	LGN	lateral geniculate nucleus
GAP	glyceraldehyde-3-phosphate	LH	luteinizing hormone
GCS	Glasgow Coma Scale	LHRH	luteinizing hormone-releasing hormone
GDP	guanosine diphosphate	LMWH	low-molecular-weight heparin
GFR	glomerular filtration rate	LPS	lipopolysaccharide
GH	growth hormone (somatotropin)	LTA	lipoteichoic acid
GHRH	growth hormone-releasing hormone	LTD	long-term depression
		LTP	long-term potentiation
GI	gastrointestinal	MALT	mucosa-associated lymphoid tissue
GIP	gastric inhibitory peptide	MAO	monoamine oxidase
GLUT	glucose transporter	MCAD	medium-chain acyl-CoA dehydrogenase
GM-CSF	granulocyte-macrophage colony stimulating factor	MCS	multiple cloning site
GMP	guanosine monophosphate	MHC	major histocompatibility complex
GnRH	gonadotropin-releasing hormone	MI	myocardial infarction
GP	glycoprotein	MIC	minimal inhibitory concentration
GR	glucocorticoid receptor	MMC	migratory motor complex
GRP	gastrin-releasing peptide	Mr	relative molecular mass
GSH	glutathione	mRNA	messenger RNA
GSK3	glycogen synthase kinase 3	MSH	melanophore-stimulating hormone
GTP	guanosine triphosphate		
Hb	haemoglobin	NAD	nicotinamide adenine dinucleotide
HbF	foetal haemoglobin	NADH	reduced nicotinamide adenine dinucleotide
HbS	sickle cell haemoglobin	NADP	nicotinamide adenine dinucleotide phosphate
hCG	human chorionic gonadotropin	NADPH	reduced nicotinamide adenine dinucleotide phosphate
HDL	high-density lipoprotein	NANC	non-adrenergic, non-cholinergic
HIV	human immunodeficiency virus		
HLA	human leucocyte antigen	NCC	Na^+-Cl^- co-transport
IBD	inflammatory bowel disease	NFκB	nuclear factor κB
IBS	irritable bowel syndrome	NK	natural killer (cell)
ICF	intracellular fluid	NKCC	Na^+-K^+-$2Cl^-$ co-transport
ICM	inner cell mass	NMDA	N-methyl-D-aspartate
IFN	interferon	NMJ	neuromuscular junction
Ig	immunoglobulin (e.g. IgA, IgE, IgG, IgM)	NO	nitric oxide
		NREM	non-rapid eye movement
IGF	insulin-like growth factor	NSAID	non-steroidal anti-inflammatory drug
IL	interleukin		
IMM	inner mitochondrial membrane	OHCM10	Oxford Handbook of Clinical Medicine, tenth edition
IP_3	inositol-1,4,5-triphosphate		
IPSP	inhibitory postsynaptic potential		
IVF	in vitro fertilization		
LAL	lysosomal acid lipase		

OHCS11	*Oxford Handbook of Clinical Specialties*, eleventh edition		rRNA	ribosomal RNA
OHPDT2	*Oxford Handbook of Practical Drug Therapy*, second edition		RT-PCR	reverse transcription polymerase chain reaction
ORF	open reading frame		RV	residual volume
p.d.	potential difference		SAN	sinoatrial node
PAH	p-aminohippurate		SCAD	short-chain acyl-CoA dehydrogenase
PAI	plasminogen activator inhibitor		SCN	suprachiasmatic nucleus
PAL	physical activity level		snRNA	small nuclear RNA
PCR	polymerase chain reaction		SOD	superoxide dismutase
PDGF	platelet-derived growth factor		SR	sarcoplasmic reticulum
PDH	pyruvate dehydrogenase		T_3	tri-iodothyronine
PDK	pyruvate dehydrogenase kinase		T_4	thyroxine
			TAF	TBP-associated factor
PEPCK	phosphoenolpyruvate carboxykinase		TALH	thick ascending loop of Henle
			TBP	TATA box-binding protein
PFK	phosphofructokinase		TCA	tricarboxylic acid
PG	prostaglandin (e.g. PGE_2)		TGF	transforming growth factor
PIP	phosphatidylinositol-4,5-bisphosphate		TGN	trans-Golgi network
			TLR	toll-like receptor
PK	protein kinase		TNF	tumour necrosis factor
PMF	proton motive force		t-PA	tissue plasminogen activator
PMN	polymorphonuclear leucocyte		TPP	thiamine pyrophosphate
PNS	peripheral nervous system		TRH	thyrotropin-releasing hormone
PP	protein phosphatase		tRNA	transfer RNA
PPP	pentose phosphate pathway		TSH	thyroid-stimulating hormone
PRL	prolactin		TXA_2	thromboxane A_2
PTH	parathyroid hormone		V	flow rate
PVN	paraventricular nucleus		V/Q	ventilation/perfusion ratio (in lungs)
RE	restriction enzyme			
REM	rapid eye movement		VIP	vasoactive intestinal peptide
RER	rough endoplasmic reticulum		VLCAD	very long-chain acyl-CoA dehydrogenase
Rh	rhesus factor (D antigen)			
RNA	ribonucleic acid		VLDL	very low-density lipoprotein
ROS	reactive oxygen species		VTA	ventral tegmental area
RPF	renal plasma flow		YAC	yeast artificial chromosome
RQ	respiratory quotient (respiratory exchange ratio)			

Cellular structure and function

General principles: overview

There are four major chemical components to biological life—carbon (C), hydrogen (H), oxygen (O), and nitrogen (N):

- O, C, and H form the bulk of the dry mass—65%, 18%, and 9% respectively of the human body (e.g. carbohydrates, simple lipids, hydrocarbons)
- N represents 4% and is an essential part of life (nucleotides, amino acids, amino sugars, complex lipids)
- They are supplemented by phosphorus (P) in the form of phosphate (PO_4^{3-}) and by small amounts of sulphur (S) in the amino acids methionine and cysteine, and in thioester bonds
- There are also a number of essential trace elements (e.g. Mn, Zn, Co, Cu, I, Cr, Se, Mo) that are essential co-factors for enzymes.

Note that all biological molecules (biomolecules) must obey the basic rules of chemistry!

One important such example for C atoms is stereochemistry:

- Three-dimensional (3D) structures in biomolecules are often key to their function
- Carbon atoms can have four different groups attached to them (i.e. be tetrahedral)
- Under these conditions, the C atom is said to be chiral—it has two mirror-image forms (D- and L-) that cannot be superimposed (Fig. 1.1)
- Nature has favoured certain stereoisomers:
  - For example, naturally occurring mammalian amino acids are always in the L-form, whereas carbohydrates are always in the D-form
  - The stereoisomers of therapeutic compounds may have very different effects:
    - For example, one stereoisomer of the drug thalidomide (a racemic mix), used briefly in the late 1950s to relieve symptoms of morning sickness in pregnancy, caused developmental defects in ~12,000 babies.

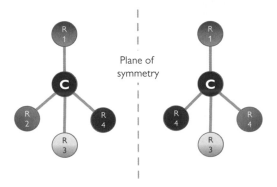

Fig. 1.1 Stereoisomers: the molecule on the left cannot be superimposed on the one on the right however it is rotated.

Biomolecules

Roughly speaking, biomolecules can be divided up into two groups: small and large (macromolecules).

- Small—relative molecular mass (Mr) <1000 (mostly <400):
  - Intermediates of metabolic pathways (metabolites), and/or
  - Components of larger molecules
- Macromolecules—Mr >1000, up to millions:
  - Usually comprised of small biomolecule building blocks, for example:
    - Proteins (made up of amino acids)
    - Polysaccharides (sugars)
    - DNA, RNA (nucleotides, themselves sugars, bases, and phosphate)
  - Macromolecules can contain more than one type of building block, e.g. glycoproteins (amino acids and sugars).

How are larger biomolecules formed from small ones?

When smaller molecules are to be combined into larger ones, often the smaller molecules are activated in some way before the joining reaction takes place. This is because firstly, often the activation will allow an otherwise energetically unfavourable reaction to take place and, secondly, it may well influence the reaction that takes place to ensure that the correct product is formed.

What roles do macromolecules play in cell structure and function?

There are four basic types of macromolecule in cells:

- Proteins—probably have one of the broadest range of functions in the body of any of the macromolecules. For example:
  - Enzymes (biological catalysts)
  - Structural
  - Membrane transporters (channels, carriers, pumps)
  - Receptors
  - Signalling molecules, e.g. hormones
- Lipids—lipid molecules do not polymerize, but can associate non-covalently in large numbers:
  - Major component of cell membranes (plasma and organelle)
  - Energy storage
  - Signalling (intracellular, hormone)
  - Insulation (electrical in nerves, thermal)
  - Components of other macromolecules (e.g. lipoproteins, glycolipids)
- Carbohydrates (polysaccharides)—also have a diverse range of functions in the body, including:
  - Structural (connective tissue)
  - Cell surface receptors (short chains linked to lipids and proteins)
  - Energy storage (glycogen)
  - Source of building blocks for other molecules (e.g. conversion to fat, amino acids, nucleic acids)
- Nucleic acids (DNA/RNA)—transfer of genetic information from generation to generation (DNA) and for determining the order of amino acids in proteins (RNA).

Proteins: overview

Proteins have a wide range of biological functions, determined ultimately by their 3D structures (in turn determined by primary sequence) and potentially by higher levels of organization (i.e. multi-subunit complexes).

Proteins fall into a number of functional types.

Structural

- Can be extracellular ('fibrous'), e.g. in bone (collagen), skin, hair (keratin):
  - May use disulphide bonds to stabilize structure → strength
- Intracellular structural proteins, e.g. cytoskeletal proteins such as actin, fibrin (blood clots).

Globular

Describes most non-structural proteins:
- Generally compact due to folding of polypeptide chain
- Usually bind other molecules (ligands), e.g. haemoglobin and O_2, antibody and epitope, ferritin and iron:
  - Binding is specific and involves a particular part of the protein (binding site):
    ○ Can be tight (high affinity) or loose (low affinity) or anywhere in between
  - Ligand binding can alter the function of the protein.

Enzymes are a specific type of globular protein:
- Bind substrate(s) and catalyse conversion into product(s) (reactants)
- Enzymes are catalysts, i.e. are chemically identical before and after the reaction they catalyse
- Enzymes do not change reaction equilibrium, but increase the rate at which equilibrium is reached:
  - This is achieved by lowering the reaction activation energy.

Plasma membrane proteins

- Integral membrane proteins, e.g. transporters, receptors:
  - Generally, compact globular proteins
  - Have hydrophobic domains in the membrane lipid and hydrophilic domains in the cytoplasm/extracellular fluid
- Membrane-anchored proteins:
  - Attached by covalent bonds to membrane elements, e.g. phospholipid head group.

Regulatory proteins

These are proteins which affect the activity of other proteins, for example:
- Protein kinases (PKs), which phosphorylate proteins and affect their function, e.g. PKA, PKC, tyrosine kinase
- Calmodulin, a Ca^{2+} binding protein, can be part of multi-subunit enzyme complexes and regulate their activity.

Amino acids and the peptide bond

Amino acids are the building blocks from which proteins are formed, by the joining of these simple subunits into polymers (Table 1.1).
All amino acids have the same basic structure (Fig. 1.2).

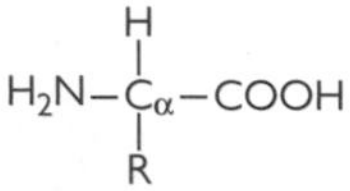

$$H_2N-\underset{\underset{R}{|}}{\overset{\overset{H}{|}}{C_\alpha}}-COOH$$

Fig. 1.2 Basic structure of amino acids; R is the side-chain.

There are 20 common side-chains (R) in the amino acids that are found in mammalian proteins, and they fall into five broad categories (NB: amino acids can be in more than one category; Fig. 1.3):
1. Hydrophilic, e.g. serine, glutamine
2. Hydrophobic, e.g. phenylalanine, valine
3. Basic—arginine, lysine
4. Acidic—aspartate, glutamate
5. 'Structural', e.g. proline.
- Cysteine and methionine contain sulphur
- The amino group of the side-chain of arginine and lysine has a pK of ~14, so is always positively charged under physiological conditions
- The carboxyl group of the side-chain of aspartate and glutamate has a pK of ~4, so is always negatively charged under physiological conditions
- Histidine side-chain has a pK of ~6 so can be either protonated (basic) or unprotonated (neutral) under physiological conditions
- Proline is known as a 'structural' amino acid as it puts a kink in polypeptide chains:
  - Glycine also allows flexibility in polypeptide chains due to the small size of its R group (i.e. a proton).

All the amino acids in mammalian proteins are the L stereoisomer.
- Bacteria make use of D-isomers:
  - Some antibiotics mimic D-amino acid and thus interfere with bacterial metabolism.

Amino acids are joined together in polypeptides by peptide bonds.
- Methionine is the starting amino acid of all proteins
- The peptide bond (–CO–NH–) is formed by a condensation reaction
- The peptide bond is planar and rigid (*trans* configuration), with no rotation around C–N bond
- Rotation around C_α–C and N–C_α bonds allows polypeptide chains to form 3D shape of proteins.

Protein function is determined from the primary amino acid sequence.
- The order of the amino acids will give the protein its shape
- The 3D shape of the protein is essential to its function, e.g. the shape of the substrate binding site of an enzyme—this includes the position of the R groups, which is determined by the stereoisomer (L vs D).

Table 1.1 Amino acids

Name	3-letter code	Single-letter code	Side group (except for proline where whole amino acid is shown)
Glycine	Gly	G	–H
Alanine	Ala	A	–CH$_3$
Valine	Val	V	–CH(CH$_3$)(CH$_3$)
Leucine	Leu	L	–CH$_2$–CH(CH$_3$)(CH$_3$)
Isoleucine	Iso	I	–CH(CH$_3$)–CH$_2$–CH$_3$
Serine	Ser	S	–CH$_2$–OH
Cysteine	Cys	C	–CH$_2$–SH
Threonine	Thr	T	–CH(CH$_3$)–OH
Methionine	Met	M	–CH$_2$–CH$_2$–S–CH$_3$
Proline	Pro	P	$^+$H$_2$N–CH–COO$^-$ with H$_2$C–CH$_2$–CH$_2$ ring
Phenylalanine	Phe	F	–CH$_2$–C$_6$H$_5$
Tyrosine	Tyr	Y	–CH$_2$–C$_6$H$_4$–OH
Tryptophan	Trp	W	–CH$_2$–(indole)
Histidine	His	H	–CH$_2$–(imidazole)
Lysine	Lys	K	–CH$_2$–CH$_2$–CH$_2$–CH$_2$–NH$_3^+$
Arginine	Arg	R	–CH$_2$–CH$_2$–CH$_2$–NH–C(NH$_2^+$)(NH$_2$)
Aspartate	Asp	D	–CH$_2$–C(=O)O$^-$
Asparagine	Asn	N	–CH$_2$–C(=O)NH$_2$
Glutamate	Glu	E	–CH$_2$–CH$_2$–C(=O)O$^-$
Glutamine	Gln	Q	–CH$_2$–CH$_2$–C(=O)NH$_2$

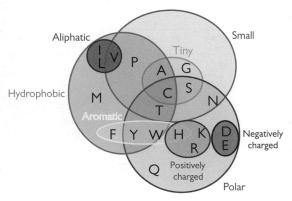

Fig. 1.3 Amino acids grouped by physicochemical characteristics (see Table 1.1 for the single letter amino acid codes).

Principles of protein structure

Proteins are held together at a molecular level by a number of forces:
- Van der Waals':
  - The weakest interactions, also known as London forces
- Hydrogen bonding:
  - Non-covalent; occurs between electronegative atoms (N or O) and an electropositive proton, e.g. =N ... NH=, =N ... HO⁻, C=O ... HN=
- Hydrophobic forces:
  - Interactions between groups of hydrophobic amino acids to reduce the amount of water associated with them and thus to increase thermodynamic stability of structure
  - Strong driving force for globular protein formation
  - Can also occur between hydrophilic amino acids in non-aqueous surroundings
- Ionic interactions:
  - Between R groups of oppositely charged amino acids, e.g. arginine—glutamate, aspartate—lysine
- Disulphide bonds:
  - The only covalent bonding involved in protein structure
  - Occur between the –SH groups of cysteines, to give –S–S– bonds. These bonds can be broken by denaturing the protein, e.g. with heat or solvents. Denaturation can be reversible or irreversible.

Proteins have four levels of organization

1. Primary—the order of the amino acids, e.g. Gly–Ala–Val.
2. Secondary—the formation of regions of structure in the polypeptide chain. These are stabilized by hydrogen bonds between the C=O and N–H moieties of peptide bonds in the polypeptide chain. Examples of such regions of local structure within the protein are the:
- α-helix (Fig. 1.4):
  - Most stable of several possible helical structures which amino acids can adopt due to least strain on the inter-amino acid hydrogen bonding
  - R groups point outwards from the helix. Usual form protein adopts when crossing lipid bilayers
  - Stabilized by hydrogen bonds between every fourth peptide bond; the peptide bonds are adjacent in 3D space (periodicity of helix = 3.6 residues)
- β-sheet (Fig. 1.4):
  - Polypeptide chain is in a fully extended conformation
  - Can run parallel or anti-parallel
  - Stabilized by hydrogen bonds between peptide bonds of adjacent polypeptide chains
- Loop/turn or β-turn (Fig. 1.4):
  - Region where polypeptide chain makes a 180° change in direction, often at surface of globular proteins
  - β-turns are stabilized by hydrogen bonds between residues 1 and 4 of this four-amino acid structure. Residues 2 and 3 are at the end of the loop and are often Pro or Gly, and do not contribute stabilizing bonds
  - Loops can have a more complex structure.

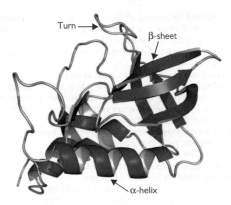

Turn →

β-sheet

α-helix ←

Fig. 1.4 Diagram of the structure of staphylococcal nuclease protein.

Reproduced with permission from Papachristodoulou, D., *Biochemistry and Molecular Biology* 6e, 2018, Oxford University Press.

3. Tertiary—the overall 3D structure formed by a polypeptide chain.
- Describes the overall arrangement of the regions of secondary protein structure:
  - Recognizable combinations of secondary structures (super secondary structures, or motifs) may be present in different proteins, e.g. α-helix–turn–α-helix is often a DNA binding motif
  - Certain more stand-alone structural regions (domains) may be identified whose function is known, e.g. an ATP binding domain
  - Families of functionally related proteins may have a similar tertiary structure. For example, ATP-binding cassette (ABC) proteins have two domains, each of six membrane-spanning (α helical) regions and an ATP-binding site.
- Stabilized by the forces described earlier (van der Waals', hydrogen bonding, hydrophobic forces, ionic interactions and potentially disulphide bonds) including between the R groups
- The primary structure, i.e. the amino acid sequence, determines the tertiary (3D) structure of a protein. This was shown by Christian Anfinsen in 1961 when the enzyme ribonuclease spontaneously refolded into its active native structure after chemical denaturation[1]
- Proteins fold in an at least partially defined pathway with intermediates between the unfolded and dully folded forms—they fold far too quickly to try all possible conformations
- In aqueous globular proteins, one of major determinants of 3D structure is the packing of the hydrophobic residues in the interior with water excluded (hydrophobic forces):

---

1 Anfinsen CB *et al.* (1961). The kinetics of formation of native ribonuclease during oxidation of the reduced polypeptide chain. *Proc Natl Acad Sci USA* **47**, 1309–14.

- The C=O and NH groups of the peptide bonds are all hydrogen bonded to each other in α-helices and β-sheets, and van der Waals' bonds stabilize the hydrocarbon backbones
- The outer surface has the hydrophilic residues, including those found in the turns/loops
- The amino acid sequence often leads to amphipathic structures, such as α-helices and β-sheets where one side is hydrophilic and one side hydrophobic, allowing these to interact with two environments
- Proteins in hydrophobic environments, such as membrane proteins, need to have the opposite orientation.

4. Quaternary—the interaction of a number of polypeptide chains to form a multimeric protein complex.

- Many proteins are made up of more than one subunit, with the two or more polypeptide chains held together usually by the weak non-covalent interactions previously described:
  - The resulting oligomers can be either homo-oligomers (made up of a number of identical subunits) or hetero-oligomers (made up of different subunits)
- Misfolding of proteins is associated with some neurological diseases, e.g. Parkinson's disease, Huntington's disease, and Creutzfeldt–Jakob disease (CJD):
  - Misfolded forms of normally soluble proteins accumulate as amyloid fibrils or plaques (hence the diseases are collectively known as amyloidoses)
  - The insoluble amyloid fibrils are rich in β-sheets.

Two of the best studied and understood proteins are the monomeric myoglobin and the hetero-tetrameric haemoglobin (Fig. 1.5):

- These are 3D structures solved by X-ray crystallography in the late 1950s/early 1960s[2,3]
- Myoglobin has a single polypeptide chain with a haem group non-covalently associated (an organic group attached to a protein is known as a prosthetic group; Fig. 1.5a):
  - Binds O_2 with a simple hyperbolic affinity curve
- Haemoglobin has four polypeptide chains ($\alpha_2\beta_2$ in adults; Fig. 1.5b), each with a haem group and very similar to myoglobin:
  - O_2 affinity curve is sigmoidal due to cooperation between the four O_2 binding sites. Binding of the first O_2 causes an allosteric change in the protein shape that makes the binding of the next O_2 easier, and so on until all four sites are occupied. See Figs 6.17, 6.18; ⊃ pp.411, 413.

Proteins can also be modified after synthesis:

- Post-translational modification:
  - Disulphide bonding—as described previously
  - Cross-linking—formation of covalent cross-links between individual molecules

---

2 Kendrew JC et al. (1958). A three-dimensional model of the myoglobin molecule obtained by X-ray analysis. *Nature* **181**, 662–6.

3 Perutz MF et al. (1960). Structure of haemoglobin: a three-dimensional Fourier synthesis at 5.5-Å resolution, obtained by X-ray analysis. *Nature* **185**, 416–22.

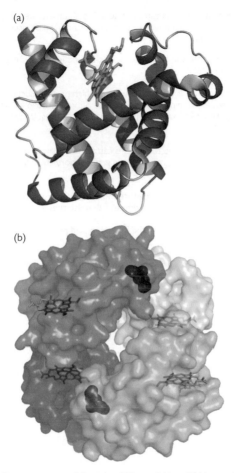

Fig. 1.5 Computer-generated diagrams of (a) myoglobin and (b) haemoglobin.

Reproduced with permission from Papachristodoulou, D., *Biochemistry and Molecular Biology* 6e, 2018, Oxford University Press.

- Peptidolysis—enzymic removal of part of the protein after synthesis
- Attachment of non-peptidic moieties
- Glycosylation—addition of carbohydrate groups
- Phosphorylation—addition of a phosphate group to specific residue(s) of a protein, e.g. serine, tyrosine
- Adenylation—addition of an AMP group to a protein

- Farnesylation—proteins can be attached to an unsaturated C15 hydrocarbon group (known as a farnesyl anchor) which inserts into the plasma membrane.

These modifications can have effects on the functioning of the protein, affecting, for example:
- Regulation:
  - Phosphorylation is a common way by which protein function is modified, e.g. turning an enzyme on or off with protein kinase A (PKA). This is an example of one protein (PKA) regulating the function of others
- Targeting:
  - Molecules conjugated to proteins can act as signals to the intracellular sorting and targeting machinery, e.g. phosphorylation can affect targeting
- Turnover:
  - Glycosylation levels can regulate the half-life of proteins in the circulation, e.g. by affecting the rate at which they are taken up and degraded by liver cells
  - Labelling of proteins with the 74-amino acid protein ubiquitin marks them for breakdown by proteasomes
- Structural:
  - For example, cross-linking of individual molecules in collagen greatly increases the strength of the fibril.

Structural proteins

Collagen

- Collagen is a structural fibrous protein of tendons and ligaments; it is also the protein component onto which minerals are deposited to form bone
- There are 13 types of collagen reported to date, with type I making up 90% of the collagen found in most mammals (and up to a third of the total protein in humans)
- Type I collagen is made up of three chains, two α_1 and one α_2 in a triple helix (Fig. 1.6):
  - Each individual chain is a left-handed helix formed from a repeating amino acid sequence: –Gly–Pro–X– and –Gly–X–OHPro– which occur >100 times each, accounting for ~60% of the molecule:
    - Only Gly has a small enough R group to be at the centre of such a triple helical structure
    - Collagen contains two unusual amino acids—hydroxyproline and hydroxylysine
    - Formed by post-translational enzymatic modification
  - These chains form a right-handed triple helix (tropocollagen molecule). The triple helix is stabilized by inter-chain hydrogen bonding between the peptidyl amino and carbonyl groups of the glycine and the hydroxyl groups of hydroxyproline
  - Tropocollagen molecules form bundles of parallel fibres 50nm in diameter and several millimetres long:
    - The tropocollagen molecules are staggered in their assembly
    - Tropocollagen molecules are cross-linked between lysine residues and hydroxylysine residues.

There are clinical disorders arising from errors in collagen synthesis.

1. Osteogenesis imperfecta—there are two possible causes:
- Abnormally short α_1 chains: these associate with normal α_2 chains but cannot form a stable triple helix and, therefore, nor fibres
- Substitution of the Gly residue: replacement with either Arg or Cys also blocks formation of a stable triple helix → symptoms—brittle bones, repeated fractures, and bone deformities in children.
2. Ehlers–Danlos syndrome (⊕ OHCS11, Figs 14.6 and 14.7, p.847)— structural weakness in connective tissue.
- Deficient cross-linking due to a defective enzyme resulting in lower numbers of hydroxylysine residues
- Can also be due to failure to process precursor precollagen into tropocollagen → symptoms: hyperextensible skin and recurrent joint dislocation.
3. Marfan's syndrome (⊕ OHCM10 p.706)—inherited disorder of weakened tissues.
- Extra amino acids near C-termini of α_2 chains results in reduced cross-linking as residues no longer line up
- Also reduced levels of fibrillin (a small glycoprotein which is an important element of the extracellular matrix) → symptoms: weakening, especially of cardiovascular system, causing aortic rupture; skeletal and ocular muscle, lungs, and nervous systems also may be affected.

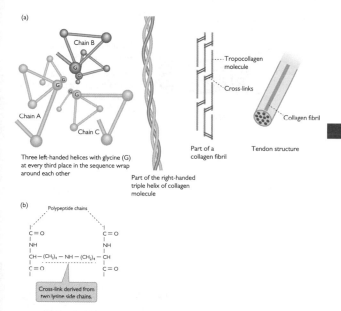

Fig. 1.6 (a) Arrangement of collagen fibrils in collagen fibres; (b) one type of cross-link formed between two adjacent lysine residues.

Reproduced with permission from Papachristodoulou, D., *Biochemistry and Molecular Biology* 6e, 2018, Oxford University Press.

Histones

Histones are involved in the packaging of DNA (Fig. 1.7) in a space-efficient way (they increase the packing factor by ~7-fold).

- Histone proteins are globular with a cationic surface which neutralizes the phosphates of the DNA
- Octamer of two each of H2A, H2B, H3, and H4 proteins has DNA wound around it, forming a nucleosome, which is often stabilized by the structural H1 protein:
  - Nucleosomes are further organized into helical arrays called solenoids which, in turn, are arranged around a central protein scaffold, allowing ~2m of DNA in the human nucleus to form the 46 chromosomes with a total length of just 200μm.

(a)

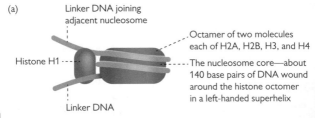

Linker DNA joining
adjacent nucleosome

Octamer of two molecules
each of H2A, H2B, H3, and H4

Histone H1

The nucleosome core—about
140 base pairs of DNA wound
around the histone octomer
in a left-handed superhelix

Linker DNA

(b)

Nucleosome

Linker DNA—about
30–40 base pairs

200 base pairs

50 nm

(c)

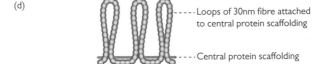

30 nm

30nm fibre with
nucleosomes arranged
in a zigzag fashion

(d)

Loops of 30nm fibre attached
to central protein scaffolding

Central protein scaffolding

Fig. 1.7 Order of chromatin packing in eukaryotes.

Reproduced with permission from Papachristodoulou, D., *Biochemistry and Molecular Biology* 6e, 2018, Oxford University Press.

Concepts of biochemical reactions and enzymes

Classes of common biochemical reactions:
- Hydrolysis: splitting with water, e.g. breaking of peptide bond
- Ligation: joining of two compounds, e.g. two pieces of DNA
- Condensation: forms water, e.g. synthesis of peptide bond
- Group transfer: movement of a biochemical group from one compound to another
- Redox: reaction of two compounds during which one is oxidized and the other reduced
- Isomerization: physical conversion from one stereoisomer to another.

Biochemical reactions rarely occur in the absence of an enzyme.
- An enzyme is defined as a biological catalyst:
  - Enzymes increase the rate at which a reaction occurs (usually by at least 10^6-fold)
  - Like a chemical catalyst, the enzyme is unchanged by the reaction that it catalyses, nor does it alter the equilibrium of the reaction (i.e. the forward and reverse reactions are speeded up by the same factor)
  - Enzymes are very specific for the reaction that they catalyse, and often enzyme activity is regulated
- Enzymes achieve their catalytic effect by reducing the size of the activation energy step for the reaction (Gibbs' free energy of activation or $\Delta G^{\ddagger}$; Fig. 1.8):
  - $\Delta G^{\ddagger}$ is the difference in the free energy between the free substrates and the transition state
  - The substrates binding to the enzyme have a lower transition state energy (i.e. a lower $\Delta G^{\ddagger}$) but can still react to form the same product as they could in free solution
  - → the speed of the reaction is increased.

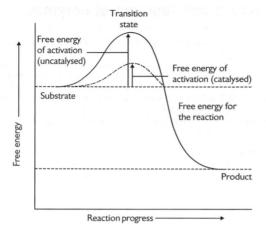

Fig. 1.8 Enzymes decrease the activation energy.

Structure and function of enzymes

As well as being effective catalysts, one of the most striking observations concerns enzyme specificity, both in terms of the reaction catalysed and their substrates.

- Usually, only a single reaction is catalysed (or a few very closely related reactions), with a very high, if not absolute, choice of substrate(s)
- The active site is essential for both the catalysis and specificity of enzymes. It is defined by the structure of the enzyme protein, both at the primary level (for essential residues for the reaction) and at higher levels of protein organization (to put these residues in the correct place in the 3D structure).

Serine proteases

- So named because they have a serine residue, which is rendered catalytically active by an aspartate and histidine residue closely adjacent in 3D space (the 'catalytic triad')
- These three residues form a charge relay network on the enzyme binding a substrate (Fig. 1.9)
- Family of enzymes includes chymotrypsin, trypsin, elastase, thrombin, and subtilisin.

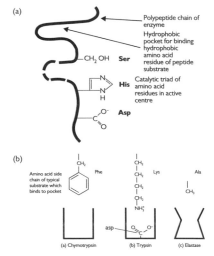

Fig. 1.9 Serine proteases: schematic representations of (a) the catalytic triad; (b) the binding site and how substrate specificity is achieved.

Carboxypeptidase A

- Contains a Zn molecule coordinated by two histidine residues, a glutamate, and a water molecule
- These destabilize the substrate and allow it to be attacked by another catalytic glutamate residue, resulting in bond cleavage.

Lysozyme

- Cleaves glycosidic bonds between modified sugars of the polysaccharide chains that make up part of the bacterial cell wall
- The substrate is bound in the correct orientation by a number of hydrogen bonds
- This allows a specific aspartate residue to catalyse the cleavage reaction.

How are proteins arranged to form enzymes?

- Enzymes can be single proteins, or multimers
- Different tissues can have different forms of the same enzyme (isozymes); these can be used for diagnostic testing. For example, finding the heart muscle isozyme of lactate dehydrogenase (LDH; ➔ p.188) in the blood is indicative of a heart attack (➔ *OHCM10* pp.118, 119, 688)
- As well as multimeric enzymes, there are also protein complexes that contain multiple enzyme activities, e.g. pyruvate dehydrogenase (PDH; ➔ pp.145, 148).

How are enzymes regulated?

- Allosteric effectors bind to the enzyme at a site distinct from the active site and modify the rate at which the reaction proceeds:
  - These changes are brought about by a change in the shape of the enzyme
  - Can result either in either increased or decreased rates, e.g. PDH
- Covalent modification, usually by addition of a phosphate group to specific residue(s)
- Subunit dissociation, e.g. cAMP-dependent protein kinase (PKA; Fig. 1.10).
  - Consists of two catalytic (C) subunits and two regulatory (R) subunits
  - The R subunit has a pseudosubstrate site for the catalytic unit, and so in the absence of cAMP binds to and inactivates the C subunit
  - cAMP binds to the R subunit, allosterically abolishing the pseudosubstrate site so that the R subunit dissociates and leaves the C subunit free to act.

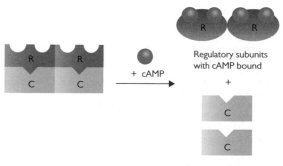

Fig. 1.10 Activation of cAMP-dependent protein kinase (PKA) by cAMP (R regulatory subunit of PKA and C catalytic subunit of PKA).

Enzyme co-factors

In addition to their protein subunit(s), many enzymes also have essential non-protein components known as co-factors or co-enzymes.
- These co-factors play a vital role in the reaction catalysed by the enzyme
- They are usually either trace elements or derivatives of vitamins.

Trace elements and enzymes

- Small but essential amounts of minerals are absorbed from the diet and used as co-factors:
  - Zinc in lysozyme enzymes (e.g. superoxide dismutase (SOD))
  - Manganese in, e.g. isocitrate dehydrogenase in tricarboxylic acid cycle (TCA) cycle
  - Cobalt (constituent of vitamin B_{12}) essential for methionine biosynthesis
  - Selenium in, e.g. glutathione peroxidase
  - Molybdenum in oxidation/reduction reactions
  - Copper in, e.g. cytochrome oxidase.
  - Manganese in Mn-SOD

Vitamins as precursors of co-enzymes

- Vitamins ('vital amines') are small organic molecules (Fig. 1.11) that cannot be made by the body and so must be obtained from the diet:
  - In most cases, vitamins must be chemically modified to form the co-enzyme. Modifications range from minor (e.g. phosphorylation) to substantial (e.g. incorporation into much larger molecules)
  - Lack of vitamins results in metabolic disorders and diseases associated with deficiency, e.g. vitamin C (ascorbic acid) and scurvy (Ⓢ *OHCM10* p.268):
    ○ Most vitamins are only required in very small amounts in the diet
    ○ Water-soluble vitamins are usually co-enzyme precursors.

Examples of vitamins and their active roles

- Biotin → covalently bonded to phosphoenolpyruvate carboxykinase (PEPCK): gluconeogenesis
- Pyridoxine (vitamin B_6) → pyridoxal phosphate: transamination
- Pantothenic acid → co-enzyme A: acyl transfer across the internal mitochondrial membrane (IMM)
- Riboflavin (vitamin B_2) → FAD: oxidation reduction, e.g. TCA cycle and electron transport chain (ETC)
- Niacin → NAD: oxidation reduction, e.g. TCA cycle and ETC
- Thiamine (vitamin B_1) → thiamine phosphate: PDH and α-ketoglutarate dehydrogenase in glycolysis
- Folic acid → tetrahydrofolate derivatives: biosynthetic reactions.

Examples of water-soluble vitamins

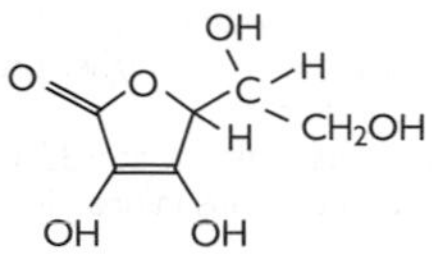

Vitamin C

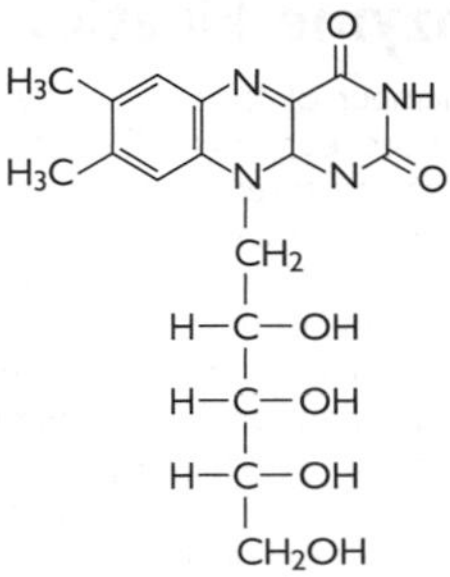

Vitamin B$_2$
(riboflavin)

Examples of fat-soluble vitamins

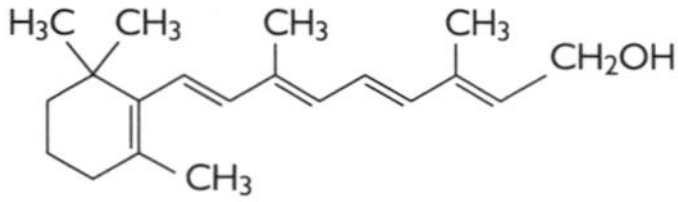

Vitamin A (retinol)

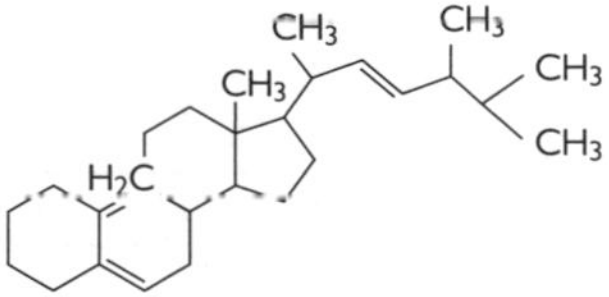

Vitamin D$_2$

Fig. 1.11 Some vitamin structures.

Enzyme kinetics

A number of factors can affect the rate at which enzyme-catalysed reactions proceed (Figs 1.12), including:

- Temperature: generally, reactions proceed faster with increased temperature until a point is reached at which the crucial higher orders of protein structure are destroyed ('denaturation')
- In humans, most (but not all) enzymes are most efficient at ~37°C
- pH: the level of protonation of amino acid side-chains in proteins is dependent on the environmental pH:
  - Most cytosolic enzymes have maximal activity at pH 7.4 whereas, for example, those in the acid environment of the stomach have a much lower pH optimum
- Amount of enzyme: the more enzyme, the more active sites and, therefore, the more reactions can be catalysed. Enzyme activity is often expressed as either the total activity (units of activity per volume of enzyme solution) or the specific activity (units per amount of protein)
- Concentration of substrates and products: enzymes are catalysts and, as such, only increase the speed at which reactions achieve equilibrium, and do not change the equilibrium itself.

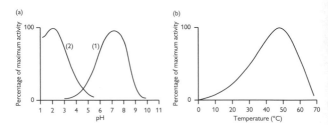

Fig. 1.12 Effect of (a) pH and (b) temperature on enzyme activity. In (a), (1) represents the majority of enzymes, and (2) gastric enzymes.

Enzyme activity can be measured by looking at the rate at which a product appears (or a substrate disappears) under defined reaction conditions:

- The interaction between enzyme rate (V) and substrate concentration ([S]), as seen in Fig. 1.13, can be described for most enzymes by the Michaelis–Menten equation:

$$V = \frac{[S]V_{max}}{[S] + K_m}$$

- This can also be drawn as a linear plot, either a Lineweaver–Burke (1/V vs 1/[S]), Eadie–Hoftee (V vs V/[S]), or Hanes–Woolf ([S]/V vs [S]) plot
- The Hanes–Woolf plot can be considered the most reliable as it gives greatest weight to the most robust data (i.e. at highest [S], when there is highest activity).

Useful constants that can be determined from enzyme assay studies include:
- K_m: the affinity of the enzyme binding site for its substrate
- V_{max}: the maximum rate at which the enzyme can operate (given unlimited substrate)
- Turnover number: the number of reactions that the enzyme can perform per unit time.

Kinetic analysis allows inhibitors of enzymes to be defined into different classes:
- Irreversible: the inhibitor covalently modifies the enzyme and so permanently inhibits it
- Reversible: these inhibitors do not permanently affect the enzyme. These can be usefully analysed kinetically to determine whether they are competitive or non-competitive:
  - Competitive: the inhibitor and the substrate compete for the same binding site (change in kinetic parameters: K_m increased, V_{max} unchanged)
  - Non-competitive: the inhibitor binds at a site distinct to the substrate binding site (K_m unchanged, V_{max} decreased)
  - Uncompetitive: inhibitor binds at a site distinct to the substrate, but only when the enzyme also has substrate bound (K_m decreases, V_{max} decreases).

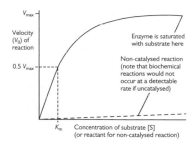

Fig. 1.13 Effect of substrate concentration on the reaction velocity catalysed by a classical Michaelis–Menten type of enzyme.

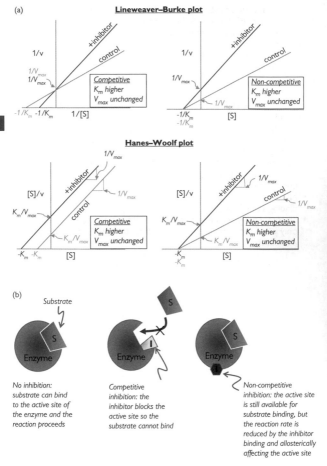

Fig. 1.14 (a) Plots of enzyme reaction rate in the presence of competitive and non-competitive inhibitors, shown as a double reciprocal (Lineweaver–Burke) plot and a single reciprocal (Hanes–Woolf) plot. (b) Schematic of different types of inhibition (S = substrate, I = inhibitor).

Membrane transporter proteins: structure and function

(See also ➔ pp.52–53.) Membrane transporter proteins (Fig. 1.15) can be split into:

1. Channels: aqueous-filled pores that form a pathway for small hydrophilic substrates (usually ions) to cross lipid bilayers. The passage of substrates is nearly always gated in some way, with the channel being opened by, for example, a compound binding (ligand-gated), a change in membrane potential (voltage-gated), stretching the membrane (stretch-activated). Any of the activators result in a conformational change in the channel that allows the passage of substrate.
- Examples: nicotinic acetylcholine receptor (nAChR, neuromuscular junction), voltage-gated Na^+ channels (nerve), stretch-activated Ca^{2+} channels (smooth muscle).
2. Transporters.
- Primary active—use ATP hydrolysis to energize movement of substrates:
  - Examples: Na^+/K^+-ATPase (aka the sodium pump), Ca^{2+}-ATPase
- Secondary active—use the energy from an (electro)chemical gradient set up by a primary active transporter:
  - Examples: Na^+-glucose co-transporter (SGLT), Na^+-Ca^{2+} exchange
- Facilitated diffusion—speeds up the equilibration of substrates across cell membranes but, unlike the two previous classes, cannot concentrate substrates above equilibrium:
  - Examples: facilitated glucose transporter (GLUT).

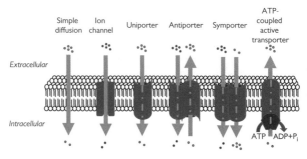

Fig. 1.15 The main types of membrane transport systems.

Structure

- Despite the wide range of functions of membrane transporter proteins, they have remarkable similarities regarding their molecular structures, although it should be noted that only a few membrane transporter proteins have been structurally characterized by X-ray crystallography and, thus, much is based on prediction
- Both channels and transporters have multiple membrane spanning domains (transmembrane domains (TMs)), usually α-helixes, connected by intra- and extracellular loop regions. As the TMs pass through the hydrophobic core of the lipid bilayer, they tend to be made up of largely hydrophobic amino acids.

Channels
- The aqueous pore in the channel is formed from a number of TMs, which have a polar amino acid residue once every turn of the α-helix, giving a polar side to the TM. A number of such TMs then come together, forming a pore lined with hydrophilic residues which thus allows water to enter and substrates to pass through
- The specificity of an ion channel is often determined by the charges on residues at the 'mouth' of the channel. For example, the nAChR is a channel for cations and has a ring of negatively charged residues at the mouth of the channel to attract cations and repel anions:
  - Experimentally reversing these charges with site-directed mutagenesis can turn it into an anion-selective channel.

Transporters
- Many are predicted to have 12 TMs (although estimates can vary between 10 and 14)
- Although the amino acid sequences of the many families of transporters have little in common, there are a few motifs/features:
  - Proteins with the motif for ATP binding/hydrolysis are known as ABC transporters
  - Some transporters appear to have two similar halves, suggesting they evolved from gene duplication
- Function by binding substrate then undergoing a conformational change, resulting in the substrate-binding site being re-orientated to face the opposite side of the membrane:
  - Binding site is usually very specific for a small number of very closely related substrates
  - Binding of substrate to transporter is the basis behind transporter kinetics (🔿 p.52).

Examples of ABC transporters

1. 'Flippases': ATP-driven transporter proteins which 'flip' membrane lipids from one side of the lipid bilayer to the other. Roles in creating/maintaining asymmetric distribution of lipids in membranes, cellular signalling.
2. pGlycoprotein.
- Widely expressed ATP-driven transporter with a wide substrate range of small lipophilic compounds
- Known substrates include a large number of drug molecules, which are exported from cells and out of the body:
  - This can reduce absorption rates (intestine), increase excretion rates (liver, kidney), and affect tissue distribution (e.g. blood–brain barrier)
  - Genetic variations between individuals in pGlycoprotein expression will affect drug efficacy between patients. Gene profiling of patients may allow tailoring of drug prescribing in future (ethically contentious).
3. Cystic fibrosis (🔿 *OHCM10* p.173) transmembrane conductance regulator (CFTR).
- CFTR is the channel involved in Cl^- exit across the apical membrane in secretion (e.g. sweat glands, pancreatic ducts)
- The majority of CF cases are caused by a mutation → the deletion of a single Phe amino acid (ΔF_{508}). The ΔF_{508} mutant protein misfolds and fails to traffic to the apical membrane.

Roles of lipids

Lipids play a wide range of roles in the body—from energy storage to hormone signalling, plasma membrane structure to heat insulation. They are a structurally diverse class of macromolecules that have the common feature of being poorly soluble in water (hydrophobic).

Energy

- Stored triacylglycerides ('fat') makes up about 20% of the body mass of an average person
- Efficient way to store energy:
  - Higher specific energy (kJ g^{-1}) than glycogen (➔ p.126) or protein
  - Much lower hydration level than glycogen due to the inherent hydrophobicity of lipids
  - To store as much energy in glycogen would require approximately a doubling of body mass!

Structural roles

- As discussed elsewhere, lipids (phospholipids, sphingolipids, cholesterol) are the main structural components of the lipid bilayer membranes (➔ p.36):
  - The phospholipids arrange themselves into a bilayer with the polar headgroups facing outwards towards the aqueous environment and the hydrophobic 'tails' pointing inwards
  - The hydrophobic interior of the bilayer acts as a diffusion barrier to prevent ions or water-soluble (hydrophilic) molecules from crossing to enter or leave, e.g. the cell (plasma membrane) or organelles
  - To be able to cross the membrane, such compounds require a specific pathway (e.g. a channel or transporter, ➔ p.30)
- Lipids also stabilize fat–water interfaces:
  - Bile salts in the intestine: act as detergent-like compounds to emulsify dietary lipid and allow it to be absorbed as chylomicrons
  - Phospholipid in membrane: as mentioned previously, the polar headgroups allow the formation of the bilayer
  - Cholesterol in membrane: as a lipid-soluble molecule, cholesterol inserts into the hydrophobic portion of the membrane. In doing so, it makes the membrane less fluid by sterically inhibiting the movement of the fatty acyl chains
  - Pulmonary surfactant: this complex mixture of phospholipids and proteins reduces the surface tension at the interface of the alveolar lining fluid and the air in the lung, preventing alveolar collapse.
  - Premature babies (born before surfactant production begins) are unable to inflate their lungs (infant respiratory distress syndrome). Artificial surfactant can assist in these cases
  - Insulation: lipids can be either electrical insulators, as in the myelination of nerves (➔ p.237), or thermal insulation, e.g. subcutaneous fat.

Signalling molecules

- Extracellular—a number of classes of hormones are synthesized from cholesterol:
  - Steroid hormones
  - Eicosanoids (prostaglandins, thromboxanes, leukotrienes)—derived from arachidonic acid, an abundant component of plasma membrane phospholipids (e.g. as a fatty acid tail of phosphatidylcholine, phosphatidylethanolamine). Released by phospholipases A_2 in response to hormone and inflammatory signals
- Intracellular—second messengers can be derived from the breakdown of the phospholipid phosphatidylinositol-4,5-bisphosphate (PIP_2). Phospholipase C cleaves the headgroup to leave two components with signalling roles:
  - Inositol-1,4,5-triphosphate (IP_3) enters cytoplasm where it causes release of calcium from intracellular stores
  - Diacylglycerol (DAG) remains in the membrane:
    - Activates PKC, which has become membrane associated due to the rise in intracellular calcium
    - PKC phosphorylates target proteins, affecting their activity.

Fatty acids and triacylglycerides

Fatty acids have the general structure of a long hydrocarbon chain (usually 14–24) with a terminal carboxyl group (Fig. 1.16):
- Most animal fatty acids have an even number of carbon atoms (➔ p.128)
- Fatty acids can display different levels of saturation, depending on the number of C=C bonds in the chain.

What is generally thought of as 'fat' (i.e. the form that lipids are stored as in the body) are triacylglycerides (triglycerides), consisting of three fatty acids esterified to a glycerol molecule (Fig. 1.17):
- Triacylglycerides are good for energy storage due to their high energy per gram and low hydration level
- Major site of storage: adipocyte cells of adipose tissue.

Sources of fatty acids

Fatty acids can be either obtained from the diet or made in the body *de novo*:
- Triacylglycerides from the diet are absorbed across the intestinal epithelium and transported in the blood by chylomicrons to the liver and adipose tissue, where they are hydrolysed to free fatty acids by lipoprotein lipase and cross into the cell by diffusion
- Triacylglycerides synthesized *de novo* in the body (mainly in the liver) are carried by very low-density lipoprotein (VLDL) to other tissues.

Essential fatty acids

The body does not have the metabolic pathways to make all fatty acids and, therefore, some must be obtained from the diet (essential fatty acids). Mammals cannot introduce C=C bonds past C9 of the hydrocarbon chain:
- Linoleic (C=C at C9 and C12) and linolenic acids (C=C at C9, C12, and C15) are the essential fatty acids. Good dietary source: sunflower seed oil.

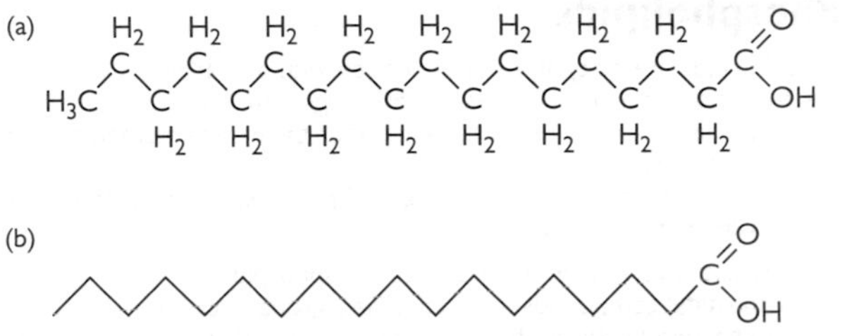

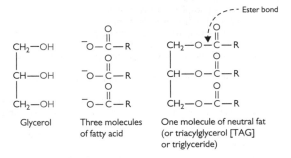

Fig. 1.16 (a) Structure of stearic acid (C18); (b) simple representation of fatty acids.

Glycerol

Three molecules of fatty acid

One molecule of neutral fat (or triacylglycerol [TAG] or triglyceride)

Ester bond

Fig. 1.17 A triacylglycerol and its component parts.

Phospholipids

Phospholipids are diacylglycerol molecules with a headgroup attached to the 3-position via a phosphodiester bond (Figs 1.18, 1.19):
- Phospholipids are the major constituent of the plasma membrane lipid bilayer (➔ p.32)
- Phospholipid is the active component of pulmonary surfactant, essential for maintenance of lung structure.

There are several different classes of phospholipids:
- Phosphatidyl compounds—common headgroups include:
  - No headgroup except for the glycerol → phosphatidylglycerol. The phosphatidylglycerol can be further modified to cardiolipin
  - Serine → phosphatidylserine
  - Ethanolamine → phosphatidylethanolamine
  - Choline → phosphatidylcholine (lecithin)
  - Inositol → phosphatidylinositol (can be phosphorylated and plays a role in intracellular signalling—➔ p.33)
- Sphingolipids have a backbone derived from sphingosine rather than glycerol:
  - Sphingosine is an amino alcohol with a long (12C) unsaturated hydrocarbon chain
  - Sphingomyelin is the only non-phosphatidyl membrane lipid; it has a choline headgroup and a fatty acid linked to the sphingosine backbone by an amide bond (Fig. 1.20)
- Many membranes also contain glycolipids, which have a sugar unit for the headgroup:
  - In animals, glycolipids are also derived from sphingosine:
    ○ Cerebrosides are the simplest, having the same basic structure as sphingomyelin except with either a glucose or a galactose ester linked in place of the choline group (Fig. 1.20)
    ○ Gangliosides have more complex sugars as the headgroup, with a branched chain of up to seven sugar residues (Fig. 1.20).

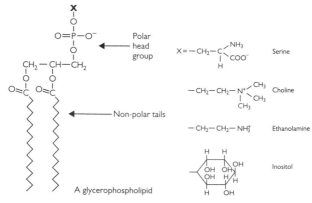

Fig. 1.18 The structure of a molecule of phosphatidic acid.

Fig. 1.19 Sample headgroup structures.

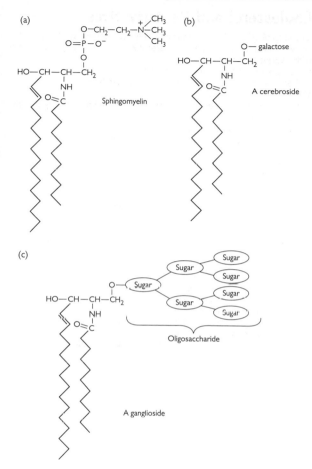

Fig. 1.20 Diagram of a molecule of (a) sphingomyelin; (b) a cerebroside; and (c) a ganglioside.

Cholesterol and its derivatives

Cholesterol (Fig. 1.21) is a four-ring, 27C compound, synthesized from acetyl-CoA (mainly in liver). Cholesterol is an important lipid in animals, with a variety of roles:

- Component of plasma membranes:
  - Structural role: affects the fluidity of the membrane
- Precursor of steroid hormones:
  - All steroid hormones are synthesized by a common pathway starting with cholesterol
  - Important steroids include:
    - Sex hormones (female: progesterone, oestrogens; male: testosterone)
    - Aldosterone (regulates sodium balance)
    - Cortisol (stress)
- Precursor of bile salts:
  - Conjugated with glycine (glycocholate) or taurine (taurocholate)
  - Secreted from liver into intestine to emulsify dietary lipid
- Precursor of vitamin D:
  - Key role in calcium regulation
- Component of plasma lipoproteins (chylomicrons, VLDL, intermediate-density lipoprotein (IDL), low-density lipoprotein (LDL), high-density lipoprotein (HDL)):
  - Involved in transport of cholesterol between tissues
  - LDL cholesterol (➲ *OHCM10* p.690) is known as 'bad cholesterol' as it is associated with increased risk of atherosclerosis:
    - Leads to deposition of cholesterol in plaques which can ultimately block blood vessels. Such blockage of coronary arteries is the cause of coronary artery disease
    - Treated with HMG-CoA reductase inhibitors ('statins'; ➲ *OHPDT2* p.22).

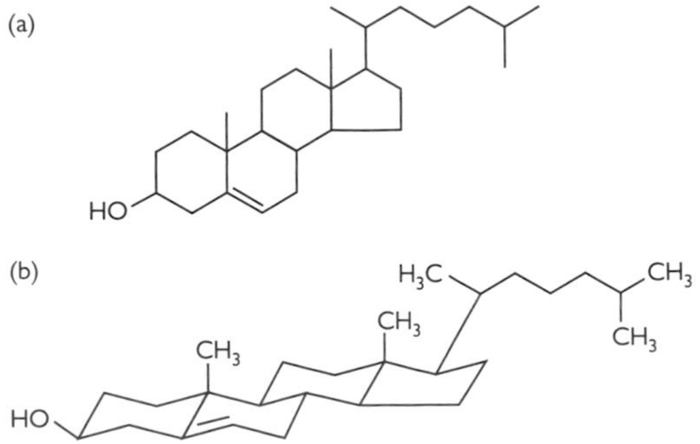

(a)

(b)

Fig. 1.21 (a) Conventionally drawn structure of cholesterol; (b) structure drawn to indicate actual conformation of cholesterol.

Carbohydrates: general principles

Carbohydrates are the most abundant form of organic matter on earth:
- Contain the elements C, H, and O
- Produced by the fundamental pathway of photosynthesis combining CO_2 and H_2O:
  - All animals are ultimately reliant on this source of new organic material.

Monosaccharides (sugars) have the general formula $(CH_2O)_n$. Most common form are the hexoses (n = 6), e.g. glucose, galactose, fructose:
- Sugars have chiral carbon atoms, so exist as stereoisomers:
  - D-glucose is the naturally occurring form, sometimes known clinically as dextrose
- As well as stereoisomers, as sugars are cyclisized, a second asymmetrical centre is created, giving two forms:
  - The α form is when the hydroxyl group is below the ring
  - The β form is when it is above the ring
  - α and β forms can interconvert in solution (mutorotation): equilibrium 66% β, 33% α, 1% open chain.

Although sugars may be found as monosaccharides, they can also exist as:
- Disaccharides, e.g. sucrose (glucose-α,1–2-fructose), lactose (galactose-β,1-4-fructose), maltose (glucose-α,1–4-glucose)
- Small multimers (oligosaccharides), e.g. on sphingolipids, glycoproteins
- Mainly found as large multimers (polysaccharides).

The monosaccharide building blocks are joined by glycosidic (C–O–C–) bonds (Fig. 1.22) which are formed by dehydration reactions:
- They are named from the numbers of the C atoms in the sugars, e.g. a 1,4 bond is a joining of C1 in the first sugar to C4 in the second
- The type of glycosidic bond can have major effects on the final structure (and thus function) of the molecule formed
- This can be seen in the following examples of commonly occurring polysaccharides (Fig. 1.22), all made up of glucose building blocks:
  - Glycogen (animal energy storage):
    - Very large polymer of thousands of glucose monomers
    - Predominantly joined by α,1–4 glycosidic bonds, with branches via α,1–6 linkages every 8–12 residues, forming a branched tree-like structure
  - Starch (plant energy storage):
    - Mixture of two glucose polymers: amylose is a linear α,1–4 glucose polymer (forms a helix); amylopectin is structurally very similar to glycogen, although the branching is less frequent (every 24–30)
  - Cellulose (plant structural molecule):
    - Unlike glycogen and starch, has β,1–4 glycosidic bonds between the glucose monomers to form long straight chains. Chains line up in parallel, stabilized by hydrogen bonding to form fibrils and then fibres. These fibres have the high tensile strength required for their structural role in plant tissue
    - Mammals lack the enzyme to digest cellulose (cellulase) and thus it passes undigested through the intestine (known as dietary fibre). Ruminants (e.g. cows) have cellulase-producing bacteria in their digestive tracts.

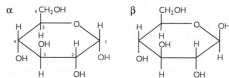

(a) Glucose

(b) 1–4 glycosidic bond

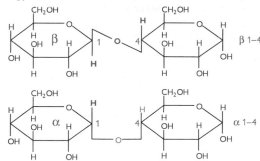

(c) glycogen

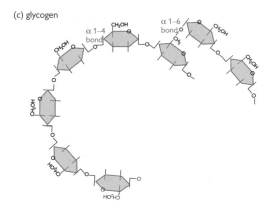

(d) Cellulose

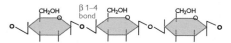

Fig. 1.22 Diagrams of (a) glucose structure; (b) 1–4-glycosidic bond; (c) glycogen (or starch); and (d) cellulose.

Structural carbohydrates

As well as using polysaccharides for energy storage in the form of glycogen, animals also employ structural carbohydrates in the extracellular matrix. Polysaccharide units constitute 95% of proteoglycan, with the remainder being a protein backbone.

- The polysaccharide chains are known as glycosaminoglycans (GAGs; Fig. 1.23). The chains are made up of repeating disaccharide subunits containing either glucosamine or galactosamine (amino sugars):
  - At least one of the sugars in the disaccharide has a negatively charged group (either carboxylate or sulphate).

Examples include:
- Hyaluronic acid (hyaluronate; Fig. 1.23)
- Chondroitin sulphate
- Keratan sulphate (Fig. 1.24):
  - Found in cartilage extracellular matrix
  - Large (around 10^5kDa), highly hydrated molecule, which acts as a cushion in joints (Fig. 1.25)
- Dermatan: found in skin
- Heparin: present on blood vessel walls where it plays a role in preventing inappropriate blood clotting (● p.490).

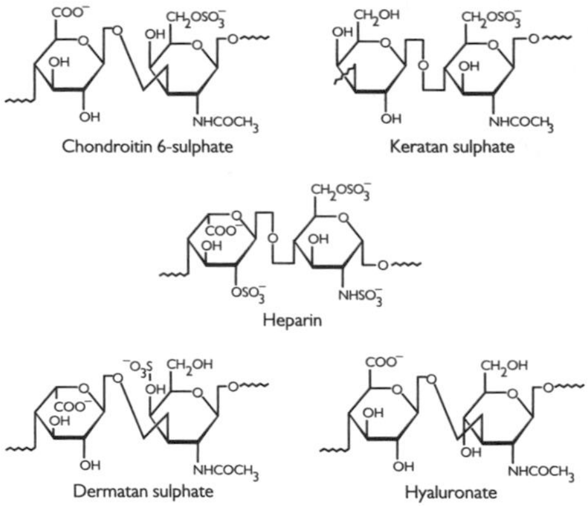

Fig. 1.23 Structural formulae for five repeating units of important glycosaminoglycans.

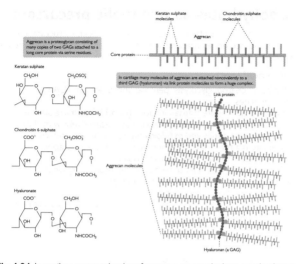

Fig. 1.24 In cartilage many molecules of aggrecan are attached non-covalently to a third GAG (hyaluronan) via link protein molecules to form a huge complex.

Reproduced with permission from Papachristodoulou, D., *Biochemistry and Molecular Biology* 6e, 2018, Oxford University Press.

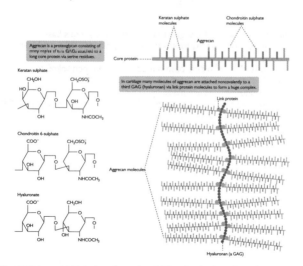

Fig. 1.25 Large, highly hydrated complex GAG molecules form the cushion in cartilage in joints.

Reproduced with permission from Papachristodoulou, D., *Biochemistry and Molecular Biology* 6e, 2018, Oxford University Press.

Carbohydrates as metabolic precursors

While polysaccharides play roles as energy sources and structural components, monosaccharides are also important as precursors in biosynthetic pathways.

Amino acid synthesis

- May be made from glycolysis, pentose phosphate, or TCA cycle intermediates
- Simplest reactions are transaminations (◑ p.164) to make alanine and aspartate (from pyruvate and oxaloacetate respectively, with glutamate as the nitrogen donor and pyridoxal phosphate as the co-factor)
- Aspartate is a precursor for other amino acids (→ asparagine, methionine, threonine (→ isoleucine, lysine))
- α-ketoglutarate → glutamate (→ glutamine, proline, arginine)
- 3-phosphoglycerate → serine (→ cysteine, glycine)
- Pyruvate → alanine, valine, leucine
- Phosphoenolpyruvate + erythrose-4-phosphate → phenylalanine (→ tyrosine), tyrosine, tryptophan
- Ribose-5-phosphate → histidine.

Fatty acid synthesis

- Fat is a major energy storage form
- Excess carbohydrate enters the TCA cycle as normal, but as cell does not need to make ATP, the level of citrate builds up:
  - Citrate leaves the mitochondrial matrix and enters the fatty acid synthetic pathway (◑ p.132).

Nucleotide synthesis

- Nucleotides consist of a sugar moiety and a nitrogenous base
- The sugar ribose-5-phosphate is the starting place for the synthetic pathway of the purine bases
- In contrast, the pyrimidine base is made first, and then linked to ribose-5-phosphate
- Deoxyribonucleotides are made by reduction
- The ribose 5-phosphate is synthesized from glucose 6-phosphate by the pentose phosphate pathway (◑ p.150).

Carbohydrates as conjugates

Many proteins and lipids have carbohydrate moieties attached, e.g. most secreted proteins (e.g. antibodies), many integral membrane proteins, and membrane lipids.

Glycoproteins

There are two ways in which carbohydrates can be attached to proteins:
- O-glycosidic link (O-linked) to the hydroxyl group of a serine or threonine
- N-glycosidic link (N-linked) to the amine group of asparagines:
  - Consensus sequence Asn–X–Ser or Asn–X–Thr where X is any amino acid, except proline:
    ○ Motif necessary but not always used due to protein 3D structure constraints
  - First two sugars added always *N*-acetylglucosamines, followed by three mannose residues that form a 'core'. Addition of further monosaccharides gives rise to a huge diversity of oligosaccharide structures of two major forms:
    ○ High mannose—additional mannose residues added to the core described
    ○ Complex—variety of (less common) monosaccharide units added to core.

The wide diversity of protein-linked oligosaccharide suggests a variety of functionally important roles, although these are not well understood at present. One important role is as recognition signals, of which blood groups are a good example:
- There are four major blood groups in humans: A, B, AB, and O (❷ p.32)
- These correspond to the presence of certain oligosaccharide residues on the erythrocyte integral membrane proteins (e.g. glycophorin A):
  - There is a basic core oligosaccharide, which can have a different terminal sugar or none at all (Figs 1.26, 1.27)
- Some individuals do not have the enzymes (glycosyl transferases) to add the terminal sugar (type O), whereas others can make type A, some type B, and some both (type AB).

Glycolipids

The main glycolipids are based on the sphingolipids, with one or more sugars replacing the sphingosine headgroup of the phospholipids (❷ p.36). There are two groups:
- Cerebrosides: simple glucose or galactose residue. Important in brain cell membranes
- Gangliosides: more complex branched chain of several monosaccharide groups. Also involved in the blood group types in erythrocytes.

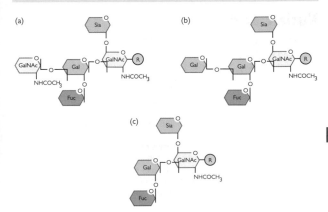

Fig. 1.26 The molecular basis of ABO blood groups: (a) type A; (b) type B; and (c) type O.

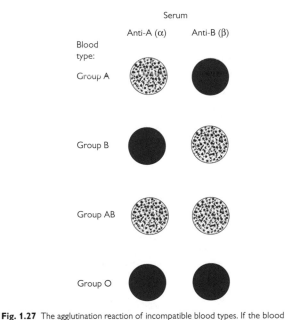

Fig. 1.27 The agglutination reaction of incompatible blood types. If the blood sample is compatible, the mixed blood sample appears uniform. If the blood is incompatible with the serum, it aggregates and precipitates as shown.

Nucleic acids

Molecular structure of nucleic acids

DNA and RNA are nucleic acids:
- Consist of a long polymer of nucleotides (also known as bases)
- Nucleotides are made up from a nitrogenous base, a sugar (deoxyribose in DNA, ribose in RNA), and a phosphate group
- There are two types of nitrogenous base (Fig. 1.28):
  - The purines: guanine (G) and adenine (A)
  - The pyrimidines: cytosine (C) and thymine (T, only in DNA) or uracil (U, only in RNA)
- DNA has a polarity, in that it is always read 5′→3′:
  - The terms 5′ and 3′ refer to the free carbon atom from the sugar which is free at the end of the chain in question:
    ○ The 5′ end has a phosphate on the C5 of the (deoxy)ribose
    ○ The 3′ end finishes with a hydroxyl group on the C3
- DNA strands associate into pairs and run in opposite directions (anti-parallel) (Fig. 1.29)
- Base pairing rules are always followed:
  - C pairs with G with three hydrogen bonds
  - A pairs with T (or U in RNA) with two hydrogen bonds
- The two strands of anti-parallel DNA form a double helix (Fig. 1.30):
  - Elucidated from X-ray diffraction images of crystallized DNA
  - Structure solved by Watson and Crick (1953).[4]

DNA replication

DNA replication is semi-conservative (⊃ p.70).

Amino acid coding

The sequence of the nucleotides codes for amino acids in proteins:
- Codons are triplets of bases that code for an amino acid
- $4^3 = 64$ potential triplets, so some of the 20 amino acids in proteins are coded for by more than one triplet (Table 3.1, ⊃ p.197)
- ATG is the start codon for almost all proteins
- TAG, TGA, and TAA are all stop codons.

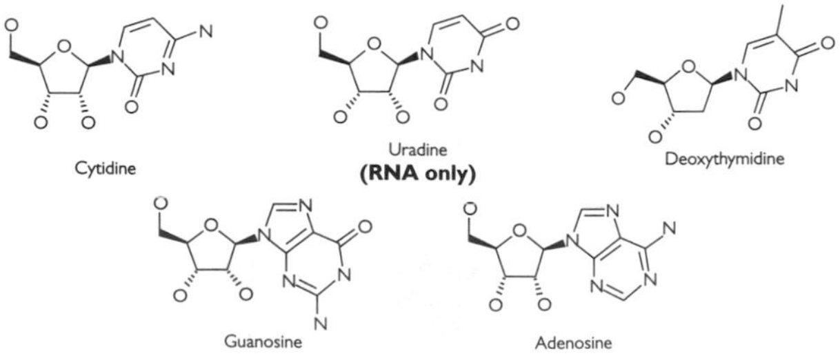

Cytidine	Uradine **(RNA only)**	Deoxythymidine

Guanosine	Adenosine

Fig. 1.28 Types of nitrogenous base.

4 Watson JD, Crick FH (1953). Molecular structure of nucleic acids; a structure for deoxyribose nucleic acid. *Nature* **171**, 737–8.

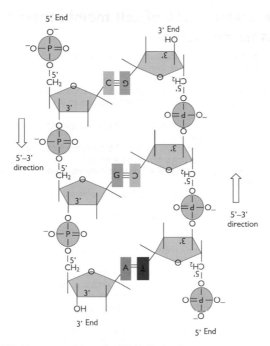

Fig. 1.29 Two anti-parallel strands of DNA (B = base).

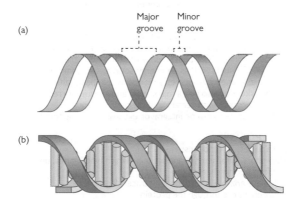

Fig. 1.30 Outline of the backbone arrangements in the DNA double helix, showing the base pairs in the centre of the helix.

Reproduced with permission from Papachristodoulou, D., *Biochemistry and Molecular Biology* 6e, 2018, Oxford University Press.

The organization of cell membranes: the plasma membrane

Plasma membranes establish discrete environments (Fig. 1.31). The cell membrane bounds the cell contents; other membranes establish cell inclusions—the nucleus, mitochondria, endoplasmic (sarcoplasmic in muscle) reticulum, Golgi vesicles, lysosomes, and endosomes.

- Membranes are fluid mosaics: there is a lipid bilayer, in which proteins are embedded
- The fluidity derives from free movement of both lipids and proteins within the membrane
- Membranes can be from 25% protein:75% lipid (myelinated nerves) to 75% protein:25% lipid (mitochondrial membrane)
- Membrane proteins and lipids are often also glycosylated and contribute to the glycocalyx
- There are three lipid classes: phospholipids, sphingolipids, and cholesterol (➲ pp.36–9)
- Bipolar phospholipids are most abundant:
  - They comprise a charged headgroup and two uncharged hydrophobic tails
  - The tails face inwards, thereby forming the bilayer
  - Polar headgroups face the extra and intracellular environments
  - Lipid tails can have kinks due to double bonds
- For most phospholipids, there are four chemical components: fatty acids, glycerol, phosphate, and one other species
- Principal lipids are phosphatidylcholine (lecithin), phosphatidylserine, and phosphatidylethanolamine
- Sphingolipids have a similar structure and can be glycosylated
- Cholesterol, an unsaturated alcohol (sterol), inserts between phospholipids and influences membrane fluidity. This allows lateral diffusion of membrane proteins, important for hormone-receptor coupling.

Membrane composition is asymmetrical: the inner half of the bilayer contains more phospholipids with negatively charged headgroups (e.g. phosphatidylserine). A 'flippase' enzyme (an ABC transporter family member, ➲ p.31) uses ATP as an energy source, sweeping the membrane to preserve asymmetry. Asymmetry maintains cell shape (although proteins and the cytoskeleton also play a major role) and ensures correct association of particular proteins with certain lipid species.

Membrane proteins can be integral or extrinsic (or peripheral).

Integral proteins

- Have a variety of functions:
  - Receptors—transmembrane proteins which bind ligands, undergo conformational changes to:
    - Initiate enzyme activity on an intracellular domain
    - Open a pore through the protein
    - Activate intracellular signalling cascades
  - Transporters: channels or carriers
  - Enzymes
  - Adhesion molecules: integrins for extracellular matrix, cadherins for cells

- Can span the membrane once or may have multiple, closely packed transmembrane segments (e.g. ten spans in the Na$^+$-K$^+$-ATPase). Transmembrane spans are typically 25 amino acid α-helices—hydrophobic residues face the lipid environment of the membrane
- In some cases can only partially cross the membrane or sit at the extra or intracellular faces and link to phospholipids by oligosaccharides (GPI-linked proteins)
- Can be multimers made of non-covalent bound subunits. Multimers may comprise multiple copies of a single protein or two or more different ones
- Can combine in non-mobile clusters or act as anchors for non-covalent binding of extrinsic cytoskeletal proteins.

Extrinsic proteins

- Form non-covalent bonds with integral proteins
- May be components of the cytoskeleton (e.g. spectrin, ankyrin), that bind transmembrane channels, carriers, and adhesion molecules. They define cell shape, strength, and polarity.

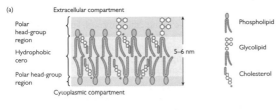

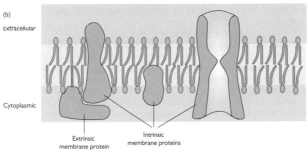

Fig. 1.31 The structure of the plasma membrane: (a) the basic arrangement of the lipid layer; (b) a simplified model showing the arrangement of some of the membrane proteins.

Reproduced with permission from Pocock G, Richards CD (2009). The Human Body 3e, p20 Oxford University Press.

Transport across membranes

Membranes form selective barriers to maintain the composition of different compartments. Without proteins, only small non-polar solutes would permeate the lipid bilayer. Even when lipid permeability exists, movement can be augmented by protein pathways:

- The simplest pathway is solubility diffusion through lipid
- Gases such as oxygen can dissolve in the lipid and then diffuse passively through Brownian motion along concentration gradients according to Fick's law: $J = -D \times A \times \Delta C/\Delta x$

  where J = diffusion flux
  D = diffusion coefficient of solute
  A = surface area of membrane
  ΔC = concentration gradient
  Δx = membrane thickness

  - Diffusion is a passive process: there is no direct energy expenditure involved
- Water also diffuses across membranes by osmosis, from regions of high water concentration (hypo-osmotic) to regions of low water concentration (hyperosmotic):
  - Although water can dissolve in and diffuse through lipids, most cell membranes contain selective aquaporin proteins that convey a high permeability to water and mean that osmotic gradients cannot be sustained. Water movements cease when intracellular and extracellular solutions become isotonic.

Other solutes, including ions and some organic osmolytes such as taurine also move across the membrane by passive diffusion through channels which are water-filled protein pores:

- Channels can be constitutively open (leak channels) or gated, i.e. opened by:
  - Change in membrane potential
  - The extracellular binding of a ligand to the channel itself or to a G-protein-linked receptor associated with the channel
  - The intracellular binding of a second messenger (such as cGMP)
  - Membrane deformation
- Channels can be selective for the ions that can permeate. There are families of channels specific for ions including Na^+, K^+, Ca^{2+}, and Cl^-
- Gap junctions are channels connecting cytoplasm of two cells. Each membrane contains connexons, comprising six connexins which form a pore. The two pores are aligned to make a patent, non-selective channel for electrical and chemical cell–cell communication.

Other transport processes are mediated by carrier proteins (Fig. 1.32):

- Carriers undergo a conformation change to present the solute at the opposing face. The binding and conformation change processes mean that the process is:
  - Slower than channel-mediated transport
  - Temperature sensitive
  - Saturable, with V_{max} and K_m

- Passive carriers mediate facilitated diffusion, e.g. the GLUT carrier for glucose:
  - Like simple diffusion, there is no direct input of energy and the solute gradient is dissipated by the transport process
  - Movement of solute stops when equilibrium is reached. For uncharged solutes, this is when the inward and outward chemical gradients are equal. Charged solutes reach equilibrium when the electrochemical gradients are equal, although an asymmetry of solute concentration at either side of the membrane will still exist.

Active transport processes also exist. These accumulate the transported solute to a level above that predicted by passive equilibration:
- Active transporters undergo conformational changes energized directly or indirectly by ATP hydrolysis
- Primary active transporters are ATPases—the hydrolysis of ATP by the protein initiates a conformational change that translocates bound ions across the membrane:
  - For example, Na^+/K^+-ATPase, keeping intracellular $[Na^+]$ low in cells, the Ca^{2+}-ATPase, which extrudes Ca^{2+} and the H^+/K^+-ATPase in the stomach, which secretes gastric acid
- Secondary active transporters use gradients made by primary systems to energize transport of other solutes. An inward Na^+ electrochemical gradient is most commonly used. A conformational change upon solute binding translocates the two solutes across the membrane. The driver ion gradient dissipates as it energizes accumulation of the substrate. The process can be:
  - Co-transport (symport), e.g. Na^+-glucose co-transport by the SGLT protein
  - Exchange (antiport), e.g. $Na^+ \times H^+$ exchange by the NHE protein.

Intracellular organelle membranes also possess transport pathways. Notable examples include H^+-ATPases in lysosomes, Ca^{2+}-ATPases in mitochondria and in endoplasmic reticulum (ER)/sarcoplasmic reticulum (SR), and Ca^{2+} channels in ER/SR.

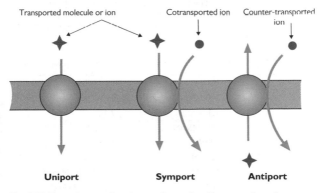

Fig. 1.32 The main types of carrier proteins employed by mammalian cells.
Reproduced with permission from Pocock G, Richards CD (2009). The Human Body 3e, p36 Oxford University Press.

The intracellular milieu

In an average 70kg adult, total body water volume is 42L: there are 14L of extracellular fluid and 28L of intracellular fluid. The intracellular and extracellular compartments differ markedly in their composition, and it is the cell membrane separating the compartments that is responsible for the maintenance of the differences.

Extracellular fluid comprises:
- Blood plasma - 3L
- Interstitial fluid (including bone and connective tissues) - 10L
- Transcellular fluid (considered to be fluid found in enclosed epithelial-lined cavities; for example, synovial fluid, cerebrospinal fluid, pleural fluid, ocular fluid) - 1L

Plasma and interstitial fluid are essentially identical with the exception of proteins, which are unable to exchange across capillary walls and so are largely absent from interstitial fluid. The composition of transcellular fluid is variable.

Typical values for the composition of the extracellular and intracellular compartments are listed in Table 1.2.
- The most important difference is the separation of Na^+ and K^+ ions:
  - The Na^+-K^+-ATPase membrane transporter protein actively extrudes Na^+ ions from and accumulates K^+ ions in the cell.
  - The asymmetry underlies the establishment of the resting membrane potential and the generation of action potentials in excitable cells
- Free $[Ca^{2+}]$ is kept low inside cells by active extrusion of Ca^{2+} ions by a Ca^{2+}-ATPase and by secondary active $Na^+ \times Ca^{2+}$ exchange, and sequestration within the ER by another Ca^{2+}-ATPase (SERCA):
  - Low steady-state $[Ca^{2+}]$ levels can be exploited for cell signalling, when hormones initiate second messenger cascades, which liberate Ca^{2+} from intracellular stores
- Intracellular pH is maintained close to neutrality by active extrusion of H^+ ions by a primary H^+-ATPase or, more commonly, by secondary active Na^+-H^+ exchange. Na^+-driven HCO_3^- uptake can also contribute to acid extrusion by adding HCO_3^- ions to buffer the cytoplasm:
  - Without these active acid extruders, the negative membrane potential would mean that H^+ ions would be passively equilibrated at approximately pH 6.6

Table 1.2 Typical extracellular and intracellular compartment composition values

	Extracellular	Intracellular
Osmolarity	290mOsm	290mOsm
pH	7.4	7.1
$[Na^+]$	140–145mM	5–15mM
$[K^+]$	4.5mM	120mM
$[Cl^-]$	120mM	20–50mM
$[Ca^{2+}]$	1–2mM	1–2mM (10^{-7}M free)
$[Mg^{2+}]$	1mM	18mM (1mM free)
$[HCO_3^-]$	22mM	15mM

- Cl^- ions are accumulated inside cells by Na^+-driven active processes, although the effect is restricted by passive efflux of Cl^- through channels
- A large proportion of intracellular osmotic potential is derived from impermeant structural proteins, the concentration of which can be 300g L^{-1}:
  - These ensure that the intracellular solution, like the extracellular solution, has similar numbers of anions and cations, and hence exhibits bulk electroneutrality (despite varying contributions from different ions)
- The high water permeability of the cell membrane that results from water channels (aquaporins) means that cell volume must change in response to changes in extra- or intracellular osmolarity:
  - Extracellular osmolarity is tightly regulated, although there are regions in the body—notably in the kidneys—where cells can be subjected to varying osmotic conditions
  - Intracellular osmolyte content can vary with metabolic changes or following uptake or loss of solutes across the cell membrane
  - Cells can limit these changes by activating channels and carriers to lose or gain solutes and hence water. These systems rely on ion gradients established by the Na^+-K^+-ATPase.

Cell signalling pathways

Cell function can be regulated by a variety of external chemical factors that interface with the cell at receptors on the cell membrane or, if lipid soluble, at cytoplasmic receptors.

- These factors include:
  - Peptides such as antidiuretic hormone (ADH)
  - Amines such as adrenaline (epinephrine)
  - Steroids such as aldosterone
  - A diverse array of small signals (nucleotides, ions, gases)
- These regulators can be derived from:
  - The cell itself (termed autocrine regulation)
  - Nearby cells (paracrine regulation)
  - Cells located at some distance and delivered by the bloodstream (endocrine)
  - From nerve endings (neurocrine).

The agonist or ligand binds to the receptor protein and initiates a conformational change in the protein. Thereafter, a number of direct or indirect responses can be elicited (Fig. 1.33):

- Ligand-gated ion channels (ionotropic receptors) are receptor proteins that operate as ion channels upon occupancy. Exemplified by the nicotinic receptor of skeletal muscle cells, which operates as a cation channel when acetylcholine binds
- Catalytic receptors act as enzymes or are associated with enzyme complexes which are activated upon occupancy:
  - Insulin and many growth factors initiate serine/threonine or tyrosine kinases within the receptor protein, or associated with it, in this way
  - Alternatively, receptor guanylate cyclases (e.g. the ANP receptor) can convert GTP to cGMP which, in turn, activates PKG (a serine/threonine kinase)
- Receptors can also be coupled, through a GTP-binding protein (G-protein), to effectors:
  - A G-protein is a heterotrimeric complex (Fig. 1.34). Agonist binding promotes interaction of the receptor with a G-protein, which undergoes a conformational change. GTP binding ensues, which causes the trimer to split into α and $\beta\gamma$ units
  - G-protein-linked receptor occupancy can initiate stimulatory (G_s) or inhibitory (G_i) effects, largely through α subunit actions on effector enzymes to generate second messenger signalling molecules:
    - Adenylate cyclase: converts ATP to cAMP. cAMP activates PKA, which phosphorylates serine and threonine residues on proteins to alter their conformation and modulate their function. Effects are reversed by phosphatases that dephosphorylate target proteins. The activation of glycogenolysis by adrenaline is mediated by PKA activation of phosphorylase kinase
    - Phospholipase C: converts membrane phosphatidyl inositols to IP_3 and diacylglycerol (DAG). IP_3 is released into the cytoplasm where it binds to a receptor on the ER, which acts as a channel for release of Ca^{2+} ions into the cytoplasm. Ca^{2+} exerts its effects through Ca^{2+}-binding proteins such as calmodulin which activate serine/threonine kinases. Myosin light chain kinase can be activated in

this way. DAG remains in the membrane where it associates with cytoplasmic PKC, a Ca^{2+}-dependent serine/threonine kinase, and increases its affinity for Ca^{2+}. PKC mediates growth factor actions such as cell shape changes and proliferation

o Phospholipase A_2 (PLA_2): converts membrane phospholipids to arachidonic acid. Agonists activating PLA_2 include serotonin and glutamate. Arachidonic acid is a precursor for eicosanoids. The eicosanoids are involved in inflammatory responses and modify blood vessel diameter, platelet activity, and cell membrane permeability. Cyclo-oxygenase (COX; inhibited by aspirin) generates prostaglandins, prostacyclins, and thromboxanes; 5-lipoxygenase generates leukotrienes

- After interaction with effectors, the G-protein trimer is reformed by α-catalysed hydrolysis of GTP to GDP, which is displaced for GTP when an agonist–receptor complex next interacts
- Each subunit exhibits a number of isoforms, with differences in tissue distribution and function apparent. The huge variety of subunit combinations which can be assembled provides for enormous diversity in signalling possibilities
- G-proteins can also have direct actions on effectors: an α subunit directly mediates the activation of L-type Ca^{2+} channels following β-adrenoreceptor occupancy in the heart. Likewise, a $\beta\gamma$ subunit complex can directly activate the muscarinic M_2 acetylcholine receptor-linked K^+ channel in the heart
- A family of small G proteins also exists, which are structurally similar to the α subunit. The Ras isoforms regulate cell growth by regulating kinase cascades from the cell membrane to the nucleus. Mutations in Ras lead to oncogenes which encode constitutively active Ras pathways and induce malignant transformation of the cell
- Intracellular or nuclear receptors. These proteins bind to lipid-soluble agonists (e.g. steroid hormones such as aldosterone) which can cross the plasma membrane. Receptors may be located in the cytoplasm or in the nucleus. Cytoplasmic ones translocate to the nucleus after receptor occupancy and dissociation from heat shock chaperone proteins. The receptor conformational change ultimately leads to a change in DNA conformation which initiates transcription; responses are therefore slower than those of other receptor types.

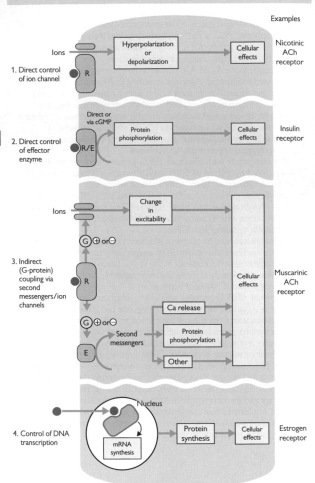

Fig. 1.33 The principal ways in which chemical signals affect their target cells. Examples of each type of coupling are shown. (R = receptor; E = enzyme; G = G-protein; + indicates increased activity; − indicates decreased activity.)

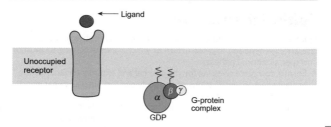

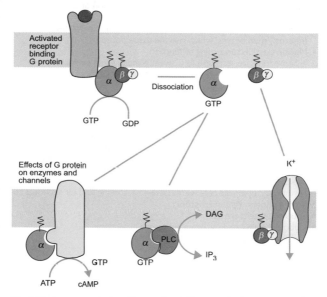

Fig. 1.34 Receptor activation of heterotrimeric G-proteins leads to activation of enzymes and ion channels.

Reproduced with permission from Pocock G, Richards CD (2009). The Human Body 3e, p54 Oxford University Press.

Epithelial structure

Epithelia are planar sheets of cells and associated connective tissue, the precise arrangement and microstructure of which varies according to location and specialization. Epithelia:

- Form a complete covering for the internal and external surfaces of the body (including invaginations of the body surfaces such as sweat glands and the exocrine pancreas)
- Separate the external environment from the internal milieu
- Form the major functional component of several organs (e.g. gastrointestinal (GI) tract, urogenital system, skin)
- Their principal function is to mediate the selective transfer of substances and, in so doing, control the composition of the internal environment
- Possess cells that have discrete apical (external) and basolateral (internal) membranes, which give the tissue polarity
- May be classified according to function:
  - Some mediate secretion (e.g. sweat glands), some perform absorption (e.g. the epithelium of the GI tract), others fulfil a barrier role (e.g. skin)
  - Most epithelial cells exhibit more than one of the earlier listed three functional characteristics and certain specialized epithelia also play a role in sensation and movement (mucociliary escalator).

Epithelial cells are invariably associated with a connective tissue layer at their basal surface which they synthesize. This basement membrane contains specialized collagen protein (type IV) and provides structural integrity to the epithelial cell layer above. In certain locations, the basement membrane itself is highly specialized (e.g. in the glomerulus of the kidney).

Historically, epithelial cells were classified according to their sectional appearance under a light microscope:

- Squamous: cells are flat
- Cuboidal: cells are approximately square in section
- Columnar: cells significantly taller than they are wide.

Epithelia are arranged into single or multiple layers of cells:

- Simple epithelium: single layer of epithelial cells
- Stratified: several layers of cells
- Pseudo-stratified: single layer of cells which has the appearance of a stratified epithelium since nuclei are arranged at different heights.

Such distinctions are still in use, but they only provide limited information as to the functional specializations of different epithelia.

The arrangement of epithelial cells into layers accounts for many of their functional specializations. This arrangement is maintained by cell junctions between epithelial cells which bind them together and provide epithelia with their gross structure. Broadly speaking, three types of cell junction between epithelial cells exist:

- Occluding or tight junctions: contribute to the barrier function of epithelia by linking individual cells tightly together. They also maintain the polarity of epithelial cells by separating the apical and basolateral membranes. They are formed from membrane proteins (occludins and claudins) that adhere epithelial cells together near to their apical surfaces. The extent to which they do this varies between epithelia and determines whether an epithelia is 'tight' or 'leaky'

• Anchoring junctions: provide epithelia with the ability to resist shearing and tensile forces by linking cytoskeletal elements between epithelial cells as well as to the extracellular matrix. In this way they bind epithelial cells and associated extracellular matrix together into a functional unit. Anchoring junctions comprise a component of the cytoskeleton, a cytoplasmic link-protein, and a cell–cell adhesion molecule bound in series:

  • Adherens junctions: bind the actin networks of epithelial cells together via catenins (link-protein) and cadherins (cell–cell adhesion molecule). They are particularly associated with columnar epithelial cells, where they form an adhesion belt
  • Desmosomes: connect cytoskeletal intermediate filaments between adjacent epithelial cells. The cytoplasmic link proteins are desmoplakins and the cell–cell adhesion molecules are desmogleins. The formation of antibodies to desmosomal proteins results in widespread blistering of the skin and mucous membranes—an autoimmune condition known as pemphigus
  • Hemidesmosomes: connect intermediate filament networks to the extracellular matrix. They resemble desmosomes except that instead of desmogleins, desmoplakins are bound to connective tissue receptors called integrins. They are important for the attachment of epithelial cells to the underlying basement membrane

• Communicating (gap) junctions: mediate cell–cell communication by allowing selective diffusion of small molecules and ions between adjacent cells. They comprise proteins called connexins, which combine to form a conducting pore (connexon) in cell membranes. Connexons in adjacent epithelial cells align, allowing direct cell–cell communication. In epithelia, they are particularly important in embryogenesis, where they play a role in organization of developing sheets of cells. Gap junctions are regulated by intracellular pH and Ca^{2+}.

Often, several types of epithelial junction are found in close proximity to each other. This is known as a junctional complex.

Epithelial cell specializations

Structural cell surface specializations occur on certain epithelial cells, normally to increase the surface area of the cell or to move foreign particles:

• Cilia: long cytoplasmic extensions (5–10µm long, 0.25µm diameter) from the surfaces of some epithelial cells. They are motile and are important in moving fluid over cells. They are particularly important in respiratory epithelia
• Microvilli: small, cytoplasmic projections (1µm long, 0.08µm wide), found on the cell surface of certain epithelial cells. In the intestine, for example, the mass of microvilli on the cell surface forms a brush border, aiding the absorption of nutrients
• Basolateral folds: deep folding of the basal or lateral surface of the epithelial cells. Important in fluid or ion transport functions of cells (e.g. renal tubular cells).

Epithelial function

Epithelia are layers of cells that isolate the internal from the external environment. They regulate the movement of solutes and water to and from the body. Examples include the skin, the linings of the respiratory tract, alimentary canal, and kidney tubules.

Epithelia can be divided into two categories:

- Absorptive: active Na^+ transport drives solute and water reabsorption
- Secretory: active Cl^- transport drives fluid secretion.

The cells of epithelia exhibit considerable diversity of form and function but have fundamental properties in common. They are formed into sheets, which may be multilayered, held together by tight junctions at their luminal edge. They are separated from neighbouring cells by lateral intercellular spaces.

Vectorial transport

- The ability to translocate ions from one compartment to another (unidirectional transport) is the cardinal property of epithelia
- It is achieved by asymmetry of the cell membranes at the two faces. The cells are termed 'polarized' (but not in the sense that a nerve cell is 'polarized'):
  - The external-facing membrane is the apical (or luminal or mucosal) membrane
  - The internal-facing membrane is the basolateral (or contraluminal or serosal) membrane
- Membranes have different morphology (villi), biochemistry (protein distribution), and function (ion selectivity), and remain separated by the tight junctions which form a barrier (to varying degrees) to solutes and water.

Tight and leaky epithelia

Epithelia can be tight or leaky (Table 1.3):

- In tight epithelia, tight junctions prevent significant movement of molecules between cells
- In leaky epithelia, the tight junctions form imperfect seals and are a low-resistance, leak pathway ('shunt') for ions and water:
  - Leaky epithelia perform bulk handling of isosmotic solutions (either for absorption or secretion) = 'valves'. Located in proximal parts of the kidney and GI tract (e.g. proximal tubule, small intestine)
  - Tight epithelia withstand large osmotic gradients; they are more selective in the way they handle the load with which they are presented. Located distally (e.g. collecting duct, colon)
- By placing the two different types of epithelia in series, bulk absorption, followed by fine control, is achieved.

Epithelial solute transport can be transcellular (though the cells) or paracellular (between the cells). The former is ultimately dependent on active transport processes, while the latter occurs passively, by diffusion or convection.

Direction of transport depends on electrical and chemical gradients for ions, and osmotic and hydrostatic pressure gradients for water. Ion movements lead to charge separation—establish a potential difference (p.d.)

Table 1.3 Tight vs leaky epithelia

	Tight	Leaky
Tight junctions	Complex	Simple
Paracellular ion permeability	Low	High
Electrical resistance	High	Low
Transepithelial p.d.	High (30mV)	Low (5mV)
Water permeability	Low*	High
Apical entry of sodium	Channels	Carriers

* May be raised in collecting duct by ADH-induced insertion of water channels.

across the epithelium. Orientation of p.d. depends on which ions move, and in which direction. Magnitude of p.d. depends on whether the epithelium is leaky (so that charge dissipates or 'shunts').

Tight epithelia have low water permeability: it may be upregulated by channel insertion (e.g. collecting duct). The water permeability of leaky epithelia is high and unregulated, and can be transcellular—through water channels (aquaporins)—or paracellular. Note that high water permeability does not necessarily derive from the leaky tight junction.

Absorptive epithelia

- In absorptive epithelia, active transport of Na^+ ions is the fundamental event
- Described by the Ussing model (Fig. 1.35):
  - Na^+ concentration is kept low in cells by the basolateral Na^+-K^+-ATPase

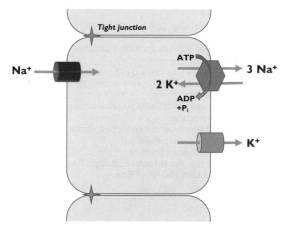

Fig. 1.35 Ussing model.

- Na$^+$ ions move down an electrochemical gradient into the cell across the apical membrane
- The transepithelial movement of Na$^+$ ions leaves the lumen negative with respect to the contraluminal side
- The vectorial transport of sodium and associated solutes depends on the specific permeability properties of individual membranes, which are in turn determined by a 'pick and mix' from a selection of transport proteins
- Basolateral membranes have properties in common with 'regular' cells (nerve, muscle) and possess:
  - Na$^+$-K$^+$-ATPase
  - K$^+$ leak channels = high permeability to K$^+$ (P$_K$)
  - Low permeability to Na$^+$ (P$_{Na}$)
  - Ca^{2+}-ATPase
  - Cl$^-$-HCO$_3^-$ exchanger ('band 3')
  - Hormone receptors
- Apical membranes all have high P$_{Na}$:
  - In tight epithelia, this is due to Na$^+$ channels (ENaC), inhibited by amiloride and regulated by aldosterone
  - In leaky epithelia, carrier-mediated symports and antiports are also present, appropriate for specific tissue functions (e.g. Na$^+$-glucose for sugar absorption in small intestine)
  - Passive movement via paracellular route down electrochemical gradients established by sodium-coupled movements also occurs in leaky epithelia
- Note that all absorptive processes depend in some way on the sodium gradient, established by the Na$^+$-K$^+$-ATPase in the basolateral membrane.

Secretory epithelia

In secretory epithelia (e.g. lung, pancreas), the underlying process is active Cl$^-$ transport:

- Cl$^-$ ions enter the epithelial cells across the basolateral membrane on a carrier: often Na$^+$-K$^+$-2Cl$^-$ symport
- Cl$^-$ ions accumulate within the cell and exit passively down a chemical gradient through Cl$^-$ channels (which are incorrectly inserted or regulated in cystic fibrosis)
- The exit of Cl$^-$ sets up a Cl$^-$ diffusion potential across the membrane = lumen negative
- Na$^+$ ions move passively across the epithelium via the paracellular pathway, driven by the transepithelial p.d.
- H$_2$O follows along an osmotic gradient
- Note that this process is, like absorption, is dependent on the asymmetrically distributed Na$^+$-K$^+$-ATPase.

Organelles: overview

Eukaryote cells are distinguished from prokaryotic bacterial cells by the presence of organelles which are distinct membrane-bound compartments within the cell:

- 90% of total cell membrane is intracellular
- Many important metabolic reactions occur within or on these organelle membranes.

Metabolic compartmentalization is an essential role of the organelles, allowing both oxidative and reductive conditions in the cell, separating anabolic and catabolic reactions, and containing toxic compounds to prevent widespread cellular damage.

Major organelles in the cell include:

- Nucleus:
  - Contains the genetic material (chromosomes) encased in a double membrane perforated with nuclear pores which allow movement of macromolecules in and out of the nucleus
  - Also contains the nucleoli, assemblies of RNA and protein involved in ribosome production
- Mitochondria:
  - Responsible for the production of ATP—the cellular fuel
  - Bounded by a double membrane. The inner membrane is highly folded into cristae (increases surface area)
- Endoplasmic reticulum (ER)—interconnected tubular membranes (cisternae) in the cytoplasm:
  - Rough ER (RER):
    - Generally, found closer to the nucleus
    - The 'rough' refers to the ribosomes, which give a studded appearance and are actively involved in protein synthesis
  - Smooth ER:
    - Involved in packaging and delivery of proteins to the Golgi apparatus
    - Site of membrane lipid synthesis
    - Contains cytochrome P450, which plays a role in detoxification of drugs and toxic compounds, especially in liver (➲ p.608)
- Golgi apparatus:
  - Along with RER, a prominent Golgi is associated with actively secreting cells
  - Consists of a stack of flattened membrane bound vesicles, which can be distinguished into cis-, median-, and trans-Golgi regions
  - As proteins made in the RER pass through the Golgi complex, they are modified and processed (e.g. glycosylation) before entering the trans-Golgi network (TGN) for sorting and delivery to the appropriate target
- Ribosomes:
  - Protein and RNA aggregates that catalyse the manufacture of proteins
  - Many are found attached to the RER for making secreted proteins; plus there are also free cytoplasmic ribosomes

- Lysosomes:
  - Bounded by a single membrane and containing lytic enzymes
  - Involved in digestion of ingested macromolecules and turnover of intracellular components
- Peroxisomes:
  - Contain enzymes involved in oxidative metabolism which use molecular oxygen and generate toxic hydrogen peroxide (H_2O_2)
- Cytoskeletal elements—different types of contractile proteins:
  - Microfilaments:
    - Mostly made up of actin with regulatory proteins
    - Involved in cell movement
  - Intermediate filaments:
    - Form α-helical structures
    - Contribute to mechanical strength and stability of cells
  - Microtubules:
    - α- and β-tubulins forming hollow tubes
    - Involved in chromosome separation in cell division (mitosis), intracellular transport of vesicles and organelles, and movement of cilia.

NB: organelles are not independent of each other. Indeed, there is a large amount of vesicular movement between organelles, especially between the ER, the Golgi, and the TGN involved in protein trafficking.

The nucleus

The presence of a nucleus defines a cell as eukaryotic, and thus all cells in the human body have a nucleus, except mature red blood cells, which are enucleate (although the precursor cells they are derived from are nucleated).

Features of the nucleus

- Approximately 7–8μm in diameter
- The nuclear contents are kept separate from the cytoplasm by the nuclear membrane. This double membrane contains pores that allow macromolecules to cross. This is a two-way process. For example, nucleotides need to enter the nucleus, and messenger RNA (mRNA) leaves the nucleus to be translated in the cytoplasm:
  - The space between the inner and outer membrane is called the perinuclear space
  - The inner nuclear membrane (the nuclear lamina) consists mainly of a scaffold-like network of protein filaments (lamins or intermediate filament type V):
    - Proposed roles of the nuclear lamina include maintenance of nuclear shape and spatial organization of nuclear pores, as well as in transcription and DNA replication
- The nucleus contains 46 chromosomes:
  - DNA–protein complexes representing the hereditary material
  - 22 homologous pairs, plus the sex chromosomes (XX = female, XY = male)
  - The chromosomes are only visible as distinct entities during cell division (**◉** mitosis p. 78 (cell cycle) or meiosis p. 86 (meiosis)), when maximally condensed. The rest of the time they are unwound and dispersed (chromatin):
    - Heterochromatin (appears darker)—relatively more condensed form
    - Euchromatin (lighter region) is less dense and contains most of the active genes
    - The Barr body is the inactive X-chromosome in female cells which appears as a darkly stained mass of chromatin
- There are also several nucleoli, which are dense regions of RNA and protein (nucleoprotein):
  - These are regions involved in the production of new ribosomes for export into the cytoplasm through the nuclear pores
  - They are associated with particular chromosomes which have the genes for ribosomal RNA (in humans—chromosomes 13, 14, 15, 21, and 22).

Functions of the nucleus

- Gene replication and repair:
  - As well as DNA replication during cell division, there are also mechanisms to repair DNA to maintain the integrity of the hereditary material
- Genetic transcription:
  - Production of mRNA that will be translated into proteins in the cytoplasm by ribosomes
- Ribosome production:
  - Production of ribosomes in the nucleoli.

Gene replication

DNA replication needs to occur before cell division (⊃ mitosis p. 78) so that each daughter cell has a complete set of hereditary material. It is obviously important that the process should be highly faithful.

- Newly synthesized DNA requires packaging into nucleosomes and chromosomes
- Thus DNA replication also requires significant protein synthesis
- DNA synthesis requires a supply of nucleotides bases—the dNTPs:
  - Anti-cancer drugs (⊃ OHCM10 p.376) (e.g. methotrexate, 5-fluorouracil) stop cell division by interfering with dNTP supply.

DNA replication occurs by a semi-conservative process

- Each daughter molecule has one DNA strand from the parent and one newly copied strand
- Shown by Meselson and Stahl[5] in a classic experiment (Fig. 1.36):
  - *Escherichia coli* were grown in media containing ^{15}N, which was incorporated into their DNA
  - These cells were suddenly switched into ^{14}N-containing media. First-generation cells had DNA which was 50% ^{15}N, 50% ^{14}N.

Mechanism of DNA replication

- Helical double-stranded DNA has to be unwound by DNA helicase:
  - DNA gyrase (a topoisomerase enzyme) stops unwound single DNA strands from getting tangled by a breaking and rejoining mechanism
  - Single-stranded DNA binding proteins bind to and stabilize the single strands
  - There are now two strands: the leading and the lagging strand
- DNA is copied by DNA polymerases in a 5'→3' direction (Fig. 1.37):
  - These enzymes cannot initiate a new strand of DNA
  - RNA primase (an RNA polymerase) makes a short RNA primer (10–20 bases)
  - For the leading strand, DNA polymerase can extend this directly 5'→3'
  - However, the lagging strand has a 3'→5' direction and cannot be copied directly:
    - ○ It is copied in small pieces (Okazaki fragments), each primed with RNA
    - ○ When the DNA polymerase encounters the RNA of a previously made fragment, a 5'→3' RNase H removes the RNA and DNA polymerase replaces it with DNA
    - ○ DNA ligase joins the two adjacent DNA fragments together
- DNA polymerases use one strand of the original DNA as a template, sequentially adding the correct nucleotide using the base pair rules (C&G, A&T). These pairings need to very accurate for faithful DNA replication:
  - DNA polymerases have a proof-reading facility to enhance accuracy.

---

5 Meselson M, Stahl F W (1958). The replication of DNA in *Escherichia coli*. *Proc Natl Acad Sci U S A* **44**, 671–82.

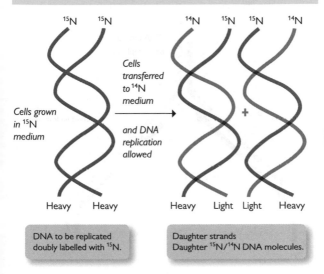

Fig. 1.36 Demonstration of semi-conservative DNA replication by Meselson and Stahl.

Reproduced with permission from Pocock G, Richards CD (2009). *The Human Body* 3e, p20 Oxford University Press.

DNA damage and repair

- Factors such as UV light and chemical agents can damage DNA, → modified bases and/or base mismatches
- Enzymes recognize damage and repair it by nicking the strand (endonuclease), removing the incorrect/damaged bases (exonuclease), filling in the gap (DNA polymerases), and then joining up the strand (DNA ligase).

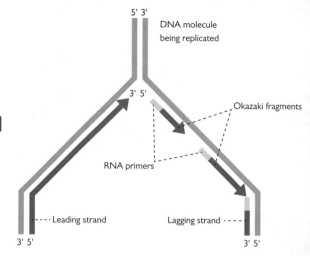

Fig. 1.37 Diagram of a replicative fork. The leading strand is synthesized continuously, while the lagging strand is synthesized as a series of short (Okazaki) fragments.

Reproduced with permission from Pocock G, Richards CD (2009). *The Human Body* 3e, p20 Oxford University Press.

Trafficking

All protein synthesis takes place on ribosomes in the cytoplasm (either free or associated with RER), yet the final destination of proteins is very varied (e.g. secreted, integral membrane, intracellular organelles, cytoplasmic). Trafficking is the term given to the movement of compounds from their site of manufacture to their target site (Fig. 1.38).

Vesicle trafficking routes

- Proteins which are destined for secretion, insertion into the membrane, or targeting to organelles (e.g. lysosomes) are made on the RER:
  - Those which are destined for secretion or an intraorganelle space have a signal sequence at their N-terminus which allows them to enter the RER lumen
  - Integral membrane proteins also have a signal sequence and are inserted into the RER membrane
  - Often, the signal peptide is cleaved after translation
- Vesicles bud off from the RER and move into the Golgi apparatus, where processing and glycosylation takes place

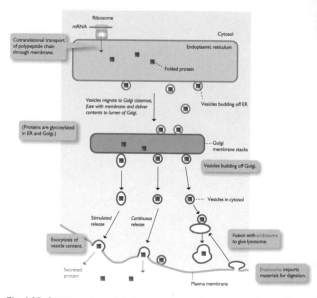

Fig. 1.38 Overview of protein trafficking: how proteins are secreted from cells and how enzymes are delivered to lysosomes.

Reproduced with permission from Pocock G, Richards CD (2009). *The Human Body* 3e, p20 Oxford University Press.

- After passing through the three parts of the Golgi (cis-, median-, and trans-regions), vesicles again bud off and enter the TGN, where they are sorted and targeted to the correct destination (e.g. plasma membrane, lysosomes)
- The arrival and fusion of the vesicle at the plasma membrane either adds the new proteins to the membrane or allows the contents of the vesicle to be released into the extracellular medium. There needs to be membrane retrieval at the same rate as addition, otherwise the cell would increase in size
- Proteins also can enter the nucleus and mitochondria, but do so after synthesis:
  - These are synthesized in the cytoplasm on free ribosomes
  - Signal sequences allow proteins to enter the mitochondria post-translationally
  - Nuclear proteins can enter via the nuclear pores—again a signal sequence is responsible for the targeting
- Chaperonins are required for assembly of large oligomeric proteins into functional active complexes in mitochondria and ER lumen
- Examples of inherited disorders of trafficking:
  - Lysosomal storage diseases → secretion of harmful degradative enzymes into the bloodstream rather than targeting into lysosomes (e.g. I-cell disease)
  - Primary hyperoxaluria (⊃ OHCM10 p.118) → peroxisomal enzyme mistargeted to mitochondria, resulting in an inability to metabolize oxalate

Secretion

There are two basic types:
- Constitutive—where the proteins are secreted as soon as they are synthesized and processed
- Regulated—proteins are synthesized, processed, and stored in vesicles before being released when a particular signal is received.

Receptor-mediated endocytosis

- Macromolecules which cannot pass through the plasma membrane may enter cells through receptor-mediated endocytosis (Fig. 1.39):
  - Receptors are present in the plasma membrane, often in clusters
  - Clathrin coated pits → vesicles (endosomes)
  - Lysosomes are the usual recipient of the vesicles, where their hydrolytic activity acts on the endocytosed contents
- Low pH is essential for enzyme activity—achieved by H^+-ATPase activity.

Transcytosis (pinocytosis)

- Way for macromolecules to cross cells (especially endothelial cells). Compounds are taken up in vesicles on one side, cross the cell in the vesicle, and are released by exocytosis on the other.

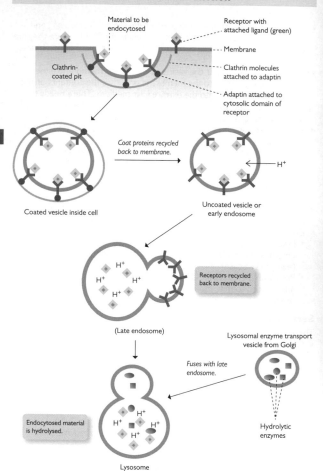

Fig. 1.39 Formation of lysosome by receptor-mediated endocytosis.
Reproduced with permission from Pocock G, Richards CD (2009). *The Human Body* 3e, p20
Oxford University Press.

Cell cycle

Many cells in the body divide to replace cells that are lost through maturation and apoptosis or to respond to increased work imposed on a tissue. Multiple processes need to be carried out before a cell can divide:
- Replication of chromosomes
- Segregation of chromosomes into two diploid sets
- Division of cytoplasm and cell membrane.

These processes have many potential problems because they involve large-scale replication and segregation of the genome. The cell cycle (Fig. 1.40) is a useful way of controlling the process of cell division into sequential steps which can be policed to ensure integrity of the cell progeny. There are several important checkpoints during the cycle which ensure that all necessary actions have been performed before progression to the next stage (e.g. that all chromosomes have duplicated before mitosis). These are detailed in the descriptions of the phases. These checkpoints are all controlled by the levels of checkpoint-specific cyclin-dependent kinase proteins.

Phases of the cell cycle

G1 (gap 1) phase
- Variable length ∴ major determinant of overall length of cell cycle
- Most important phase for growth ∴ high metabolic requirement
- Contains the restriction point = cellular 'decision' as to whether to progress to S phase, and thus irreversibly to cell division, or to enter resting phase (G0).

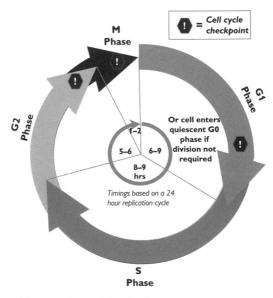

Fig. 1.40 Schematic diagram of the cell cycle.

S (synthesis) phase
- Phase of DNA replication.

G2 (gap 2) phase
- Phase of chromosome packaging
- Synthesis of proteins required for mitosis
- Checkpoint at the end of G2 to ensure that all DNA has been replicated before mitosis and that environmental conditions are favourable.

M (mitosis) phase
- Actual physical division into two cells
- Visible as mitotic figures in histological sections
- Subdivided by the morphology of the chromosomes seen in histological sections (Fig. 1.41):
  - Prophase—chromosomes begin to condense to discretely visible structures
  - Metaphase—chromosomes line up on the equator of the nuclear spindle
  - Anaphase—chromosomes begin to pull apart into the two separate clusters
  - Telophase—chromosomes are now in the two tight clusters which will form the daughter cell nuclei
- The nuclear spindle controls the segregation and movement of the chromosomes:
  - Spindle is composed of microtubules
  - Kinetochore attaches the centromere of each chromosome to the spindle
  - The spindle-attachment checkpoint only allows progression from metaphase to anaphase when all the chromosomes are attached to the nuclear spindle
  - Drugs such as vinblastine and colchicine destabilize the nuclear spindle microtubules and arrest mitosis, sometimes for days.

G0 (gap nought) phase
- Not strictly part of the cell cycle but the resting state outside the cell cycle
- Some cells remain in G0 until they die through senescence
- Other cells remain in G0 for a variable length of time before re-entering the cell cycle.

Cell cycle control in human disease and disease treatment
Neoplasia
- Loss of cell-cycle checkpoint control points (= mutator phenotype) → accumulation of DNA damage and disarray → development of further malignant characteristics
- Increased throughput through cell cycle → increase in number of neoplastic cells in relation to normal cells → overgrowth of neoplastic cells.

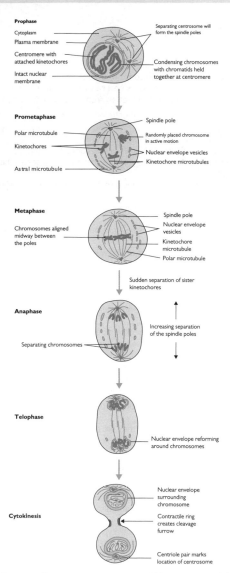

Fig. 1.41 Diagrams of the subprocesses within the M (mitotic) phase of the cell cycle.

Reproduced with permission from Pocock G, Richards CD (2009). *The Human Body* 3e, p24 Oxford University Press.

Atherosclerosis

(⊅ OHCM10 p.657.)
- Proliferation of smooth muscle cells in atherosclerotic plaques →
  increased vascular stenosis
- Therapies are aimed at slowing or stopping cell cycle in these cells at
  time of angioplasty or stent placement to prevent restenosis.
  (e.g. Sirolimus-coated stents)

Alzheimer's disease

(⊅ OHCM10 p.488.)
- May be due to re-entry of quiescent neuronal cells into the cell cycle
  with abortive attempt at DNA replication → cell death by apoptosis
  after failure to pass checkpoint control.

Radiotherapy

(⊅ OHCM10 p.526.)
- Induces DNA damage so entry into mitosis through the checkpoint at
  the end of G2 will either be delayed, while the DNA is repaired, or
  permanently prevented with apoptosis of the cell, if the DNA damage
  is irreparable
- This is one of the mechanisms of therapeutic radiotherapy for the
  treatment of cancer.

Drug therapy of cancers

(⊅ OHCM10 p.524.)
- Destabilization of the microtubules of the nuclear spindle (e.g.
  vinblastine, colchicine)
- DNA damage, e.g. alkylating agents, such as chlorambucil, which
  covalently bind to DNA by alkyl groups
- Drugs which bind to topoisomerase II (e.g. doxorubicin) which prevent
  this protein from functioning as the cleavage complex during DNA
  replication → prevention of entry to mitosis at the checkpoint at the
  end of G2.

Cell growth

The term 'cell growth' is used rather misleadingly in many texts, since strictly it refers to increase in size of cell without division, but it is usually used to mean increase in size of a tissue or organ by cell division. The cell cycle is the basic mechanism by which proliferation occurs, but it requires regulation by other factors, including growth factors and growth inhibitors.

Growth factors may arrive at a cell by three different routes:
- *Autocrine*—when the factor is produced by the cell itself but acts back on it to stimulate growth. This may sound odd, and a positive feedback loop could lead to overproliferation, but it is an indicator of the metabolic well-being of the cell, which suggests that there is 'room' for more similar cells
- *Paracrine*—when the factor is produced by a cell in close proximity to the cell it is affecting, mediated by short-range soluble molecules
- *Endocrine*—when the factor is produced at some distance to the affected cells and carried to it by the blood. For example, thyroid-stimulating hormone (TSH), produced by the pituitary, which stimulates the growth of thyroid epithelial cells and the production of thyroxine.

Growth factors

There are a huge number of identified growth factors with generally long and complicated names that suggest where they were first discovered (e.g. vascular endothelial growth factor, transforming growth factor β). These names are confusing and often specifically named growth factors have a proliferative action on many different tissues.

The mechanism of action of these growth factors has a generic pattern:
- The growth factor binds to a specific transmembrane cell surface receptor
- The cell surface receptor either has an intrinsic enzyme (usually kinase) activity on its intracellular domain or is linked to a second messenger molecule which has such an activity
- This sets off a signal transduction chain to the nucleus, where there is activation of transcription regulation factors → the transcription of more proteins → cells to pass into and through the cell cycle.

The specifics of each pathway are complex: it is enough to know if therapeutic intervention is planned in that specific area.

Growth inhibitory factors

In addition to positive growth factors, there are also a series of counterbalancing inhibitors of growth. These act by similar mechanisms to growth factors but result in the increased transcription of genes which code for inhibitors of the cell cycle, such as p27.

Apoptosis

Definition

Genetically regulated form of cell death affecting individual cells. Distinguish from necrosis, which is the death of many adjacent cells due to some factor extrinsic to them (e.g. ischaemia). Apoptosis is derived from the Greek word meaning 'dropping off'. Importantly, unlike necrosis, apoptosis is not pro-inflammatory.

Morphology

- Membrane blebbing
- Cell shrinkage
- Condensation of chromatin
- Fragmentation of DNA
- Expression of apoptotic markers on the cell surface to mark for phagocytic clearance
- Phagocytosis by macrophages.

Initiators of apoptosis

- Deprivation of survival factors (e.g. interleukin (IL)-1)
- Proapoptotic cytokines (e.g. Fas, tumour necrosis factor)
- Irradiation—gamma and ultraviolet
- Anti-cancer drugs.

Intracellular regulators of apoptosis

- bcl-2—suppresses the apoptotic pathway
- p53.

Effectors of apoptosis

Caspases—a family of enzymes that cleave proteins close to aspartate residues. They have specific intracellular targets such as proteins of the nuclear lamina and cytoskeleton.

Physiological roles of apoptosis

- *Growth and development*—loss of redundant tissue during organ development, e.g. interdigital webs, the majority of human neuronal cells produced during development also die during development
- *Control of cell number in adult life*—regulation of balance between proliferation and cell death.

Role of apoptosis in disease

- *Increased apoptosis*—acquired immune deficiency syndrome (AIDS), neurodegenerative diseases, post-ischaemic injury, hepatitis, graft-versus-host disease
- *Decreased apoptosis*—many malignancies, autoimmune disorders (e.g. systemic lupus erythematosus).

Medical therapies which modulate apoptosis

- Aspirin and other non-steroidal anti-inflammatory drugs (NSAIDs)—protect against colorectal adenomas and cancer by inducing apoptosis of colorectal epithelium through inhibition of COX-2
- *Anticancer drugs and radiotherapy* (⊃ OHCM10 pp.524, 526)—induce apoptosis in tumour and normal tissues; both p53-dependent and independent mechanisms are recognized.

Differentiation

Humans are all developed from single cells, but the adult body contains about 250 specific cell types arranged into a large multiorgan system. The process by which a single progenitor cell produces these millions of specialized cells is called differentiation. The overall pattern of organization within a cell, which is synonymous with differentiation, is often referred to as its phenotype. Phenotype can be described at many different levels from its appearances by light microscopy, through to the pattern of proteins in its cytoplasm defined by mass spectrometry. The phenotype is often contrasted with the genotype of a cell (the description of its genetic material) but this boundary is becoming increasingly blurred as more and more interactions are being discovered between the genome and the rest of the cell.

Morphology

The change in a cell from an undifferentiated to differentiated phenotype can be visualized by light microscopy and by electron microscopy (for tissue-specific organelles). Different types of epithelia are easily distinguished, but other specialized cells, especially lymphocytes, may require immunohistochemistry of cell surface proteins since their basic morphology is similar for different types (e.g. B vs T lymphocytes).

Genetic mechanism of differentiation

The process of differentiation is controlled by the regulation of the expression of genes. The detail of the mechanism for specific cell types has yet to be fully worked out but it will involve the known methods of gene regulation including promoters, repressors, and DNA methylation.

Metaplasia

This is the change of a cell from one fully differentiated phenotype to a different, fully differentiated phenotype. It must be distinguished from dysplasia, which is a change to a less differentiated phenotype and is a precursor of cancer. Metaplasia occurs in a number of important sites in the human body:

- In the distal oesophagus, there may be a change from squamous to glandular epithelium if there is gastro-oesophageal reflux, which exposes the squamous epithelium to the acidic contents of the stomach. The metaplasia results in Barrett's oesophagus (OHCM10 p.695)
- In the bronchi, there is often a metaplasia from ciliated glandular epithelium to squamous epithelium in cigarette smokers due to the irritant effect of the smoke. This results in the loss of the mucociliary escalator, the mechanism by which inhaled and secreted debris is removed from the bronchi, which leads to the development of chronic bronchitis. This metaplasia may precede dysplasia of the metaplastic squamous epithelium and the development of cancer
- In the uterine cervix at puberty, the influence of hormones causes outgrowth of the cervix so that glandular endocervical epithelium is exposed to the vaginal environment, which stimulates metaplasia to squamous epithelium.

Meiosis and gametogenesis

The normal adult cell contains two copies of each chromosome (one inherited from the mother, the other from the father). In order to produce an oocyte or sperm, this needs to be reduced to a single copy. Thus, a special type of cell division is required (Fig. 1.42), which is called meiosis (from the Greek word meaning diminution). However, if this division simply segregated the chromosomes into pairs and put one of each pair into a cell that would become a gamete, then this gamete would contain individual chromosomes that came wholly from the mother or father. Since chromosomes contain many thousands of genes, this whole group of genes would always be passed on together as a single unit of inheritance. This would thwart evolutionary selection of the genome. Thus, within the process of meiosis there is a phase of recombination between maternal and paternal chromosomes before division, which leads to a much more thorough mixing of the genomes.

Division I of meiosis
- Each chromosome replicates to produce two sister chromatids which are tightly linked along their length
- Each duplicated chromosome pairs with the equivalent chromosome from the other parent (the homologous chromosome) to form a structure called a bivalent. The sex chromosomes also participate in this process, even if male, because there is homology between some regions of the X and Y chromosomes
- There is genetic exchange between the homologous chromosomes by crossing over of segments of the chromosomes (genetic recombination)
- The bivalents line up on the mitotic spindle
- Division leads to two cells each with a sister chromatid, with mixtures of maternal and paternal genes (i.e. haploid cell but with diploid amounts of genetic material).

Division II of meiosis
- This is a simple segregation to divide the sister chromatids into genuine haploid cells.

Things that can go wrong during meiosis
- *Non-disjunction*—in some divisions, a bivalent (in meiosis division I) or a sister chromatid (in meiosis division II) may not separate and so one cell will end up with no copies of that chromosome and another will have three. This occurs in around 10% of meiotic divisions. The embryos that develop from such cells are usually non-viable and account for the majority of spontaneous first-trimester abortions ('miscarriages') but some are viable, with recognized phenotypes (e.g. trisomy 21, producing Down syndrome)
- *Translocation, deletion, etc.*—any of these can occur during the meiotic process and the crossing over of chromosomes makes this a more frequent occurrence in meiosis compared with mitosis.

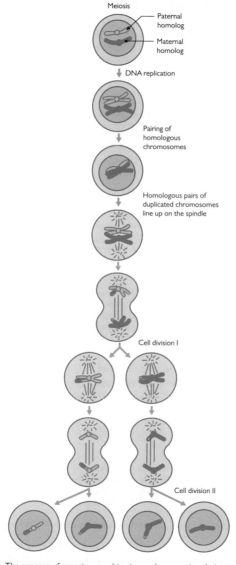

Fig. 1.42 The process of genetic recombination and segregation during meiosis.

Reproduced with permission from Pocock G, Richards CD (2009). *The Human Body* 3e, p25 Oxford University Press.

Receptors and ligands

Pharmacology is the study of drugs and their actions. In order to fully understand how drugs work, it is first necessary to grasp some basic concepts of the chemical and physical principles that underlie drug–receptor interactions, as well as some of the nomenclature.

Receptors

Cell surface receptors are proteins that are designed to recognize and respond to specific transmitter molecules, often known as ligands. The active site of a receptor is shaped and charged in such a way as to facilitate binding of its specific ligand, just as the active site of an enzyme is specific to its substrate.

Receptors can be crudely divided into two categories:
• Those linked to ion channels
• Those linked to G-proteins (➔ p.56).

NB: receptors are not only found on the cell surface; enzymes and carrier proteins in the cytoplasm can also act as 'receptors' in the sense that their function is modified by binding of specific ligands. Furthermore, receptors that modulate gene expression are also found in or around the nucleus.

Irrespective of the protein receptor target, binding of a specific ligand instigates a chain of events that ultimately results in modification of cellular function (➔ p.58).

Ligands

Ligand is a general term used to describe any molecule that binds to a specific receptor. Endogenous ligands range from very simple inorganic molecules (e.g. nitric oxide (NO)) to complex organic molecules, including amino acids (e.g. glutamate), peptides, and lipid derivatives. Drugs are exogenous ligands that either mimic (agonists) or block (antagonists) the effects of their natural counterparts.

Ligand binding—general principles

Binding of a ligand (A) to a specific receptor (R) is a chemical interaction that results in the formation of a ligand–receptor complex (AR), often represented by the following equation:

$$A + R \underset{K_{-1}}{\overset{K_{+1}}{\rightleftharpoons}} A$$

A number of important chemical principles apply to this equation:
• Law of mass action—the rate of reaction is proportional to the concentration of the reactants (Fig. 1.43a)
• Equilibrium—when the rate of formation of AR equals the rate of dissociation of AR to A + R
• The proportion of A that is complexed at equilibrium is determined by its affinity for the receptor, characterized by the affinity constant, k_{+1}. The dissociation constant (k_{-1}) is inversely proportional to k_{+1}.
• k_{-1}/k_{+1} is known as the equilibrium constant (K_A) and is characteristic for a particular drug for a given receptor. K_A equals the amount of drug required to occupy 50% of the receptor population at equilibrium

- The proportion of receptors occupied at equilibrium is therefore dependent on both the equilibrium constant and the concentration of ligand present and is described by the Hill–Langmuir equation (Fig. 1.43b):

$$P_A = \frac{[A]}{[A]+K_A}$$

where P_A = the proportion of receptors occupied at equilibrium.

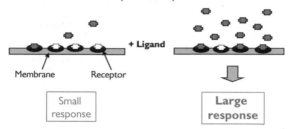

(a)   The proportion of receptors occupied at equilibrium is a function of the concentration of ligand present and its affinity for the receptors

The size of the response achieved at equilibrium is a function of receptor occupancy and drug efficacy

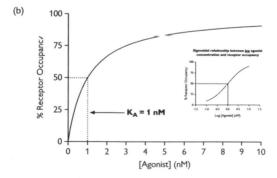

Fig. 1.43 Ligand binding. (a) Law of mass action. (b) The relationship between agonist concentration and receptor occupancy for an idealized agonist. This relationship is specific for a given drug and receptors in a particular tissue.

Concentration–response relationship

Although receptor occupancy is an important consideration for pharmacologists, it is the response that a given concentration of ligand evokes in the body, or in an isolated piece of tissue, that is ultimately of greatest interest when exploring drug action.

At first glance, one might assume that ligand concentration would be related to response in the same way that it is to receptor occupancy. Theoretically, this is true for a ligand that evokes a maximum response when all the receptors are occupied. However, there are several factors that ensure that this is rarely the case in reality:

- Availability—the ligand concentration delivered does not necessarily reflect the concentration at the receptors. This is especially true *in vivo*, when drug absorption and distribution, sequestration by plasma components, and breakdown by enzymes have a large impact on the concentration reaching the target receptors (◐ p.96)
- Efficacy—the size of the response obtained for a given concentration of ligand is determined by the efficiency with which the second messenger system is evoked by its binding to the receptor. It may be the case, therefore, that binding of different ligands to an equal number of the same population of receptors produces different sized responses. Efficacy is the term given to describe the ability of a ligand to cause a physiological or cellular response and determines whether a ligand is an agonist (has efficacy) or antagonist (no efficacy)
- A full agonist will cause a maximal response on binding all, or sometimes only a proportion of the receptors (the remainder is known as the receptor reserve) (see A and B in Fig. 1.44)
- A partial agonist has lower efficacy than a full agonist and is unable to cause a maximum response, even when the concentration of agonist is sufficiently high to have 100% receptor occupancy (see C in Fig. 1.44)
- An antagonist has very little or no efficacy. This means that binding of an antagonist to a receptor population fails to evoke any response, even when receptor occupancy is 100%.

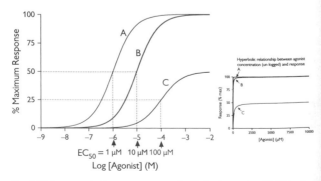

Fig. 1.44 Log concentration response curves for idealized examples of full (A and B) and partial (C) agonists.

NB: it is important to recognize that efficacy is unrelated to affinity. Thus, an antagonist can have high affinity (binds avidly to receptors) but zero efficacy (fails to produce a response).

Potency is a complex, non-specific pharmacological term sometimes misused to describe affinity on the basis of a drug's ability to evoke a physiological response. As highlighted previously, response amplitude is not a good measure of affinity and the term can be misleading. That said, equipotent molar ratio (EPMR) is sometimes a useful measure to establish the relative amount of a given drug that is required to generate the same size of response as another drug in the same tissue (Fig. 1.45). The EPMR will be independent of the chosen response size (between ~25% and 75% of maximum) providing the concentration response curves are parallel. It should not be used in cases where slopes vary.

The EC_{50} of a drug is the effective concentration required to cause 50% maximal response for that drug (see Fig. 1.44). This value is related to agonist affinity but does not provide any information about efficacy. ED_{50} is the equivalent measure when using doses (e.g. mg/kg) instead of concentration (e.g. µM). IC_{50} is a measure to describe the concentration of an inhibitor required to cause 50% inhibition of a response.

The pD_2 is −log of the EC_{50} (molar concentration). Thus, the pD_2 for a drug with an EC_{50} of 1µM (10^{-6} M; see A in Fig. 1.44) is 6. The higher the pD_2, the less drug is required to produce a 50% response.

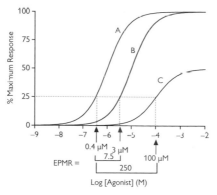

Fig. 1.45 Equipotent molar ratios (EPMRs) for drugs A, B, and C.

Antagonists

Antagonists fall into several different classes, depending on their mode of action.

Competitive antagonists

- Have affinity for the binding site of a specific receptor
- Compete with the endogenous agonist for that site: the higher the affinity of the antagonist for the receptor, the lower the concentration that will be required to compete effectively with the agonist for receptors and reduce agonist receptor occupancy
- The inhibitory effect of a given concentration of a competitive antagonist can be overcome by increasing the concentration of agonist and is, therefore, said to be reversible (Fig. 1.46)
- A truly competitive antagonist must cause a parallel shift in concentration response curve without affecting the maximum response (Fig. 1.46).

The dissociation constant (K_D) of a competitive antagonist can be established from experiments using an established agonist in the presence of different antagonist concentrations to determine the dose ratio (DR)—the ratio of agonist required to generate a given response in the presence of a known antagonist concentration compared to that in the absence of antagonist. Using the Schild equation or the Arunlakshana and Schild plot (Fig. 1.47), the K_D (sometimes also referred to as the K_B) can be determined. The K_D for a true competitive antagonist is independent of antagonist concentration and, therefore, the gradient of the plotted line is equal to 1.0 (Fig. 1.47).

Irreversible competitive antagonists

- Bind irreversibly but competitively to the agonist binding site, preventing agonist binding
- Cause inhibitory effects that cannot be reversed by increasing the agonist concentration
- Reduce the maximum response that can be achieved with an agonist

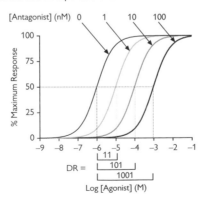

Fig. 1.46 Competitive antagonism and dose ratio (DR) for several concentrations (1–100nM) of an idealized competitive antagonist.

- The irreversible nature of the effect means that the antagonistic action develops with exposure time, as an increasing number of receptors become irreversibly bound.

Non-competitive antagonists

- Bind to a region of the receptor (or associated ion channel) other than the binding site for the endogenous agonist. Non-competitive antagonist binding can either alter the conformation of the agonist binding site to reduce the affinity for agonists, or prevent activation of the transduction mechanism required to evoke a response, or block ion channel opening
- The effects cannot be reversed by increasing the agonist concentration and, therefore, the inhibitory action is reflected in a reduction in maximum response (Fig. 1.48).

Physiological antagonists

Agonists that cause the reversal of a physiological response via a different receptor and second messenger system (e.g. an agent that causes relaxation of a contracted muscle).

Chemical antagonists

React with the agonist before it binds to a receptor, either reducing the affinity of the agonist for the receptor or completely preventing its ability to bind.

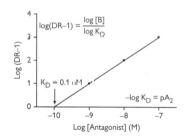

Fig. 1.47 Schild equation and plot to establish the dissociation constant (K_D) for the competitive antagonist (B) shown in Fig. 1.46.

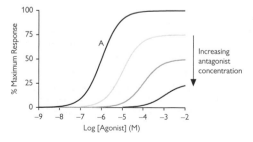

Fig. 1.48 Effect of an irreversible (non-competitive) antagonist on the actions of an idealized agonist (A).

Administration of drugs

In order for a drug to have a physiological effect *in vivo*, it must first be de-livered in such a way that the correct amount of drug reaches the relevant receptors in the target tissue.

The chemical characteristics of the drug will determine:
- How well it is absorbed into the bloodstream
- How it will be distributed in the body
- How quickly it will be metabolized and excreted from the body.

These are essential criteria in determining the chemical composition, dose, and frequency of dosing required to make for a useful pharmaceutical agent.

A number of routes are available for drug delivery. The route of adminis-tration is ultimately dependent on the chemical properties of the drug, but can usually be tailored for the particular application. For example, the most effective means of delivering a therapeutic agent to an asthmatic patient is directly to the bronchial tree by means of an inhaler or nebulizer. However, a cream would be the best form of drug delivery for a patient with eczema, eye drops for an eye infection, while a pill is most suitable in the treatment of stomach ulcers.

Intravenous delivery (injection or infusion)
- The fastest means of delivery into the bloodstream
- Avoids some of the problems associated with oral administration
- Equally applicable in conscious and unconscious patients
- Patients in hospital can receive chronic drug treatments via pump-driven or drip infusions
- Generally unsuitable for self-administration.

Subcutaneous (under the skin) and intramuscular injections
- Less rapid in onset than intravenous administration
- Dependent on drug diffusion at the site of injection as well as local blood flow—an important consideration in patients in 'shock', where perfusion pressure is greatly reduced because of blood loss or vascular collapse
- Generally, only implemented by medical staff, but patients with diabetes are often trained to inject themselves with insulin subcutaneously.

Intrathecal delivery
- Used to deliver drugs directly into to the CSF
- Lumbar puncture injections are technically challenging
- Necessary for central nervous system (CNS) drugs that are unable to cross the blood–brain barrier, or that have undesirable side effects elsewhere in the body if injected systemically.

Delivery via the gastrointestinal tract (oral and rectal)
- Oral administration is the most convenient because it does not necessarily require special equipment or medical supervision
- Carries the disadvantage that the drug will be exposed to peptic enzymes and very acidic conditions in the stomach

- Drugs that are not absorbed in the stomach will encounter different enzymes and alkaline conditions in the lower GI tract. Extreme changes in pH can lead to chemical changes that inactivate compounds and alter the rate at which they diffuse through cell membranes. For this reason, some agents (e.g. peptides like insulin that will be broken down by pepsin in the stomach) do not lend themselves to oral delivery. In such instances, alternative delivery routes or protective capsules can be used
- Not appropriate in patients before operations and in those who are unable to swallow or are vomiting. In some instances, drugs can be delivered rectally instead
- The rate at which orally ingested drugs are absorbed varies greatly in different patients and with different types of drug. Although some (e.g. aspirin, alcohol) are absorbed in the stomach, most are absorbed in the small intestine: gut motility is therefore a factor that determines the latent period between ingestion and the drug reaching its site of absorption. For this reason, taking an oral drug just after a meal is likely to slow its absorption by delaying its progress through the intestine. Once there, the rate at which it diffuses across the epithelium is determined primarily by its physico-chemical properties (e.g. particle surface area and equilibrium constant). Finally, the rate at which it enters the bloodstream is determined by the blood flow to the gut (splanchnic blood flow); rapid blood flow constantly removes drug from the gut wall and maintains the concentration gradient for diffusion. In general, however, peak absorption is usually achieved after ~1h and most of the drug has been absorbed within 4h.

Locally acting preparations

Where the desired site of action of a drug is accessible, it is desirable to deliver it directly to the target area to minimize (but not necessarily eliminate) systemic effects. The following are some examples of locally acting drug preparations:
- Eye drops
- Ear drops
- Dermatological creams
- Inhaled preparations for bronchial complaints.

Absorption, distribution, and clearance of drugs

Drug absorption

Drug absorption involves the passage of drug molecules across the epithelial barrier layer (e.g. intestinal epithelium, lung epithelium). The cells that constitute the barrier are tightly connected to one another, so the only means of passage is across the cell membrane (lipid bilayer). The most important means of diffusion of drugs is through the lipid membrane itself, requiring drug molecules to dissolve in both lipid and aqueous (water) environments.

Partition coefficient

The partition coefficient is a measure of the relative solubility of a compound in aqueous and lipid environments and is a critical determinant of drug absorption.

Ionization

One of the key factors in determining the lipid solubility of a molecule is its readiness to ionize, generating a charged species that is repelled by uncharged lipid molecules. Most drugs can be categorized as acids or bases and, in general, the weaker the acid or base (i.e. less readily ionized), the greater the lipid solubility of the drug and the more rapid the diffusion across membranes.

pH partition

Acids and bases ionize as follows:

$$\text{Acid}: AH \underset{}{\overset{K_a}{\rightleftharpoons}} A^- + H^+$$

$$\text{Base}: BH^+ \underset{}{\overset{K_a}{\rightleftharpoons}} B + H^+$$

where K_a is the equilibrium constant for ionization.

Because drugs are generally weak acids and bases, the equilibrium is heavily weighted towards the non-ionized form, when the pH of the environment is neutral (~7.0).

Ionization is, however, affected by the pH of the environment in which the drug is dissolved:

- A weak acidic drug ionizes more readily under alkaline conditions (pH >7)
- A weak basic drug ionizes more readily under acidic conditions (pH <7).

The dissociation constant (pK_a) is a term which allows for the environmental pH and is derived using the Henderson–Hasselbalch equation:

$$\text{Weak acid}: pK_a = pH + \log_{10} \frac{[AH]}{A^-}; \quad \text{Weak base}: pK_a = pH + \log_{10} \frac{[BH^+]}{[B]}$$

Remembering that drugs generally only cross membranes when they are not ionized, orally administered acidic drugs will be absorbed primarily in the stomach (pH ~3), while oral basic drugs will be absorbed in the small intestine (pH ~9).

Ion trapping

The rate of diffusion of a non-ionized molecule through a membrane is determined by the concentration gradient for that molecule across the membrane (➔ p.52). As molecules diffuse down the concentration gradient, their concentration will tend to equalize between the two compartments. At equilibrium, the net diffusion of molecules will be zero because a concentration gradient no longer exists.

In reality, most drugs show a degree of ionization, as determined by the pK_a of the drug and the pH of the compartment in which it is dissolved. Nevertheless, the non-ionized form of the drug will equilibrate across a membrane (Fig. 1.49). If the pH is the same in both compartments, the concentration of drug at equilibrium will be the same on both sides of the membrane. This is how drugs diffuse from cell to cell throughout a tissue.

However, the situation is more complicated when the pH is different between the two compartments. For example, a weakly acidic drug such as aspirin (pK_a = 3.5) is absorbed in the stomach (pH ~3), where it passes first into the epithelial barrier layer and then into the blood (both of which pH ~7.4). The pK_a for aspirin determines that it is mainly in the non-ionized form at this low pH, and can therefore readily diffuse across the epithelial cell membrane. Once inside the cell, however, the higher pH will cause a large proportion of the drug to ionize, rendering it unable to diffuse back through the membrane into the stomach compartment (Fig. 1.49). As a result, when the non-ionized form of the drug reaches equilibrium between the two compartments (stomach and intracellular), a large amount of drug is effectively 'trapped' within the cell in the ionic form.

Concentration and amount

When considering drug distribution between compartments, the relative size of the compartments is another important factor. At equilibrium, the concentration (i.e. amount per unit volume) (e.g. moles l^{-1}, mg mL^{-1}) of non-ionized molecules is the same in each of the compartments. However,

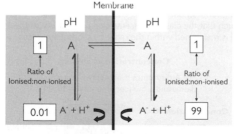

Ion Trapping for a weak acid:

Total drug conc ratio = 1.01:

Concentration of drug is ~100 x higher in pH 7.4 compartment

Fig. 1.49 Ion trapping for a weak acid in the body compartment in preference to the acidic stomach compartment.

by definition, the amount of drug (e.g. moles or mg) will only be the same if the volume of the compartments is the same. In reality, the volume of the compartments might be very different (Fig. 1.50). At equilibrium, under these conditions, although the concentration is the same in the compartments, there is considerably more of the drug in the body compartment. This is an important aspect in volume of distribution.

Drug distribution and clearance

Unless drugs can be applied directly to the target tissue (e.g. topical application for skin conditions, inhalation for asthma), we ultimately rely on the blood to deliver the drug to the relevant tissue. The amount of drug that reaches the tissue is dependent on absorption kinetics, blood flow to and from the target tissue, the rate of de-activation by metabolic processes, and subsequent excretion of the drug—collectively termed elimination.

Elimination of the vast majority of drugs occurs according to first-order kinetics. This is a term that describes the rate at which the plasma concentration of a drug falls over time. The characteristics of first-order kinetics are that, while the rate of drug elimination is dependent on its concentration, the time it takes for half of the drug to be eliminated is constant (the half-life). Thus, a plot of concentration against time falls exponentially on a linear scale (Fig. 1.51) but is linear on a semi-logarithmic scale. The half-life of a drug is a critical tool used by pharmacologists to predict plasma concentrations of a drug and to establish the interval required between dosing to maintain its therapeutic effects (i.e. to maintain a level above the effective concentration (Fig. 1.51)).

Dosing, steady state, and loading doses

A sound understanding of the pharmacokinetics of a particular drug is essential for determination of a suitable dosing regimen. For example, a drug with a very short plasma half-life will need to be administered frequently to maintain a therapeutic dose, while a bigger dose of a slowly absorbed drug might have to be administered to attain a therapeutic effect. It is a feature of drugs with first-order elimination kinetics that repeated doses at consistent intervals will ultimately result in generation of 'steady-state conditions', where the inter-dosing plasma concentration fluctuates between consistent maximum and minimum levels (Fig. 1.52). Ideally, a dosing strategy can be derived such that the plasma concentration rises rapidly to the therapeutic range and is maintained within the range by subsequent doses.

Concentration vs amount

Drug A distributes evenly between stomach,
plasma and urine (concentration = 10 nM):

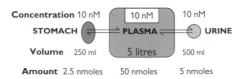

Fig. 1.50 The relationship between concentration and amount.

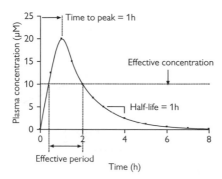

Fig. 1.51 An example of the plasma concentration profile for a single oral drug dose.

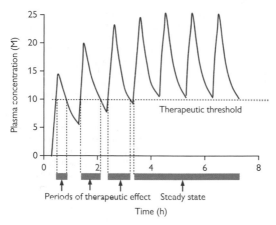

Fig. 1.52 An example of the plasma concentration profile for repeated oral drug dosing (1h intervals) of a drug with a half-life of ~30min. Note how repeated doses progressively increase the duration of the effective periods until steady state is reached.

Zero-order kinetics: alcohol and salicylate

Alcohol and salicylate (the metabolite of aspirin) belong to a small group of compounds for which the rate of clearance is independent of its concentration (zero order). A plot of plasma alcohol concentration against time on a linear scale gives a straight line and the half-life is not a constant, as it is for most drugs (Fig. 1.53). The reason for the zero-order (or saturation) kinetics of these compounds is that the enzymes required to metabolize the drugs become saturated at very low levels; the enzyme is unable to increase its rate of activity in the face of increasing drug concentrations. Drugs that fall into this category are difficult to administer effectively because they do not attain steady state conditions.

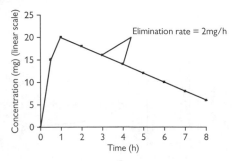

Fig. 1.53 Zero-order kinetics.

Volume of distribution
The plasma concentration of a drug in the body is not only dependent on the amount of drug administered and the rate at which it is absorbed and eliminated, but is also dependent on the volume in which the drug is diluted. The plasma concentration of a drug that is confined to the bloodstream will accurately reflect the amount of drug in the body (Fig. 1.54a). However, the plasma concentration of a different drug, which is heavily absorbed in the aqueous component of tissue (Fig. 1.54b) and/or body fat (Fig. 1.54c), will vastly underestimate the amount of drug in the body because only a small fraction of the total drug is found in the blood. The apparent volume of distribution of a drug is calculated using the amount of drug administered and the concentration of the drug measured in the blood; by estimating the expected blood concentration if the drug distributes evenly throughout the water compartments of the body (i.e. blood, extracellular fluid, cellular fluid), we can determine whether the drug accumulates in the blood or distributes to other, non-aqueous compartments (e.g. subcutaneous fat; Fig. 1.54).

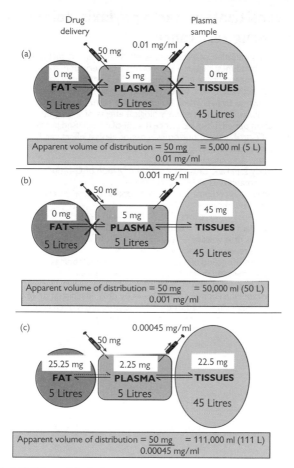

Fig. 1.54 Volume of distribution.

Desensitization, tachyphylaxis, tolerance, and drug resistance

Continual stimulation of receptors leads to the use of metabolically ex-pensive second messengers. As a means of conservation, many receptor-mediated processes have in-built mechanisms to gradually damp down the effects of activation, → a time-dependent loss of cellular effects, some-times known as tachyphylaxis. Tolerance is a term often used to describe a gradual loss of the desired physiological effects of a drug during the ad-ministration period. Drug resistance is a specific term used primarily for the loss of effect of anti-tumour drugs. Loss of agonist effect can be due to a number of reasons:

- Receptor desensitization:
  - Usually rapid and caused by phosphorylation of G-proteins or conformational change in ion channels
  - Reversed by receptor removal and replacement with newly synthesized receptor
- Adaptive loss of receptors:
  - A cellular response to continuous stimulation
  - Reduces drug efficacy
  - Can be reversed by removal of stimulus
- Exhaustion of second messengers or essential metabolites:
  - Usually reversed during a period without stimulus
- Physiological adaptation:
  - Relevant when homeostasis is upset by drug administration (e.g. a rapid drug-induced reduction in blood pressure is often compensated for by increased heart rate or blood volume).

Drug interactions

Before a drug can be prescribed, it is important to establish whether it might interact with any other medications that a patient is taking (↔ *OHCM10* p.757).

Interactions can take many forms:
- Antagonistic: drugs can act to cancel each other out as a result of physiological, competitive, or non-competitive antagonism
- Additive: drugs cause the same physiological effect and the total amplitude is the sum of the effects of the individual drugs
- Synergistic: drugs cause the same physiological effect but the total amplitude of the response is considerably greater than the sum of the effects of the individual drugs. This can be related to a secondary interaction of drugs at the metabolic level, resulting in increased duration of action. Alcohol can have this effect because its metabolism depletes substances in the liver that are essential for metabolism of other drugs, slowing their clearance and prolonging their activity.

Indications, contraindications, and side effects
- Indications are the clinical conditions for which a drug is prescribed
- Contraindications are clinical conditions or other considerations (e.g. patient age, pregnancy) that preclude the use of a particular drug. For example, β-blockers (↔ *OHCM10* p.114), which are often prescribed for hypertension should not be given to patients with asthma as they can exacerbate airway constriction, and aspirin should not be prescribed to children under 12 years or to breastfeeding mothers. Furthermore, many drugs are unsuitable for patients with liver or kidney impairment because they are not cleared rapidly enough, → toxic effects
- Side effects: unfortunately, very few drugs (if any) are so specific that they only cause the desired effect; most have secondary (off-target) effects (side effects) that might be undesirable and are responsible for the contraindications. As mentioned previously, β-blockers have the side effect of inhibiting β-adrenoceptors in the lung, where they can exacerbate bronchoconstriction in asthmatic patients and counteract β-adrenoceptor agonists prescribed to help alleviate asthma.

Cellular metabolism

General principles: overview

- All animals need food, both for energy and for the (precursors to) building blocks for growth:
  - This food comes from hunting and scavenging in a competitive environment
  - The pathways of metabolism have evolved to waste as little as possible of these precious resources
  - Metabolic pathways are regulated to avoid inappropriate actions and opposing pathways running simultaneously ('futile cycling')
- Humans are no different except, for most of us, work and shopping have replaced hunting
- Eating is intermittent and so the body needs to have mechanisms to store energy in easily releasable forms (e.g. fat, glycogen).

What is metabolism?

A series of chemical reactions catalysed by enzymes:
- Catabolism—breakdown of compounds to release energy; usually involves oxidation
- Anabolism—biosynthesis of more complex compounds from small precursors, usually consumes energy and involves reduction.

Entropy and (Gibbs) free energy (G)

- Entropy is the degree of chaos or randomness in a system:
  - Reactions can only proceed if there is an overall increase in entropy
- Free energy is the energy in a system available for useful work, usually expressed at standard conditions
- The difference in free energy for a reaction (ΔG) gives an idea of whether a reaction will proceed:
  - A negative ΔG means that a reaction will proceed
  - A ΔG of zero means that a reaction is at equilibrium
  - Reactions with a positive ΔG can be driven by linking to another reaction, so that the sum of the total ΔGs is negative.

Cellular energy

It is not feasible to directly, physically link reactions that produce and consume energy, and so it is necessary to have a short-term way to trap that energy. The molecule ATP is the universal cellular currency of energy (Fig. 2.1). ATP is present in cells complexed with Mg^{2+}.

Structure of ATP

- There are 30.5kJ mol^{-1} of energy stored in each of the phosphate–phosphate bonds:
  - The phosphate–ribose sugar bond only has 14.2kJ mol^{-1}, and so the hydrolysis of AMP to adenosine is not used as a source of energy in metabolic reactions
- ATP can be hydrolysed in a number of ways:
  - $ATP \rightarrow ADP + P_i$
  - $ADP \rightarrow AMP + P_i$
  - $ATP \rightarrow AMP + PP_i$ (NB: energy release comes from the hydrolysis of $PP_i \rightarrow 2P_i$)

- In all cases, the energy released can be used to drive reactions with a positive ΔG
- ATP is only present in cells in small amounts:
  - At rest, the daily ATP turnover is ~40kg (and can be as high as 0.5kg min^{-1} during exercise)
  - Therefore, there must be mechanisms to quickly regenerate ATP to match cellular usage.

Electrons can also be stored in reduced intermediates such as NADH and FADH$_2$ which are important in both the generation of ATP by the electron transport chain and biosynthetic reactions.
- Some biosynthetic reactions use NADPH instead of NADH, e.g. fatty acid synthesis.

Oxidation is the end-point for metabolic fuels.
- Ultimately, compounds enter the TCA cycle, resulting in the production of NADH and FADH$_2$ and CO_2. NADH and FADH$_2$ are reoxidized in the electron transport chain to produce ATP.

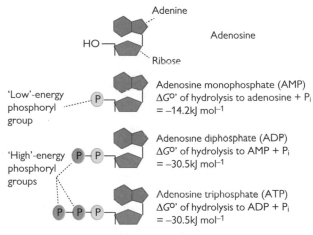

Fig. 2.1 Diagrammatic representation of adenosine and its phosphorylated derivatives.

Metabolic control

Being able to control the rate at which metabolic pathways proceed is essential to ensure that:
- Supply meets demand
- Forward and reverse pathways are not run simultaneously ('futile cycling').

Control over enzyme activity can be both:
- Short term:
  - Allosteric effects: the binding of molecules to enzymes affecting their catalytic rate (time of onset: milliseconds)
  - Covalent modification such as phosphorylation (seconds to minutes)
- Long term:
  - Changes in enzyme protein levels (induction/suppression) (hours to days).

Points in a metabolic pathway which are suitable for control:
- Regulated reactions are often those which are essentially irreversible (for energetic reasons)
- Regulation often takes place (Fig. 2.2):
  - Early in a linear pathway
  - At branch points
  - Reciprocally, at bi-directional points, where there is a different enzyme for the forward and reverse directions.

Metabolic control can be described as being either:
- Intrinsic, i.e. brought about by changes in intracellular levels of (almost always) an allosteric regulator of an enzyme *or*
- Extrinsic, i.e. brought about by signals originating outside the cell, e.g. hormones:
  - Small changes in plasma hormone levels often have large effects on cell functions due to amplification cascades.

Cycles between organs

In addition to cellular compartmentalization being an important feature of metabolic pathways, there is also a sharing of metabolic load between organs of the body. One good example of this is the Cori cycle (Fig. 2.3).
- These types of inter-organ cycles have multiple levels of potential control, including:
  - The delivery of substrates to different organs by the circulation
  - The rate at which substrates cross cell membranes via membrane transporters
  - Regulation of enzyme function in the cells.

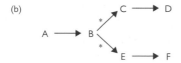

* = regulated reaction.

Fig. 2.2 Metabolic control: regulation. * = regulated reaction.

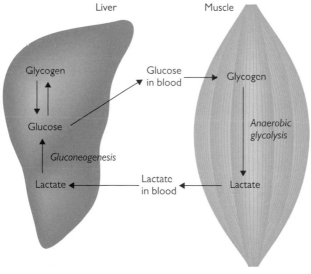

Fig. 2.3 The Cori cycle.
Reproduced with permission from Papachristodoulou, D., *Biochemistry and Molecular Biology* 6e, 2018, Oxford University Press.

Oxidation and reduction

Generally speaking:
- Breakdown (catabolic) reactions involve oxidation
- Synthetic (anabolic) reactions involve reduction
- Rather than being directly linked, intermediate molecules store/donate the electrons. For example, in glycolysis, glyceraldehyde-3-phosphate is oxidized while the intermediate NAD^+ is reduced to NADH.

There are three main biological intermediates (Figs 2.4, 2.5):
- Nicotinamide adenine dinucleotide (NAD^+)
- Flavin adenine dinucleotide (FAD)
- Nicotinamide adenine dinucleotide phosphate ($NADP^+$).

Compartmentalization allows controlled oxidation and reduction reactions to occur in the same cell.
- Most of the NAD^+ and FAD are unreduced and in the mitochondria (ideal for oxidative reactions)
- In contrast, most $NADP^+$ is in the reduced form (NADPH) in the cytosol, where it participates in reactions involving reduction.

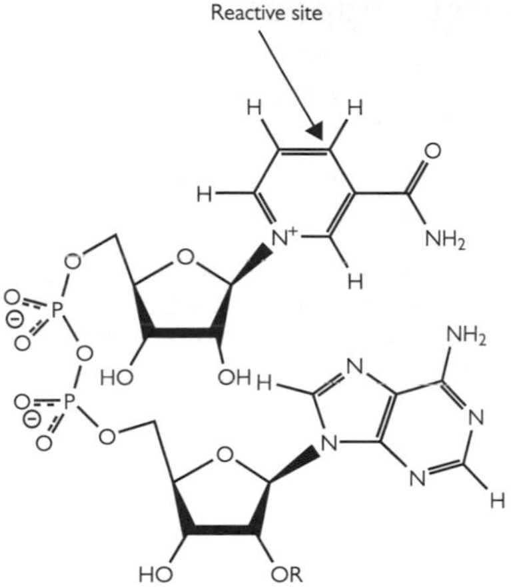

Fig. 2.4 Structures of the oxidized forms of nicotinamide-derived electron carriers: NAD^+ (R = H) and $NADP^+$ (R $= PO_3^{2-}$).

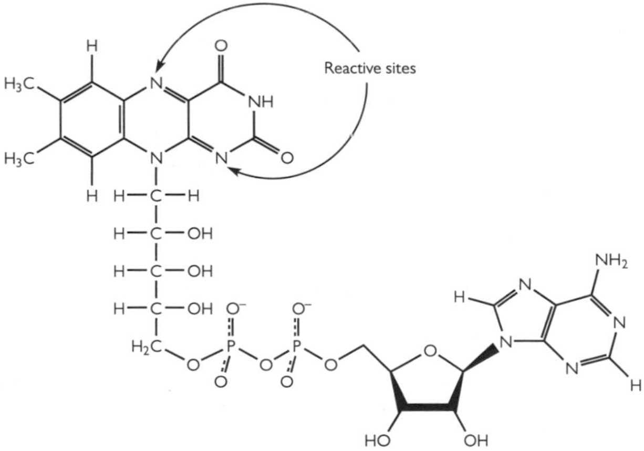

Fig. 2.5 Structure of the oxidized form of flavin adenine dinucleotide (FAD).

Body energy supplies

Food intake (➲ *OHCM10* p.244) is not a continuous process, and so the body must be able to store energy. Energy can be stored in a variety of ways, listed in order of use:

- Carbohydrate:
  - Glucose in plasma (3L with an average concentration of ~5mM)
  - Glucose is stored as the polymer, glycogen, in all cells but the two major sites are:
    ○ Liver (10% of liver mass): used to maintain blood glucose during short periods of fasting; enough stores for about 24h at rest
    ○ Skeletal muscle (2% of muscle mass): only used by muscle itself during exercise
- Lipid (fat):
  - The majority of lipid is stored in adipose tissue (fat cells):
    ○ Fat accounts for about 15kg of a typical 70kg man
    ○ Fat is a highly compact energy store (>300-fold more energy in body fat stores than in liver glycogen)
    ○ Enough fat stores for about 3 months
- Protein:
  - Protein is not a classical energy store but is used in starvation conditions to supply carbon skeletons for gluconeogenesis (➲ p.158)
  - Skeletal muscle is the major store of mobilizable protein
  - Loss of protein from heart, kidney, and liver compromises their function and will ultimately lead to death.

When we do eat, what is our relative dietary intake?

- Carbohydrate:
  - Constitutes 35–45% of our daily energy intake: 60% from starch, 35% in Western diet from sucrose ('sugar')
- Fat:
  - 40–50% of our daily energy requirement: mostly ingested as triacylglycerides
- Protein:
  - 7–10% of energy requirements needed from protein intake (Western diet supplies about 15%)
  - About 35g minimum required per day to maintain nitrogen balance (i.e. to ingest as much nitrogen as we excrete)
- Carbohydrate-free diets (such as the 'Atkins diet') work by tricking the body into starvation mode (➲ p.178), with energy obtained from protein and fat metabolism due to the prevailing glucagon signal. As with all diets, for it to be effective, energy intake must be less than energy expenditure.

Central metabolic pathways

Tricarboxylic acid cycle and its control

The TCA cycle is the common pathway for the oxidation of fuel molecules.
- Also known as the Krebs cycle (after its discoverer) or the citric acid cycle
- It is a cyclic pathway (Fig. 2.6): intermediates are regenerated so that net amounts of each remain the same after each turn of the cycle:
  - Intermediates present in relatively small amounts, and essentially play a catalytic role
  - Many intermediates are starting points for biosynthetic pathways. Anapleurotic reactions fill up the cycle to replace any of the intermediates used in this way.

The TCA cycle

- The reactions of the TCA cycle take place in the mitochondrial matrix:
  - Pyruvate enters the mitochondria on a specific transporter in the inner mitochondrial membrane (IMM)
- Entry point into the TCA cycle is the compound acetyl-CoA:
  - Acetyl-CoA is formed from pyruvate (the end-point of glycolysis; ⟴ p.140) by the link reaction in Fig. 2.7
  - Acetyl-CoA can also come from fatty acid breakdown (⟴ p.128) or the carbon skeletons of amino acids (⟴ p.163)
  - PDH is inhibited directly by high levels of acetyl-CoA and NADH, and indirectly by ATP, acetyl-CoA, and NADH (via activation of PDH kinase, which phosphorylates PDH and inactivates it; PDH kinase is inactivated by PDH substrates pyruvate, CoA-SH and NAD$^+$). Inhibition of PDH is relieved through dephosphorylation by phosphoprotein phosphatase.

Regulation of TCA cycle

- The rate of TCA cycling matches the cellular demand for ATP, and not the availability of substrates.
  - The main regulated enzymes are isocitrate dehydrogenase (inhibited by ATP, NADH; activated by ADP) and α-ketoglutarate dehydrogenase (inhibited by ATP, NADH, succinyl CoA):
    - TCA cycle is therefore inhibited when the cell has no need for further ATP synthesis, and activated when it needs to make more ATP
  - The regulated TCA enzymes are also activated by a rise in intra-mitochondrial Ca^{2+}:
    - PDH (via dephosphorylation by Ca^{2+}-activated phosphoprotein phosphatase), isocitrate dehydrogenase, and α-ketoglutarate dehydrogenase directly
    - The rise in Ca^{2+} could be caused by adrenaline ('flight or fight' response) or increased muscle contraction. Both of these situations will increase ATP consumption, so the TCA cycle is stimulated to increase ATP synthesis.

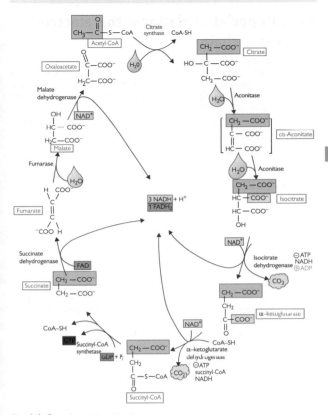

Fig. 2.6 Complete citric acid cycle.

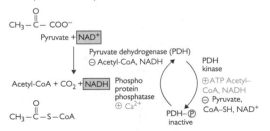

Fig. 2.7 Formation of acetyl-CoA from pyruvate.

Mitochondrial oxidation—the electron transport chain

The ETC converts the electrons stored as reduced intermediates NADH and $FADH_2$ into a proton motive force (PMF) across the IMM. Reactions occur through sequential oxidation/reduction centres containing transition metals (Fe in haem, Fe-S, Cu^{2+} in cytochrome oxidase) embedded in large protein complexes (Fig. 2.8).

- The large protein complexes will only move relatively slowly in the lipid bilayer:
  - Coenzyme Q and cytochrome C are small, highly mobile electron carriers that transport electrons from one complex to another
- The reaction centres have increasing redox potential
- Three of the four complexes are also proton pumps: for each pair of electrons, complex I extrudes $4H^+$ from the matrix; complex III, $4H^+$; and complex IV, $2H^+$. Complex II does not pump protons when it transfers electrons from $FADH_2$ to coenzyme Q:
  - Thus, for each NADH oxidized, $10H^+$ are extruded; and, for each $FADH_2$, $6H^+$
- The ultimate electron acceptor is molecular oxygen, which is reduced to water.

Although most NADH is formed in the mitochondria during the TCA cycle (⊙ p.116), it is also formed cytoplasmically, e.g. in glycolysis (⊙ p.140).

- There is no direct pathway for NADH to cross the IMM to enter the ETC. If there was, this would destroy the distinct oxidative/reductive compartments of the cell
- NADH can effectively cross the membrane by means of the malate/aspartate shuttle (Fig. 2.9)
- When cytoplasmic NADH is low, the glycerol-3-phosphate shuttle may be used (Fig. 2.10):
  - Electrons enter the ETC at the level of $FADH_2$ and so get less ATP per original NADH than with the malate/aspartate shuttle.

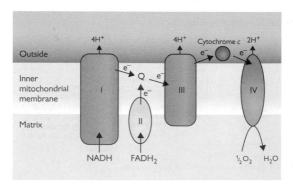

Fig. 2.8 The electron transport chain.
Reproduced with permission from Papachristodoulou, D., *Biochemistry and Molecular Biology* 6e, 2018, Oxford University Press.

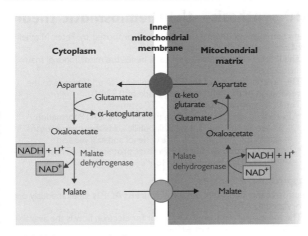

Fig. 2.9 Malate–aspartate shuttle.

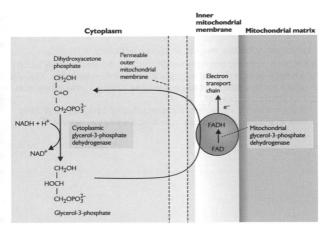

Fig. 2.10 Glycerol-3-phosphate shuttle.

ATP synthesis—the chemiosmotic theory

The chemiosmotic theory (Fig. 2.11) was proposed by Peter Mitchell[1] in 1961 and is based on the following premise:
- The IMM is impermeable to protons, hence the mitochondrial matrix is a closed environment
- The proton pumping of the ETC complexes (➲ p.116) leads to the generation of the PMF (total magnitude of 0.224V)
- The PMF provides the energy for ATP synthesis:
  - Evidence—agents that collapse this PMF inhibit ATP formation:
    - These compounds are weak lipophilic acids that carry protons across the IMM, e.g. 2,4-dinitrophenol and salicylic acid
  - Proton flow through the ATP synthase protein (F_0F_1-ATPase, complex V) drives ATP synthesis (➲ p.118).

Respiratory control

Electrons cannot flow through the ETC unless ADP is simultaneously phosphorylated to ATP.
- The most significant controlling factor for electron flow is the availability of ADP for conversion to ATP. In this way, the ADP concentration exercises what is known as 'respiratory control'
- With no ADP available, protons cannot move through the ATP synthase, and with the maximal possible PMF built up, electrons cannot flow through the ETC. Therefore, unless ATP needs to be synthesized, electrons are not accepted from NADH or FADH by the ETC.

Hypothesis for the evolution of mitochondria

It has been proposed that mitochondria originated from free-living bacteria which became incorporated into cells in a symbiotic relationship. This idea is supported by the fact that bacteria also use a PMF to drive uptake of nutrients across their cell wall.
- Some antibiotics are proton ionophores that kill bacteria by collapsing their PMF. One such example is the topical antifungal, nystatin (➲ *OHPDT2* p.440).

1 Mitchell P (1961) Coupling of phosphorylation to electron and hydrogen transfer by a chemiosmotic type of mechanism. *Nature*, **191**, 144–8.

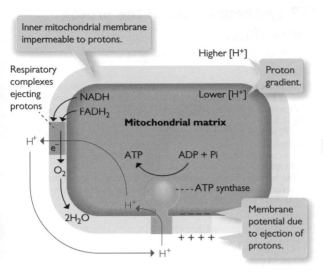

Fig. 2.11 Generation of ATP in mitochondria by the chemiosmotic mechanism.
Reproduced with permission from Papachristodoulou, D., *Biochemistry and Molecular Biology* 6e, 2018, Oxford University Press.

ATP synthesis—uses of the proton motive force

The PMF across the IMM can be used to drive a number of processes.

1. ATP synthesis

The impermeability of the IMM to protons, except through the protein responsible for ATP synthesis, is a key feature of the chemiosmotic theory.

- This protein is known as the F_0F_1-ATPase, ATP synthase, or complex V (Fig. 2.12)
- The F_0 subunit is an integral membrane protein which forms a proton channel
- F_1 is a complex (α_3, β_3, γ, δ, and ε) that has the catalytic site for ATP synthesis:
  - The F_0 and F_1 subunits are functionally linked, such that protons can only flow when ATP is being synthesized (dependent on [ADP] = 'respiratory control').

Mechanism of ATP synthesis

The movement of protons through the F_0 subunit induces the F_1 subunit to physically rotate.

- This is proposed to propel the binding sites through their different transition states of loose (ADP + P_i), tight (ADP + P_i), and ATP release. Therefore, it takes three protons to make one ATP by the ATP synthase (Fig. 2.13).

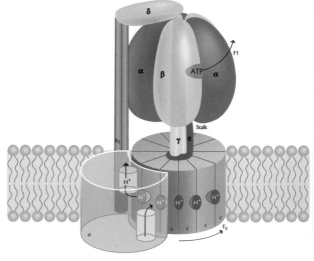

Fig. 2.12 F_1F_0-ATPase.

Reproduced with permission from Papachristodoulou, D., *Biochemistry and Molecular Biology* 6e, 2018, Oxford University Press.

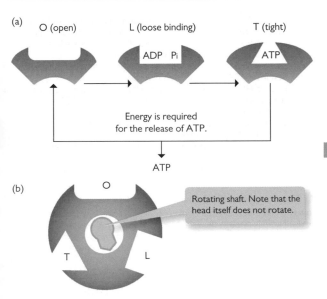

Fig. 2.13 The catalytic sites of ATP synthase as proposed in the Boyer model: (a) the changes that occur in a single site of one β subunit of F₁ during the synthesis of ATP; (b) the three β subunits work in a cooperative manner and the conversion in one site is coordinated with that in the other two sites.

Reproduced with permission from Papachristodoulou, D., *Biochemistry and Molecular Biology* 6e, 2018, Oxford University Press.

The F_1 subunit can be dissociated from the F_0 subunit by protease activity.
- When not linked to the F_1, it can act as an ATPase:
  - ATP hydrolysis will drive the rotation of the F_1 subunit. This is shown experimentally by attaching a fluorescent actin filament and seeing it rotate ('the world's smallest motor')
- In the absence of a PMF (e.g. tissue hypoxia) the ATP synthase is prevented from hydrolysing ATP by inhibitory factor 1 (IF1):
  - IF1 is over-expressed in cancer and may play a role in the change in tissue energy generation from oxidative phosphorylation to glycolysis (the Warburg effect).

2. Inner membrane transport

The proton gradient is also used to drive the movement of compounds through specific transporters in the IMM.
- Most ATP is made in the mitochondrial matrix, yet is needed in the cytoplasm; conversely, most ADP is formed in the cytoplasm, but regenerated in the matrix:

- Surely ADP/ATP exchanger (the adenosine nucleotide translocase (ANT)) is present in the IMM. Although not proton-coupled, due to the fact that ATP is more negative than ADP (4^- vs 3^-), it is driven by the membrane potential component of the PMF
- P_i is also required in the matrix for ATP synthesis (although most will be released from ATP hydrolysis in the cytoplasm):
  - There is a H^+/P_i co-transporter in the IMM, which effectively means that each ATP formed uses four protons. Hence, 1 NADH = 2.5 ATP, 1 $FADH_2$ = 1.5 ATP
- Pyruvate needs to cross the IMM to enter the TCA cycle:
  - There is an IMM pyruvate/H^+ co-transporter
- Mitochondria also take up Ca^{2+} in response to a rise in intracellular levels:
  - Uptake will be electrogenically favourable due to the PMF
  - Plays a part in regulating the TCA cycle (⊃ p.114).

3. Thermogenesis in brown adipose tissue

So far, it has been stressed that the only natural route by which the PMF can be dissipated is through the F_0F_1-ATPase and the synthesis of ATP.

- The only tissue for which this is not true is brown adipose tissue ('brown fat'—brown due to its high mitochondria content)
- There is an uncoupling protein (thermogenin) in brown fat mitochondria that allows the PMF to be dissipated without making ATP:
  - The energy is released as heat
  - This is important in human babies, who cannot shiver to generate heat.

ATP synthesis—control

The daily turnover of ATP in the average 70kg man is ~40kg, yet cells contain relatively little ATP at any one time—it cannot be stored and ATP molecules have a half-life in the order of seconds. Therefore, ATP production must match usage. Intracellular ATP usually remains virtually constant.

Relative concentrations of:
• ATP—high
• ADP—low
• AMP—very low.

ADP and AMP as controls of ATP synthesis

• ADP controls the rate of ATP synthesis at a mitochondrial level through the process of respiratory control (⊃ p.118). Unless there is ADP to make into ATP, the ETC does not run
• AMP is an important intracellular signal. As the normal intracellular concentration is very low, cells are very sensitive to even a small change:
  • When energy levels become lower, $2ADP \rightarrow ATP + AMP$
  • A rise in AMP activates the glycolysis pathway via the PFK enzyme (⊃ p.144) and stimulates cellular ATP production.

Fat as a fuel: overview

Fat is the long-term energy store for mammals.
- Fat constitutes about 20% of body weight of a well-nourished average sized individual (15kg out of 70kg)
- Fat has double the energy per gram dry weight than glycogen (39 vs 18kJ g^{-1}, respectively)
- In addition, fat has a low hydration level due to its hydrophobic nature:
  - If all the fat energy was stored as glycogen, body weight would be almost double!
- Energy release from fat is nowhere near as rapid as it is from glycogen due to the number of metabolic processes needed before it can enter the TCA cycle:
  - This is reflected in the way fat is used
  - Fat contributes about 35% of total daily energy production.

Tissue use

As mentioned previously, release of energy stored in fat is not as rapid as from glycogen.
- Fat is a suitable energy supply for tissues with steady energy requirements.

Cardiac muscle
- Cardiac muscle is almost exclusively aerobic (supported by high mitochondrial content)
- It has virtually no glycogen stores
- Cardiac muscle uses fatty acids (plus ketone bodies and lactate) for energy.

Skeletal muscle
- Despite having large stores of glycogen, skeletal muscle uses fatty acids for about 85% of its energy needs while resting:
  - Glycogen breakdown provides glucose to generate energy during bursts of activity.

Renal cortex
- The kidney has a very high energy requirement for the size of the organ
- Most energy is required in the cortex for reabsorption of filtered nutrients from the proximal convoluted tubule:
  - Fatty acids are the favoured fuel.

Assimilation of dietary fat

- Triacylglycerides from the diet are incorporated into micelles formed with the aid of bile salts in the intestinal lumen:
  - The triacylglycerides are broken down by luminal lipases into the constituent fatty acids and monoacylglycerides
- These can then cross the enterocyte cell wall via membrane transport proteins in the lipid bilayer
- Inside the enterocyte, they are reassembled and packaged into chylomicrons:
  - These are a mix of triacylglycerides, proteins (apolipoproteins, principally apoB-48), and other lipids such as cholesterol
  - Chylomicrons are also the vehicle for carrying fat-soluble vitamins (e.g. vitamin E)
- Chylomicrons pass into the lymph system, which in turn drains into the venous circulation (into the vena cava near the heart)
- Peripheral tissues, especially adipose tissue and muscle, have membrane-bound lipoprotein lipases which once again break down the triacylglycerides into free fatty acids and monoacylglycerides:
  - These can then be taken up into the cells (by a combination of diffusion and mediated transport)
  - In adipose tissue, they will be resynthesized into triacylglycerides for storage
  - In muscle, they will be oxidized for energy
- Free fatty acids can also be carried in the bloodstream, bound to albumin.

Hormone regulation

- Lipase activity in adipose tissue is regulated by the hormones glucagon, adrenaline (epinephrine), noradrenaline (norepinephrine), and adrenocorticotrophic hormone (ACTH). These hormones bind to G-protein-coupled receptors (membrane proteins with seven transmembrane domains):
  - These, in turn, activate adenylate cyclase, raising intracellular cAMP and activating PKA
  - PKA phosphorylates triacylglycerol lipase, activating it to break down triacylglycerides
  - The released fatty acids leave the cell and are transported to peripheral tissues bound to albumin in the plasma
- Conversely, insulin activates a phosphatase which dephosphorylates the lipase, thus inactivating it.

Plasma fatty acid levels

- Fatty acids are usually at a fairly low concentration (sub mM) under normal conditions:
  - Levels rise to about 1mM during starvation
- Ketone bodies are virtually absent under fed conditions:
  - This rises greatly (to around 5mM) during starvation
  - In type 1 diabetes, the uncontrolled production of ketones can cause metabolic acidosis (→ p.182; see also → *OHCM10* p.670).

β-oxidation

Once they have entered the cell, fatty acids are oxidized in the mitochondrial matrix.

- On diffusing across the plasma membrane, the hydrophobic fatty acids associate with a cytoplasmic binding protein:
  - This complex moves to the mitochondrial membrane for uptake into matrix
  - Medium-chain (C8–C10) fatty acids can cross the IMM directly.

Long-chain fatty acids need to be activated before they can cross the IMM.

- A CoA group is joined by a thioester linkage to the carboxyl group of the fatty acid:
  - Reaction driven by ATP hydrolysis
  - Catalysed by acyl-CoA synthetase (also known as fatty acid thiokinase):

$$R\text{-}COO^- + ATP + CoA\text{-}SH \rightarrow R\text{-}CO\text{-}S\text{-}CoA + AMP + PPi$$

Activated fatty acids cannot cross the IMM unaided (Fig. 2.14).

- First, they are conjugated to carnitine, a zwitterionic alcohol, to form acyl carnitine:
  - This reaction is catalysed by carnitine acyl transferase
- Acyl carnitine crosses the IMM on a specific carrier, acyl carnitine translocase:
  - Acyl carnitine is exchanged for free carnitine

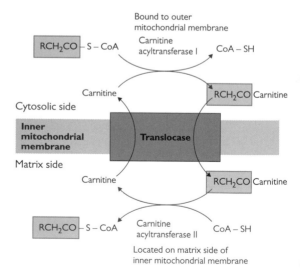

Fig. 2.14 Mechanism of transport of long-chain fatty acyl groups into mitochondria where they are oxidized in the mitochondrial matrix.

- This process is strongly inhibited by malonyl-CoA (thus preventing simultaneous fatty acid breakdown and synthesis; ⊃ p.132)
- The acyl carnitine is reconverted to acyl-CoA and free carnitine in the mitochondrial matrix:
  - Reaction in matrix catalysed by carnitine acyl transferase II
- A number of diseases are linked to carnitine, acyl carnitine translocase, or acyl carnitine transferase deficiencies:
  - Carnitine deficiency leads to muscle weakness during long-term exercise (when fatty acids are an important source of energy):
    ○ Heart and kidney are also affected as they use fatty acids for the majority of their energy supply
    ○ Symptoms range from mild muscle cramps to severe weakness and even death.

Once in the mitochondrial matrix, acyl-CoA can be oxidized by the process known as β-oxidation (Fig. 2.15).

- Four-step cyclic reaction removes a C2 subunit in the form of acetyl-CoA. This can enter the TCA cycle → ATP
- There are different isoenzymes for reaction 1 depending on the length of the fatty acid being metabolized: very long-chain acyl-CoA dehydrogenase (VLCAD), long-chain (LCAD), medium-chain (MCAD), and short-chain (SCAD).

Not all fatty acids in our diet are of an even chain length.

- Although animals have even chain lengths (i.e. C_{2n}), plants have an odd number of fatty acids:
  - β-oxidation eventually leaves a C3 unit (propionyl CoA). This is converted into the TCA cycle intermediate, succinyl-CoA.

Fatty acids can have differing degrees of saturation (C–C vs C=C bonds).

- One extra enzyme is required for monounsaturated fatty acid:
  - Normal rounds of β-oxidation occur until there is a *cis* double bond between the C_3 and C_4 atoms
  - An isomerase then rearranges the C=C bond so that it is *trans*-double bond between C_2 and C_3:
    ○ This has formed the *trans*-enoyl-CoA compound on the β-oxidation pathway, which can continue as normal
- Any polyunsaturated fatty acid requires two extra enzymes, the isomerase plus a reductase:
  - β-oxidation rounds occur with the help of the isomerase until a fatty acid chain with a –C=C–C=C– (*trans*-double bond between C_4 and C_5 and *cis*-double bond between C_2 and C_3) is formed after the fatty acyl-CoA dehydrogenase step of β-oxidation
  - This cannot be processed further without a reductase enzyme:
    ○ The reductase utilizes NADPH to reduce this to –C–C=C–C– (*trans*-double bond between C_3 and C_4)
    ○ This can then be isomerized to the trans-enoyl CoA (i.e. *cis*-double bond between C_2 and C_3) and metabolized (as previously for a monosaturated fatty acid).

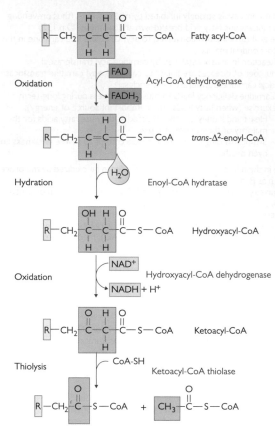

Fig. 2.15 One round of the four reactions of β-oxidation by which a fatty acyl-CoA is shortened by two carbon atoms with the production of a molecule of acetyl-CoA.

Diseases of fatty acid oxidation

- Known to be inherited diseases related to deficiencies in all of the acyl-CoA dehydrogenases
- Best characterized is deficiency in MCAD:
  - Thought to be one of the most common inborn errors of metabolism
  - Symptoms include lethargy, vomiting, and often coma after fasting for more than 12h:
    ○ Ketogenesis is blocked in liver by lack of β-oxidation of fatty acids

- ○ This in turn slows gluconeogenesis
- ○ Failure to be able to metabolize fat in muscle causes increase use of glucose, exasperating the hypoglycaemia
- ○ Medium-chain fatty acids metabolized by alternative pathways and excreted in urine (the disease can be diagnosed by urine analysis)
- Disorder can be managed by avoiding fasting:
  - ○ May be the cause of some cases of sudden infant death syndrome.

Biosynthesis by the liver

During times of plenty, the body will store energy. After the glycogen stores have been replenished (to ~10% of liver weight), the liver switches to fat biosynthesis. Both excess sugars and amino acid carbon skeletons can be used to make fatty acids.

Fatty acids are made in the cytosol by a complex of enzymes, collectively known as fatty acid synthase:
- Dimer of identical 260kDa subunits
- Each monomer has three domains joined by flexible linker regions:
  - Total of seven catalytic sites per subunit. The proximity of these sites allows intermediates to be handed efficiently from one active site to another without leaving the complex.

The reactions of fat synthesis are distinct from those of breakdown.
- Fatty acid synthase is located in the cytoplasm (breakdown in mitochondrial matrix)
- The intermediates of synthesis are covalently bound to the enzyme (rather than to CoA).

The committed step of fat synthesis is the carboxylation of acetyl-CoA to malonyl-CoA. This reaction is driven by ATP hydrolysis and thus effectively irreversible (Fig. 2.16).
- Biotin is an essential co-factor for acetyl-CoA carboxylase
- Allosterically activated by citrate.

The reaction scheme is as follows (Fig. 2.17):
1. For the first round only, an acetyl-CoA is covalently linked to the acyl carrier protein (ACP), part of the fatty acid synthase protein monomer 1, via a flexible linker molecule (phosphopantetheine). It is then passed to the condensing enzyme (CE) in the other monomer (2).
2. Malonyl-CoA is covalently joined to the ACP of monomer 1.
3. There follows a series of four reactions: condensation, reduction (with NADPH as the reductant), dehydration, and a final reduction (again using NADPH) (Figs 2.17, 2.18).
4. The elongated chain is transferred to the CE of monomer 1, and another malonyl-CoA is covalently linked to the ACP of monomer 2.
   - Further rounds continue until a palmitoyl (C16) unit is formed
   - This is released by hydrolysis to give free palmitate.
Longer chain and unsaturated fatty acids are synthesized in the smooth ER.
- Palmitoyl CoA is the starting substrate
- Four similar reactions occur as previously (i.e. condensation, reduction, dehydration, reduction)
- >60% of fatty acids are >C18, with C20, C22, and C24 being the most common
- Unsaturated fatty acids are also common:
  - Catalysed by desaturase, cytochrome b5, and cytochrome b5 reductase
  - Most common in animals are the C16 palmitoleic and C18 oleic acids which have a single C=C bond at C9
  - As mammals cannot introduce double bonds past C9, such fatty acids have to come from the diet (essential fatty acids)

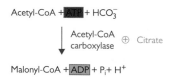

$$\text{Acetyl-CoA} + \boxed{\text{ATP}} + \text{HCO}_3^-$$

Acetyl-CoA carboxylase   ⊕ Citrate

$$\text{Malonyl-CoA} + \boxed{\text{ADP}} + \text{P}_i + \text{H}^+$$

Fig. 2.16 The committed step of fatty acid synthesis: acetyl-CoA to malonyl-CoA.

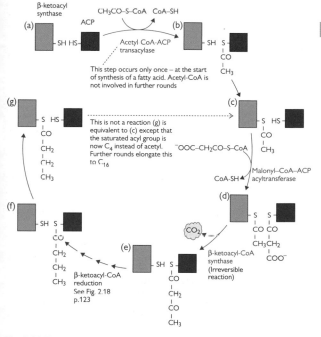

Fig. 2.17 The steps involved in the synthesis of fatty acids.

- The rate of synthesis and breakdown of fatty acids reflects the energy state of the cell
- When ATP levels in the cell are high, mitochondrial citrate rises as the ETC and the enzymes of TCA cycle are inhibited (Fig. 2.19):
  - Citrate leaves the mitochondria on a specific carrier in exchange for malate
  - In the cytosol, citrate is split into acetyl-CoA and oxaloacetate:
    - Acetyl-CoA is converted into malonyl-CoA for fatty acid synthesis
    - Oxaloacetate is converted back into pyruvate

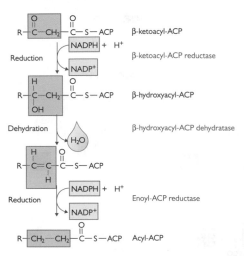

Fig. 2.18 Reductive steps in fatty acid synthesis, (e)–(f) in cycle shown in Fig. 2.17.

- Pyruvate can return into the mitochondrion, where it is converted into oxaloacetate by pyruvate carboxylase
- This process generates one NADH and one NADPH:
  - Each cycle of the fatty acid synthase reaction results in the oxidation of two NADPH, the second of which comes from the pentose phosphate pathway (PPP; ➲ p.150)
- Acetyl-CoA carboxylase is regulated by phosphorylation:
  - An AMP-sensitive kinase (AMPK) inactivates acetyl-CoA carboxylase when energy levels are low in the cell, thus inactivating fatty acid synthesis. This inhibition can be partially overcome allosterically by citrate. This effect of citrate is antagonized by high levels of palmitoyl-CoA, indicating an excess of fatty acids. Palmitoyl-CoA also inhibits the mitochondrial citrate exporter and the production of NADPH by the PPP
- Acetyl-CoA carboxylase is also under hormonal control:
  - Insulin activates it by dephosphorylation (via protein phosphatase 2A). Glucagon and adrenaline inactivate the protein phosphatase 2A via PKA
- Malonyl-CoA inhibits carnitine acyltransferase I, preventing substrates for β-oxidation entering the mitochondria.

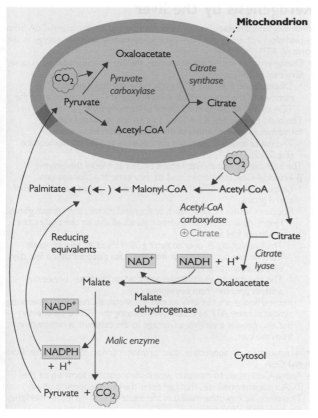

Fig. 2.19 Source of acetyl groups (acetyl-CoA) and reducing equivalents (NADPH) for fatty acid synthesis. The other NADPH comes from the PPP.

Ketogenesis by the liver

With a balanced metabolism of carbohydrate and fat, the acetyl-CoA from β-oxidation will enter the TCA cycle to ultimately produce energy in the form of ATP.

- During times of fasting and starvation, the liver maintains blood glucose levels by gluconeogenesis:
  - Oxaloacetate from the TCA cycle is the starting substrate. The removal of this intermediate prevents acetyl-CoA from entering the TCA cycle ('fat burns in the flame of carbohydrate')
- The build-up of acetyl-CoA leads to a greatly increased rate of formation of ketone bodies in the mitochondria (ketogenesis):
  - The major ketone bodies are acetoacetate and β-hydroxybutyrate (Fig. 2.20)
- The liver cannot metabolize ketone bodies as it lacks the enzyme β-ketoacyl-CoA transferase, and so they enter the bloodstream:
  - Ketone bodies are effectively a water-soluble, transportable form of acetyl groups
  - Recipient tissues include heart, brain, renal cortex, and adrenal glands:
    - Heart, renal cortex, and adrenal glands all use ketone bodies as a preferred fuel source
    - The brain switches over to getting 50–75% of its energy needs from ketone bodies (rather than the usual glucose) after a few days of starvation (⊃ p.178)
    - This reduces the gluconeogenesis load on the body, preserving protein (muscle) from breakdown
  - Ketone bodies are not only an efficient metabolic resource (generating almost as many ATP as acetyl-CoA entering the TCA cycle directly) but also provide a survival advantage to the tissues that receive them from the liver.

It is important to appreciate that animals cannot make glucose from acetyl-CoA.

- Acetyl-CoA needs to combine with oxaloacetate to form any of the TCA cycle intermediates that can enter the gluconeogenic pathway. Therefore, no new intermediates are created (TCA cycle intermediates are essentially catalytic).

The levels of ketone bodies act as signals for availability of energy substrates.

- High levels of acetoacetate acts as a signal for abundantly available acetyl groups. This inhibits the further breakdown of fat in adipose tissue.

Disease conditions can cause confused signals. Most common is diabetes mellitus (⊃ p.182; see also ⊃ *OHCM10* p.206).

- Lack of insulin secretion means that the liver and skeletal muscles do not absorb glucose; the lack of hepatic carbohydrate leads to ketogenesis. This is made worse by the lack of signals to adipose tissue to inhibit fat breakdown
- Ketone bodies (⊃ *OHCM10* p.832) are acidic and their accumulation (up to 200-fold the normal concentration), and the ensuing metabolic acidosis, can be severe enough to impair CNS function:
  - Acetoacetate is unstable and spontaneously decays to acetone. This can be smelt on the breath of people with uncontrolled diabetes ("pear-drops").

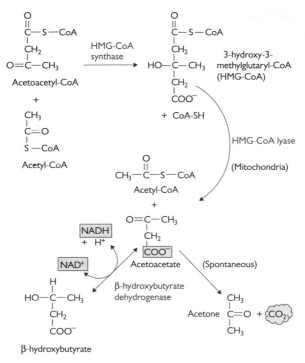

Fig. 2.20 Ketone body production in the liver during excessive oxidation of fat in starvation or diabetes.

Integration

In all metabolic pathways, their regulation is the key to integrating their functions. Extrinsic regulation of fat metabolism is controlled by a number of hormones: insulin, glucagon, adrenaline, and thyroxine.

- Insulin is the hormone that signals the well-fed state:
  - Turns on lipid synthesis
  - Activates acetyl-CoA carboxylase by dephosphorylation (via activating protein phosphatase 2A)
  - Also increases the amount of lipoprotein lipase ('clearing factor' lipase) on the endothelial cells in adipose tissue:
    - Increases the breakdown of circulating triacylglycerols and thus their uptake and storage in adipocytes
- Glucagon and adrenaline signal the need for energy release from fat:
  - Activate PKA via cAMP which:
    - Inhibits fat synthesis by phosphorylating acetyl-CoA carboxylase and, thus, inactivates it
    - Promotes breakdown of triacylglycerides in adipose tissue and release of fatty acids by activating lipases
    - Reduces the amount of lipoprotein lipase so that circulating triacylglycerols are available to other tissues (and not all taken up by adipose tissue)
- Thyroxine, the active thyroid hormone, is a long-term signal of growth and development:
  - Increases the oxidative metabolism of both carbohydrate and fat:
    - The mechanism of thyroxine action involves up-regulation of gene transcription of the relevant metabolic enzymes.

Glucose as a fuel: overview

Glucose intake after a meal is usually more than enough to meet the immediate energy needs of the body, and so the excess energy needs to be stored.

- Storage is primarily in the form of the glucose polymer, glycogen:
  - Main sites of storage are the liver (10% of organ weight) and skeletal muscle (2%)
  - Glycogen is a rapidly mobilizable storage form
- When these stores are replete, further excess glucose will be stored as fat.

Glucose is the primary fuel of a number of tissues when the body is in a fed state, with a daily consumption at rest of 160g.

- The brain uses ~75% (120g) of the total glucose used per day:
  - Ketone bodies fulfil much of this need during starvation
- Red blood cells have no mitochondria and so can only make ATP by glycolysis. They therefore have an obligate need for glucose, which they convert into lactate
- Renal medulla has a high energy need to power the membrane transport that occurs in reabsorption from ultrafiltrate compared to its perfusion
- Skeletal muscle uses glucose for immediate energy:
  - This comes from its store of glycogen:
    - Muscle cannot release glucose into the blood, and so its glycogen stores are for its own use only
  - Muscle also obtains glucose generated in the liver via the Cori cycle (Fig. 2.3; ⊃ p.109)
- In pregnancy, the foetus uses glucose as its main energy source.
  - Glucose transporters in the brush-border and basolateral membranes of the placenta ensure that the foetus always has a good supply of glucose:
    - The foetal demand for glucose can even result in maternal hypoglycaemia
  - The placenta itself also uses glucose for glycolytic energy production, and the lactate is either released into the circulation or taken up by the foetus and used as an energy source.

Glycolysis

Glycolysis (Greek for 'splitting sweetness') is a sequence of enzyme-catalysed reactions occurring in the cytoplasm of all cells.

- These reactions split a 6-carbon glucose into two 3-carbon pyruvate molecules
- Oxygen is not required for glycolysis and two ATP are produced per glucose
- Under aerobic conditions, these pyruvate molecules can enter the TCA cycle (◑ p.114) and the two NADH, the ETC (◑ p.116).

Glycolysis can be divided into two phases (Fig. 2.21):
- An energy-investment phase
- An energy-generation phase.

The energy-investment phase

- Glucose enters the cell on a facilitated glucose transporter (GLUT) and is immediately phosphorylated by hexokinase (glucokinase in liver):
  - This irreversible reaction traps the glucose in the cell and maintains the glucose gradient for entry
- Glucose-6-phosphate (G-6-P) is isomerized to fructose-6-phosphate
- Fructose-6-phosphate is phosphorylated to fructose-1,6-bisphosphate (F-1,6-P) by phosphofructokinase (PFK)
  - This irreversible reaction is known as the 'committed step' of glycolysis, as the only fate for F-1,6-P is to enter the next reaction in the glycolytic pathway
  - PFK is the primary regulated step of the glycolysis pathway. PFK responds to both cellular energy needs and hormonal regulation
- Fructose-6-phosphate is cleaved into two 3-carbon molecules—glyceraldehyde-3-phosphate (GAP) and dihydroxyacetone phosphate (DHAP):
  - Only GAP is a substrate for further steps in glycolysis
  - DHAP is isomerized into GAP.

The energy-generation phase

- The oxidizing power of NAD^+ is used to form a high-energy bond between an inorganic phosphate molecule and the aldehyde group of GAP, to give 1,3-bisphosphoglycerate
- The hydrolysis of the acyl phosphate group at C1 gives enough energy to drive the formation of ATP from ADP + P_i. This is known as substrate-level phosphorylation
- The two-step rearrangement of 3-phosphoglycerate into phosphoenolpyruvate creates another high-energy phosphate group—this time an enol phosphate at C2
- A second substrate-level phosphorylation reaction produces an ATP and pyruvate.

Points to note

- Glycolysis generates a net two ATP plus two NADH and two pyruvates:
  - Four ATP per glucose are formed in the energy-generation phase two for each C3 GAP molecule, but two ATP per glucose are invested in the energy-investment phase

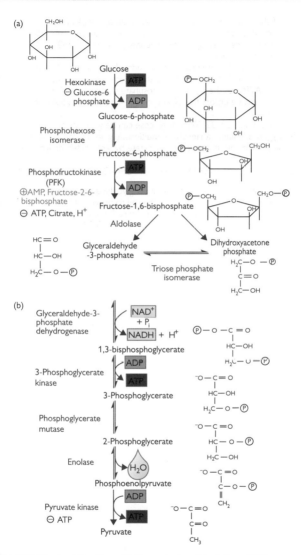

Fig. 2.21 The glycolytic pathway. (a) The energy-investment phase. (b) The energy-generation phase, which only glyceraldehyde-3-phosphate can enter.

- All of the reactions (except those catalysed by hexokinase, phosphoglycerate kinase, and pyruvate kinase) are reversed when glucose is made from pyruvate (gluconeogenesis; ⮕ p.158)
- Under aerobic conditions:
  - The two NADH molecules are reoxidized to NAD^+ by the ETC to give an extra five ATP per glucose (if the malate-aspartate shuttle is used, or three ATP with the glycerol-3-phosphate shuttle; ⮕ pp.116–17)
  - The pyruvate enters the TCA cycle (giving a further 25 ATP per glucose)
- Thus a total of 30–32 ATP per glucose can be generated
- If oxygen is limiting (anaerobic conditions, e.g. in vigorously exercising muscle), pyruvate is converted to lactate by lactate dehydrogenase:

$$CH_3\text{-}CO\text{-}COO^- + NADH + H^+ \leftrightarrow CH_3\text{-}CHOH\text{-}COO^- + NAD^+$$

  - This regenerates the NAD^+ to allow glycolysis to continue
  - Lactate may be exported from the cell to prevent it becoming acidotic
  - Lactate is the usual fate of pyruvate in red blood cells, which lack mitochondria and therefore have no ETC or TCA cycle.

Control of glycolysis

Glycolysis is regulated by the energy needs of the cell. There are three main points of regulation.

Hexokinase

- Has high affinity (K_m <0.1mM) and shows strong end-product inhibition by G-6-P in most tissues
- Inhibition by G-6-P is important, because in the presence of high glucose concentrations and low rates of glycolysis, it prevents cellular depletion of P_i by hexokinase:
  - Liver has glucokinase, which has lower affinity (K_m ~7mM) and is not inhibited by G-6-P. This is a problem in fructose intolerance because, in the absence of end-product inhibition, the liver generates large quantities of fructose-6-phosphate, which cannot be metabolized further (⊃ p.146). This causes liver ATP levels to drop, compromising hepatocyte cellular function.

Phosphofructokinase (PFK)

- PFK is the major site of regulation of glycolysis
- There are several important allosteric regulators, both positive and negative.

Positive regulators

- AMP: a rise in the cellular AMP level indicates low ATP (as 2ADP → ATP + AMP, catalysed by adenylate cyclase)
- Fructose-2,6-bisphosphate (F-2,6-bisP): formed by phosphofructokinase-2 (PFK-2) phosphorylating F-6-P to F-2,6-bisP:
  - PFK-2 is inhibited by ATP, and strongly activated by AMP
  - The opposing fructose-2,6-phosphatase (F-2,6-Pase), which converts F-2,6-bisP to F-6-P, is inhibited by AMP. These two enzyme activities are found in the same protein (a 'bi-directional enzyme').

Negative regulators

- H^+ ions: if lactic acid builds up, then PFK will be inhibited by the increased protons:
  - This is a form of end-product negative feedback
  - It is especially likely to occur when blood flow is inadequate, e.g. extreme exercise or an attack of angina pectoris (⊃ *OHCM10* p.116) in the heart
- ATP: high levels of ATP inhibit PKF as they indicate that the cell does not need glycolysis
- Citrate: fatty acids and ketone bodies are often the favoured cellular fuel, and their oxidation produces citrate, which inhibits PFK and decreases glucose utilization
- PFK is also under hormonal control, with the exception of the brain PFK isoform:
  - In liver, increases in intracellular cAMP inhibit glycolysis via a decrease in F-2,6-bisP, by inhibiting PFK-2/stimulating F-2,6-Pase:
    - Glucagon and adrenaline activate adenylate cyclase by binding to G-protein-linked receptors in the hepatocyte plasma membrane

- The heart also has adrenaline receptors which cause an increase in cAMP but, in this case, the heart isozyme of PFK-2 is activated, increasing F-2,6-bisP and stimulating glycolysis:
  - Increases ATP production to match adrenaline-signalled increased work load
- Muscle PFK is stimulated by insulin and adrenaline:
  - PFK regulation is especially important in type IIB (glycolytic) skeletal muscle fibres. These have a low capacity for oxidative phosphorylation and no triacylglycerol stores
  - Nevertheless, during long periods of strenuous activity, muscle fibres must adapt to use fatty acids as a fuel. The rise in fatty acids inhibits:
    - Hexokinase (via G-6-P increase due to decreased glycolysis)
    - PFK (via increased cytoplasmic citrate as a result of increased mitochondrial acetyl-CoA)
    - PDH (via rise in matrix acetyl-CoA causing pyruvate dehydrogenase kinase (PDK) to phosphorylate and inactivate PDH)
    - PDH inactivation reduces pyruvate entry rate into TCA cycle
    - Instead, increased amounts of pyruvate are converted to oxaloacetate, allowing the acetyl-CoA from fatty acids to be oxidized by the TCA cycle.

Pyruvate kinase

- Strongly inhibited by ATP
- Different tissues express slightly different forms of the same enzyme (known as isozymes):
  - For example, hexokinase and glucokinase are isozymes
  - This can be useful in diagnostic testing, e.g. if a heart muscle enzyme isoform is found in the plasma, it indicates that the individual has had a heart attack (➲ p.22; see also ➲ OHCM10 p.118).

Use of other monosaccharides

Although glucose is the major carbohydrate fuel, both galactose and fructose are important, with the latter making up a significant part of dietary carbohydrate.

Galactose

Galactose (from the disaccharide, lactose) is metabolized by converting it into the glucose metabolite, G-6-P (Fig. 2.22). This is a four-reaction process:
- Galactose is phosphorylated by galactokinase
- Galactose-1-phosphate is converted into UDP-galactose by reaction with UDP-glucose, giving glucose-1-phosphate. This is then isomerized to G-6-P
- UDP-galactose is isomerized back to UDP glucose for re-use.

Galactosaemia is a rare, inherited inability to metabolize galactose.
- A mild form is seen when galactokinase is deficient
- In the severe form, galactose-1-phosphate uridyl transferase enzyme is absent:
  - High blood and urine levels of galactose
- Infants fail to thrive, with symptoms including:
  - Vomiting/diarrhoea after milk
  - Enlargement of the liver and jaundice—even cirrhosis:
    ○ These are due to toxic effects of galactose-1-phosphate
  - Cataracts:
    ○ Due to build-up of reduced form of galactose (galactitol) in the lens
  - Lethargy:
  - Mental retardation (often delayed language skill acquisition):
    ○ Still persists even if patient has a galactose-free diet.

Fructose

Fructose has a simpler entry pathway into metabolism (Fig. 2.23).
- In the liver, it is converted into fructose-1-phosphate by fructokinase
- This is then split into DHAP and glyceraldehyde:
  - Glyceraldehyde is converted into GAP by triose kinase
  - DHAP and GAP are both intermediates of the glycolysis pathway (➲ p.140)
- In adipose tissue, hexokinase converts fructose into fructose-6-phosphate, which can continue directly through glycolysis.

Hereditary fructose intolerance prevents cleavage of fructose-1-phosphate to DHAP and GAP.
- Deficiency in fructose-1-phosphate aldolase
- Characterized by hypoglycaemia after fructose ingestion, and death in young children after prolonged ingestion
- Fructose-1-phosphate accumulates intracellularly, effectively depleting the cells of free P_i and therefore reducing their ability to make ATP.

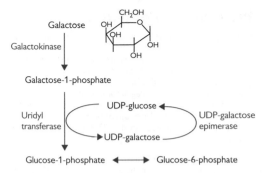

Fig. 2.22 Galactose metabolism.

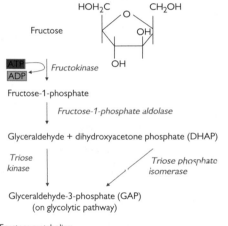

Fig. 2.23 Fructose metabolism.

Aerobic oxidation of glucose

PDH is a key regulatory enzyme for aerobic oxidation of glucose, as it commits pyruvate to acetyl-CoA to enter the TCA cycle. Other potential fates of pyruvate are conversion to lactate (anaerobic conditions), oxaloacetate (to replenish TCA cycle intermediates), or alanine (by transamination).

The PDH complex consists of a large number of subunits, with multiple copies of three catalytic and two regulatory enzymes, with five co-factors (all derived from water-soluble vitamins). Regulation is at two levels:
• Feedback inhibition by acetyl-CoA and NADH
• More important is the regulation of the PDH enzyme complex by phosphorylation.

PDH is inactivated by phosphorylation.
• The kinase responsible for PDH phosphorylation is itself part of the PDH complex:
  • The kinase is activated by ATP, acetyl-CoA, and NADH
  • It is inhibited by CoA-SH, NAD^+, pyruvate, and ADP
• The phosphatase that activates PDH is also part of the complex:
  • It is Mg^{2+} and Ca^{2+} dependent. Ca^{2+} is important during muscle contraction, as it will cause the activation of PDH when energy is required
  • Adrenaline activates PDH in cardiac muscle via G-protein receptors
  • Insulin activates PDH via Ca^{2+} in adipose tissue (increasing acetyl-CoA conversion into fat)
• These regulatory factors allow PDH activity to reflect the metabolic state of the mitochondrion:
  • An increase in the $NADH/NAD^+$ or acetyl-CoA/CoASH ratio signals that the ETC is not operating fast enough to match NAD^+ reduction to NADH. This could be due to lack of oxygen or a high ATP level (respiratory control is in operation)
  • The result is an inactivation of PDH → a reduction in the rate of pyruvate entry into the TCA cycle.

The brain has a high energy requirement that is normally satisfied by aerobic glucose oxidation.
• Deficiencies in PDH correlate with severe neurological defects:
  • Often result in childhood death
  • Raised blood levels of lactate, pyruvate, and alanine (with resulting acidosis)
• Alcoholism causes a reduction in thiamine (vitamin B_1) absorption from the diet, and thiamine phosphorylation to thiamine pyrophosphate (TPP) in the liver:
  • TPP is one of the co-factors for PDH
  • Reduction in PDH activity causes mental disorder (with memory loss, partial paralysis) known as Wernicke–Korsakoff syndrome (➲ OHCM10 pp.714, 786).

Pentose phosphate pathway

The pentose phosphate pathway (PPP) has several functions (Fig. 2.24), including:

- Generation of NADPH for biosynthetic reducing power, e.g. fat synthesis
- Production of ribose-5-phosphate for nucleic acid synthesis.

Reactions take place in the cytosol of cells involved in biosynthesis, e.g. adipose tissue.

Reaction scheme

- The need of the cell for NADPH or sugar intermediates will determine the flow through the PPP:
    - When more NADPH is needed than ribose-5-phosphate, one G-6-P is oxidized to CO_2 for every five that are regenerated:

$6 \text{ G-6-P} + 12 \text{ NADP}^+ + 7 \text{ H}_2\text{O} \rightarrow 5 \text{ G-6-P} + 6 \text{ CO}_2 + 12 \text{ NADPH} + 12 \text{ H}^+ + P_i$

    - When balanced amounts of ribose-5-phosphate and NADPH are needed:

$\text{G-6-P} + 2\text{NADP}^+ + \text{H}_2\text{O} \rightarrow \text{ribose-5-phosphate} + 2 \text{ NADPH} + 2\text{H}^+ + \text{CO}_2$

    - When only ribose-5-phosphate is needed:

$5 \text{ glucose-6-phosphate} + \text{ATP} \rightarrow 6 \text{ ribose-5-phosphate} + \text{ADP}$

NADPH is important in recycling of the antioxidant glutathione.
- Glutathione (GSH) is important in detoxifying harmful peroxides (–OOH) including H_2O_2 (catalysed by glutathione peroxidase) (🔁 p.174):
    - Oxidized GSH is regenerated by glutathione reductase using NADPH as its source of reducing power
- GSH is also important in red blood cells for keeping the methionine residues of haemoglobin in a reduced state.

Deficiency in glucose-6-phosphate dehydrogenase (G6PD or G6PDH) (🔁 OHCM10 p.338) reduces the NADPH availability.
- Anything that increases the oxidative stress in cells will then cause problems due to the lack of reduced GSH. Examples include antimalarial drugs, e.g. primaquine (🔁 OHPDT2 p.445) and flavobeans (broad beans)
- Symptoms include black urine, jaundice, haemolytic anaemia
- NADPH is important in maintaining the erythrocyte membrane integrity. Deficiencies in G6PD lead to weakened red blood cells that are more susceptible to haemolysis:
    - There are >300 known mutations in this enzyme
    - Frequency varies from <1% in northern Europeans, 10% in African Caribbeans, to up to 25% in southern Europeans:
        - The high prevalence in southern Europeans is due to its protective effects against malaria
        - Selective advantage may be due to malaria parasite needing PPP products and/or the extra stress due to the parasite causing the red blood cell host to lyse before the parasite matures.

(a) If cell has balanced need for ribose-5-phosphate and NADPH

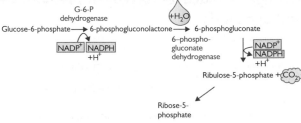

(b) If cell requires more NAPH than ribose-5-phosphate, then the excess ribose-5-phosphate formed in part (a) can be converted into the glycolytic/gluconeogenic intermediate fructose-6-phosphate by the overall reaction

6 ribose-5-phosphate C_5 → 5 fructose-6-phosphate C_6 + Pi

The reaction scheme is shown schematically

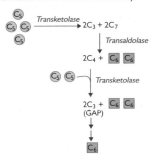

(c) If cell requires more ribose-5-phosphate than NADPH

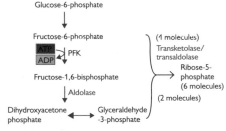

Fig. 2.24 Pentose phosphate pathway.

Storage of glucose—glycogen breakdown and synthesis

Glycogen is a readily mobilized storage form of glucose.
* Glycogen is a very large, branched polymer of glucose:
* It has mainly α-1,4 glycosidic bonds, with branches about every tenth residue caused by α-1,6 bonds (◉ p.40)
* The many free 4-OH ends allow for rapid breakdown to release glucose.

There are separate pathways for breakdown (Figs 2.25, 2.26) and synthesis of glycogen.

Glycogen breakdown (glycogenolysis)

Glycogen is broken down by the liberation of a glucose-1-phosphate (G-1-P) molecule, leaving the glycogen chain one residue shorter.
* The reaction is catalysed by glycogen phosphorylase
* The α-1,4 glycosidic bond is cleaved by phosphorolysis (cleavage of bond by orthophosphate), rather than by hydrolysis
* In most tissues, G-1-P is converted into G-6-P by phosphoglucomutase, which can then enter the glycolytic pathway to form energy:
  * As the main site of gluconeogenesis, the liver has the enzyme glucose-6-phosphatase. This converts G-6-P into glucose, which is released into the bloodstream.

Glycogen phosphorylase can only remove glucose residues from free chain ends until it is four residues from a branch point.
* Three residues are moved by a transferase to an adjacent chain for future breakdown by glycogen phosphorylase
* The remaining single residue is hydrolysed by α-1,6 glucosidase (debranching enzyme) to give glucose, leaving a linear chain for continued breakdown by glycogen phosphorylase
* Debranching enzyme and transferase activity are present in the same 160-kDa polypeptide chain.

McArdle's disease (◉ *OHCM10* p.704)—absence of muscle glycogen phosphorylase:
* Patients have a limited ability to perform strenuous exercise due to painful muscle cramps
* This is caused by a failure to utilize the (larger than normal) stores of glycogen to make ATP.

Glycogen synthesis (glycogenesis)

Glycogen is synthesized by the addition of glucose molecules to the 4-OH end of an existing chain of glycogen, using an activated form of glucose—UDP-glucose.
* Glycogen chain extension is catalysed by glycogen synthase:

$$G\text{-}1\text{-}P + UTP \rightarrow UDP\text{-glucose} + PP_i$$

$$Glycogen_n + UDP\text{-glucose} \rightarrow Glycogen_{n+1} + UDP$$

- There is also a branching enzyme:
  - When at least 11 residues have been added, it breaks off a chain of about seven glucose units and rejoins them to a free 6-OH group (i.e. as an α-1,6 linkage)
- The new branch must be at least four residues from the nearest existing branch.

The storage of glucose as glycogen is an energy-efficient process.
- One ATP equivalent is used in generating UDP-glucose
- Approximately one in ten glucose monomers released on glycogen breakdown will be a branch point and released as glucose (→ G-6-P for further metabolism except in liver):
  - Therefore, the metabolic cost is ~1.1 ATP per glucose
  - Overall, this represents around 97% efficiency (based on one G-6-P yielding 31 ATP).

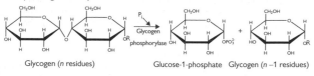

Glycogen (n residues)   Glucose-1-phosphate   Glycogen (n −1 residues)

Fig. 2.25 Breakdown of glycogen is by the sequential liberation of a glucose-1-phosphate molecule.

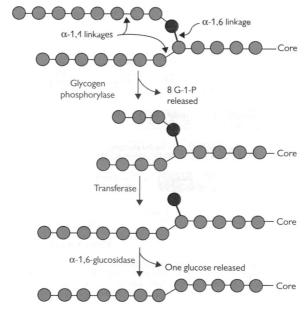

Fig. 2.26 Schematic representation of glycogen breakdown, including of a branch point.

Regulation of glycogen synthesis and breakdown

The two separate pathways of glycogen synthesis and breakdown must be regulated, both to maintain suitable plasma glucose concentrations and also to avoid futile substrate cycling. There are two potential forms of regulation: intrinsic and extrinsic.

- Intrinsic—allows cells to respond to their own energy needs by breaking down glycogen when cell ATP and glucose levels fall, and to switch on glycogen synthesis when these concentrations rise
- Extrinsic (Fig. 2.27)—mediated by hormones or other stimuli:
  - Increases in intracellular levels of Ca^{2+} or cAMP will promote glycogen breakdown and inhibit synthesis, e.g. to prepare muscle cells for action or liver to release glucose for other tissues
  - Insulin signals the fed state and enhances glycogen synthesis and inhibits breakdown, thus storing energy for use in the future. These effects are mediated via reversible phosphorylation of the synthesis/breakdown enzymes.

Regulation of glycogen breakdown

The enzyme directly responsible for glycogen breakdown, i.e. glycogen phosphorylase, can exist in two interconvertible forms (Fig. 2.28a):

- The *a* form is active
- The *b* form is usually inactive

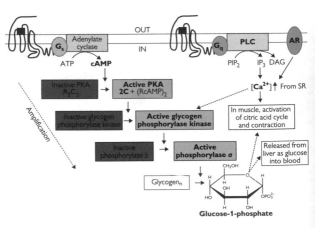

Fig. 2.27 Overall mechanism of activation of glycogen breakdown from the activation of membrane receptors (note how each step acts to amplify the signal). Glucagon acts via Gs-protein linked receptor, adrenaline via Gs or Gq-linked receptor (tissue dependent) and 'AR' is the nicotinic acetylcholine receptor (nAChR) at the neuromuscular junction.

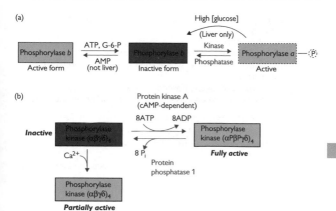

Fig. 2.28 Mechanisms of activation/inactivation of (a) glycogen phosphorylase and (b) glycogen phosphorylase kinase.

- The usually inactive b form can be converted into the active a form by phosphorylation:
  - Catalysed by glycogen phosphorylase kinase
  - Glycogen phosphorylase a is deactivated by dephosphorylation by protein phosphatase 1 (PP1). This is the mechanism behind hormonally exerted extrinsic control
- Although glycogen phosphorylase b is usually inactive, it can be activated allosterically by molecules that signal the energy charge of the cell.
  - This represents the intrinsic control
  - In muscle cells, high (AMP) will activate glycogen phosphorylase b, whereas high (ATP) and (G-6-P) inactivate it. All three compounds act at the same allosteric regulatory site
- The liver isoform of glycogen phosphorylase is different, in that active glycogen phosphorylase a is deactivated by the binding of glucose, but the b isoform is insensitive to AMP levels. This difference reflects the role of liver glycogen stores in supplying glucose for the rest of the body:
  - Glycogen breakdown is prevented when plasma glucose concentrations are high.

What regulates the regulator (i.e. what regulates phosphorylase kinase)? Glycogen phosphorylase kinase is a very large protein (1200kDa), made up of $(\alpha\beta\gamma\delta)^4$ subunits, and it can be controlled in two ways (Fig. 2.28b):
- It is converted from a low to a high activity form by phosphorylation by PKA:
  - As PKA is activated by cAMP, this is makes glycogen phosphorylase kinase sensitive to hormones such as adrenaline

- Glycogen phosphorylase kinase is phosphorylated on a serine residue on subunits α and β
- Glycogen phosphorylase kinase can be partially activated by Ca^{2+} at levels of ~1μM because the δ subunit is calmodulin:
  - This is important in muscle, where contraction is triggered by Ca^{2+} release from the SR
  - It will also make glycogen phosphorylase kinase sensitive to hormones which raise cytoplasmic Ca^{2+} (especially relevant in liver).

Regulation of glycogen production

It is clearly important that glycogen synthetase is switched off when glycogen phosphorylase is activated and vice versa (i.e. that they are regulated reciprocally).

- Glycogen synthase also exists in two forms—the active *a* form and the inactive *b* form
- Conversion from the active *a* to the inactive *b* form requires phosphorylation
- The three most important kinases responsible are PKA, phosphorylase kinase, and the Ca^{2+}-calmodulin CaM kinase II
- Thus, the hormones which turn on glycogen breakdown simultaneously turn off glycogen synthase (Fig. 2.29):
  - Those which acted via cAMP, through activating PKA and phosphorylase kinase
  - Those which raised $[Ca^{2+}]$, via phosphorylase kinase and CaM kinase II.

There must be a cellular mechanism present to reverse the effects of the phosphorylation steps (i.e. activation of glycogen phosphorylase kinase and glycogen phosphorylase, and inactivation of glycogen synthase).

- PP1 is the most important cell phosphatase regulating glycogen metabolism:
  - It dephosphorylates and thus inactivates glycogen phosphorylase kinase (and hence glycogen phosphorylase)
  - It also dephosphorylates glycogen synthase, and thus activates it.

The activity of PP1 itself is regulated by phosphorylation.

- PP1 has two subunits—the catalytic 37kDa and the 160kDa glycogen-binding subunit
- The glycogen-binding subunit is phosphorylated by PKA, rendering it unable to bind the catalytic subunit, thus inactivating it.

Further inhibition of PP1 is brought about by an inhibitor protein, known as inhibitor 1. When phosphorylated by PKA, this small protein blocks the catalytic subunit of PP1.

Thus, cAMP not only activates the kinase cascade (Fig. 2.29), but also prevents PP1 from dephosphorylating the enzymes involved in glycogen metabolism.

What happens in times of plenty when glycogen synthesis needs to be switched on?

- The hormonal signal of the fed state, insulin, activates glycogen synthetase and inhibits glycogen phosphorylase (Fig. 2.30):
  - Insulin binds its plasma membrane receptor, itself a tyrosine kinase

- This catalyses the auto-phosphorylation of the receptor and the initiation of a protein kinase cascade:
  - PKB is activated, which in turn phosphorylates and inactivates glycogen synthase kinase 3 (GSK3). (GSK3 normally keeps glycogen synthase phosphorylated and inactive.)
  - PP1 is activated by phosphorylation, and so glycogen synthase is dephosphorylated and activated. Simultaneously, glycogen phosphorylase kinase and glycogen phosphorylase will be dephosphorylated and inactivated
  - The net result will be increased glycogen synthesis and decreased glycogen breakdown.

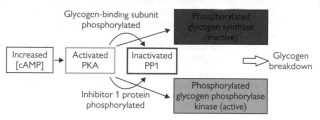

Fig. 2.29 Regulation of the process of glycogen breakdown.

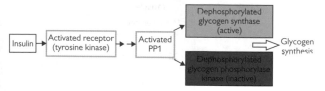

Fig. 2.30 Regulation of the process of glycogen synthesis.

Gluconeogenesis

Gluconeogenesis (Figs 2.31, 2.32) is the synthesis of glucose from non-carbohydrate precursors. This process plays different roles depending on the nutritional state and the tissue in question.

- In tissues that are generating sufficient energy and have surplus nutrients, glucose is produced to be stored as glycogen
- The liver synthesizes glucose for export to other glucose-dependent tissues (especially brain, red blood cells) during starvation and intense exercise:
  - Renal cortex also contributes about 10% of normal glucose production, which is largely used for energy by the renal medulla
- NB: mammals cannot convert fatty acids into glucose, as there is no enzyme to catalyse the reaction of acetyl-CoA into oxaloacetate:
  - The glycerol backbone of triacylglycerols is a gluconeogenic substrate
  - The last unit of β-oxidation of an odd-chain fatty acid, propionyl-CoA, can also enter gluconeogenesis.

The gluconeogenesis pathway is not simply a reversal of glycolysis.

- Thermodynamics favour glycolysis direction of glucose → pyruvate
- There are three essentially irreversible reactions in glycolysis to be bypassed.

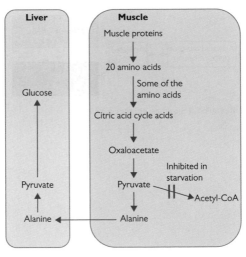

Fig. 2.31 Mechanism by which breakdown of muscle proteins supplies the liver with a source of pyruvate for gluconeogenesis during starvation.

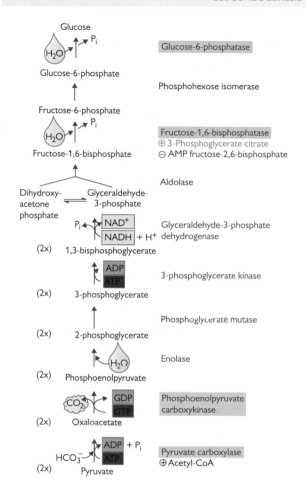

Fig. 2.32 The complete gluconeogenesis pathway from pyruvate to glucose. The highlighted enzymes are not shared with the glycolytic pathway.

Bypass reactions

Bypass reaction 1: pyruvate → phosphoenolpyruvate

Two enzyme-catalysed reactions involved:

Pyruvate + CO_2 + ATP → oxaloacetate + ADP + Pi *(pyruvate*

carboxylase)

- Reaction occurs in the mitochondrial matrix
- Requires biotin as co-factor
- Reaction driven by ATP hydrolysis
- Allosterically stimulated by acetyl-CoA
  - Signals cell is energy-replete, turning on storage pathway.

Oxaloacetate + GTP → phosphoenolpyruvate + CO_2 + GDP

(*phosphoenolpyruvate carboxykinase*)

- Can occur in either matrix or cytoplasm
- Reaction is driven by GTP hydrolysis.

Bypass reaction 2: fructose-1,6-bisphosphate → fructose-6-phosphate (fructose-1,6-bisphosphatase)

Allosterically regulated:
- Stimulated by an increase in 3-phosphoglycerate and citrate
- Inhibited by increased AMP and fructose-2,6-bisphosphate:
  - These are the opposite to the signals to regulate glycolysis
  - Prevents futile cycling.

Bypass reaction 3: glucose-6-phosphate (G-6-P) → glucose + Pi (glucose-6-phosphatase)

- Enzyme only found in liver and renal cortex
- Located in the smooth ER
- G-6-P has to enter across the smooth ER membrane, and products (glucose, phosphate) have to leave—there are specific transporters for each
  - Deficiencies in either transporters or enzymes cause von Gierke's disease:
    - Failure to break down G-6-P during hypoglycaemia leads to inappropriate production of glycogen (stimulated by rise in G-6-P).

Although these three bypass reactions render gluconeogenesis energetically favourable, they do so at a cost.
- Conversion of pyruvate to glucose uses more ATP than glycolysis releases
- Gluconeogenesis bestows two major advantages that outweigh this:
  - The ability to store excess non-carbohydrate nutrients as glycogen in time of plenty
  - The role of the liver to effect nutrient redistribution between tissues, e.g. Cori cycle (Fig. 2.3; ⊃ p.109).

In addition to local (intrinsic) control of the bypass reactions mentioned previously, gluconeogenesis is also under extrinsic hormonal control.
- Acute regulation:

- Glucagon stimulates gluconeogenesis (cAMP leads to reduction in the levels of fructose-2,6-bisphosphate, an allosteric activator of fructose-1,6-bisphosphatase):
  - The rise in fructose-6-phosphate also inhibits glucokinase
- Insulin has the opposite effect
- Long-term regulation:
  - Both glucagon and insulin affect glycolysis and gluconeogenesis by induction and repression of key enzymes in the pathways:
    - A high plasma glucagon/insulin ratio increases the liver capacity for gluconeogenesis
    - A high insulin/glucagon ratio has the opposite effect
  - Corticosteroids have the same effect as glucagon.

Alcohol inhibits gluconeogenesis in the liver.
- Detoxification of ethanol creates high levels of NADH in the cytosol
- This in turn promotes the formation of lactate from pyruvate, and malate from oxaloacetate, which effectively limits the availability of substrates for gluconeogenesis.

Amino acid metabolism: overview

Protein from the diet (~70g per day) is broken down by sequential enzyme digestion.

- The endopeptidase (protease) pepsin is secreted in the stomach in an inactive form (pepsinogen) which is activated by cleavage of a peptide fragment from its amino terminus:
  - Either autoactivated when the pH <5 or by active pepsin
  - Endopeptidases cleave internal peptide bonds and release large peptide fragments
  - Acidity of stomach lumen also denatures proteins and makes them more susceptible to hydrolysis
- Pancreatic secretion of the endopeptidases trypsin, chymotrypsin, and elastase (serine proteases) and the exopeptidases carboxypeptidases A and B:
  - Work in neutral conditions of small intestine lumen, achieved by secretion of bicarbonate-rich pancreatic juice
  - All the pancreatic proteases are secreted as inactive precursors (trypsinogen, chymotrypsinogen, proelastase, procarboxypeptidase A and B):
    - Trypsinogen is activated by enteropeptidase (enterokinase) released from the epithelial cells of the small intestine (enterocytes) and by active trypsin
    - The others are all activated by cleavage with trypsin
  - Exopeptidases remove the last (carboxypeptidase) or the first (aminopeptidase) amino acid from a peptide chain
  - Results in a mixture of amino acids and small peptides up to six amino acids long (oligopeptides):
    - Diseases which interfere with pancreatic secretion (e.g. pancreatitis, CFTR (⊕ *OHCM10* pp.173, 636)) prevent proper protein digestion and thus lead to protein malabsorption and malnutrition. This can be overcome by either supplying preparations of exogenous pancreatic enzymes or dietary supplements of easily digested proteins
- The brush-border membrane of the enterocytes contains enzymes that continue digestion:
  - Endopeptidases, aminopeptidases, and dipeptidases continue the digestion to dipeptides, tripeptides, and free amino acids.

A mixture of dipeptides, tripeptides, and amino acids is taken up by enterocytes.

- Di- and tripeptides are taken up by a proton-coupled co-transporter:
  - Also responsible for absorption of some drugs such as β-lactam (aminopenicillin) antibiotics
- Amino acids are absorbed by a number of mainly sodium-coupled transport systems
- Di- and tripeptides are cleaved by intracellular peptidases into free amino acids
- Amino acids leave the enterocytes via the basolateral membrane and enter the circulation:
  - A few hydrolysis-resistant peptides (and drugs such as antibiotics) may leave the cell intact.

Amino acids fall into different categories and can have different fates (Fig. 2.33).
- Essential amino acids (those the body cannot make): arginine, histidine, isoleucine, leucine, lysine, methionine, phenylalanine, threonine, tryptophan, and valine:
  - Arginine can be synthesized in the body, but not in large enough quantities, especially during periods of growth
  - Lack of an essential amino acid will result in an inability to synthesise proteins containing that amino acid
- Non-essential amino acids can be made by the body: alanine, aspartate, asparagine, cysteine, glutamate, glutamine, glycine, proline, serine, and tyrosine
- Amino acids can be used for a variety of purposes:
  - Protein synthesis
  - Hormones, e.g. adrenaline
  - Neurotransmitters, e.g. 5-hydroxytryptamine (5-HT)
  - Deaminated, and then the remaining carbon skeleton is:
    - Oxidized via the TCA cycle, or
    - Converted into glucose via gluconeogenesis, or
    - Turned into fatty acids
- Not all amino acids can entertain all of these fates:
  - Those which can be degraded to pyruvate or TCA cycle intermediates are termed glucogenic
  - Those which are converted to acetyl-CoA or acetoacetyl-CoA are termed ketogenic:
    - Only leucine and lysine are solely ketogenic
  - Those which can do either are termed mixed.

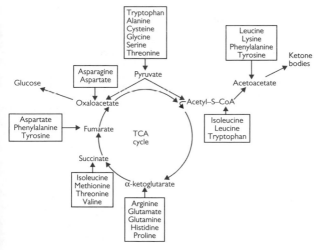

Fig. 2.33 The points of entry of amino acid carbon skeletons into the citric acid cycle and into ketone body synthesis.

Oxidation

Western diets are generally relatively high in protein, and excess amino acids cannot be stored.

- The carbon skeletons of amino acids can be used as an energy source
- First, the amino group is removed and excreted as urea (➲ p.166)
- This is done by transamination (class of enzyme: aminotransferases)
  - Each amino acid has its own specific aminotransferase (Fig. 2.34):
    ○ All have pyridoxal phosphate (a vitamin B_6 derivative) as a co-factor
  - Reactions are easily reversible and require no energy input
- The most common acceptor of the amino group is α-ketoglutarate:
  - Forms glutamate, which provides a pool of amino groups for making other non-essential amino acids or for deamination
  - Other amino group acceptors include pyruvate (→ alanine) and oxaloacetate (→ aspartate).

Glutamate is deaminated by glutamate dehydrogenase:

- The pooling of excess amino groups into glutamate means that only one deamination pathway is required
- The deamination reaction regenerates α-ketoglutarate and a free ammonium (NH_4^+), plus an NADH:
  - Glutamate dehydrogenase is allosterically regulated by increases in ADP and GDP. These compounds signal that amino acids need to be used as an energy source
- The deamination reaction takes place in the mitochondria of liver cells:
  - The major fate of NH_4^+ is incorporation into urea for excretion (➲ p.166).

Other sites of ammonium production include:

- Brain: breakdown (and therefore inactivation) of the neurotransmitter γ-aminobutyric acid (GABA) to succinate and an ammonium ion:
  - The ammonium ion is combined with α-ketoglutarate to produce glutamate, and then another ammonium ion is incorporated to form glutamine. This is transported to the liver for deamination and urea production
- Muscle: formed from natural protein turnover, muscle catabolism during starvation, and from breakdown of excess ADP during severe exercise ($2ADP → ATP + AMP$; $AMP → IMP$ (inosine monophosphate) $+ NH_4^+$):
  - An ammonium ion combines with α-ketoglutarate to form glutamate:
    ○ Glutamate is used to transaminate pyruvate to form alanine and regenerate α-ketoglutarate
    ○ The alanine is released into the bloodstream and taken up by the liver

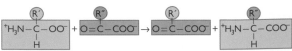

Amino acid (R′) + α-keto acid (R″) → α-keto acid (R′) + amino acid (R″)

Fig. 2.34 Amino acid aminotransferase.

○ Following deamination, the pyruvate released can be oxidized (TCA cycle) or used for gluconeogenesis
- Intestinal cells: glutamine serves as an energy source (Fig. 2.35).

Excess nitrogen can be excreted from the body either as urea or as ammonium ions.
- Urea is generated in the liver as a soluble, non-toxic way of eliminating excess ammonia
- The renal cortex can also deaminate glutamine:
  - The ammonium is used to assist with acidifying the urine (➲ p.526)
  - This mechanism also conserves HCO_3^- which would otherwise need to be used in urea synthesis and would exacerbate any acidosis.

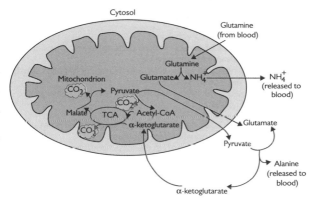

Fig. 2.35 Glutamine catabolism in the mitochondria of rapidly dividing enterocytes.

Urea cycle

Under all nutritional states, the body needs to excrete amino groups (i.e. ammonia).

- Protein intake >> need (generally true on a Western diet):
  - Cannot store excess amino acids
  - Will use/store carbon skeletons as energy source and excrete unwanted amino groups
- Protein intake << need:
  - Protein catabolism will occur to free carbon skeletons for energy
  - Excess amino groups will need to be excreted.

Free ammonia is toxic, and blood levels need to be kept low (25–40μM).

- If ammonium ion levels rise, NH_4^+ reacts with α-ketoglutarate to form glutamate—at high levels in the brain this reduces the rate at which ATP can be formed → cells starved of energy.

Healthy adults are in 'nitrogen balance':

- Approximately 80% of excess nitrogen excreted as urea (the remainder is as free ammonium ions and creatine)
- Most urea is synthesized by the liver, with the rate of synthesis strictly controlled to prevent ammonia build-up
- Urea is also used in the kidney as part of the urinary concentrating mechanism (⊙ p.526).

The urea cycle (Fig. 2.36) describes the formation of urea from one free ammonium ion and one donated from aspartate.

- The urea cycle takes place partly in the mitochondrial matrix, partly in the cytoplasm
- It involves two amino acids not found in proteins: ornithine and citrulline.

Control of the urea cycle

Control of the urea cycle is at two levels—acute and chronic:

- Acute regulation is via carbamoyl-phosphate synthetase:
  - Regulated by the concentration of N-acetyl-glutamate
  - N-acetyl-glutamate is formed by N-acetyl-glutamate synthase, the activity of which is stimulated by arginine, itself an intermediate of the urea cycle
- Chronic regulation: urea cycle enzymes are induced over 24–36h:
  - This is in response to increased levels of ammonia in liver cells
  - Ammonia levels can vary 10–20-fold with diet and under starvation conditions when muscle is broken down:
    - Under prolonged or severe starvation conditions, the ability for enzyme (protein) synthesis may be compromised.

What happens when these pathways do not function correctly?

- General strategy for treatment is to reduce the protein level in the diet, and to give a compound which aids nitrogen excretion (either by stimulating the urea cycle or another compensatory pathway)

- Ornithine transcarbamylase deficiency is the most common urea cycle problem:
  - X-linked, so generally males more seriously affected
  - Causes mental retardation and even death
  - Symptoms: raised ammonia and amino acid levels, high blood orotate levels
  - Treatment—give large quantities of benzoate and phenylacetate:
    - Benzoyl-CoA reacts with glycine to form hippurate, phenylacetyl-CoA reacts with glutamine to form phenylacetylglutamine
    - These excretable conjugates substitute for urea in the disposal of nitrogen
- Argininosuccinate synthetase/argininosuccinase deficiency:
  - Symptoms: usually benign—excrete high levels of citrulline/argininosuccinate respectively
  - Treatment: arginine supplements to replace ornithine, thus allowing urea cycle to continue.

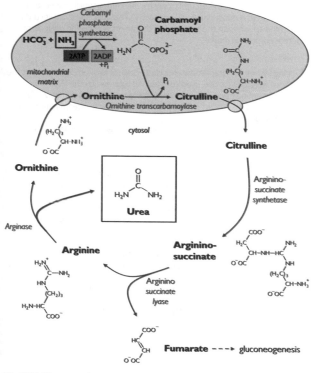

Fig. 2.36 The urea cycle.

Tissue-specific metabolism

Not all tissues metabolize amino acids in the same way, and the metabolism
by any one tissue often depends on the metabolic status of the body.

Liver

- The liver is the main site of amino acid degradation (deamination). It is
  also the major site of urea synthesis for nitrogen excretion
- During fasting, the liver is the main site of gluconeogenesis, using carbon
  skeletons from amino acids
- The liver plays a major role in the synthesis of the tripeptide, glutathione
  (➲ p.174).

Intestine

- Enterocytes take up glutamine and release it as alanine. This enables
  them to generate energy from it (Fig. 2.35; ➲ p.165)
- Enterocytes are the only cells to contain glutamate reductase, the
  synthetic enzyme for citrulline. Citrulline produced in the gut is released
  into the plasma and taken up largely by the kidney which converts it
  to arginine (the intestinal–renal pathway). This arginine is converted to
  ornithine to increase the capacity of the urea cycle during periods of
  increased protein intake.

Skeletal muscle

- During fasting and starvation, muscle protein is broken down so that the
  carbon skeletons can be used for gluconeogenesis by the liver
- The main amino acids released are alanine and glutamine (Fig. 2.37):
  - Alanine is transported by the blood to the liver for deamination and
    gluconeogenesis
  - Glutamine is taken up by enterocytes for energy and released as
    alanine.

Renal cortex

- The renal cortex is the only tissue other than liver that can perform
  gluconeogenesis. It has a capacity of up to 10% of total glucose
  generation, which is largely consumed by the renal medulla
- The renal cortex converts citrulline via arginine to creatine, which
  is used by skeletal muscle to store high-energy phosphate bonds as
  creatine phosphate:
  - Creatine phosphate spontaneously forms creatinine
  - Creatinine (➲ *OHCM10* p.669) is excreted by filtration by the
    kidneys, and its level in the blood can be used to assess renal function
    (➲ p.510; see also ➲ *OHCM10* p.298)
- The kidney is the major site of carnitine synthesis, with liver to a lesser
  extent:
  - Carnitine is important in fatty acid metabolism (➲ p.128).

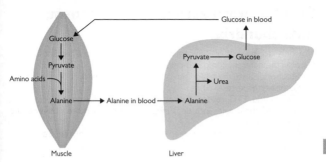

Fig. 2.37 The glucose–alanine cycle for transporting nitrogen to the liver as alanine, and glucose back to the muscles.

Reproduced with permission from Papachristodoulou, D., *Biochemistry and Molecular Biology* 6e, 2018, Oxford University Press.

Cellular organization of metabolism: mitochondria

Not only do the cells of different tissues have specialized metabolic roles, but also the different compartments in individual cells.

- Mitochondria are the major site of energy (ATP) production in all cells except erythrocytes:
  - TCA cycle, ETC, oxidative phosphorylation all take place there
- Other metabolic roles include β-oxidation of fats, synthesis of urea, and haem synthesis
- Mitochondria have their own separate genome (~4% of the total cell DNA):
  - The genome encodes 13 proteins, including some components of the electron transport chain and ATP synthase
  - Also 12S and 16S ribosomes and 22 unique tRNAs
- Evolutionary origin of mitochondria suggests that they may have originally been free-living bacteria that became incorporated in cells in a symbiotic relationship ('symbionts')
- Mitochondria can replicate in cells:
  - They have their own protein-synthesizing apparatus
  - Mitochondrial density can vary both up and down, e.g. up to tenfold increase in resting skeletal muscle if it is repeatedly stimulated to contract over a prolonged period:
    - Density also increases in hypoxia
- Mitochondrial DNA is inherited almost exclusively from the maternal side as the egg has several hundred thousand mitochondria compared to only a few hundred in the sperm
- Mitochondrial pathologies tend to be complex:
  - Not all mitochondria are affected to the same extent, and there can be large variations in severity and time of onset of diseases
  - Eventually, the energy-generating capacity of the mitochondria falls below the level required to sustain the cellular function
  - The nervous tissue and heart are highly dependent on oxidative phosphorylation, and therefore most susceptible to mitochondrial mutations:
    - The first disease discovered to be caused by mutations in mitochondrial DNA was Leber hereditary optic neuropathy (◆ *OHCM10* p.71), which causes blindness, with onset usually in adulthood (early to mid-life). Caused by mutations in the DNA encoding NADH-Q reductase (complex I)
  - In addition, the continued presence of developmental isoforms of cytochrome c oxidase (complex IV) in neonates can lead to severe respiratory distress or 'floppy baby syndrome'. Recovery occurs after several months in a high-oxygen environment.

Endoplasmic reticulum and Golgi apparatus

Endoplasmic reticulum

- Smooth ER contains the cytochrome P450 class of enzymes:
  - These play an important role in detoxification of both endogenous and exogenous compounds
  - They are also involved in the production of steroids
- Smooth ER is the site of elongation of fatty acid chains over C16 during the biosynthesis of lipids (⮕ p.132)
- The final step in the gluconeogenic pathway in liver and renal cortex takes place in smooth ER (site of the enzyme glucose-6-phosphatase; ⮕ p.158)
- Protein synthesis takes place on ribosomes attached to the cytoplasmic side of RER:
  - Protein modification and processing also occur in the RER lumen.

Golgi apparatus

- Site of modification of proteins, e.g. glycosylation
- Site of protein sorting for appropriate trafficking:
  - Proteins can be targeted, e.g. to the plasma membrane, to organelles, or for secretion
- Manufacture site (along with ER) of new lipid membrane for incorporation into the plasma membrane or for other organelles, e.g. lysosomes, peroxisomes.

Lysosomes

- Lysosomes are the site of intracellular digestion of extracellular and intracellular components
- They are a membrane-bound compartment that is acidified to ~pH 5:
  - This is the appropriate acidity for activity of the enzymes within the lysosome lumen
  - These enzymes are collectively known as hydrolases:
    ○ Wide range of bonds broken, including C–C, C–N, C–S, O–P in lipids, proteins, carbohydrates, and nucleic acids
- The lysosomal membrane is largely impermeable:
  - Specific mechanisms exist to move compounds in/out of the lysosome
- Extracellular materials are taken into the cell by endocytosis, where the material is encapsulated and internalized in membrane-bound vesicles:
  - Phagocytosis involves the uptake of foreign particles such as microorganisms
  - Pinocytosis is the uptake of material suspended in extracellular fluid
  - These vesicles fuse with primary (inactive) lysosomes to form secondary lysosomes, also known as digestive vacuoles
  - If the membrane of the primary lysosome is disrupted, the enzyme contents can be released into the cytoplasm. This causes digestion of cellular components and, ultimately, cell lysis:
    ○ Occurs in gout when crystals of uric acid are phagocytosed
- Intracellular components undergo routine breakdown and resynthesis, including proteins, lipids, nucleic acid, and mitochondria:
  - Those identified by the cell as due for breakdown are taken up by lysosomes in a process known as autophagy
- Most breakdown products are released from the liposome back into the cell cytoplasm for re-use:
  - Non-digestible material in vesicles ('residual bodies') can be removed from the cell by exocytosis
  - Some residual bodies are pigmented (lipofuscin, or 'age' or 'wear and tear' pigment)
- Sometimes the enzymic contents of lysosomes are secreted from the cell to break down extracellular material in connective tissue or the prostate gland
- Inappropriate release of lysosomal enzymes occurs in I cell disease when the enzymes are mis-targeted and released from the cell on synthesis, damaging the extracellular matrix
- Other diseases are associated with missing individual lysosomal enzymes, resulting in a build-up of the substrate for the missing enzyme:
  - These are collectively known as lysosomal storage diseases
  - The build-up of indigestible matter causes enlarged lysosomes, which can interfere with the normal cellular functions. Examples include lysosomal acid lipase (LAL) deficiency, characterized by impaired cholesterol metabolism (rare):
    ○ Cholesterol ester storage disease: low LAL activity (<5%) → hypercholesterolaemia, hepatomegaly, early onset of severe atherosclerosis
    ○ Wolman's disease: no detectable LAL activity → usually fatal by 1 year of age.

Peroxisomes

- Peroxisomes (or microbodies) are small spherical or oval organelles with a fine network of tubules in their lumen
- Over 50 peroxisomal enzymes have been identified:
  - Some use or produce hydrogen peroxide (H_2O_2)—hence the name peroxisome
  - They play essential role in lipid breakdown (especially oxidation of very long-chain fatty acids C_{24} and C_{26}), bile acid synthesis, synthesis of glycerolipids, glycerol ether lipids (plasmalogens), and isoprenoids
  - Peroxisomes also contain enzymes for metabolizing D-amino acids, uric acid, and 2-hydroxy acids using molecular oxygen to form H_2O_2:
    - Compounds are oxidized by the H_2O_2, which is itself then broken down to oxygen and water by catalase
    - By both forming and breaking H_2O_2 in the same organelle, the potential for cellular damage is limited
- Peroxisome biogenesis disorders (PBDs) are rare and associated with insufficiencies in the peroxisomal enzymes:
  - Tissues affected include liver, brain, kidney, and skeletal system
  - Symptoms include low plasmalogens, high levels of very long-chain fatty acids, and build-up of bile acid precursors
  - Most severe is Zellweger's syndrome (⊃ *OHCM10* p.716):
    - Failure to traffic enzymes properly → non-functional peroxisomes; usually fatal by 6 months of age.

Protection of cells against reactive oxygen species

- Peroxides are highly reactive oxygen species that can damage membranes and other biomolecules
- As a protectant, cells contain a high level (~5mM) of the tripeptide, glutathione (GSH) (→ p.150):
  - GSH is kept in its reduced form by glutathione reductase. The ratio of reduced glutathione (GSH) to oxidized (GS-SG) is ~500:1 in healthy cells
  - Peroxides are detoxified by the glutathione peroxidase catalysed reaction:

$$2GSH + ROOH \rightarrow GS\text{-}SG + H_2O + ROH$$

  - Glutathione peroxidase is very unusual in containing a modified amino acid containing a selenium (Se) atom
- The enzyme SOD is also responsible for detoxifying superoxide (O_2^-)
- Oxidized SOD coverts superoxide into O_2 and itself becomes reduced:

$$SOD_{ox} + O_2^- \rightarrow SOD_{red} + O_2$$

- The reduced SOD then reacts with a second superoxide and two protons to form hydrogen peroxide, reforming oxidized SOD:

$$SOD_{red} + 2H^+ + O_2^- \rightarrow SOD_{ox} + H_2O_2$$

- The hydrogen peroxide formed by SOD is broken down by catalase or GSH peroxidase:

$$2H_2O_2 \rightarrow O_2 + H_2O$$

- Antioxidant vitamins C and E are further cellular defences against oxidative damage:
  - Being lipophilic, vitamin E is especially useful in protecting membrane lipids and those in low-density lipoproteins from peroxidation
  - Vitamin C (ascorbate) is found in very high (mM) concentrations in cells and is maintained in the reduced state by GSH-mediated recycling.

Integration and regulation of metabolism: overview

Obviously not all of the metabolic processes that have been discussed in the preceding sections will be occurring at the same time in any one individual. Indeed, if this were the case, then there would be the danger of futile cycles that would be wasteful of energy. Therefore, it is important to have some concept of when and which pathways are largely active/inactive, and how this coordination of metabolism is brought about.

The major situations that will be considered are:

- Feeding
- Starvation
- The response to exercise
- Pregnancy and lactation
- Diabetes mellitus.

Cellular metabolic response to feeding

Humans eat intermittently, so need to consume calories in excess of their immediate need and store energy for later in the form of glycogen and triacylglycerides.

- In the affluent Western world, excess food consumption → obesity (➔ *OHCS11* p.830) is the most common form of malnutrition.

What happens to the major digestion products of food (i.e. glucose, amino acids, and triacylglycerides) on ingestion?

General points

- Glucose and amino acids are carried to the liver in the portal vein before they enter the main circulation
- Lipids are absorbed via the lymph system which drains into the vena cava, i.e. they are not subjected to first-pass metabolism by the liver:
  - Lipids are transported in lymph/plasma as chylomicrons (➔ pp.126, 587)
- A rise in blood glucose triggers the release of insulin from pancreatic β-cells (➔ p.636).

In the well-fed state

Glucose

- Taken up by the liver and stored as glycogen (glycogen synthesis, ➔ p.152):
  - Glucose can be metabolized to pyruvate en route to fat synthesis:
    - Triacylglycerides synthesized in the liver are carried in the blood in the form of VLDLs (➔ p.34)
- Continues into main circulation:
  - Used by brain and testis (almost solely dependent on glucose), red blood cells, and renal medulla (obligatory glucose users)
  - Taken up by muscle cells and stored as glycogen:
    - Insulin triggers insertion of GLUT4 transporters to increase uptake capacity
  - Taken up by adipocytes for conversion into fat:
    - Insulin triggers insertion of GLUT4 transporters to increase uptake capacity.

Protein

- Taken up by liver:
  - Used by liver for protein synthesis
  - Deaminated to give α-keto acid (ultimately pyruvate) and ammonia:
    - Pyruvate used for fat and glucose/glycogen synthesis; ammonia excreted via urea cycle
  - Pass into main circulation:
    - Protein synthesis by all tissues.

Lipid

- Triacylglycerides are absorbed, via the lymph, and enter the main circulation:
  - Taken up by adipose tissue and stored
  - Used by muscle as an energy source.

On re-feeding after fasting

In fasting conditions, the liver will have exhausted its glycogen stores (and will have been gluconeogenic), and these stores need to be replenished.

- Due to the insulin-stimulated insertion of higher-affinity GLUT4 (K_m ~5mM) glucose transporters into the membrane of muscle and adipose cells, these tissues will preferentially take up glucose
- The liver takes up relatively little glucose via GLUT2 (K_m ~17mM) and remains gluconeogenic for several hours after:
  - It can also replenish its glycogen stores from deaminated dietary amino acids and lactate from glycolytic tissue (e.g. red blood cells).

Cellular metabolic response to fasting and starvation

In comparison to the fed state, when all tissues use exogenous glucose as a metabolic fuel, in a fasting state the body needs to use the energy it has stored as glycogen and fat. In starvation, protein will also be broken down to provide energy.

General points

- The pancreatic α-cells release glucagon, triggered by a fall in blood glucose
- Skeletal muscle is unresponsive to glucagon
- During fasting, the liver no longer uses glucose as a fuel source
- During prolonged fasting/starvation, glucose use by other tissues falls as well:
  - Within 24–36h, muscle has almost entirely switched to other fuel sources (fatty acids, ketone bodies), and the brain starts using ketone bodies
  - By ~3 weeks, the brain has largely switched to ketone bodies:
    - Red blood cells, renal medulla, and, to a diminished extent, brain, are the only major tissues still using glucose
- Protein is not an inert energy store like fat or glycogen—breakdown of proteins such as muscles and enzymes is a last resort of starvation.

Early fasting state

- During the initial unfed state of fasting, liver glycogen is broken down (glycogenolysis) and glucose released into the circulation:
  - This glucose is used by tissues such as the brain, red blood cells, and muscle
  - The alanine cycle becomes important:
    - Alanine is generated in muscle cells by amination of pyruvate and released into the bloodstream
    - The alanine is taken up by the liver and deaminated; the nitrogen is excreted as urea, while the pyruvate is converted into glucose by gluconeogenesis
  - The Cori cycle (Fig. 2.3; ⊕ p.109) operates:
    - Similar to the alanine cycle, except involves lactate rather than alanine.

Later fasting state/starvation

- In addition to being gluconeogenic, the liver becomes ketogenic and proteolytic (liver glycogen stores will have been exhausted)
- Ketone bodies are formed from fatty acids released by lipolysis in adipose cells:
  - Circulating fatty acids can be used directly as fuel by tissues, e.g. muscle
  - Ketone bodies are used as fuel by brain and muscle
  - Glycerol from triacylglyceride breakdown is used by liver for gluconeogenesis

- Protein hydrolysis takes place in muscle, with alanine and glutamine being the main amino acids released:
  - Alanine participates in the alanine cycle
  - Glutamine is metabolized by enterocytes, and alanine released:
    - Glutamine is also an important fuel for cells of the immune system
  - Liver proteins are also hydrolysed, with the amino acids used as substrates for gluconeogenesis
- The metabolic changes to promote the use of fat as an energy source minimizes the need for protein breakdown during starvation; once fat stores are depleted then protein breakdown will increase, → organ failure (heart, liver, kidney) and death:
  - Some muscle breakdown in starvation must occur to supply enough glucose (via gluconeogenesis) to obligate glycolytic tissues: this is about 20g per day in starving humans
  - Therefore, the time of survival of starvation is largely dependent on the size of fat stores at the onset.

Cellular metabolic response to exercise

There are two metabolically distinct forms of exercise: anaerobic (e.g. sprinting) and aerobic (e.g. marathon running).

Anaerobic exercise

- Tends to be of short duration but very intense
- Energy for muscle contraction comes from intramuscular stores (e.g. glycogen, phosphocreatine)
- Due to oxygen delivery not matching demand, muscle cells are glycolytic:
  - High levels of lactic acid in cells will cause muscle weakness and cramp, and also the fall in intracellular pH (pHi) will inhibit glycolysis, thus limiting further lactate production.

Aerobic exercise—short to moderate term

- Glucose from muscle glycogen is again the main source of energy:
  - Glucose uptake from the blood is also increased by non- insulin-dependent insertion of GLUT4 into the muscle cell plasma membrane:
    - Glucose is released from liver glycogenolysis
  - Muscle glycogen stores can be increased by depletion with exhaustive exercise, then rest and a high carbohydrate diet ('carbohydrate loading')
- Muscle also increases oxidation of branched-chain amino acids, ammonium production, and alanine release.

Aerobic exercise—long term

- There is not enough stored glycogen or a large enough capacity for glucose uptake into muscle cells to provide the necessary energy for running long distances
- Metabolic adaptation is similar to that for fasting:
  - Increase in lipolysis as glycogen stores become depleted (stimulated by glucagon)
  - Fatty acids used directly as fuel by muscle or following conversion to ketone bodies (by ketogenesis in liver):
    - Little rise in blood ketone concentrations (unlike in fasting) as use by muscle matches liver production
  - Progressive switch over to preferential fatty acid oxidation by muscle:
    - Acetyl-CoA carboxylase inhibited by increase in AMP and long-chain acyl-CoA esters; reduced malonyl-CoA level stimulates carnitine palmitoyltransferase I activity and fatty acid oxidation
    - Any glucose still used is metabolized to lactate which enters the Cori cycle.

Pregnancy and lactation

Pregnancy

During pregnancy, the foetus can be thought of as, metabolically speaking, simply extra tissues that will require nutrients.
- Mainly uses glucose for energy, but can also use lactate, fatty acids, ketone bodies, and amino acids.

The placenta requires nutrients itself, but also plays a major metabolic role in pregnancy:
- Metabolizes glucose to lactate:
  - Either taken up by the foetus and used as energy source or returned to circulation to establish a placenta–liver Cori cycle
- Synthesizes placental steroids (oestradiol and progesterone):
  - Precursor is maternal cholesterol (LDL)
  - Induces an insulin-resistant state in pregnant woman
- Secretes polypeptide hormone placental lactogen (similar in structure to growth hormone):
  - Stimulates lipolysis in adipose tissue
- Combination of lactogen and steroid hormone effects are to maintain plasma nutrient levels so that the foetus is well catered for
- Perturbed fed–starve cycle in pregnant woman:
  - Foetal consumption of nutrients results in faster maternal return to starved state after eating:
    - Plasma glucose, amino acid, and insulin levels fall rapidly, and glucagons and placental lactogen levels rise, stimulating lipolysis and ketogenesis
    - Foetal consumption can cause maternal hypoglycaemia
  - In the fed state, maternal insulin and glucose levels are increased (and there is insulin resistance)
  - These dramatic swings in nutrient and hormone levels are accentuated by maternal diabetes:
    - Controlling blood glucose in such cases is difficult, but important
    - Maternal hyperglycaemia adversely affects foetal development.

Lactation

During lactation, the mammary gland takes up nutrients from the circulation:
- Glucose for making lactose (the major sugar in milk) and for triacylglyceride synthesis
- Amino acids for protein synthesis
- Fatty acids for triacylglyceride synthesis:
  - Directly from diet (chylomicrons)
  - From circulating VLDL.

If these compounds are not taken in the diet in sufficient amounts, then they will be sourced from gluconeogenesis, proteolysis, and lipolysis.
- In the long term, this can cause maternal malnutrition and, ultimately, the quality of the milk will fall.

Diabetes mellitus

Type 1 diabetes mellitus

(⊖ p.640; see also ⊕ *OHCM10* pp.206–13.)

- Formerly known as insulin-dependent diabetes mellitus (IDDM)
- Sometimes also referred to as juvenile-onset diabetes, although not limited to onset in childhood
- Caused by defective or absent pancreatic β-cells:
  - β-cells are destroyed by an autoimmune response
  - Glucagon from α-cells is the only output from the endocrine pancreas
  - Body sens it is in a continuously starved state irrespective of nutritional input
- Symptoms include hyperglycaemia, hyperlipoproteinaemia (raised plasma chylomicrons, VLDL), and severe ketoacidotic episodes:
  - Hyperglycaemia results both from the failure of muscle and adipose tissue to take up glucose (no insulin-dependent GLUT4 insertion) and liver being continuously gluconeogenic:
    - Liver gluconeogenesis is fuelled by protein breakdown including muscle wasting (hence type 1 diabetes is sometimes referred to as 'fed starvation')
    - If untreated, will result in death in the same way as starvation
  - Hyperlipoproteinaemia is from low levels of insulin-sensitive lipoprotein lipase activity in adipose tissue capillaries
  - Ketogenesis results from increased lipolysis in adipose tissue and fatty acid oxidation in liver:
    - Ketogenesis can lead to potentially fatal ketoacidosis in type 1 diabetes on account of prolonged or repeated hypoglycaemic episodes
- Symptoms reversible by administration of exogenous insulin:
  - Regular injections and monitoring of blood glucose levels necessary:
    - Variable diet and exercise levels make this more challenging for the patient.
    - Continuous glucose monitoring by in-dwelling sensors and insulin delivery by insulin pumps is becoming more common

Type 2 diabetes mellitus

(⊖ p.640; see also ⊕ *OHCM10* p.206–13.)

- Formerly known as non-insulin-dependent diabetes mellitus (NIDDM)
- Accounts for up to 90% of diagnosed diabetes
- Usual onset is middle to older age
- Strong correlation with obesity:
  - Much concern that Western diet of 'junk' food and accompanying obesity is resulting in sharp rise in type 2 diabetes and lowering of average age of onset, to even as low as childhood
- Patients have insulin resistance, i.e. their insulin receptors are not as sensitive to circulating insulin:
  - Pancreatic β-cells do not produce enough insulin to overcome this resistance ('β-cell failure')
- Patients show hyperglycaemia, hypertriglyceridaemia (from liver synthesis rather than adipose tissue lipolysis), and other obesity symptoms:
  - Tend not to see ketoacidosis in type 2 diabetes.

Clinical aspects: energy balance

Components of energy balance

- Energy is taken in the form of food
- Energy is expended as 'basal metabolism' (maintenance and repair of the organism), thermic effect of food (energy expenditure rises after meals), physical activity ('exercise'), and non-planned physical activity ('fidgeting'). Typically, 5–10% of dietary energy is lost in faeces
- Any difference between energy intake and expenditure is reflected in a change in the body's energy store.

Positive energy balance (intake > expenditure) is a normal part of growth or anabolism, e.g. during recovery from surgery or trauma. Positive energy balance beyond the needs of growth leads to fat accumulation and, ultimately, overweight and obesity (⊅ p.192).

Negative energy balance occurs during dieting or in anorexia, in people who become physically active for long periods (e.g. Mike Stroud and Ranulph Fiennes lost almost all their bodily energy reserves during their Antarctic crossing in 1992–1993), and during periods of catabolism following major trauma or during severe infection.

Regulation of energy intake

Energy intake is regulated by endocrine and neuroendocrine mechanisms.
- Leptin is a peptide hormone secreted from adipocytes in response to the amount of fat stored:
  - Increasing fat storage leads to increasing circulating leptin concentrations
  - Leptin acts through hypothalamic receptors and a complex neuroendocrine system to reduce appetite. However, in humans this system is directed more towards avoiding starvation (leptin deficiency is associated with intense hunger)
  - Variation in leptin levels within the normal range seems not to have major effects on appetite
  - Treatment with recombinant human leptin has been remarkably successful in rare patients with complete leptin deficiency (characterized by massive childhood obesity, sexual immaturity, and T-cell dysfunction) but has little effect in the normal obese patient.

Energy intake is determined, to some extent, by diet composition. It is easier to overconsume energy when the diet is 'energy dense' (high kJ per 100g).
- Energy-dense foods are generally those with little water and a high fat or sugar content ('junk foods' for instance)
- Fibre-rich (more 'natural') foods are generally low in energy density
- The relationship between dietary fat intake and obesity is not clear-cut within populations, although it is clear when comparing one population with another.

Regulation of energy expenditure

The major component of energy expenditure is normally basal metabolism, accounting typically for 60% of daily energy expenditure.

- Basal metabolic rate is determined almost entirely by lean body mass
- Thyroid activity is also a determinant (hyperthyroidism (→ OHCM10 p.218) increases basal metabolic rate)
- Physical activity is the major component of energy expenditure that can be manipulated
- Daily energy expenditure may be expressed as a ratio to resting metabolic rate—the 'physical activity level' (PAL):
  - Sedentary people have a PAL around 1.4; highly active people (e.g. soldiers in field training) have PALs up to 2.5; elite endurance athletes may maintain even higher PAL values, e.g. 3–4 in Tour de France cyclists.

Nutrition

We ingest nutrients that yield energy and nutrients that are essential for health.

- Energy-providing nutrients are also called macronutrients (fat, carbohydrate, and protein)
- Micronutrients include vitamins and minerals
- The distinction between energy-yielding and other essential nutrients is not absolute. We need to ingest specific ('essential') fatty acids and some essential amino acids. Water and oxygen are also essential but are not usually thought of as nutrients.

Nutritional disorders

Nutritional disorders may involve a deficiency, an excess, or an imbalance of nutrients.

- In developed countries, the most common nutritional disorder by far, in both humans and their pet animals, is obesity (➲ OHCS11 p.830)
- Under-nutrition is still prevalent in many parts of the world:
  - This includes British hospitals: a number of surveys of inpatients have shown alarming degrees of malnutrition, especially in elderly patients. Poor nutrition in sick patients may increase morbidity and mortality.

Macronutrients (energy-yielding nutrients)

Dietary fat

- Fat is mainly ingested as triacylglycerol (triglyceride) (95% of dietary fat) but includes some phospholipids (4–5%) and cholesterol (typically 500mg/day)
- Fat provides 9kcal/g. It contributes an average of 35% of dietary energy in the UK:
  - This has fallen over the last few years and is now in line with the recommendations of the UK's Committee on Medical Aspects of Food Policy (COMA) made in 1994
  - Lowering fat intake decreases serum cholesterol and may reduce energy intake
- The fatty acids of dietary fat may be saturated (typical of animal fat but present in all fats), monounsaturated (animal fat and vegetable oils, especially olive and rapeseed oils), or polyunsaturated (typical of sunflower and safflower oils; these are called n-6 or ω-6 fatty acids):
  - The energy content is almost identical but the effects on serum cholesterol differ: saturated fat raises and monounsaturated and polyunsaturated fats lower serum cholesterol
  - Average UK intake of saturated fatty acids (13% of dietary energy) is still greater than COMA's recommendation of not more than 11% of energy
  - The characteristic polyunsaturated fatty acids of fish oils (➲ OHCM10 p.244) belong to the n-3 or ω-3 family. These lower serum triglyceride levels, inhibit platelet aggregation, and have generally anti-inflammatory properties.

Dietary carbohydrate

Carbohydrates are diverse. The energy yield is 4kcal/g.

- Simple sugars are energy dense (➲ p.184)

- Complex carbohydrates are absorbed more slowly and, because they may be hydrated in foods (e.g. potato is 80% water), are low in energy density
- Polysaccharides that are not digested in the small intestine (typically plant cell wall material) are classed as fibre or non-glycaemic carbohydrate
- Bacterial fermentation in the colon produces gas and short-chain fatty acids that may have beneficial effects on colonic function including cancer protection.

Dietary protein
Most people in the Western world are not short of protein.
- Typical protein intake is 60–90g/day, whereas 40–50g/day is probably sufficient for life
- Vegetarians not consuming a range of protein sources may be at risk of deficiency of particular essential amino acids. Wheat (low in lysine) and legume protein (low in cysteine, methionine) complement each other.
- Protein yields 4kcal/g but its importance is not primarily as an energy source.

Alcohol
Alcohol provides energy (7kcal/g) and in some groups of people makes a substantial contribution to energy intake.

Micronutrients

Micronutrients have diverse functions in the body. Anyone eating a balanced and varied diet that meets energy requirements is unlikely to suffer from vitamin deficiencies and, for most people, there is no clear evidence of benefit from supplementation. There are exceptions.
- High folic acid (➲ OHCM10 p.332) intake in the periconceptual period reduces neural tube defects in babies and fortification of flour with folic acid is mandatory in the USA and Australia (not yet in Europe)
- Women with heavy menstrual losses may be iron deficient
- In the developing world it is estimated that billions of people suffer from micronutrient deficiency (iron, zinc, vitamin A)
- Some vitamins (A, D, E, and K) are fat-soluble and are only absorbed in the presence of dietary fat
- Vitamin B_{12} (a water-soluble vitamin) requires a specific protein, intrinsic factor, for its absorption:
  - Intrinsic factor is secreted from the gastric mucosa and impaired secretion will result in vitamin B_{12} deficiency and pernicious anaemia (➲ OHCM10 p.334)
- In some parts of the world there are mineral deficiencies in the soil, e.g. selenium in China, some parts of the USA, and Finland; iodine in many upland areas of Africa, Asia, and South America. Public health campaigns are then needed to encourage supplementation.

Assays: diagnostic enzymology

Measurement of enzyme activity plays an important and continually increasing role in medical diagnosis. There are basically two applications of diagnostic enzymology—measurement of the enzyme itself, or exploitation of enzyme specificity to measure the concentration of its substrate.

Enzyme activity measurement

This is important in two areas: assessment of tissue damage (in which enzymes from the affected tissue are released into the circulation) and diagnosis of specific enzyme deficiencies.

- In the case of tissue damage, the enzyme is being measured in an abnormal site, usually in serum, and results are dependent on the half-life of the protein after release from the cells of origin:
  - Enzymes released from tissues are cleared relatively rapidly from the circulation (half-lives of the order 10–20h) and this provides current information about the damage
  - Many enzymes are present in a variety of different tissues and simple measurement of a single enzyme may not be sufficient to identify the site of origin:
    - Measurement of several different enzymes or additional analysis of isoenzyme forms which may be preferentially expressed in particular tissues may provide the necessary specificity
    - Isoenzyme forms of lactate dehydrogenase, creatine kinase, and alkaline phosphatase are most widely used in this regard (⮷ *OHCM10* p.688).
- Enzyme activity is determined with saturating substrate concentrations and can be measured by following the rate of change in the concentration of substrate or product or by measuring the amount of substrate remaining or product formed after a specified time:
  - To avoid problems due to substrate depletion, reverse reaction, and product inhibition, conditions should be chosen to measure the initial rate, and all relevant conditions such as temperature and buffer composition and pH should be standardized:
    - In some cases, it is necessary to couple the enzyme reaction being measured to a second enzyme-catalysed reaction in order to generate a readily detectable product
    - In this situation, it is necessary to ensure that the coupled enzyme and any additional substrates are not rate-limiting
- For the investigation of specific enzyme deficiencies, the same technical considerations are relevant:
  - However, these investigations are required much less frequently and are usually performed in specialist centres with the necessary experience and expertise
  - While enzyme measurement for assessment of tissue damage is usually performed on serum samples, diagnosis of enzyme defects requires direct analysis of affected tissues
  - Some defects are expressed in readily accessible samples, such as blood cells or cultured fibroblasts, but others may require tissue biopsy

- With increasing identification of the genes responsible for many enzyme deficiencies, direct mutation analysis, avoiding the need for biopsy and complex enzymological studies, is becoming more widely available.

Measurement of metabolite concentrations using enzymes

The specificity of many enzyme-catalysed reactions can be exploited to measure the concentration of their substrate. This is particularly important when dealing with the complex mixtures of related compounds found in biological samples such as blood or urine:

- For example, there are several enzymes which oxidize glucose with the necessary specificity for measuring blood glucose concentration
- For this type of analysis, it is important that the equilibrium of the reaction lies far to the right and that there is sufficient enzyme to ensure almost complete conversion of substrate to product within a short time.

Inborn errors of metabolism

There are >300 defined enzyme deficiencies affecting the function and regulation of many different metabolic pathways.

- Most present in the first years of life—many within the newborn period
- Although identification of specific enzyme deficiencies is performed by a small number of specialist laboratories, clues to the nature of the underlying biochemical defect can usually be found on the basis of a number of widely available screening tests.

Inborn errors of metabolism can be divided into three main groups based on the consequences of the biochemical defect, which in turn is related to the type of metabolic pathway involved.

Group 1

Conditions due to accumulation of toxic metabolites. These include the disorders of amino acid oxidation and the related organic acidurias, defects of the urea cycle, and the various forms of carbohydrate intolerance.

Clinical presentation

- The reaction to toxic intermediates is non-specific
- Most patients present in the newborn period, but there may be delayed onset, often with milder symptoms
- Clues to the diagnosis are a period of normality after birth before symptoms commence, precipitation of symptoms by feeding or intercurrent illness, and a positive family history
- Common symptoms include feeding difficulties, vomiting, irritability, hypotonia, drowsiness, and seizures.

Investigation

- A small number of inborn errors fulfil criteria for universal newborn screening. However, this is widely performed only for phenylketonuria
- When there is already an affected child in a family, early diagnosis in subsequent siblings, before the onset of symptoms, is associated with much better prognosis. Specific tests will be indicated depending on the prior diagnosis
- In all cases, rapid assessment is essential once symptoms of toxicity appear
- The most widely available and useful screening tests are for urine organic acids and urine and plasma amino acids:
  - Specific measurement of acid–base status, blood ammonia, glucose, and ketones, and liver function tests are usually required in addition.

Group 2

Conditions with impaired energy generation. Disorders of glycogen mobilization, pyruvate metabolism, gluconeogenesis, fatty acid oxidation, and the mitochondrial electron transport chain.

Clinical presentation

- Symptoms may be precipitated by fasting, rather than feeding
- Neurological dysfunction is again prominent, but there may also be specific problems related to energetic failure in other organs such as heart, liver, kidney, and skeletal muscle.

Investigation

- Most patients will be screened as for patients in group 1, although blood lactate and pyruvate concentrations are often more relevant, and the lactate concentration in CSF may be particularly helpful in patients with predominantly neurological dysfunction
- Fasting tests and muscle biopsy for morphology and enzymology may be indicated, but are usually performed in specialist units.

Defects in metabolism of macromolecules

These include the lysosomal storage diseases and peroxisomal diseases.

Clinical presentation

- These conditions are generally of later onset and are often characterized by organomegaly, with or without progressive neurodegeneration
- Appearance of features such as characteristic facial appearance, skeletal deformities, and corneal clouding may provide additional clues.

Investigation

- Apart from analysis of urine, glycosaminoglycans (GAGs) for the diagnosis of the mucopolysaccharidoses, there are no screening tests for this group of conditions
- Vacuolated lymphocytes or foamy macrophages in bone marrow may provide a clue to a lysosomal storage disease, and plasma very long-chain fatty acids are often elevated in peroxisomal diseases. However, diagnosis is usually based on specific enzyme assays
- Once an inborn error of metabolism has been suggested on the basis of the tests outlined, further investigation and management is generally undertaken by specialist centres where expertise in laboratory diagnosis, monitoring, and long-term treatment (where possible) is available
- As many inborn errors of metabolism present with non-specific symptoms which can mimic other conditions, such as infection, the key role for the primary physician is to suspect that an inborn error may be the cause of the patient's problems and to arrange for the appropriate screening and routine laboratory tests to be performed as quickly as possible
- While the course of many inborn errors of metabolism is unaltered by any therapy, there are continuing advances in management, especially of conditions due to accumulation of toxic metabolites:
  - It is in this group, particularly, that delayed diagnosis may result in a much poorer prognosis, with a high risk of permanent brain damage.

Obesity and treatment

Obesity is an excessive accumulation of body fat.
- A useful measurement is the body mass index (BMI) (⊜ *OHCM10* p.245):

$$\frac{\text{weight}\,(\text{kg})}{\text{height}\,(\text{m})^2}$$

- A BMI of 25–30kg/m² is usually taken to represent overweight; a BMI >30kg/m² obesity
- However, a weight lifter (for instance) may have a BMI of 30kg/m² without excessive fat accumulation
- More specific measurements of body fat content can be made from skinfolds (calipers), electrical impedance, body density (usually involving underwater weighing), dual-energy X-ray absorptiometry (DXA), or isotopic techniques to measure body water.

Obesity results from an excess of energy intake over energy expenditure (⊜ p.172).
- The imbalance may be very small but, over a long period, a large change in fat accumulation occurs:
  - An extra pat of butter (12g) each day could, in principle, lead to accumulation of almost 50kg of body fat over 10 years
- Measurements of energy expenditure in obese people universally show this to be increased compared with lean people (hence energy intake must also be high):
  - There is no evidence for 'slow metabolism' maintaining obesity except in particular medical conditions such as hypothyroidism (⊜ *OHCM10* p.220), although it is almost impossible to know what caused the obesity to develop initially.

Obesity has a strong inherited component.
- Nevertheless, there is clearly an interaction between genes and environment because the present alarming increase in obesity incidence cannot represent a sudden change in the gene pool:
  - Rather, it represents the effects of an 'obesogenic' environment (plentiful food, sedentary lifestyle) on a genetic background evolved to protect the organism from food shortages
- Almost all adult obesity is polygenic:
  - A small number of single-gene mutations causing obesity are known, mostly in children and young people
  - In those cases in which the gene has been identified, these are all in pathways involved with appetite regulation, not energy expenditure.

Obesity is a medical, not a cosmetic problem.
- Obesity strongly increases the risk of developing type 2 diabetes, coronary heart disease, hypertension, gallstones, polycystic ovaries, various cancers, sleep apnoea, and osteoarthritis.

The distribution of body fat may be almost as important in this respect as the total amount.

- Upper-body ('abdominal') obesity poses far more of a health risk than does lower-body obesity
  - This is simply estimated by measurement of waist circumference:
    - A waist circumference of >102cm (men) or 88cm (women) is a call for action.

Treatment involves increasing energy expenditure (physical activity), decreasing energy intake, or both.

- Treating obesity is notoriously difficult, with poor long-term results—prevention would be a better public health strategy
- Despite many 'fad diets', ultimately they must act by reducing energy intake
- Pharmacological treatments include orlistat, which reduces dietary fat absorption by inhibition of pancreatic lipase. Metformin is useful in the obese type 2 diabetic patient (→ p.640). Sibutramine, which acts on central appetite-regulating pathways and rimonabant, a novel cannabinoid-receptor blocker which helps both smoking cessation and weight loss, have both been withdrawn because of side effects. 'Incretin'-based therapies (mimicking or increasing the action of glucagon-like peptide 1), used in type 2 diabetes to augment insulin secretion (→ *OHCM10* p.208), also have a useful weight-reducing effect.
- 'Bariatric' surgery, which reduces the size of the stomach or bypasses sections of the small intestine, can be very effective but because of cost and significant morbidity is restricted to the grossly obese or those with other medical complications.

Molecular and medical genetics

Gene expression

The term 'gene' is used to represent an inherited unit of information, as reported by Mendel[1] in his experiments on the garden pea, and can be interpreted in two different ways:

- At a gross level, genes determine the phenotype, e.g. morphological characteristics such as eye or hair colour
- At a molecular level, a gene is a stretch of nucleotides that codes for a polypeptide, so determines the amino acid sequence and, therefore, the function of a protein:
  - Each amino acid is coded for by a three-base-pair sequence (known as a codon; Table 3.1)
  - As there are 22 pairs of homologous chromosomes, plus the sex chromosomes, most genes are present twice. The copies are known as alleles.

There are different types of mutations in genes that will be of varying seriousness depending on the nature of the change and whether they occur in the coding (exon) or non-coding (intron) regions.

Types of mutation and their effect when in exons

Point mutation: a change in a single nucleotide

If this happens in the coding sequence, the result may be either:

- No change in the amino acid sequence of the polypeptide:
  - Due to the fact that there are four bases and the codons are triplets, there are $4^3 = 64$ codons for 20 amino acids. Therefore, a number of codons code for each individual amino acid, known as redundancy
- A change in the amino acid coded for by the changed triplet:
  - This may or may not affect the function of the protein, depending on the amino acid changed, what it is changed to, and its role in the protein function. For example, many of the amino acids in a protein can be regarded as scaffolding, simply holding the catalytic residues in the correct 3D space
- The change in the base pair may result in the triplet coding for a stop codon (non-sense mutation):
  - This will result in a truncated protein, which is unlikely to have the function of the wild-type.

Single base insertion or deletion

This is likely to have a serious effect on the protein structure and function:

- Will cause a mis-sense or frame-shift mutation, where the amino acid coding will be disrupted by the inserted/deleted base:

Wild-type	Base insertion	Base deletion
CAT CAT CAT CAT CAT	CAT ACA TCA TCA TCA	CAT ATC ATC ATC AT
His His His His His	His Thr Ser Ser Ser	His Ile Ile Ile

- Usually will result in a stop codon being coded for, but even if not, it is virtually impossible that a mis-sense protein will be able to substitute for a wild-type one.

1 Mendel JG (1865). Translated by Druery CT and Bateson W (1901). Experiments in plant hybridization. *J R Horticultural Soc* **26**, 1–32. Available at: htttp://www.esp.org/foundations/genetics/classical/gm-65.pdf.

Multiple base insertion or deletion
• Insertions or deletions of anything other than multiples of three bases will cause frame-shift mutations

Table 3.1 The genetic code

5' base	Middle base				3' base
	U	C	A	G	
U	UUU Phe	UCU Ser	UAU Tyr	UGU Cys	U
	UUC Phe	UCC Ser	UAC Tyr	UGC Cys	C
	UUA Leu	UCA Ser	UAA Stop*	UGA Stop*	A
	UUG Leu	UCG Ser	UAG Stop*	UGG Trp	G
C	CUU Leu	CCU Pro	CAU His	CGU Arg	U
	CUC Leu	CCC Pro	CAC His	CGC Arg	C
	CUA Leu	CCA Pro	CAA Gln	CGA Arg	A
	CUG Leu	CCG Pro	CAG Gln	CGG Arg	G
A	AUU Ile	ACU Thr	AAU Asn	AGU Ser	U
	AUC Ile	ACC Thr	AAC Asn	AGC Ser	C
	AUA Ile	ACA Thr	AAA Lys	AGA Arg	A
	AUG Met†	ACG Thr	AAG Lys	AGG Arg	G
G	GUU Val	GCU Ala	GAU Asp	GGU Gly	U
	GUC Val	GCC Ala	GAC Asp	GGC Gly	C
	GUA Val	GCA Ala	GAA Glu	GGA Gly	A
	GUG Val	GCG Ala	GAG Glu	GGG Gly	G

* Stop codons have no amino acids assigned to them.

† The AUG codon is the initiation codon as well as that for other methionine residues.

• Insertion/deletion of multiples of three will insert/delete amino acids:
  • As with point mutations, the severity of this will depend on the protein involved and which amino acids are affected
  • The importance of even losing a single amino acid is highlighted by the condition cystic fibrosis (⊃ OHCM10 p.173), where the most common genotype is ΔF_{508}—the loss of the Phe residue at position 508 out of 1480. This results in the trafficking failure of the CFTR protein to the epithelial membrane
  • Dynamic mutations are when nucleotide repeats expand generation by generation:
    ○ When a certain threshold of repeats is reached, the disease phenotype will become apparent, e.g. Huntington's disease (⊃ OHCM10 p.702)

- ○ This accounts for the phenomenon of anticipation, where a disease phenotype becomes progressively more severe from one generation to the next
- Most mutations are associated with a loss of function of the protein coded for by the mutated gene. Rarely, there can be a gain in function, e.g. a mutation in a growth factor receptor (FGFR3) that is constitutively active.

Mutations in introns

The previously described types of mutations can also occur in non-coding regions of genomic DNA (the introns).

- The role of the introns is not understood—they are often naively assumed to be mostly 'junk' DNA with no role
- If the mutation occurs in a regulatory element such as a repressor or promoter, this can cause inappropriate down- or up-regulation of gene expression, respectively
- Mutations in splice sites may prevent the correct splicing of exons together in the processing of RNA, leading to the inclusion or exclusion of exons:
  - There is a strong possibility this will also cause frame-shift problems and create premature stop signals.

Gene structure

- Genes consist of nucleic acid (DNA), which itself is a long polymer of nucleotides
- Genes are made up of exons and introns:
  - Exons code for the protein that the gene encodes:
    - Exon sequence is highly conserved between individuals
  - Introns will be spliced out during processing to the mRNA:
    - The length of the introns usually far outweighs the exons for any particular gene
    - The exact function of the introns is not clear. The intron sequence is not well conserved between individuals.

How was DNA identified as the hereditary material?

- Originally it was thought that, as proteins were quite complex molecules, the genetic material must be of equal or greater complexity, i.e. other proteins
- Experiments with pneumococcus disproved this:
  - Two types of *Diplococcus pneumoniae* (now called *Streptococcus pneumoniae*) were used:
    - Wild-type ('smooth' after its appearance under a microscope), which is virulent and kills mice
    - Avirulent ('rough') mutant, which is harmless as it lacks the appropriate surface polysaccharide needed to infect cells
    - Wild-type can be rendered avirulent by heat treatment, which is then harmless but remains smooth ('smooth avirulent')
- The first key experiment was performed by Griffith[2] in 1928—if a mouse was injected with smooth avirulent + rough bacteria → dead mouse (Fig. 3.1):
  - Bacteria isolated from the dead mouse were smooth and virulent
  - Therefore, something in the smooth avirulent bacteria had enabled the mutant rough bacteria to produce the polysaccharide and become virulent
- To answer whether this something was protein or DNA, Avery[3] conducted the following experiment in 1944:
  - DNA was extracted from wild-type bacteria
  - Injected mouse with mutant (rough) bacteria + wild-type (smooth) DNA → dead mouse
  - However, if the DNA extracted from the wild-type bacteria was treated with:
    - DNAase → live mouse
    - With the protease trypsin → dead mouse
  - This indicates that DNA is the key component

2 Griffith F (1928). The significance of pneumococcal types. *J Hygiene* **27**, 113–59.

3 Avery TO *et al.* (1944). Studies on the chemical nature of the substance inducing transformation of pneumococcal types. *J Exp Med* **79**, 137–58.

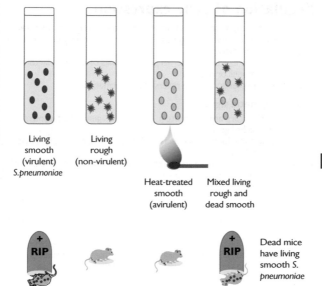

Living
smooth
(virulent)
S.pneumoniae

Living
rough
(non-virulent)

Heat-treated
smooth
(avirulent)

Mixed living
rough and
dead smooth

Dead mice
have living
smooth *S.
pneumoniae*

Fig. 3.1 Schematic of Griffith's 1928 experiment demonstrating the principle of transformation, with DNA later shown to be the material which mediated this effect.

- Hershey and Chase's experiment (1952)[4] using phage (which infect bacteria) gave further support to the findings of Griffith and Avery:
  - When a similar type of experiment was done with ^{32}P labelled phage (i.e. with labelled DNA), 30% of progeny phage were labelled
  - When the phage were ^{35}S labelled (i.e. protein labelled), <1% of progeny were labelled
  - This confirms Avery's finding that DNA is the genetic material, not protein.

4 Hershey AD, Chase M (1952). Independent functions of viral protein and nucleic acid in growth of bacteriophage. *J Gen Physiol* **36**, 39–56.

Regulation of gene expression

Some genes are expressed ubiquitously in cells at relatively consistent levels (often known as housekeeping genes). However, others vary in expression levels much more dynamically.

- Control can be exerted by internal genetic signals or external factors such as hormones or hypoxia
- Control of expression can be reversible or permanent.

Internal genetic signals

Transcription factors are the protein products of genes that control the expression of other genes.

- These tend to be long-term effects, including permanent changes that occur in cell differentiation
- These gene regulatory factors are enhancer proteins which have a DNA binding motif and a transcription activation motif. The most common type of DNA binding motif is the helix-turn-helix (→ p.68):
  - Enhancer proteins are said to act in a *trans* manner because they can affect genes far away in the genome
  - They act by recruiting other regulatory proteins expressed by the cell:
    - These complexes bind to the gene and are thought to act by increasing the accessibility of the initiation site to RNA polymerase and/or exposing other regulatory sites. In this way, enhancers perturb the chromatin structure of a gene that they regulate (→ p.68).

External signals that regulate gene expression

One of the best examples of an external factor affecting gene expression is the action of steroid hormones.

- These effects can be transient or longer term, but are generally reversed when the hormone signal is removed
- These lipid-soluble hormones diffuse from the extracellular fluid into cells where they bind to receptor proteins
- These hormone-receptor complexes in turn diffuse into the nucleus where they bind to receptor elements in the DNA and either up- or down-regulate gene expression:
  - Hormones can also act via second messenger systems to affect gene expression, e.g. there are cAMP-sensitive response elements
- The binding of a transcription factor to promoter elements upstream from the transcription start affects the rate at which the gene is transcribed
- Transcription factors can either increase or inhibit the transcription of genes:
  - These factors are known as enhancers and silencers, respectively
  - Enhancers bind to promoter regions: the GC box, the CAAT box, and the TATA box:

- These three elements are within 100 base pairs of the transcription start site
- They are said to be *cis* acting, as they act on the gene immediately following the promoter region
- Transcription factors binding to the GC and CAAT boxes enhance transcription by increasing the basal activity of the TATA box
- Silencers have a negative effect on gene transcription.

NB: control of gene expression can also occur at the level of mRNA processing, translation, and mRNA stability (i.e. changes in breakdown rates relative to formation).

Most of the information about regulation of gene expression has been obtained from studies of the *lac* operon (Fig. 3.2) in bacteria.
- An operon is a group of genes with related functions which are under the control of a single promoter:
  - They are only found in prokaryotes
  - The mRNA produced contains the coding sequence for all the proteins (i.e. it is polycistronic, compared to eukaryote mRNAs which are always monocistronic)
- Expression of the *lac* operon allows bacteria to use lactose as an energy source
- In the absence of lactose, the *lac* repressor protein binds to the operator and blocks expression
- When an inducer molecule is present (in this case, the metabolite of the sugar lactose, allolactose), it binds to the repressor protein and allosterically changes its shape:
  - The repressor protein no longer binds to the operator and so RNA polymerase can transcribe the genes involved in bacterial lactose metabolism
  - This has been shown experimentally using the artificial substrate, IPTG (isopropyl-β-D-1-thiogalactopyranoside)
  - The *lac* repressor protein has been studied by crystallography and its structure supports the proposed mechanism of action.

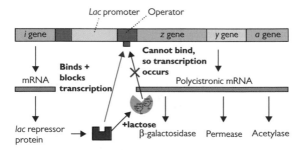

Fig. 3.2 The lac operon.

RNA transcription

In order for a eukaryotic cell to make a new protein, a number of RNAs need to be transcribed in addition to the one for the protein itself, the mRNA.

- These are transcribed from their genes by different isoforms of RNA polymerase:
  - Ribosomal RNA (rRNA) *(polymerase I)*
  - Transfer RNA (tRNA) *(polymerase III)*
  - Small nuclear RNA (snRNA) involved in RNA processing *(polymerase II)*
- These are not translated into proteins, but are catalytically active in the RNA form.

The first step in gene expression is the transcription of the gene of interest: this takes place in three stages—initiation (Fig. 3.3), elongation, and termination.

Initiation

The first step is the binding of the TATA box-binding protein (TBP) and TBP-associated factors (TAFs, collectively known as the TFIID complex) to the TATA box.

- This causes the sequential recruitment of other transcription factors for RNA polymerase II (TFIIs)—namely TFIIA, TFIIB, and TFIIF
- This is followed by RNA polymerase II itself and, finally, TFIIE and TFIIH

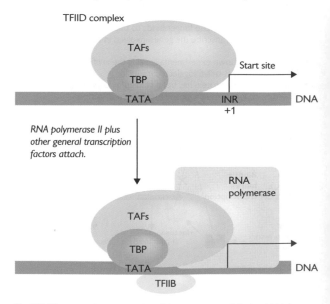

Fig. 3.3 Diagrammatic representation of the components of the basal initiation complex.

- On completion of its assembly, the initiation complex (known as the basal transcription apparatus) unwinds a short stretch of the double helix to reveal the single-stranded DNA that it will transcribe and interacts with transcription factors which regulate the rate of transcription:
  - The rate of transcription initiation by the basal transcription apparatus is quite slow in the absence of enhancing transcription factors.

Elongation

RNA polymerase II selects the correct ribonucleotide triphosphate and catalyses the formation of the phosphodiester bond.
- The RNA molecule does not require a primer and is synthesized in the $5' \rightarrow 3'$ direction:
  - Normal nucleotide base pairing rules apply, except uracil (U) is paired with adenine (A) when it occurs in the template, e.g.

DNA template	CCG	ATT	TTA	GCG
RNA strand	GGC	UAA	AAU	CGC

- This process is repeated many times and the enzyme moves unidirectionally away from the promoter region along the DNA template
- The polymerase is completely processive, i.e. a single enzyme transcribes a complete RNA
- Studies have shown that RNA polymerases do have proof reading abilities, and can correct misincorporated bases as part of the elongation process:
  - The error rate after proof reading is estimated at ~1 in every 10^3–10^5 bases, significantly higher than that for DNA polymerase (~1 in 10^7)
  - A lower fidelity is acceptable as RNA is not inherited, and as many copies of RNA are made, the majority of proteins synthesized will be correct.

Termination

As RNA polymerases are so processive, it is important that there are termination signals to indicate when the RNA transcription should stop.
- The simplest stop signal is a palindromic GC region immediately followed by a T-rich region:
  - The palindromic GC-rich region forms a hairpin structure in the RNA due to base pairing
  - The T-rich region causes an oligo(U) sequence immediately after the hairpin. This hairpin–oligo(U) complex is thought to destabilize the relatively weak association between the newly synthesized RNA and the DNA template:
    - This leads to their dissociation, and that of the RNA polymerase II complex
    - The DNA strands reform into a double helix
- There are also proteins that assist in terminating RNA transcription:
  - One such protein discovered in prokaryotes is the rho (ρ) protein
  - This binds to the newly made RNA at C-rich G-poor regions and scans along the RNA towards the RNA polymerase in an ATP-dependent manner
  - When it catches up with the RNA polymerase, it breaks the DNA/ RNA association and thus terminates transcription.

In both cases, the signal for termination lies in the newly formed RNA, not in the DNA template.

RNA processing

The RNA, as it is transcribed by RNA polymerase II, is known as the pre-mRNA and requires processing (Fig. 3.4).

- First, the pre-mRNA is capped:
  - This involves the addition of a G residue to the 5' end of the pre-mRNA by a very unusual 5'–5' triphosphate bond, which is then methylated on the N7 of the terminal G and possibly also on adjacent riboses
  - The role of the cap is to protect the 5' terminus of the mRNA against degradation and to improve ribosomal recognition for translation
- Second, a poly(A) tail is added (polyadenylation):
  - The polyadenylation consensus sequence (AAUAAA) signals that the pre-mRNA should be cleaved 20 bases or so downstream by an endonuclease
  - A poly(A) polymerase then adds the poly(A) tail, which can consist of hundreds of residues. Like the cap, the poly(A) tail is thought to enhance translation and improve mRNA stability
- Third, the pre-mRNA contains sequences for both the introns and the exons, whereas the mature mRNA only has those of the exons:
  - The introns are spliced out at specific recognition site at the ends of introns by spliceosomes. Spliceosomes consist of a number of snRNA molecules, associated proteins called splicing factors, and the pre-mRNA being processed:
    - The snRNA and protein complexes alone are known as small nuclear ribonucleoproteins (snRNPs or 'snurps')
  - mRNA can also be edited, i.e. its base sequence changed:
    - For example, the lipid and cholesterol transport protein apolipoprotein B exists in two forms—apoB-100 (formed in the liver) and apoB-48 (small intestine)
    - ApoB-48 is a truncated form of apoB-100, formed not by protein cleavage of the larger form, nor by alternative splicing, but by tissue-specific mRNA editing inserting a stop codon into the transcript.

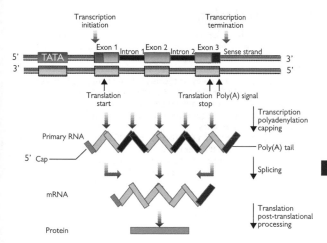

Fig. 3.4 Transcription, post-transcriptional processing, translation, and post-translational processing.

RNA translation

Once the pre-mRNA has been processed to form the mature mRNA, it leaves the nucleus through the pores in the nuclear membrane and moves to the cytoplasm where translation takes place (Fig. 3.5).

- Translation is carried out by ribosomes:
    - These are situated on the cytoplasmic face of rough endoplasmic reticulum (RER) for proteins that will be secreted, targeted to organelles, or inserted in the plasma membrane
    - Proteins that will remain in the cytoplasm are synthesized on free cytoplasmic ribosomes
    - Ribosomes consist of a 60S (large) and a 40S (small) subunit, giving the total 4200kDa 80S complex
    - The subunits are ribonucleoproteins, i.e. they are made up of a large number of proteins and one or more RNA molecules:
        - The key reactive sites in ribosomes are made up almost entirely of RNA
        - The 60S subunit contains three RNAs: 5S, 5.8S, and 23S
        - The 40S contains an 18S RNA
- Protein synthesis begins at the amino terminus and runs through to the carboxyl terminus:
    - The mRNA is translated in a 5'→3' direction
    - The 40S subunit attaches to the cap at the 5' end of the mRNA and scans along the mRNA looking for the start codon:
        - Methionine (AUG) is always the first amino acid in eukaryotic proteins
        - As eukaryotic mRNAs are monocistronic (i.e. only code for one protein), the most 5' AUG in the mRNA is usually the protein start
        - Bound to the 40S subunit are initiation factor proteins (eIFs), GTP, and the Met-tRNA$_i$ (the specific tRNA for starting a new polypeptide chain)
        - When they find the initiation codon, the initiation factors dissociate and the 60S subunit is recruited
    - The 80S ribosome complex has three sites to which tRNAs can bind—the E, P, and A sites:
        - Met-tRNA$_i$ binds to the P site
        - The next tRNA, as specified by the mRNA, template binds to the A site
        - A peptide bond is then formed between the two amino acids by peptidyl transferase
        - The ribosome translocates along the RNA towards the 3' end by one codon so that the amino acids formerly bound to the P and A sites are now bound to the E and P sites, respectively, driven by GTP and enzymes known as elongation factors
        - The newly liberated A site can accept the next tRNA
        - This sequence is repeated many times until the stop codon is reached in the mRNA template. There are no tRNAs with anticodons complementary to the codons coding for the stop signal. Instead, the codons are recognized by the release factor (eRF1), which causes dissociation of the ribosome from the completed polypeptide chain

- Prokaryotic ribosomes and translation processes are slightly different from eukaryotic ones. These differences have been exploited therapeutically by antibiotics (→ OHCM10 p.386) that inhibit protein synthesis such as tetracycline, aminoglycoside (e.g. streptomycin), and macrolide (e.g. erythromycin) antibiotics (→ OHPDT2 pp.398–402).

Proteins synthesized by ribosomes of the RER will either be inserted into the RER membrane or enter the RER lumen.

- For the majority of secreted proteins, there is a signal sequence that allows the protein to enter the RER lumen as it is synthesized
- This signal sequence is usually cleaved during subsequent protein processing.

For membrane proteins, there is also thought to be some sort of signal sequence that allows the N-terminus to cross the membrane in the case of type II membrane proteins or the first transmembrane region to enter the membrane in the case of type I proteins.

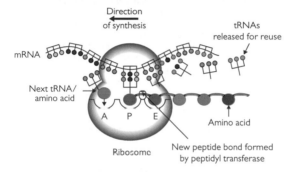

Fig. 3.5 Representation of the way in which genetic information is translated into protein.

Organization of the genome

The human genome contains approximately 3.2×10^9 base pairs.
- First draft of the complete sequence was released in early 2001
- Predicted to be around 30,000 genes (open reading frames (ORFs) coding for proteins)
- Exons make up only 1.5% of the total DNA:
  - The rest are regulatory elements and introns (non-coding regions)
  - The average length of an exon is 145 base pairs, with 8.8 exons per gene.

DNA sequences can be present in the genome in different amounts.
- Single-copy DNA represent almost half of the human genome. Only a small fraction of this codes for proteins, the rest being unique intron and regulatory sequence
- Moderately repeated DNA is that which is repeated from a few to a thousand times:
  - The sequences vary from a hundred and many thousand base pairs
  - Makes up about 15% of the human genome
- Highly repeated DNA is made up of short sequences (usually <20 nucleotides) repeated millions of times
- Inverted repeat DNA are a structural motif of DNA, with an average length of 200 base pairs. In humans, they form hairpin structures:
  - Inverted repeats can be up to 1000 base pairs
  - These can be repeated 1000 or more times per cell.

As can be seen from Table 3.2, the size of the genome does not necessarily correlate to the number of ORFs:
- For example, wall cress (*Arabidopsis*) has a genome of very similar size to the fruit fly *Drosophila*, but almost double the number of genes
- The human genome is almost 10^3-fold larger than that of *Escherichia coli* but has less than ten times the number of genes.

The relatively large size of the human genome is due to the large amount of non-coding regions (introns).
- The exact function of introns is not known:
  - They are often regarded as 'junk' DNA
  - The converse argument is that they must have a function as there is a large metabolic penalty to maintain them
- Much of the non-coding sequence consists of highly repeated sequences (~53% of the total genome):
  - Prokaryotes do not have introns to the same extent, hence their smaller genomes.

Table 3.2 Comparison of genome size with number of genes for various organisms

Organism	Size of genome	Number of genes
Escherichia coli (bacteria)	4.6×10^6	4289
Arabidopsis thalania (wall cress plant)	142×10^6	~26,000
Caenorhabditis elegans (nematode worm)	97×10^6	~19,000
Drosophila melanogaster (fruit fly)	137×10^6	~14,000
Homo sapiens (human)	3.2×10^9	~30,000

Medical genetics: general concepts

With the sequencing of the human genome, the concepts of molecular and medical genetics are coming closer together. The distinction is purely pragmatic—any molecular genetics which have been identified to cause human disease are then called medical genetics. The specialty of medical genetics serves to investigate patients with diseases which may have a genetic basis and then to counsel them on the implications of this for their own futures and that of any actual or potential offspring. Currently, the prospect of treating genetic diseases through replacement of defective genes by gene therapy remains a dream, although there have been some encouraging results from attempts to treat inherited immunodeficiencies by this route.

Most diseases represent a combination of genetic and environmental influences and the relative contribution of either varies. A genetic disease is usually taken to be one in which the disease is always manifest in 'standard' environmental conditions if the genetic abnormality is present. Some genetic diseases have complete expression whatever the environment (e.g. trisomy 21 producing Down's syndrome; ➋ OHCS11 p.273), while others may only be revealed in environments which are now 'normal' but would not have been in the recent evolutionary past, e.g. certain types of congenital hyperlipidaemia (➋ OHCM10 pp.690–1) which are now manifest as ischaemic heart disease in populations with a higher-calorie, higher-fat diet. Diseases with an overwhelming environmental factor and a minor genetic predisposition would not usually be regarded as genetic disease, e.g. development of lung cancer (➋ OHCM10 pp.174–5) in cigarette smokers.

Human genomic analysis

Sequencing a relatively small stretch of DNA has been routine since the development of the Sanger method of sequencing (➔ p.956). However, with the human genome having over 3×10^9 bases, new sequencing technologies have been developed to allow large-scale, high-throughput sequencing.

- The Human Genome Project (HGP), declared officially 'completed' in 2001, used the technique of hierarchical shotgun sequencing (Fig. 3.6) to sequence a single entire human genome:
  - The genome was sheared into random pieces of ~150,000 base pairs whose order in the genome could be determined
  - Each of these fragments was then sheared into pieces small enough to be Sanger sequenced
  - These sequenced fragments overlapped, and with computing power advancing at the time it was possible to reassemble this jigsaw
  - These larger fragments could then also be reassembled to give the whole genome sequence
  - The publicly funded cost of the HGP was ~$3 billion
  - (Whole genome shotgun sequencing was not used for the HGP due to issues with reassembling millions of small pieces, especially difficult with the large proportion of repetitive sequence (>50%) found in the human genome.)
- As human genome sequencing becomes more commonplace, then cost is going to be a big factor
- High-throughput ('next-generation') sequencing techniques have brought the cost of sequencing a human genome to around $1000 or below, and the time taken to hours rather than years:
  - Works by parallelizing sequencing, i.e. producing thousands to millions of short reads (25–500 base pairs) concurrently which can be assembled to produce full reads
    ○ Assembly process is computationally intensive
  - Allows more ambitious projects such as the 1000 and the 100,000 Genome Projects
  - Sequencing many genomes will allow the role of polymorphisms to become clearer
- Another way to reduce cost is not to sequence the whole genome, just the exons. This gives the exosome (i.e. the expressed genes in the genome, about 1% of the total genome) and can be used to look for rare variants:
  - Will, of course, only reveal variants that are in the protein coding regions of the genome, not those in non-coding or structural regions.

Sequencing is not just restricted to genomes, for example:
- RNA-seq: allows analysis of expressed genes (the transcriptosome) in cells by sequencing all RNA:
  - Allows analysis of splice variants, single nucleotide polymorphisms (SNPs), and post-transcriptional modifications in coding RNA, as well as tRNA, ribosomal RNA, non-coding RNA
- ChIP: cross-linking of DNA binding proteins to their target, followed by purification by immunoprecipitation and then sequencing.

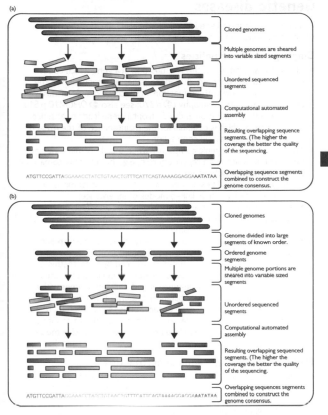

Fig. 3.6 (a) Whole genome shotgun sequencing; (b) hierarchical shotgun genome sequencing.

Reproduced under CC BY-SA 2.5 Creative Commons license from https://commons.wikimedia.org/wiki/File:Whole_genome_shotgun_sequencing_versus_Hierarchical_shotgun_sequencing.png

Genetic diseases

A number of different genetic mechanisms can influence the manifestation of genetic diseases and their modes of inheritance.

Single gene abnormalities

- The simplest example of this is a mutation in a gene which codes for a structural protein which leads to either no production of the protein or a protein with no function. A specific example of this is Duchenne muscular dystrophy (➜ p.370; see also ➜ *OHCM10* p.510) where a deletion mutation in the dystrophin gene leads to absence of this protein which is essential for the maintenance of the myocyte membrane during muscle contraction
- There is more pathogenic complexity when the abnormal gene codes for a protein which has a regulatory role, e.g. the retinoblastoma gene and hereditary retinoblastoma
- Single gene abnormalities may be transmitted by autosomal dominant, autosomal recessive, or X-linked modes of inheritance:
  - *Autosomal dominant*—the abnormal gene has an effect, i.e. it is penetrant, even when a normal gene is present on the other allele of the homologous chromosome
  - *Autosomal recessive*—the abnormal gene does not have an effect unless the allele on the homologous chromosome has the same abnormality or has been deleted or silenced by hypermethylation
  - *X-linked*—the sex chromosomes present a special case because there is incomplete homology between X and Y chromosomes with the X chromosome having a substantial amount of additional genomic material. If an abnormal gene occurs in this region, then there is no corresponding normal allele on the Y chromosome to counteract it if the abnormal gene would otherwise have a recessive model of penetrance.

Mechanisms of single gene abnormalities

- Deletion of bases within a coding region of a gene leading to misreading of all the bases within that gene after that point causing either no production of the gene protein product or an abnormally functioning protein, e.g. dystrophin in Duchenne muscular dystrophy (➜ p.370; see also ➜ *OHCM10* p.511)
- Insertion of bases within a coding region of a gene leading to misreading of all the bases in that gene after that point causing an abnormal protein product, e.g. the factor VIII gene in some types of haemophilia A (➜ *OHCM10* p.344)
- Point mutation of a base within a coding region of a gene, i.e. substitution of one base for a different base, leading to coding of an abnormal protein, e.g. sickle cell anaemia (➜ *OHCM10* p.340)
- Fusion of two genes to produce an abnormal product, e.g. Lepore haemoglobins
- Deletion, insertion, or point mutation within a non-coding (i.e. promoter or enhancer) region of a gene, which can interfere with the binding of transcription factors to this region and cause loss or reduced transcription of the coding region of the gene, e.g. some hereditary haemolytic anaemias (➜ *OHCM10* p.338).

Chromosomal abnormalities

Genetic diseases may also be caused by chromosomal abnormalities and these usually have a more complex phenotype since a much larger amount of genetic material will be involved, e.g. in trisomy 21 (Down syndrome), the phenotype includes learning difficulties, premature development of cataracts and dementia, susceptibility to certain bacterial infections, congenital heart defects, and a predisposition to leukaemia.

Mechanisms of chromosomal abnormalities

- Trisomy—occurs when a pair of homologous chromosomes fail to separate at meiosis, e.g. trisomy 18—Edwards syndrome (● OHCS11 p.846), Down syndrome (● OHCS8 p.152-3)
- Monosomy—occurs as the other daughter cell of a meiotic division which produced a trisomic cell, usually causes a non-viable foetus which spontaneously aborts during the first trimester
- Deletion—loss of a portion of a chromosome, the deleted portion will be lost at the next mitotic division, there may not be much phenotypic effect if the homologous chromosome is intact
- Ring chromosome—a special case of deletion where material is lost from each end of a chromosome and the new ends join together to form a ring, this causes many problems at mitosis
- Inversion—two breakages occur within a chromosome and the central segment rotates before rejoining
- Isochromosome—occurs when one arm of a chromosome is lost (e.g. the short arm) and the other arm duplicates to replace it (e.g. the long arm) producing genetic monosomy of the lost arm and trisomy of the duplicated arm
- Translocation—a breakage occurs and material is attached to a different chromosome. Sometimes material is exchanged between two non-homologous chromosomes, which may produce a normal phenotype in that individual but with major abnormalities in gametes if only one of the translocated chromosomes segregates into an individual cell.

Trinucleotide repeat diseases

These are a unique group of diseases that are characterized by amplification of a specific trinucleotide sequence in a gene. Strictly, they could be classified as insertions in coding or non-coding regions of the affected gene but they have special properties that merit separate discussion:

- A specific trinucleotide sequence, e.g. CGG in fragile X syndrome (● OHCS11 p.852), which normally has several repeats in a gene is amplified to have many hundreds or thousands of repeats
- This amplification may occur either during spermatogenesis (e.g. Huntington's chorea; ● OHCM10 p.702) or oocytogenesis (e.g. fragile X syndrome) depending on each specific disease
- In non-coding regions, the amplified repeat sequence interferes with transcription and can induce methylation, and thus non-function, of the whole promoter region
- In coding regions, the amplified repeat sequence produces proteins with abnormal functions, e.g. a form of huntingtin in Huntington's chorea that abnormally inactivates proteins that are normally associated with it. The diseases present earlier in life, and with a more severe phenotype, with successive generations, because of the amplification which occurs during gametogenesis. This is called anticipation.

Treatment of genetic diseases

Stem cells

Stem cells have the potential to develop into many different cell types in the body during early life and growth. Stem cells generate a continuous supply of terminally differentiated cells.

• Stem cells are not terminally differentiated and can divide without limit.

When a stem cell divides, each daughter can either remain a stem cell or go on to become terminally differentiated with a specialized function, such as a muscle or red blood cell, via a series of precursor cell divisions.

Stem cells are defined by three important characteristics:
• They are non-differentiated (unspecialized) cells
• They have the ability to divide and renew themselves by cell division for long periods (long-term self-renewal)
• Under certain physiological or experimental conditions they can be induced to become specialized cell types.

Types of stem cell

• Embryonic stem cells:
  • In the 3–5-day-old embryo (the blastocyst), the inner cells are the stem cells that give rise to the entire body (see topics in Human Embryology, Chapter 10, pp.686–687)
• Non-embryonic 'somatic' or 'adult' stem cells:
  • Discrete populations of stem cells present in adult tissues, such as the GI tract and bone marrow, generate replacements for cells lost through normal wear and tear. Typically, adult stem cells generate the cell types of the tissue in which they reside but some experiments suggest adult stem cells can give rise to cell types of a different tissue
• Induced pluripotent stem cells (iPSCs):
  • These cells arise from adult specialized cells that have been reprogrammed genetically to become stem cell-like
  • Using genetically manipulation (viral transfection) expression of just four genes is sufficient to convert adult skin fibroblasts into iPSCs (however, at a low conversion rate).[5]

In some organs (such as gut, bone marrow) stem cells regularly divide to repair and replace damaged tissues. In other organs (such as pancreas, heart) stem cells only divide under special conditions.

Normal differentiation pathways of adult stem cells

• Haematopoietic stem cells give rise to all the types of blood cells: red blood cells, B-lymphocytes, T-lymphocytes, natural killer cells, neutrophils, basophils, eosinophils, monocytes, and macrophages.
• Mesenchymal stem cells give rise to connective tissue cells such as bone cells (osteocytes), cartilage cells (chondrocytes), and fat cells (adipocytes)
• Neural stem cells give rise to the three major types of cell in the brain: nerve cells (neurones), and two types of non-neuronal cells, astrocytes and oligodendrocytes
• Epithelial stem cells in the lining of the GI tract are located in deep crypts and give rise to enterocytes, goblet cells, Paneth cells, and enteroendocrine cells

5 Takahashi K, Yamanaka S (2006). Induction of pluripotent stem cells from mouse embryonic and adult fibroblast cultures by defined factors. *Cell* **126**, 663–76.

- Skin stem cells occur in the basal layer of the epidermis and at the base of hair follicles. The epidermal stem cells give rise to keratinocytes and the follicular stem cells can give rise to both the hair follicle and to the epidermis.

Transdifferentiation

Experiments have reported that certain adult stem cell types can differentiate into cell types seen in tissues or organs other than those expected from the cell's predicted lineage. Also, certain adult cell types can be reprogrammed into other cell types by genetic modification:
- For example, insulin-producing pancreatic β-cells can be produced from pancreatic exocrine cells by introducing three key β-cell genes.[6]

Role of stem cells in disease and therapy

The regenerative properties of stem cells allow the potential for stem cell treatment of diseases ('cell-based therapies') and this is an active area of current research. Examples of potential treatments include:
- Regenerating bone using cells derived from bone marrow stroma
- Developing insulin-producing cells for type 1 diabetes
- Repairing damaged heart muscle following a heart attack with cardiac muscle cells
- Regeneration of neurones to treat neurodegenerative diseases or spinal cord injury.

The generation of personalized embryonic stem cells for therapeutic cloning requires a supply of human egg cells from women donors and is a technique fraught with ethical problems. The use and manipulation of iPSCs therapeutically bypasses the ethical problems of embryonic stem cell generation from early embryos. However, despite the potential use of iPSCs in regenerative medicine, the safety of viral vectors in patients is not clear and developing methods for growing cells and tissues in the laboratory for cell-based therapies is an area of active research.

Gene therapy

Gene therapy can be defined as 'the introduction of normal genes into cells in place of missing or defective ones in order to correct genetic disorders', often using retroviral vectors to insert the genetic material into the cell's genome.

Originally championed as the way to treat single gene disorders such as cystic fibrosis, β-thalassaemia, or Duchenne muscular dystrophy, most clinical trials have proved disappointing, for two main reasons:
1. Side effects:
   - X-linked severe combined immunodeficiency disease (X-SCID) is a disease where affected children (predominantly boys) must live in strictly controlled environments due to a lack of functional immunity (the media dubbed David Vetter[(1971–1984)] 'the boy in the bubble')
   - In a 2001 French trial, although the gene therapy was remarkably successful, a number of patients then went on to develop leukaemia: insertion of the retroviral vector was into a transcriptionally active region (i.e. where the chromosome was in an 'open' formation), activating a proto-oncogene

6 Zhou Q et al. (2008). In vivo reprogramming of adult pancreatic exocrine cells to beta-cells. Nature 455, 627–32.

- Most of the leukaemia cases were subsequently successfully treated, so with X-SCID being such a devastating disease, the reward/risk ratio may still be favourable
- Additional work continues to try to modify the retroviral vector to prevent it causing leukaemia

2. Lack of efficacy: this is the main issue seen in gene therapy clinical trials:
  - Trials continue for cancer, where the tumour cell has a different genetic make-up to the native tissue
  - Studies are also exploring correcting genetic defects in patient-derived iPSCs (see earlier in this topic), which can then be returned to the patient without any immunological issues
  - One area where gene therapy has been successful is in the treatment of progressive diseases of vision impairment, such as Leber's congenital amaurosis, choroideraemia, and age-related macular degeneration. Direct injection of the adeno-associated viral vector sub-retinally has shown long-term positive effects in these diseases.

Gene silencing

The 2006 Nobel Prize in Physiology or Medicine was awarded to Andrew Fire and Craig Mello for the discovery of RNA interference—the silencing of gene expression by double-stranded RNA.

- Hereditary amyloidogenic transthyretin amyloidosis is a rare genetic disease where a mutation in transthyretin causes a build-up of extracellular amyloid protein:
  - This affects the peripheral nervous system, leading to paralysis and ultimately death
  - These are about 50,000 sufferers worldwide
- The drug patisiran works through RNA interference to prevent the expression of transthyretin, and therefore halt the progression of the disease (and can even reverse symptoms):
  - Patisiran has recently been approved for clinical use in the USA (2018) and UK (2019).

Genome editing—CRISPR-Cas9

Genome editing is based on a bacterial defence mechanism against viruses.

- Clustered regularly interspersed short palindromic regions (CRISPR) are multiple short sequences of foreign DNA found in a specific locus in the bacterial genome. Transcripts of these sequences also contain a short 'tag' of RNA from the CRISPR locus and are known as 'guide sequences'. The CRISPR RNA then complexes with a CRISPR-associated (Cas) protein, such as the endonuclease Cas9, and the guide RNA pairs with the invading viral DNA or RNA and targets its degradation by Cas9
- Experimentally, the Cas9 nuclease can be directed against any genomic sequence by designing a guide RNA that is complementary to the desired target (Fig. 3.7):
  - Cas9 introduces a double-strand break, which can be repaired by:
    - Non-homologous end joining (NHEJ), which is error prone and so will introduce a gene disrupting mutation
    - Homologous recombination (HR) of a supplied piece of DNA allows precisely targeted insertion of a desired sequence

- In 2016, the UK Human Fertilisation & Embryology Authority (HFEA) gave permission for experimental CRISPR-Cas9 genome editing of human embryos, as long as they are destroyed after 7 days and not implanted into a human:
  - However, it was reported in 2018 that genome-edited twin girls had been born in China, with a mutation made in the *CCR5* gene to attempt to confer resistance to HIV. This represented a highly controversial and widely condemned development. It is not clear that the desired mutation was achieved, and in late 2019 the scientist responsible, He Jiankui, and colleagues were jailed and/or fined for breaching regulations
  - It remains to be seen what role CRISPR-Cas9 will play in future treatments of genetic diseases, although a clinical trial for β-thalassaemia started in late 2018 and positive results were reported in summer 2020.

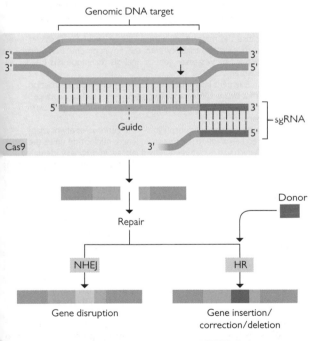

Fig. 3.7 Genome editing by adaptation of the bacterial CRISPR-Cas9 system. sgRNA, single guide RNA.

Reproduced with permission from Papachristodoulou, D., *Biochemistry and Molecular Biology* 6e, 2018, Oxford University Press. Originally adapted from Terns, R.M., and Terns, M.P. (2014). CRISPR-based technologies: prokaryotic defense weapons repurposed. *Trends in Genetics*, **30**, 111–18.

Population genetics

Most human disease is not due to single gene or chromosomal abnormalities. Rather, there is a complex interaction between a large number of genes and the individual's environment. The human genome has been subject to selections by evolutionary pressure over millions of years, but the environment in the past thousand years has been changing much more rapidly than any genetic changes. This has led to possible imbalances that may lead to the development of disease, an area which is investigated in evolutionary medicine. All these factors mean that the genetic investigation of most human diseases is directed towards finding a set of genes and environmental factors which combine to produce the disease. This requires a different set of techniques from those used in defining single gene or chromosomal abnormalities.

Polymorphisms

A genetic polymorphism is a variation between the genomes in individuals that does not appear to produce a specific deleterious phenotype, so it is not classified as a single gene inheritable disorder. This is a slightly slippery concept because the majority of human disease will be a result of interactions of such genetic variability with the environment. Polymorphisms have been defined by whatever genetic analysis technique has been available to detect them:

- Restriction fragment length polymorphisms—originally, restriction enzymes that cleave DNA at specific sites were used to produce fragments the lengths of which vary between individuals (the original DNA fingerprinting)
- Short tandem repeat polymorphisms—the genome contains many repeat stretches of di-, tri-, or variable length nucleotide units, the length of which varies between individuals, again originally identified through restriction digests. This is now the basis of genetic fingerprinting used in criminal investigations and paternity suits
- SNPs—restriction fragment length polymorphisms are just surrogate measures of SNPs and can now be identified directly through sequencing:
  - SNPs are those differences that are commonly found in the population (e.g. the same difference is found in >1% of the population)
  - The term rare variant is used for those found less commonly than SNPs, and rare functional variant for those which are known to be involved in (but not solely responsible for) disease progression
  - SNPs may not be disease causing, but can affect the way in which individuals respond to particular drugs, and form the basis of personalized medicine (➲ p.224)
- It seems that any individual will have around 3.3 million SNPs (1 in ~1000 nucleotides), so it is a large bioinformatics problem to be able to find the polymorphisms which are important in the pathogenesis of a specific disease.

Variants of uncertain significance

A variant of uncertain significance (VUS) refers to a variant that is detected by genome sequencing (➔ p.956) but for which it is unknown whether or not it has a clinical (deleterious) significance.
- The less studied the gene, the more likely differences in sequence will be labelled a VUS:
  - Even in well-studied genes, like the breast cancer-associated BRCA1 and BRCA2, new VUSs are still found
- Over time, with further study, it is likely that a VUS will be re-categorized as either deleterious or neutral:
  - This raises issues with patient management, both immediately and in case of future reclassification.

Linkage

Linkage is the statistical association between any genetic marker and a particular phenotype, which is usually a specific disease.
- If genetic markers, such as SNPs, are examined in a large population of individuals and it is known which individuals have or do not have the disease, then statistical analyses can be made to establish the strength of linkage between the marker and the disease
- If the association is very strong, this suggests that the marker is very close to, or actually in, the region of DNA that codes for proteins involved in the pathogenesis of the disease
- This task may be made easier by using populations with a relatively restricted gene pool and long-known pedigrees, which are found in geographically or socially insular communities such as the Amish in North America
- The single genes involved in cystic fibrosis and Huntington's chorea were discovered using genetic linkage studies. Defining polygenic diseases using linkage studies is much more difficult.

Genome-wide association studies (GWAS)

With genomic sequencing, the sequences of individuals can be compared to identify disease causing variants, a technique known as GWAS (or whole genome association studies (WGAS)).
- The more genomes included in the study, the more robust any conclusion about association can be
- Useful data sets include those of cohorts of twins
- GWAS can be combined with other techniques, such as metabolomics, to allow association of metabolic changes with genetic variants.

Pharmacogenetics and pharmacogenomics

Medical genetics could potentially be well employed in improving the thera-peutic benefits of administered drugs. Two inter-related terms are applied to describe studies in this area: pharmacogenetics and pharmacogenomics, although there is considerable confusion (even among scientists) as to the definition of both.

The term 'pharmacogenetics' was coined >50 years ago and relates to the study of how an individual gene might affect the body's response to a drug. Historically, this term has been applied specifically to describe the study of polymorphisms that affect enzymes involved in drug metabolism, with a bearing on the potential toxicity of a given drug. Identifying patients with particular polymorphisms, prior to drug administration, could identify those at risk of toxic side effects from a particular class of drug and would help physicians modify the treatment regimen accordingly. Much of the work to date has centred on polymorphisms among P450 enzymes, which are involved in the metabolism of many drugs in the liver.

Pharmacogenomics is a more recent term that relates to how the whole genome affects the body's response to a drug. This is a far broader dis-cipline, because it embraces not only the impact of the gene that, for ex-ample, encodes for a target protein (enzyme or receptor), but also the array of genes that might determine an individual's ability to absorb a drug, the ultimate distribution of that drug, and its clearance through metabolism and excretion. Ultimately, it is hoped that knowledge of a patient's genome will facilitate 'individualized medicines' that will be tailored to a particular patient's genetic make-up (known as "precision medicine"). This could help doctors to choose the best drug at the optimal dose for any given patient while avoiding, for example, adverse reactions determined by genetic pre-disposition. Furthermore, pharmacogenomics will help in identifying new gene product targets for drug intervention. Ethics aside, the tools are now available to us, in the post-genomic era, to take this particular aspect of medical genetics forward.

Nerve and muscle

Cells of the nervous system

Together, the peripheral nervous system (PNS) and central nervous system (CNS) comprise an integrated network of neurones, which allows rapid transmission and processing of information as nervous impulses. In conjunction with the endocrine system, this network is responsible for homeostasis of an organism, as well as for its behavioural patterns. In humans, this includes higher-level functions such as thinking and memory, as well as more basic activities.

Nervous tissue comprises two basic cell types:
- Neurones (excitable cells which transmit nervous impulses over long distances)
- Glial cells (non-neural support cells which are closely associated with neurones).

The central nervous system

The CNS includes the brain and the spinal cord and is divided into grey and white matter. Grey matter contains the cell bodies, dendrites, parts of axons, and glial cells, and makes up the cerebral and cerebellar cortex, as well as the central sections of the spinal cord. The white matter contains the nerve fibres and glial cells, but not the cell bodies, except in grey matter nuclei (islands of aggregated cell bodies within the white matter) and includes those parts of the brain which are not grey matter as well as the outer regions of the spinal cord.

The peripheral nervous system

The PNS includes all nervous tissue located outside the CNS. It includes nerves, ganglia, and nerve endings, as well as associated supporting cell types. Pns nerves may be efferent, conducting signals away from the CNS to effector organs or tissues (e.g. skeletal muscle) or afferent, conducting nervous impulses towards the CNS (e.g. from sensory receptors).

Neurones

All neurones (e.g. Fig. 4.1), whether central or peripheral, include three separate parts:
- Dendrites—multiple long processes which are specialized for receiving external stimuli from other cells
- Cell body (perikaryon)—contains the cell nucleus
- Axon—conducts a nervous impulse to other cells. Axons form synapses with other cells (neurones and non-nerve cells) via swellings called terminal boutons.

Dendrites

Dendrites are branching extensions of the nerve which, unlike axons, become thinner as they divide. They receive many synapses and are the signal reception and processing site of the nerve.

The cell body

The cell body encompasses the nucleus surrounded by cytoplasm. Free ribosomes and RER can be seen as Nissl bodies (basophilic granular areas) in the cytoplasm, and are responsible for the production of proteins, including neurotransmitters. Abundant neurofilaments (intermediate filaments) can be found within nerve cell bodies. Mitochondria are present in large numbers in the cytoplasm (as well as in axon terminals).

Axon

Axons are cylindrical processes and are often very long. The axon hillock arises from the perikaryon as a pyramidal-shaped region, from which the axon projects. The axoplasm is contained within the axon, which is surrounded by a plasma membrane (axolemma). The axons of nerves can be myelinated or unmyelinated.

There are three major axonal shapes. The cell body position determines the type of axonal shape by altering the number of cell processes formed by the dendrites and axons:

• Multipolar neurones: have more than two cell processes (e.g. motor neurones)
• Bipolar neurones: have one dendrite and one axon. Found only in specialized sensory areas such as the olfactory area and retina
• Pseudounipolar neurones: the cell body forms a T-shape, with the dendritic process and the axon 'bypassing' the cell body. Impulses avoid the perikaryon, e.g. spinal ganglia (sensory neurones).

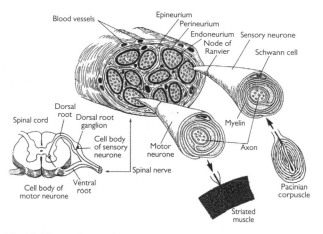

Fig. 4.1 Schematic diagram showing arrangement of motor and sensory neurones.
Reproduced with permission from Ross MH, Kaye GI, Pawlina W (2003). *Histology: A Text and Atlas*, 4th edn. Baltimore: Lippincott Williams and Wilkins.

Nerves

Bundles of nerve fibres surrounded by a sheath of connective tissue make up the peripheral nerves. Nerve fibres are ensheathed in three layers:

• The endoneurium—reticular fibres produced by Schwann cells surrounding axons
• The perineurium—a layer of flattened epithelial-like cells, which form a protective barrier surrounding a bundle of nerve fibres, preventing macromolecules from approaching the fibres
• The outermost epineurium—a dense fibrous external sheath which envelops several bundles and forms the outermost layer of the nerve.

Glial cells

There are ten times more glial cells than neurones in the brain and they have many important functions. They insulate neurones, provide structural support and help maintain the surrounding environment, play a role in repair of neurones, and form part of the blood–brain barrier. Several different types of glial cell exist and these differ between the peripheral and central nervous systems.

Astrocytes

Star-shaped cells with many processes. Their most important function is to form the blood–brain barrier. In so doing, they control metabolic exchange with the extracellular fluid and blood, thus controlling the extracellular environment of the brain.

Microglia

Small cells with multiple processes and dense drawn-out nuclei. They are the mononuclear phagocytic arm of the immune system within the CNS. They are important in inflammation and repair. When activated, the morphology of the cell changes to resemble that of a macrophage, and they perform the same functions, namely phagocytosis and antigen presentation.

Oligodendrocytes

Cells found only in the CNS. They produce myelin to insulate neuronal axons and increase the speed of conduction. They can each insulate more than one neurone, unlike peripheral myelin-producing cells (Schwann cells), and wrap around their axons. Modified membrane layers with raised lipoprotein content make up the myelin which insulates the axons.

Schwann cells

Schwann cells myelinate axons in the PNS. The Schwann cell wraps its membrane around the axon. The membranes of the cell merge to form myelin (a lipoprotein). More than one cell is required to insulate each axon. There are gaps between the Schwann cells, called nodes of Ranvier. The internodal gap is 1–2mm (the length of the Schwann cells). Unmyelinated axons are of smaller diameter. Unmyelinated cells still have a covering of Schwann cells but each Schwann cell can cover many unmyelinated axons. The Schwann cells lie adjacent to one another to form a continuous sheath, hence there are no nodes of Ranvier. In the CNS, there are many unmyelinated axons which have no sheath whatsoever.

Ganglia

Neuronal cell bodies and glial cells supported by connective tissue make up a ganglion. It is where nerves synapse with each other as one nerve enters and another exits. The ganglion cells are supported by a connective tissue framework and capsule, surrounding the cell bodies and glial cells. The axons found in ganglia tend to be pseudounipolar. There are two types of ganglion:

- Sensory ganglia receive impulses that are relayed to the CNS.
  There are:
  - Cranial nerve ganglia
  - Spinal ganglia (from the dorsal root of the spinal nerves)
- Autonomic ganglia of the parasympathetic nervous system tend to be located within the organ that they innervate such as the GI tract, as intramural ganglia. They are supported by the surrounding stroma of the organ, so have minimal or no connective tissue or supporting glia. Sympathetic ganglia are located away from target organs, e.g. the cervical ganglion or the ganglia of the sympathetic trunk.

The peripheral nervous system: somatic nervous system

The nervous system is divided into the CNS (the brain and the spinal cord; ➋ pp.764–9) and the PNS, consisting of the nerves that control skeletal muscle contraction (voluntary movements) as well as smooth muscle and organ function (involuntary activity).

Nerves that stimulate contraction of skeletal muscle belong to the somatic efferent system, while those that modulate involuntary actions belong to the autonomic system. The different functions of the somatic and autonomic nervous systems are reflected in different neuronal arrangements, neurotransmitters, and receptors associated with each system.

There are two elements to the somatic nervous system:
- Somatic motor (efferent) neurones that stimulate contraction of skeletal muscle
- Somatosensory neurones that convey information about muscle and tendon stretch to the spinal cord.

A number of key features distinguish somatic neurones from their autonomic counterparts:
- A single neurone spans between the spinal cord and the muscle—there are no intervening synapses
- The neurotransmitter released at the neuromuscular junction (NMJ) is exclusively acetylcholine (ACh)
- The postsynaptic receptors at the NMJ are exclusively nicotinic ACh receptors.

The peripheral nervous system: autonomic nervous system

The autonomic nervous system is divided into two main branches: the parasympathetic and the sympathetic nervous systems (Fig. 4.2). In some instances, these systems act in opposition to each other (e.g. increased parasympathetic stimulation slows the heart, increased sympathetic drive increases heart rate). However, some organs or tissues are only innervated by one of the systems (e.g. blood vessels usually only have sympathetic innervation).

There are a number of similarities between the sympathetic and para-sympathetic nervous systems:

- Both generally feature a single synapse, found in ganglia positioned between the spinal cord and the target organ or tissue. The neurones that stretch between the spinal cord and the ganglia are called preganglionic; those between the ganglia and the target tissue are postganglionic. The sympathetic nerves that supply the adrenal medulla are an exception to this rule, as they do not feature ganglia
- The cell bodies of the preganglionic cells are found in the ganglia; those of the postganglionic cells are generally found in the tissue
- The neurotransmitter released by the preganglionic neurone at the ganglionic synapse is exclusively ACh, which acts on nicotinic ACh receptors on the postganglionic neurone to propagate the signal
- A cocktail of neurotransmitters is often released at the synapse with the tissue. Co-transmitters often have different rates and durations of action.

There are, however, a number of very important differences between the sympathetic and parasympathetic nervous systems (Table 4.1).

Enteric nervous system

A third division of the autonomic nervous system forms a complex net around the viscera, with cell bodies embedded in the intramural plexuses of the intestinal wall. Neurones from both the sympathetic and parasympathetic systems synapse onto enteric nerves, which release a number of neuropeptides and transmitters other than noradrenaline and ACh (e.g. 5-HT, nitric oxide (NO), ATP).

Table 4.1 Differences between the parasympathetic and sympathetic nervous systems

	Parasympathetic	Sympathetic
Spinal outflow	Cranial (nerves III, VII, IX, X) and sacral	Thoracic and lumbar
Position of ganglia	Near (or in) target tissue	Near spinal cord (paravertebral sympathetic chain)
Primary neurotransmitter at target	ACh	Mainly noradrenaline (ACh in sweat glands and at adrenal medulla)
Primary receptors at tissue	ACh muscarinic receptors (M_1, M_2, M_3)	Adrenoceptors (α_1, α_2, β_1, β_2, β_3)
Co-transmitters	NO, vasoactive intestinal peptide (VIP)	ATP, neuropeptide Y

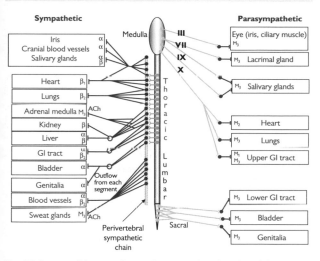

Fig. 4.2 Features of the sympathetic and parasympathetic branches of the autonomic nervous system.

Resting membrane potential

All cells have a membrane potential (E_m). E_m is the voltage difference (in mV) between the extracellular and intracellular faces of the cell membrane. E_m is usually referenced to the extracellular face. At rest, E_m is negative, meaning that the inside of the membrane is negative with respect to the outside.

The Nernst equation

E_m is established by the permeability properties of the cell membrane. At an equilibrium potential, there is not net movement of charge (no current). For a membrane which is selectively permeable to ion X, the equilibrium potential is described by the Nernst equation:

$$E_X = \frac{RT}{Z_X F}\left[\frac{\ln[X]_i}{[X]_o}\right]$$

where EX   = equilibrium potential
     R   = gas constant
     T   = absolute temperature
     F   = Faraday constant
     ZX   = the ionic charge of X
     $[X]_i$ = intracellular concentration of X
     $[X]_o$ = extracellular concentration of X.

With appropriate values for the constants, and converting natural log to $\log_{10}$, the equation for a monovalent cation at 37°C becomes:

$$E_X = 61.5 \log_{10}\frac{[X]_o}{[X]_i}$$

Ionic basis of the resting potential

- The Na^+-K^+-ATPase establishes high extracellular $[Na^+]$ and high intracellular $[K^+]$
- For an idealized cell membrane which is permeable only to K^+ ions, passive efflux of K^+ down its concentration gradient will polarize the membrane: the inside face becomes negative with respect to the outside
- The negative inside potential opposes further influx. As more ions exit the cell, so the opposing force becomes greater
- A point is reached at which chemical and electrical gradients are equal but opposing
- At this point, there is no net ion current, and equilibrium is achieved. The Nernst equation predicts $E_K = -90mV$ in mammalian nerves. Similar calculations for a membrane solely permeable to Na^+ ions predict Na^+ ion influx would cease at +60mV.

Since the resting cell membrane has high K^+ permeability, K^+ ion movements are most important for establishing E_m. However, other ions (notably Na^+, Cl^-) are also permeable, but to lesser extents. Their movements also contribute to the potential difference across the membrane. E_m is not the Nernst potential for any single ion, but rather an aggregate based on contributions weighted by membrane permeability. In a real cell, no ion is at equilibrium—although there are ionic currents, E_m is at equilibrium. These currents therefore add up to zero.

The constant field equation describes this situation:

$$E_m = -\frac{RT}{F} \ln \frac{\left(P_K\left[K^+\right]_o + P_{Na}\left[Na^+\right]_o + P_{Cl^-}\left[Cl^-\right]_i\right) + ...}{\left(P_K\left[K^+\right]_o + P_{Na}\left[Na^+\right]_o + P_{Cl^-}\left[Cl^-\right]_i\right) + ...}$$

where P = permeability of the ion and [ ] signifies concentration.

In a mammalian nerve, $E_m = -70mV$. Ion gradients are maintained by the Na$^+$-K$^+$-ATPase. This pump transporter has a 3Na$^+$:2K$^+$ pumping ratio which makes it electrogenic. Its activity makes only a very small contribution to the inside negative E_m. Note, however, that ion movements down gradients which the ATPase establishes are responsible for E_m.

The number of ions moving is miniscule in respect of the total present in the extracellular and intracellular solutions, and ion numbers do not change in bulk solution, only at the faces of the membrane.

Changing ion gradients or membrane permeability can alter E_m:
• Increasing $[K^+]_o$ reduces the outward chemical gradient for K$^+$: the contribution of K$^+$ ion movement to E_m is smaller, so E_m is less negative or depolarized
• Likewise, increasing P_{Na} will enhance the contribution of Na$^+$ ion movements. E_m again becomes less negative.

The action potential

The action potential is a rapid (1–2ms) transient electrical impulse in which E_m is displaced by up to 100mV. It is exhibited by electrically excitable cells, and is the foundation for nervous conduction and muscle contraction.

Action potentials arise from changes in membrane permeability. Electrically excitable cells possess voltage-activated channels that open in response to depolarization. The action potential is an all-or-none response: once a threshold point is passed, the events progress in a time-dependent fashion. So long as the threshold can be attained, the magnitude of the depolarizing stimulus is irrelevant.

Key events in the nerve action potential (Fig. 4.3) are:

• Initial depolarization: a relatively gentle increase in E_m. In nerve endings, this results from cation channel opening in response to chemical or physical stimuli and, within nerves, by the spread of current from neighbouring, already excited regions

• The upstroke: once threshold (generally +15mV above E_m) is reached, voltage-activated Na^+ channels open and Na^+ ions enter the cell down their electrochemical gradient. The process is regenerative: once some open, the depolarization they initiate promotes the opening of others. The increase in Na^+ permeability causes E_m to move rapidly towards, but not reach, E_{Na}. E_m overshoots zero to approximately +35mV. The increase in permeability is short-lived: the depolarization initiates fast opening of activation gates, followed by slower closing of inactivation gates

• Repolarization: this represents the inactivation of Na^+ channels and opening of voltage-activated K^+ channels. K^+ ions efflux down their electrochemical gradient. E_m recovers towards resting levels. K^+ channels are activated at the same point as Na^+ channels, but opening takes longer. Most K^+ current flows after Na^+ channels have closed

• After hyperpolarization: the closing of K^+ channels by repolarization is slow. The transient raised permeability to K^+ takes E_m closer to E_K. Repolarization also returns Na^+ channels from the inactive to closed state.

The ion fluxes occurring in this sequence are very small in comparison with the number of ions present in the extracellular and intracellular solutions. Voltage-activated Na^+ channels can be inhibited by tetrodotoxin (TTX), voltage-activated K^+ channels by tetraethylammonium (TEA).

There is a refractory period during which the neurone is resistant to firing another action potential.

• The absolute refractory period is the time during which voltage-activated Na^+ channels recover from inactivation, and during which there are insufficient channels which can be recruited to initiate an upstroke

• The relative refractory period correlates with the after hyperpolarization, where larger stimuli are required to elicit a second action potential

- Action potentials propagate by local current flow:
  - At the region of excitation, the nerve potential is reversed
  - The negative outside region acts as a current sink, drawing positive charges from areas behind and ahead of the excited region
  - The polarity of the flanking membrane regions is reduced—this is termed electrotonic depolarization
  - Na^+ channels in the membrane behind the action potential remain inactivated, preventing firing. If depolarization in the region ahead reaches threshold, however, the action potential can fire
  - This 'one-way' propagation is called orthodromic conduction
  - Electrotonic spread of depolarization decays from the site of initiation.

As nerve diameter increases, cytoplasmic resistance decreases, local current flow is more rapid, and conduction velocity is faster. Nerves can be myelinated with concentric wrappings of Schwann cell membranes, to increase membrane resistance, minimize electrotonic current decay across the nerve cell membrane, and speed up conduction. Gaps in the myelination (called nodes of Ranvier) have high densities of voltage-activated channels. The current sink at one node triggers electrotonic depolarization in the next, so the action potential 'jumps' from node to node. This is termed saltatory conduction, and speeds up conduction by up to 50 times.

The fastest conducting fibres are therefore:
- Aα (15–20μm diameter)—myelinated: innervate skeletal muscle
- Aγ (2–5μm)—myelinated: sensory (pain, temperature)
- C (0.5–1μm)—unmyelinated: sensory.

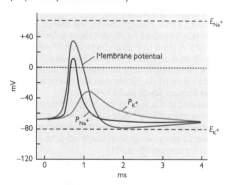

Fig. 4.3 The permeability changes that are responsible that are responsible for the action potential of a nerve.

Reproduced with permission from Pocock G, Richards CD (2006). *Human Physiology: The Basis of Medicine*, 3rd edn, p69. Oxford University Press.

Anaesthesia

Few developments have had a greater impact on medicinal practice than the discovery and development of anaesthetics from the early 1800s. General anaesthesia is routinely used to prevent patients feeling pain during surgery and has been instrumental in the development of the specialty of surgery. Increasingly, however, surgeons are turning to means of inducing local anaesthesia, in order to minimize the risk associated with anaesthetic practice (➔ *OHCM10* p.572).

General anaesthetics

It is perhaps surprising, in light of their widespread use, that the mechanism(s) by which general anaesthetics render patients unconscious and oblivious to pain are relatively poorly understood, particularly as, unlike drugs that act on specific receptors, the substances that are able to induce general anaesthesia do not have many molecular characteristics in common. Many general anaesthetics are volatile liquids or gases and are administered via inhalation. However, a number of intravenous anaesthetics are commonly used to induce rapid anaesthesia prior to prolonged administration of an inhaled anaesthetic during operations.

Inhaled anaesthetics
- Volatile liquids (e.g. those related to ether, including halothane, enflurane, and isoflurane) or gases (nitrous oxide (N_2O)—'laughing gas')
- Show a clear relationship of potency with lipid solubility: the more lipid-soluble anaesthetics are more potent. However, lipid solubility also carries the inherent danger of accumulating in body fat → a slow recovery on cessation of anaesthesia
- Are thought to act either through altering the characteristics of the lipid bilayer of cells (e.g. increased fluidity or bilayer volume) or by interacting with hydrophobic regions of critical channels and receptors to modify their function. Both mechanisms are plausible and it is possible that a mixture of the two is responsible for anaesthetic activity
- Inhibit synaptic transmission, rather than axonal conduction (compare with local anaesthetics; see ➔ pp.239–40)
- When administered at the appropriate levels for surgery, cause unconsciousness and a loss of response to painful stimuli (analgesia) and reflexes. Muscles also relax but neuromuscular blocking drugs are often administered alongside general anaesthetics to prevent muscle contraction during surgery (➔ p.244)
- Generally, result in a depression in cardiovascular and respiratory function. This is an adverse effect that requires careful monitoring: overdose can result in death because of cardiovascular and/or respiratory collapse.

Intravenous anaesthetics
- Have a much quicker action than inhaled anaesthetics: typically an intravenous anaesthetic acts within 20s of administration (the time taken to reach the brain in the bloodstream). A single injection usually induces anaesthesia for about 10min
- Are very lipid-soluble to accelerate the onset of effect: they cross the blood–brain barrier rapidly

- Are centrally acting agents such as barbiturates (e.g. thiopental, propofol), benzodiazepines (e.g. diazepam), or N-methyl-D-aspartate (NMDA) receptor blockers (e.g. ketamine)
- Can be rapidly redistributed to tissues with high blood flow and, more slowly, to poorly perfused fat. This is a problem that affects thiopental in particular, for which blood concentrations fall very rapidly, but its slow redistribution to fat prior to metabolism means that it stays within the body long after anaesthesia → a 'hangover'. The effect is exacerbated with repeated injections—thiopental is only administered to induce anaesthesia prior to maintenance by inhaled anaesthetics (⊃ p.238). However, propofol and ketamine have sufficiently long duration of action to facilitate their sole use for minor surgery.

Modern general anaesthetic practice, therefore, usually involves induction with an intravenous injection of an agent such as thiopental prior to prolonged administration of a combination of inhaled anaesthetics, such as isoflurane and nitrous oxide. A neuromuscular blocking drug such as atracurium is also administered to induce muscle paralysis.

Local anaesthetics
(⊃ OHPDT2 p.768.)

- Generally conform to a common structure consisting of an aromatic region (containing a benzene ring) linked, via an ester or amide bond, to an amine side-chain that makes the molecule basic
- Have a pKa ~8–9, meaning that they are almost entirely ionized at physiological pH, resulting in relatively slow penetration of nerves (⊃ p.96). Quaternary compounds are entirely ionized and are not able to cross lipid membranes—instead, they enter nerves through open Na^+ channels. Inflamed tissue can be resistant to local anaesthetics because it is often acidic, resulting in increased drug ionization and reduced cellular penetration (⊃ pp.106–115)
- Are rapidly hydrolysed by enzymes in the blood. Local anaesthetics that contain ester bonds are rapidly metabolized by esterase enzymes and typically have a plasma half-life of <1h. Amides are more stable (half-life ~2h) and are, therefore, longer lasting
- Act to block Na^+ channels essential to nerve conduction (⊃ pp.236–37). Some local anaesthetics have a 'use-dependent' effect—their action increases the more the channel is open. This is because the ionized form of the drug is able to enter open channels to reach its binding site within (the only means of entry for completely ionized quaternary compounds). Partially ionized tertiary local anaesthetics can reach their site of action by this route (ionized) or through the influx of non-ionized drug into the lipid bilayer
- Affect small nerve fibres more readily than large ones. This is of theoretical importance because the fibres that carry nociceptive impulses are the smallest (Aδ and C fibres), while those conducting impulses relating to other senses (e.g. touch) or motor control are much larger. In reality, however, the relative non-selectivity of local anaesthetics for Na^+ channels in all axons results in some loss of sensation and motor control in the local area

- Can be applied to the skin (eutectic mixture of local anaesthetics—EMLA®) or injected directly into the target tissue, into the bloodstream distal to a pressure cuff to prevent systemic distribution of the drug, in the vicinity of a nerve trunk for the target region, or into the subarachnoid space to block transmission in the spinal cord
- Have side-effects due to their ultimate absorption into the systemic circulation. As with general anaesthetics, systemic side-effects include centrally induced respiratory depression, as well as hypotension due to reduced sympathetic nervous system activity and depressed cardiac contractility. The cardiac effects are used to advantage in the use of lidocaine as an antiarrhythmic drug (◑ p.455). Lower systemic doses of local anaesthetics have a stimulatory effect on the CNS, resulting in agitation and tremor.

Epidural

Epidural is a specific type of local anaesthesia that involves infusion of local anaesthetic (e.g. lidocaine) into the epidural space surrounding the spinal cord. The target area is delineated by the level in the spine at which the drug is administered: infusion into the lumbar region anaesthetizes the lower abdomen and the legs; a thoracic epidural would have a more global effect but carries more risk because it might have an impact on nerves supplying the heart and muscles involved in breathing. Epidurals are used routinely to reduce the pain associated with childbirth, but their use is increasingly being advocated for abdominal surgery, and even cardiothoracic surgery.

Neuromuscular transmission

When an action potential reaches the end of a nerve, the excitation can be transmitted to another nerve or to an effector by the release of a transmitter substance. The excitation of skeletal muscle by ACh release from a motor neurone exemplifies the processes underlying chemical transmission (Fig. 4.4).

Motor nerves send processes to individual muscle fibres, which collectively form a motor unit. Each nerve process terminates at the neuromuscular junction (NMJ) or end plate, where unmyelinated bulb-like nerve endings (boutons) interface with invaginations of the muscle fibre membrane. The 50nm space between the two membranes is termed the synaptic cleft. The membrane here is highly invaginated to increase the surface area.

The unmyelinated nerve membrane contains voltage-activated Ca^{2+} channels, which are opened by the arrival of the action potential. Ca^{2+} ions flow into the cell down their electrochemical gradient and trigger fusion of ~200 vesicles containing up to 10,000 molecules of ACh. Vesicles contain 150mM ACh, which is synthesized in the nerve ending from choline and acetylcoenzyme A, and concentrated within the vesicle by a H^+-driven exchanger. Ca^{2+} ions are then extruded from the cell to terminate vesicle fusion.

The ACh released into the synaptic cleft diffuses to the post-junctional membrane of the muscle fibre where two molecules bind to each nicotinic ACh receptor. These receptors are doughnut-shaped heteropentamers and are ionotropic (ion channel) receptors. When ACh binds, a non-specific cation channel is opened, which leads to net influx of positive ions and depolarization.

This depolarization is ~40mV and is termed the end plate potential (EPP). If sufficiently large, the threshold is reached for opening of voltage-activated Na^+ channels in the muscle fibre membrane, and an action potential is initiated. The action potential involves the same sequence of events as that in the motor nerve. Small, spontaneous depolarizations (miniature EPPs) are observed in the absence of nerve stimulation. The magnitude of miniature EPPs is always a multiple of 0.4mV.

The elemental depolarization of 0.4mV represents the release of ACh contained in one vesicle or quantum.

ACh is rapidly hydrolysed in the synaptic cleft by the enzyme acetylcholine esterase (AChE), and re-uptake of choline into the nerve ending for synthesis of more ACh occurs on a Na^+-dependent transporter.

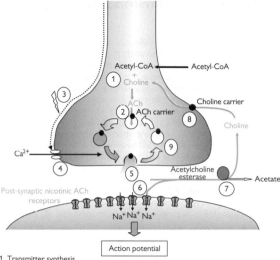

1. Transmitter synthesis
2. Transmitter uptake into vesicles
3. Action potential in presynaptic neurone
4. Voltage-gated Ca^{2+} channel activation
5. Exocytosis
6. Receptor activation in post-synaptic neurone
7. Transmitter deactivation
8. Precursor recycling
9. Vesicular recycling

Fig. 4.4 Cholinergic transmission.

Neuromuscular transmission: drug intervention

(⊘ *OHPDT2* pp.364, 670.) The ability to prevent muscle contraction is clinically important as an adjunct to anaesthesia during surgery. The global paralysis caused by neuromuscular blocking drugs means that they prevent breathing and can only be used when artificial ventilation is available.

- Tubocurarine was the first 'non-depolarizing' blocking agent to be identified. Tubocurarine is a constituent of curare, the poison that has long been used by South American Indians on their arrows and darts
- It is a competitive antagonist at nicotinic ACh receptors, where it inhibits binding of ACh in a concentration-dependent manner
- Tubocurarine has been superseded by modern alternatives with shorter action and fewer side effects (e.g. pancuronium, vecuronium, atracurium)
- The side effects of these compounds are related to their non-selective action on other nicotinic ACh receptors, including those in the autonomic ganglia. Hypotension is sometimes experienced, as well as bronchospasm due to histamine release from mast cells (⊘ p.898).

The effect of these drugs is easily reversed by inhibitors of AChE, the enzyme that breaks down ACh. Inhibition of this enzyme ultimately leads to the persistence of ACh in the synaptic cleft → an increase in its concentration and successful competition for ACh receptors.

Another class of drug that causes muscle paralysis consists of the depolarizing blocking drugs (e.g. suxamethonium).

- Unlike tubocurarine, these drugs do not act to compete with ACh for nicotinic ACh receptors; instead, they act to stimulate nicotinic ACh receptors and cause paradoxical depolarization of the motor end plate
- In the first instance, this depolarization leads to contraction—transient twitching is an early feature in depolarizing block
- Within a few seconds, this twitching (or fasciculation) gives way to paralysis as the electrical excitability of the end plate region is lost
- The major advantage of depolarizing block over non-depolarizing blocking agents is that there is less impact on ganglionic nicotinic ACh receptors. However, there are dangers, particularly bradycardia.

There are a number of other drugs that impact the presynaptic synthesis and release mechanisms of ACh:

- Vesamicol inhibits uptake of ACh into presynaptic vesicles
- The antibiotics, streptomycin and neomycin, prevent Ca^{2+} entry and can occasionally cause paralysis
- Botulinum toxin (from the bacterium, *Clostridium botulinum*, responsible for the type of food poisoning called 'botulism' (⊘ *OHCM10* p.436)) inhibits specific proteins involved in exocytosis, resulting in paralysis and inhibition of the parasympathetic nervous system. This agent is currently used 'therapeutically' to treat a number of conditions characterized by muscle spasm, including cervical dystonia, blepharospasm, hemifacial spasm associated with stroke, focal spasticity, and lower limb dysfunction associated with cerebral palsy. Injections into the skin of the armpits also relieves excessive sweating, and injections into facial skin can temporarily reduce the appearance of wrinkles (Botox®)

- β-bungarotoxin is found in cobra snake venom and has a similar action to botulinum toxin. α-bungarotoxin in the same venom inhibits the nicotinic ACh receptor
- A number of AChE inhibitors exist alongside neostigmine. Edrophonium is very short acting and is used in the diagnosis of myasthenia gravis (🔁 *OHCM10* p.512) (muscle weakness due to autoimmune destruction of ACh receptors), and physostigmine is used in the treatment of glaucoma
- Hemicholinium and triethylcholine prevent the uptake of choline into the nerve terminal, but do not have any clinical application.

Interneuronal synapses

The nervous system is not 'hard-wired'—there are spaces at connecting points called synapses. Synapses are highly complex structures that convert an electrical signal in one neurone into a chemical signal that acts, in turn, to generate (or inhibit) an electrical signal in adjacent neurones or other excitable cells (e.g. muscle cells). It is the existence of synapses and the chemical mediators that they employ that is central to much of our pharmacological interventional strategy.

Interneuronal synapses offer a means of exquisite control over the transmission of signals. The key features of synaptic transmission are summarized in Fig. 4.5.

- Action potentials arrive at the synaptic terminal of the presynaptic neurone: the greater the frequency of action potentials, the more neurotransmitter that is ultimately released
- The membrane at the synaptic terminal depolarizes (➲ p.236)
- Voltage-gated Ca^{2+} channels open to facilitate the influx of Ca^{2+} into the presynaptic terminal
- Neurotransmitter-containing synaptic vesicles migrate to, and fuse with, the presynaptic membrane in response to increased intracellular Ca^{2+}
- Neurotransmitter molecules are released into the synaptic cleft
- The neurotransmitter molecules diffuse across the cleft where they stimulate ion channel-linked receptors on the postsynaptic membrane, resulting in the altered permeability to a specific ion that causes a change in membrane potential. For example, activation of nicotinic ACh receptors is excitatory because it causes an influx of Na^+ and depolarization; activation of GABA receptors is inhibitory because it causes an influx of Cl^- ions and hyperpolarization
- The change in potential of the postsynaptic membrane caused by neurotransmitter release from a single action potential in the presynaptic terminal is called an excitatory (or inhibitory) postsynaptic potential (EPSP or IPSP; Fig. 4.6).

The ultimate effect of the synaptic transmission is dependent on a number of factors:

- The neurotransmitter released
- The receptor subtype on the postsynaptic neuronal membrane
- The ion channel linked to the receptor.

The simplest scenario for interneuronal synaptic transmission involves a single synapse between the pre- and postsynaptic neurones. In this situation, the only controlling factor that determines whether the action potential is propagated in the postsynaptic neurone is the frequency of signals arriving at the synaptic terminal of the presynaptic neurone: increased frequency causes summation of the depolarization caused by EPSPs until they exceed the necessary threshold for an action potential in the postsynaptic neurone (Fig. 4.6).

It is more likely, however, that transmission at a synapse is susceptible to modulation by a number of other signals that impact the pre- or postsynaptic neurones:

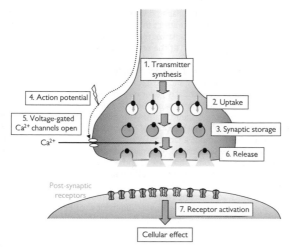

Fig. 4.5 Synaptic transmission.

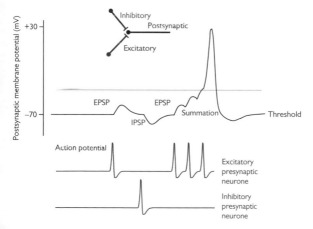

Fig. 4.6 Excitatory (EPSP) and inhibitory (IPSP) postsynaptic potentials—summation.

- Inhibitory neurones can synapse onto either the pre- or postsynaptic terminals. The neurotransmitters that are released by these neurones (e.g. GABA) act to hyperpolarize the neurone, reducing the likelihood of sufficient depolarization at the presynaptic terminal to facilitate transmitter release, or sufficient depolarization in the postsynaptic terminal to reach threshold for action potential propagation (Fig. 4.6). The balance of inhibitory and excitatory signals will ultimately determine whether an action potential is generated in the postsynaptic terminal

- Facilitatory neurones that synapse onto the presynaptic terminal act to increase the amount of neurotransmitter released by that neurone. This is generally through 'sensitization' of the presynaptic terminal to action potentials. For example, a facilitatory neurone might release 5-HT onto receptors on the presynaptic terminal. Activation of receptors stimulate G-protein-coupled phospholipase C and adenylate cyclase → longer depolarization on the arrival of an action potential at the presynaptic terminal

- Other stimulatory neurones might synapse onto the postsynaptic neurone. Here, there may be 'spatial summation' of EPSPs from a number of excitatory neurones that trigger an action potential in the postsynaptic neurone.

The huge variety of synaptic connections present in the nervous system contributes to its enormous complexity and facilitates signal integration (Fig. 4.7).

Convergence

- Neurone D is depolarized by stimulatory neurotransmitters from A and B and hyperpolarized by an inhibitory neurotransmitter from C.
- An action potential in D will only be propagated if depolarization in response to activation of A and B is sufficient to overcome hyperpolarization through activation of C.
- Neurone D integrates the information from neurones A, B, and C.

Divergence

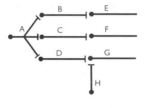

- Stimulation of Neurone A leads to release of a stimulatory neurotransmitter at synapses with B, C, and D.
- Stimulation of B and C ultimately results in stimulation of neurones E and F.
- Stimulation of D results in release of an inhibitory neurotransmitter at the synapse with G: hyperpolarization of G results in reduced sensitivity of G to stimulatory neurotransmitter released from H.

Positive feedback

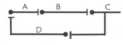

- Stimulation of Neurone A leads to release of a stimulatory neurotransmitter at synapses with B and, ultimately, C.
- An action potential in C is propagated downstream, but also feeds back to A through stimulatory neurone D.
- Stimulation of D results in further action potentials in A, and can increase the duration and/or amplitude of the response downstream of C.

Negative feedback

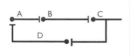

- Stimulation of Neurone A leads to release of a stimulatory neurotransmitter at synapses with B and, ultimately, C.
- An action potential in C is propagated downstream, but also feeds back to A through inhibitory neurone D.
- Stimulation of D results in hyperpolarization of A, reducing the chance of further stimulation and truncating the duration and/or amplitude of the response downstream of C.

Fig. 4.7 Signal integration through synapses.

Ganglionic synaptic transmission in the autonomic nervous system

Many of the features already described for transmission at the NMJ (Fig. 4.7) are shared by synaptic transmission in the ganglia of both the parasympathetic and sympathetic nervous systems, on account of the fact that both processes involve the neurotransmitter ACh, and postsynaptic nicotinic ACh receptors (although there are minor structural differences between neuronal and muscular nicotinic receptors). Indeed, the events leading to transmitter release at the autonomic ganglia are indistinguishable from those already described, except that far less transmitter is required to stimulate an action potential in a neurone compared to a muscle. Equally, the events at the ACh receptor are identical to those at the NMJ, except that depolarization of the postsynaptic membrane results in propagation of an action potential in the postganglionic neurone, instead of a muscle.

Drug intervention

The similarity between transmission at the NMJ and autonomic ganglia is responsible for some of the unwanted side-effects experienced with therapeutic intervention targeted at the NMJ. Nevertheless, there are some compounds with specificity for nicotinic ACh receptors at one or other of the muscular or ganglionic sites. In reality, specific ganglionic nicotinic ACh receptor stimulants or inhibitors have proved of little therapeutic use on account of their non-selective effects on both the sympathetic and parasympathetic branches of the autonomic nervous system. This lack of selectivity was responsible for the unwanted side effects of the ganglion blocker, hexamethonium—the first drug used to treat hypertension: inhibition of the sympathetic pathway was responsible for reducing blood pressure (but also caused postural hypotension), but inhibition of parasympathetic pathways caused constipation, long-sightedness, impotence, urine retention, and dry mouth. This drug is no longer used in practice, although trimetaphan is a very short-acting ganglion blocking drug that is occasionally used in specific surgical operations.

Synaptic transmission at the postganglionic parasympathetic nerve terminal

ACh is also the neurotransmitter released by postganglionic parasympathetic nerves, but it acts upon ACh muscarinic receptors on the postsynaptic membrane of the relevant tissue. To date, five subclasses of muscarinic receptor have been identified, but only three have been well characterized:

- M_1 receptors are found in the CNS, autonomic, and enteric ganglia (neural M_1), and on parietal cells that secrete gastric acid (gastric M_1)
- M_2 receptors ('cardiac') are found in the heart (primarily the atria) and on presynaptic neurones
- M_3 (glandular) receptors are found on exocrine glands, smooth muscle, and vascular endothelium.

All muscarinic receptors are G-protein coupled (⊃ pp.56–7):

- Activation of M_1 and M_3 receptors activates G-protein-coupled phospholipase C to generate IP_3 and diacylglycerol → cellular activation
- Activation of M_2 receptors inhibits G-protein-coupled adenylate cyclase → reduced rate and force of contraction in the heart.

Clinical use of muscarinic agonists and antagonists

Unlike drugs that act on ganglionic nicotinic receptors, those that act selectively on muscarinic receptors are selective for the parasympathetic nervous system. In addition, the existence of structurally distinct receptor subclasses raises the possibility of further selectivity.

In reality, however, only two muscarinic agonists are in regular clinical use:

• Pilocarpine can be used to reduce the intraocular pressure caused by glaucoma (⇨ OHCS11 p.331) by causing activation of the constrictor pupillae muscle (the iris), increasing the size of the pupil and allowing aqueous humour to flow more easily out through the canal of Schlemm. This is a long-acting drug (effects last for a day) and it is administered directly onto the surface of the eye

• Bethanechol is very occasionally used to aid bladder emptying or to ease constipation.

However, a number of muscarinic receptor antagonists are in clinical use:

• Atropine was the first muscarinic antagonist to be characterized as the active ingredient of deadly nightshade (*Atropa belladonna*). Locally delivered atropine used to be employed to dilate the pupils for ophthalmic inspections (⇨ OHCM10 p.70), but was accompanied by long-sightedness caused by inhibition of contraction of the ciliary muscle. Cyclopentolate and tropicamide (⇨ OHCM10 p.70) are preferred for this function nowadays. The antagonistic effects of systemically administered atropine are widespread, from the inhibition of secretions and sweating, to an increase in heart rate, bronchodilation, and constipation, reflecting the non-selective nature of this drug for different muscarinic receptor subtypes. These peripheral effects are accompanied by central effects that include agitation and disorientation. Hyoscine has very similar peripheral effects, but has sedative properties at low doses. It is sometimes used to treat motion sickness. Atropine methonitrate is an analogue of atropine that does not cross the blood–brain barrier, rendering it selective for peripheral parasympathetic neurones

• Ipratropium (⇨ OHCM10 p.185) is used as a bronchodilator in chronic obstructive airways disease—selectivity for bronchial receptors is imparted by delivery through an inhaler

• Pirenzepine is a relatively selective M_1 receptor antagonist that is used to reduce gastric acid secretion and help prevent gastric ulcers.

Synaptic transmission at the postganglionic sympathetic nerve terminal

With the exception of sweat glands and the adrenal medulla (Fig. 4.2) the main neurotransmitter in the sympathetic nervous system is noradrenaline. Like ACh, noradrenaline is synthesized and stored in the presynaptic nerve terminal of the postganglionic neurone and is released in response to action potentials. However, there are a number of differences between the two systems, not only in relation to the receptors that are activated but also in the re-uptake and metabolic processes.

- Synthesis from the amino acid, tyrosine, is mediated by a series of enzymes, the first of which catalyses the rate-limiting step of the process (Fig. 4.8)
- Noradrenaline is actively transported into vesicles by the monoamine transporter, to protect it from the enzyme, monoamine oxidase (MAO), which catalyses its inactivation in the cytoplasm
- Exocytosis of noradrenaline is stimulated by an influx of Ca^{2+} through voltage-sensitive Ca^{2+} channels that open in response to action potential-mediated depolarization of the nerve terminal
- Noradrenaline diffuses across the synaptic cleft, where it stimulates postsynaptic adrenoceptors, resulting in an effect that is dependent on the adrenoceptor subtype and the tissue (Table 4.2)
- Noradrenaline is reabsorbed into the presynaptic neurone via a transport protein, known as uptake 1. A similar process takes place in the surrounding tissue, primarily via a different protein—uptake 2. Uptake 1 is more selective for noradrenaline, while uptake 2 is less selective and also takes up adrenaline
- Noradrenaline reabsorbed into the presynaptic terminal is largely inactivated by MAO, although some is recycled into the vesicles. Noradrenaline and adrenaline taken into surrounding tissues via uptake 2 are inactivated by catechol-O-methyl transferase (COMT)
- As well as its action on postsynaptic receptors, noradrenaline can also activate receptors on the presynaptic terminal (α_2 adrenoceptors) that act to inhibit further exocytosis of noradrenaline (negative feedback).

Tissue effects that are mediated by sympathetic nervous activity are extremely varied and often confusing (Table 4.2). The following are examples that highlight some of the apparent contradictions:

- Noradrenaline can cause dilation of some blood vessels and constriction of others, depending on whether the vessels contain predominantly β_2 or α_1 adrenoceptors, respectively
- Stimulation of α_1 receptors in most smooth muscle causes contraction through phospholipase C activation and an increased intracellular Ca^{2+}, but in GI muscle, it causes relaxation due to activation of K^+ channels and hyperpolarization

- Stimulation of α_2 adrenoceptors located on vascular smooth muscle cells causes constriction of blood vessels. However, the same receptors located on the endothelium of blood vessels can ultimately cause relaxation of the underlying vascular smooth muscle through release of endothelium-derived relaxing factors (NO, prostacyclin). Moreover, in the GI tract, stimulation of α_2 adrenoceptors causes relaxation of smooth muscle because the receptors are located on the presynaptic terminal and their stimulation encourages inhibition of further noradrenaline release by negative feedback.

To accurately predict the effect of noradrenaline on a given tissue, it is essential to know:
- The identity of the predominating adrenoceptor subtype
- The locality of the adrenoceptors (pre- or postsynaptic, endothelium or smooth muscle cell)

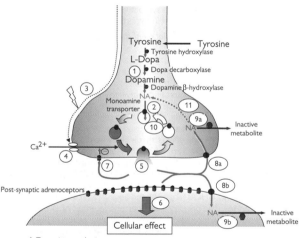

1. Transmitter synthesis
2. Transmitter uptake into vesicles
3. Action potential in presynaptic neurone
4. Voltage-gated Ca^{2+} channel activation
5. Exocytosis
6. Receptor activation in post-synaptic neurone
7. Feedback inhibition of Ca^{2+} channels (α_2 receptors)
8. (a) Uptake 1 and (b) uptake 2
9. Transmitter deactivation by (a) MAO and (b) COMT
10. Vesicular recycling
11. Transmitter recycling

Fig. 4.8 Noradrenergic transmission.

- The nature of the G-coupled second messenger system.

Cardiovascular effects

Some of the most important actions of sympathetic stimulation relate to the cardiovascular system and it is in this field that much of the pharmacological interest has centred. The main cardiovascular actions of noradrenaline and adrenaline are:

- Increased heart rate and force of contraction (β_1 receptors)
- Increased renin secretion in the kidney → elevated angiotensin II synthesis, increased plasma volume, and vasoconstriction (β receptors)
- Redistribution of blood due to vasoconstriction of blood vessels in the splanchnic vascular beds (primarily α_1 receptors) and dilatation of vessels supplying skeletal muscles in particular (primarily β_2 receptors)
- Reduced compliance (or increased stiffness) of large arteries (α receptors).

The overall effect of these changes in the cardiovascular system is to increase blood pressure, making the sympathetic nervous system a legitimate target for reducing blood pressure in hypertension.

Postsynaptic pharmacological modulation of the sympathetic nervous system—antihypertensive drugs
(➲ *OHPDT2* p.158.)

- α-adrenoceptor antagonists inhibit noradrenaline-mediated vascular constriction, but their effects are rapidly overcome by reflex increases in heart rate and force of contraction. Nevertheless, phenoxybenzamine (irreversible) and phentolamine (reversible) are non-selective α-adrenoceptor antagonists, and prazosin is a selective α_1-adrenoceptor antagonist. All have previously been used as antihypertensive drugs, but their use is limited by side effects that include postural hypotension (a dramatic fall in blood pressure on standing up due to a loss of reflex α_1-mediated vasoconstriction (➲ pp.251–2)) and nasal congestion
- β-adrenoceptor antagonists (β-blockers) (➲ *OHCM10* p.114) are effective antihypertensive (➲ *OHCM10* p.140) drugs that are also used in heart failure (➲ *OHCM10* p.136) and angina (➲ p.462; see also ➲ *OHCM10* p.116). The primary antihypertensive effect of β-blockers is now believed to be through inhibition of renin secretion in the kidney; renin is an enzyme involved in the synthesis of angiotensin II, which promotes salt and water retention together with vasoconstriction. The benefits of β-blockers in heart failure and angina, however, are primarily mediated by their inhibitory effects on β_1 receptors in the heart, → reduced heart rate and force of contraction. As a result, the cardiac workload is reduced, together with the demand for oxygen. Side effects of β-blockers include excessive slowing of the heart or even heart failure (due to inhibition of cardiac β_1 adrenoceptors), bronchospasm in asthmatic people (due to inhibition of β_2 adrenoceptors in the bronchioles), and worsening plasma lipid profile. The best-known non-selective β-blocker is propranolol; atenolol is a β_1 selective β-blocker; and labetalol is a mixed α- and β-adrenoceptor antagonist.

Table 4.2 Non-cardiovascular effects mediated by adrenoceptors

Target tissue	Effect	Adrenoceptor
Bronchioles	Dilation	β_2
GI tract	Reduced motility (relaxation)	α, β_2
Uterus: Pregnant Non-pregnant	Contraction Relaxation	α_1 β_2
Bladder: Detrusor muscle Sphincter	Contraction Relaxation	α_1 β_2
Penis	Smooth muscle contraction (impotence)	α_1
	Vas deferens contraction (ejaculation)	α_1
Eye: Pupil Ciliary muscle	Dilation (contraction of iris muscle) Relaxation	α_1 β_2
Salivary glands	Amylase secretion	β_1
Pancreas	Insulin secretion	α_2
Kidney	Renin secretion	β
Liver	Increased plasma glucose (glycogenolysis)	α_1, β_2
Fat	Lipolysis	β_3
Skeletal muscle	Tremor	β_2
	Increased speed of contraction	β_2
	Glycogenolysis	β_2

Therapeutic uses of α and β agonists

Among the most widely used adrenoceptor agonists in clinical use is the endogenous catecholamine, adrenaline. Adrenaline is more β-adrenoceptor selective than noradrenaline and it is used in the following crises:

- In cardiac arrest ($\bullet$ OHCM10 inside back cover of handbook): routinely administered during cardiopulmonary resuscitation (CPR) to help stimulate the heart (β_1 adrenoceptors)
- In anaphylactic shock ($\bullet$ OHCM10 pp.794–5): alleviates bronchoconstriction through β_2-adrenoceptor stimulation and helps prevent cardiovascular collapse, primarily through cardiac β_1-adrenoceptor stimulation.

More specific adrenoceptor agonists include:

- α_1-adrenoceptor agonists (e.g. phenylephrine) are primarily used in nasal decongestants, where they act to inhibit mucus secretion through α_1-adrenoceptor activation. There is a theoretical risk of hypertension through vascular constriction
- α_2-adrenoceptor agonists (e.g. clonidine—partial agonist) have been used in the past as antihypertensive drugs. Their primary action is through activation of the negative feedback inhibitory mechanism for noradrenaline release through presynaptic α_2 adrenoceptors (Fig. 4.7 and Table 4.3)
- β_1-adrenoreceptor agonists (e.g. dobutamine) improve cardiac contractility and are used in cardiogenic shock ($\bullet$ OHCM10 p.802). They can promote cardiac arrhythmias by increasing the heart rate ($\bullet$ p.450)
- β_2-adrenoceptor agonists (e.g. isoprenaline, salbutamol) are used in some inhalers to promote bronchodilation in people with asthma ($\bullet$ OHCM10 p.178). Isoprenaline (non-selective β agonist) has been largely superseded by salbutamol (β_2-selective agonist)
- β_3-adrenoceptor agonists (e.g. BRL 37344) are being developed as treatments for obesity. They act to accelerate β_3-mediated lipolysis (breakdown of fat; see Table 4.2).

Presynaptic pharmacological modulation of the sympathetic nervous system

Modulation of the synthesis, storage, release, or metabolism of noradrenaline is an alternative means of modifying the effects of sympathetic nervous stimulation. However, the obvious disadvantage of this approach is that it does not carry the specificity of modulation of postsynaptic receptors: the effects are global and impact both α- and β-adrenoceptor-mediated effects. As a result, this type of modulation is often associated with a wide range of more severe side effects. Nevertheless, drugs that inhibit synthesis storage and release of noradrenaline have been used in the past to treat hypertension, and some are still used in specific conditions (Table 4.3).

Table 4.3 Drugs that affect the sympathetic nervous system

Target	Effect	Note	Side effects	Examples
Synthesis	• Inhibit enzymes involved in noradrenaline synthesis (e.g. tyrosine hydroxylase) • Introduce an alternative substrate into the synthetic pathway to generate an analogue of noradrenaline that does not/activate receptors ('false transmitter')	• Former antihypertensive therapy, sometimes used in phaeochromocytoma (⊙ OHCM 0 p.228) (tumour—induced increase in adrenaline production) • Former antihypertensive therapy	• Postural hypotension • Postural hypotension	• α-methyl tyrosine • Methyl-DOPA (generating α-methyl noradrenaline, false transmitter)
Storage in vesicles	• Inhibit monoamine transporter	• Former antihypertensive agent	• Postural hypotension, depression	• Reserpine
Release into synapse	• Inhibit action potential/depolarization • Stimulate α_2-mediated negative feedback system • Inhibit α_2-mediated negative feedback system	• Former antihypertensive agent • Rarely used antihypertensive agent • Experimental tool, stimulant, aphrodisiac	• Postural hypertension, diarrhoea, impotence	• Guanethidine • Clonidine • Yohimbine
Uptake/metabolism	• Inhibit uptake 1 mechanism • Inhibit monoamine oxidase	• Prolonged activity of noradrenaline in synapse • Reduced metabolism of noradrenaline ir presynaptic neurone	• Hypertension; dependence (addiction) • Cheese reaction (hyper-tension due to sympathomimetic effects of tyramine found in fermented cheese and yeast extract)	• Tricyclic antidepressants, • Amphetamines (e.g. metamphetamine 'speed') (also inhibit uptake 1) • Selegiline

Non-adrenergic, non-cholinergic nerves in the peripheral nervous system

ACh and noradrenaline have long been recognized as the major neurotransmitters of the somatic and autonomic nervous systems. However, many of the neurones in the enteric nervous system that is associated with the GI tract, the abdominal organs, and the genitalia, do not release either of these neurotransmitters and are known as non-adrenergic, non-cholinergic (NANC) nerves. The elusive NANC neurotransmitter has now been identified as the inorganic free radical, NO, which also acts as a neuromodulator in the CNS and has important roles in the cardiovascular and immune systems (⊃ pp.468, 900).

NO as a NANC transmitter in the peripheral nervous system

- Chemically, NO is very different from noradrenaline and ACh. Its free radical status means that it reacts very rapidly with a wide range of biological molecules and has a short biological half-life
- It cannot be stored like other neurotransmitters; instead it is synthesized on demand by the neuronal isoform of the enzyme, NO synthase (nNOS). Synthesis is highly regulated by the intracellular levels of Ca^{2+} in the neurone, which is ultimately dependent on depolarization of the presynaptic terminal in response to action potentials (Fig. 4.9)
- Unlike conventional neurotransmitters, NO does not activate a receptor on the postsynaptic membrane; instead, it diffuses readily through the membrane and stimulates soluble guanylate cyclase, an enzyme which catalyses the conversion of guanosine triphosphate (GTP) to the cyclic nucleotide, cyclic guanosine monophosphate (cGMP)
- There is no specific pathway for the degradation of NO because it is rapidly inactivated in biological media through oxidation to nitrite (no_2^-) and nitrate (no_3^-)
- cGMP mediates the relaxation of smooth muscle via cGMP-dependent protein kinases, ultimately leading to phosphorylation of myosin light chains in the smooth muscle cells. The functions of NO in the PNS primarily involve relaxation of smooth muscle in the stomach, intestine, and the corpus cavernosum (penile erectile tissue)
- Sildenafil (Viagra®) is a drug that is used to overcome impotence. It prevents the deactivation of cGMP to 5´GMP by an enzyme called phosphodiesterase V, which is abundant in the corpus cavernosum.

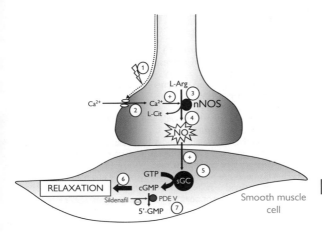

1. Action potential
2. Ca^{2+} influx
3. Ca^{2+}/calmodulin-mediated stimulation of neuronal nitric oxide synthase (nNOS)
4. NO synthesis
5. Diffusion to smooth muscle and stimulation of soluble guanylate cyclase (sGC)
6. cGMP stimulates vascular relaxation
7. cGMP inactivated by phosphdiesterase V (PDE V)

Fig. 4.9 Non-adrenergic, non-cholinergic neurotransmitter in the peripheral nervous system.

Skeletal muscle

Structure

Skeletal muscle is striated ('striped').

- Skeletal muscle cells (or myofibres) are made up of a large number of cylindrical myofibrils
- Myofibrils are divided into sarcomeres, containing interdigitating myofilaments (Fig. 4.10):
  - Sarcomeres are defined by Z lines
  - Thick filaments are ~10nm wide and 1.5µm long and comprise two heavy chains of myosin:
    - Myosin chains have entwined α helix tails, which split open at a hinge region, and globular heads. Associated with the head is the regulatory light chain of myosin
  - Thin filaments are ~5nm wide and 1µm long and made primarily of two intertwined α-helix strands of actin. Thin filaments also contain tropomyosin and troponin:
    - Two α helices of tropomyosin protein twist around the grooves formed by the actin. Tropomyosin obstructs myosin binding sites in the actin groove
    - Troponin is a trimer: troponin T binds to tropomyosin; troponin I binds to actin; and troponin C binds to Ca^{2+} ions (Fig. 4.11)
  - Z lines contain α-actinin, which anchors the thin filaments. Thick filaments lie between the thin filaments
  - The partial interdigitation confers the appearance of light and dark banding on the sarcomere:
    - Light I bands = portions of actin that do not overlap with myosin
    - Dark A bands = overlap of actin and myosin filaments
- ATP-energized movement of thin filaments over thick filaments underlies skeletal muscle contraction:
  - When sarcomeres contract, the I bands shorten but the A band length remains unchanged (Fig. 4.10).

Excitation–contraction coupling

- Skeletal muscle fibre E_m is around –90mV. Action potentials are initiated by ACh release from motor nerves and last for 2–4ms. The initiation of an action potential in the muscle fibre membrane triggers muscle contraction, by elevating intracellular $[Ca^{2+}]$
- Excitation spreads along the muscle membrane and into transverse (T) tubule invaginations at the boundary between the I and A bands which penetrate deep into the myofibre and closely abut pairs of cisternae of SR to form triads
- The muscle membrane contains voltage-activated L-type Ca^{2+} channels (also called dihydropyridine or DHP receptors) arranged in clusters of four
- These clusters are adjacent to Ca^{2+} channels in the SR membrane, called ryanodine receptors. Each DHP receptor interfaces with one of the four subunits of the ryanodine receptor

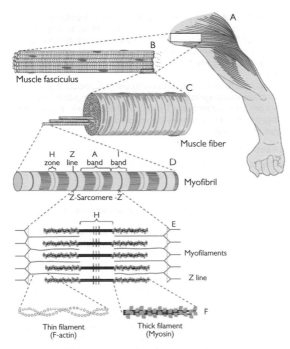

Fig. 4.10 The organization of skeletal muscle at various degrees of magnification.
Reproduced with permission from Pocock G, Richards CD (2006). *Human Physiology: The Basis of Medicine*, 3rd edn, p85. Oxford University Press.

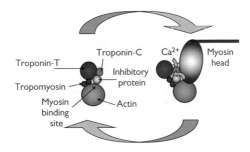

Fig. 4.11 Ca²⁺-mediated regulation of actin–myosin interaction.

- The arrival of the action potential in the T tubule membrane opens DHP receptors and Ca^{2+} ions enter the muscle fibre, although this is not essential for muscle contraction. The conformational change upon channel opening, however, initiates a conformational change in the ryanodine receptor, which opens it and releases Ca^{2+} ions from the cisternae
- Ca^{2+} ions bind to low affinity sites on troponin C, inducing a conformational change in the protein (Fig. 4.11). Troponin I moves away from actin and troponin T displaces tropomyosin. Binding sites for myosin on actin monomers are revealed, and actin–myosin cross bridges are formed.

Cross-bridge cycling

The sliding of filaments requires cross-bridge cycling (Fig. 4.12):
- ATP binds to the head of the heavy chain, disengaging myosin from actin
- ATP is hydrolysed by the myosin head. The head pivots to attain a 90° orientation to the thin filaments
- The head forms a cross-bridge with an actin monomer two positions along the filament
- Inorganic phosphate is released from myosin and the head swivels through 45°, drawing the actin past the myosin filament
- ADP is released and the cycle is complete
- The cycle repeats for as long as $[Ca^{2+}]$ remains elevated and actin sites are accessible to myosin. $[Ca^{2+}]$ levels fall when stimulation of the muscle stops, with Ca^{2+} ions taken back up into the SR, or extruded from the cell.

In the short term, ATP is rapidly resynthesized by transfer of phosphate from phosphocreatine to ADP. Longer-term recovery of ATP levels can be achieved through production of pyruvate from glycogen stores. This anaerobic process yields some ATP, which explains why muscles can cope with short-term deprivation of oxygen. Pyruvate can also be used in oxidative metabolic pathways to generate greater amounts of ATP.

Length–tension relation

- The relation between length and active tension is biphasic in skeletal muscle
- Increasing or decreasing sarcomere length from normal reduces the number of cross-bridges which can be formed: myosin and actin filaments fail to overlap at increased lengths and opposing actins obstruct each other in shortened conditions. In isometric contraction, tension is generated by the fibre without any change in its length
- In isotonic contraction, the muscle contracts against a constant load and shortens.

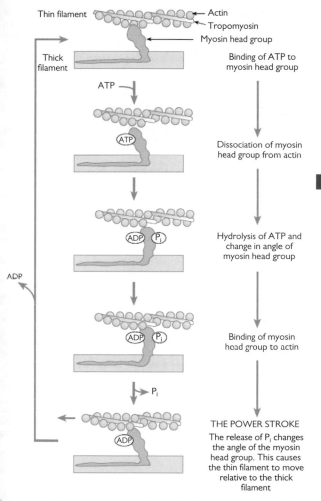

Fig. 4.12 A schematic representation of the molecular events responsible for the relative movement of the thin and thick filaments of a strained muscle.

Reproduced with permission from Pocock G, Richards CD (2006). *Human Physiology: The Basis of Medicine*, 3rd edn, p87. Oxford University Press.

Force generated in muscle can be increased by a quick-fire sequence of twitches, where a second action potential fires before Ca^{2+} levels have fallen to resting levels and the muscle fibre has relaxed (Fig. 4.13). This summation is frequency dependent: at very high frequencies a state of tetany will exist. The contraction of whole muscles is enhanced by recruitment of additional motor units.

The rate of force development can vary between fibres:

- Slow twitch ('red') fibres are prevalent in postural muscles. They have a profuse blood supply, are rich in myoglobin, and can maintain long contractions without fatigue
- In contrast, fast twitch ('white') fibres in precision muscles are glycolytic, perform oxidative metabolism, and have short twitch durations.

Motor units

Motor units are functional units of contraction, consisting of an α-motor neurone and all the fibres that it innervates. The fibres of a particular motor unit are of the same type.

Skeletal muscle contraction is controlled by α-motor neurone activity. When stimulated by a threshold action potential, fibres of a motor unit contract in an all-or-nothing fashion. Contraction of the muscle is graded in a number of ways, which include initial stretch of the muscle fibres (the length–tension relationship (Fig. 4.14) which shows an increase in developed tension with stretch, up to an optimal length); increased frequency of action potential stimulation; and recruitment of additional motor units.

Muscle spindles and *Golgi tendon organs* sense changes in length and tension of muscle fibres. Muscle fibres consist of extrafusal fibres (responsible for contraction) and intrafusal fibres (important for sensing muscle stretch). The intrafusal fibres are innervated by type Ia and type II sensory afferents and make up the muscle spindle.

Stretch of the muscle spindle results in reflex muscle contraction (Ia-α monosynaptic reflex arc). During volitional muscle contraction, resulting from somatic α-motor neurone firing, the spindle would shorten. This would result in counter-productive reflex relaxation of the muscle; to prevent this, there is concomitant γ-motor neurone activation, which contracts the spindle.

Golgi tendon organs are located within tendons of groups of muscle fibres. They are innervated by Ib sensory afferents. Increased muscle tension results in activation of Ib afferents which, via an interneurone, inhibits motor neurone activity.

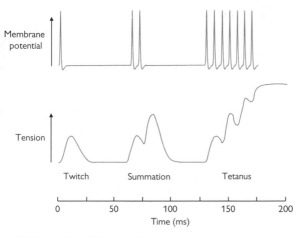

Fig. 4.13 Contraction of skeletal muscle.

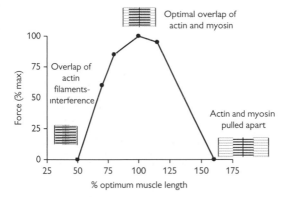

Fig. 4.14 Length–tension relationship for muscle contraction.

Cardiac muscle

Cardiac muscle is another striated muscle, but is different from skeletal muscle in a number of ways. The most striking difference is that cardiac muscle initiates contractions itself: it is myogenic. It does not require nervous stimulation (hence the viability of heart transplantation), but its activity can be modulated by nervous impulses.

Structure

- In contrast to skeletal muscles, fibres are branched and interdigitate
- At Z lines, ruffled cell membranes abut each other to form an intercalated disc, which facilitates transmission of mechanical forces from one fibre to the next
- The membranes in intercalated discs contain gap junctions. These promote the spread of electrical excitation between neighbouring fibres
- A T-tubule system is again present, with infoldings of cell membrane present at the Z lines, although there are fewer, albeit more pronounced, tubules.

Cardiac action potentials

- Excitation is initiated in a cluster of cells in the right atrium—the sinoatrial node (SAN) or pacemaker region
- This rhythmically generates spontaneous action potentials which spread throughout the atria within 40ms
- The atria are separated from the ventricles by a non-conducting fibrous ring of tissue: excitation can spread into the ventricles only at the atrioventricular node (AVN)
- After a 100ms delay, the excitation passes through the bundle of His and Purkinje fibres (30ms) and then throughout the ventricular muscle (30ms). Slowing the wave of excitation at the AVN allows time for blood to move from atria to ventricles
- E_m in cardiac myocytes during diastole is around −80mV
- The spread of depolarization through the cardiac muscle initiates action potentials
- Ventricular action potentials are far longer (200ms) and more complicated than those in skeletal muscle (Fig. 4.15). The events underlying the action potential in a ventricular cell can be summarized (Fig. 4.16):
  - Phase 0: rapid upstroke and overshoot to around +20mV. Voltage-activated Na^+ channels open at around −65mV and inward Na^+ current depolarizes the cell
  - Phase 1: an immediate incomplete repolarization towards 0mV. Na^+ channels inactivate and, at positive voltages, the background K^+ conductance of the membrane decreases (rectification)

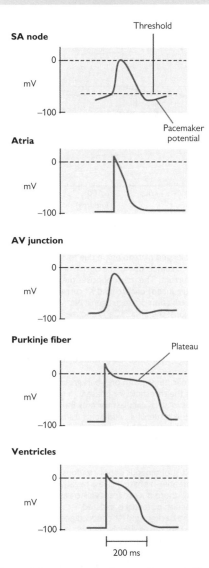

Fig. 4.15 Characteristic appearance of action potentials recorded from various types of myocardial cell. AV, atrioventricular; SA, sinoatrial.

Reproduced with the permission from Pocock G, Richards CD (2006). *Human Physiology: The Basis of Medicine*, 3rd edn, p267. Oxford University Press.

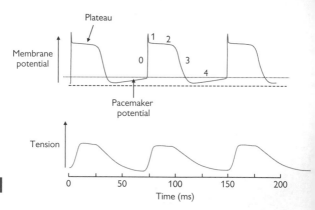

Fig. 4.16 Contraction of cardiac muscle and its association with cardiac action potentials. 0–4 (top panel) refers to phases 0–4 described in the text.

- Phase 2: a prolonged plateau phase due to slower opening of voltage-activated Ca^{2+} channels during the upstroke (at around –45mV) and inward Ca^{2+} current. The influx opposes the small hyperpolarizing effect of the (reduced) background K^+ current. The plateau is also sustained by activation of electrogenic $Na^+ \times Ca^{2+}$ exchange to extrude Ca^{2+} ions: the $3Na^+{:}1Ca^{2+}$ ratio means that each transport cycle is associated with influx of one net positive charge which has an additional depolarizing effect
- Phase 3: repolarization. Ca^{2+} channels close after 100ms and a K^+ current (through slowly activating voltage-activated K^+ channels) dominates
- Phase 4: diastolic phase, with only background currents, where E_m drifts towards the threshold for phase 0
- Action potentials in atrial cells last for only 8ms, and phase 2 is far less evident. This is the result of an additional transient K^+ current, activated at potentials positive to –30mV, which contributes to the repolarizing currents from an early stage.

Origin of the pacemaker action potential
- SAN cells exhibit spontaneous depolarizations
- The diastolic potential in phase 4 is less negative than in ventricular cells, and is not steady: there is a progressive increase in E_m from –65mV to –45mV, then an action potential is initiated
- SAN cells lack the background K^+ conductance which stabilizes E_m, making them susceptible to depolarization
- There are three contributions to the depolarization:
  - Activation of the so-called 'funny' current (Na^+ ion influx through a non-selective channel): hyperpolarization in phase 3 of the preceding action potential activates the current. This starts the depolarization but closes the channel

- • Inactivation of the outward K$^+$ current as phase 3
    hyperpolarization ends
  • Increasing inward Ca^{2+} current as depolarization opens Ca^{2+} channels.
- At threshold, a slow phase 0 upstroke occurs. This represents opening
  of activating voltage-activated Ca^{2+} channels: voltage-activated Na$^+$
  channels make little or no contribution to the upstroke
- There is no phase 2 plateau in the SAN cell action potential.
  Neurotransmitters modulate the currents determining the slope of
  phase 4 depolarization, so altering heart rate.

AVN cells have similar properties to SAN cells. They have latent rhythmicity
and can assume a pacemaker role in the absence of impulses from the SAN.
Purkinje fibres exhibit action potentials broadly similar to ventricular cells.
'Funny' currents are also present in Purkinje fibres, which can initiate slow
pacemaking when the AVN is blocked.

Excitation–contraction coupling

- Contraction initiated by the action potential lasts for around 300ms
- The Ca^{2+} -dependent sliding filament mechanism of contraction is
  identical to that in skeletal muscle
- In contrast to skeletal muscle, there is an absolute requirement for Ca^{2+}
  influx from the extracellular solution for excitation–contraction coupling.
  Ca^{2+} ions which enter during the plateau phase of the action potential
  activate Ca^{2+} channels in the SR membrane to release further Ca^{2+}: Ca^{2+}
  -induced Ca^{2+} release (CICR)
- Cross-bridge formation is not optimized in cardiac muscle: increasing
  fibre length enhances tension generated: one explanation of Starling's
  law of the heart (→ p.458). CICR can be exploited to regulate
  contraction: neurotransmitters alter the magnitude of Ca^{2+} entry and
  thereby modulate force of contraction
- Contraction is terminated by reuptake into the SR or extrusion across
  the cell membrane. The plateau phase of the action potential prevents
  tetanic contractions and ensures the heart pumps.

Smooth muscle

Smooth muscle (Figs 4.17, 4.18) is a non-striated muscle found within the walls of blood vessels, in the respiratory, GI, urinary, and reproductive tracts, in the piloerector muscles associated with hairs in the skin, and in the ciliary muscle and iris of the eye.

It is different in its structure to striated muscles:

- Although actin and myosin are present, they are not organized into parallel arrays
- Dense bodies (or attachment plaques) take the place of Z lines
- These contain α-actinin, which binds actin
- Tropomyosin is present but troponin is absent
- Intermediate cytoskeletal filaments assist in transmitting force between cells
- A SR Ca^{2+} store exists but is less highly developed.

Smooth muscle contractions are slow to establish but are sustained with low energy consumption for long periods. Innervation is derived from the autonomic nervous system. Smooth muscles have some degree of tension even at rest—contractions augment basal tone.

Smooth muscle can be divided into:

- Visceral or unitary smooth muscle:
  - Large sheets of cells with common innervation connected by gap junctions which function as low-resistance electrical connections and permit coordinated contraction
  - Spontaneous contractions can occur; stretch increases tone
  - Contractions primarily result from circulating hormones but can be modulated by nervous stimuli
  - Found in walls of visceral organs (e.g. GI tract, blood vessels, airways)
- Multi-unit smooth muscle:
  - Fibres receive individual innervation and act independently
  - Spontaneous contractions do not occur. Contractions initiated primarily by nervous triggers, modulated by hormones
  - Functions are more like those of skeletal muscle, with fine graded contractions (e.g. iris, piloerector muscles of skin).

Electrical activity

- E_m in smooth muscle is not steady: there is constant drift (a few mV). Minimum values are around $-50mV$
- Visceral smooth muscle can exhibit pacemaker activity, similar to the cardiac SAN. Pacemaker regions shift around within the muscle
- In some cases, spontaneous slow waves of graded depolarization are seen, with spike potentials superimposed at more positive potentials. Oscillating but opposing membrane currents for Ca^{2+} and K^+ probably explain slow waves
- Action potentials have slower upstroke and last longer than skeletal muscle. They can be spiked or exhibit a plateau.
  - Upstrokes represent opening of voltage-activated Ca^{2+} channels; repolarization is due to opening of voltage-activated and Ca^{2+}-activated K^+ channels
  - Multi-unit smooth muscle does not usually exhibit action potentials.

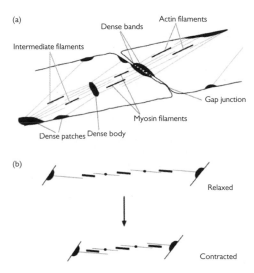

Fig. 4.17 The organization of the contractile elements of smooth muscle fibres. (a) Note the points of close contact for mechanical coupling (dense bands) and the gap junction for electrical signalling between cells. (b) A simple model of the contraction of smooth muscle. As the obliquely running contractile elements contract, the muscle shortens.

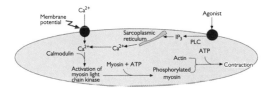

Fig. 4.18 Excitation–contraction coupling in smooth muscle. As in cardiac and skeletal muscle, the development of tension is regulated by calcium which can enter the sarcoplasm via either voltage-gated calcium channels in the plasma membrane or the intracellular calcium stores in the sarcoplasmic reticulum. Calcium binds to calmodulin, which regulates the interaction between actin and myosin via myosin light-chain kinase.

Contraction

Increases in [Ca^{2+}] initiate contraction and can originate from:

- Influx of Ca^{2+} ions during slow waves or action potentials
- IP_3-mediated Ca^{2+} mobilization from the SR in response to neurotransmitters, hormones, or E_m changes

Note that changes in E_m are not an absolute requirement for smooth muscle contraction.

Contraction is a slow sequence of events:

- Ca^{2+} ions bind to calmodulin (the smooth muscle functional homologue of troponin C)
- In the absence of Ca^{2+}, actin–myosin binding is prevented by the myosin light chain (MLC), not by tropomyosin
- Ca^{2+}-calmodulin activates MLC kinase. Phosphorylation of MLC relieves the inhibition; hydrolysis of ATP by the myosin head occurs; cross-bridge cycling begins
- A phosphatase dephosphorylates MLC when [Ca^{2+}] falls.

Some smooth muscle can exist in a latch state. Tension is generated for long periods, yet [Ca^{2+}] is lower (but still above resting levels), MLC phosphorylation is reduced, and energy consumption is minimal. The interaction of two Ca^{2+} binding proteins, caldesmon and calponin, with thin filaments is believed to alter the kinetics of actin–myosin binding.

Demyelination of neurones

Neurones are surrounded by glial cells which produce a myelin sheath around them. This sheath has two functions:

- It provides nutrition for the neurones
- It speeds the transmission of signals along the nerve since depolarization 'jumps' between the gaps between Schwann cells (the nodes of Ranvier) rather than all the way along the cell membrane.

Thus, if a neurone loses its myelin sheath, there will be severe malfunction and possible death of that neurone. There are several human diseases which are characterized by demyelination.

Central nervous system

Multiple sclerosis

($\circleddash$ *OHCM10* pp.496–7.) This is characterized by demyelination in any area of the brain and usually progresses with episodes of acute demyelination occurring at different times. The mechanism by which this demyelination occurs is not known but there are T-lymphocytes and macrophages around the demyelinated areas and, in animal models, demyelination can be induced by the transfer of T-lymphocytes from one animal to an unaffected animal.

Acute disseminated encephalomyelitis

This disease usually follows a viral infection and is characterized by a single widespread episode of demyelination affecting most of the white matter. It is possible that this represents an acute autoimmune reaction to myelin. There is a high fatality rate (~20%), but the other cases usually recover completely.

Central pontine myelinolysis

($\circleddash$ *OHCM10* p.672.) A disease with demyelination in the distribution, as described by its name. It is associated with alcoholism, severe electrolyte imbalances, and liver transplantation. One possible mechanism is rapid correction of low blood sodium (hyponatraemia). It presents as a sudden quadriplegia.

Box 4.1 summarizes the treatments for demyelination diseases.

Box 4.1 Treatments and drug therapies for demyelination diseases

Multiple sclerosis is not yet curable. The acute effects during attacks are managed by high-dose corticosteroid administration, while the use of a range of anti-inflammatory (interferon beta 1a and 1b, natalizumab), immunosuppressant (mitoxantrone), and immunomodulator (glatiramer acetate) drugs show some benefit in reducing the frequency of attacks.

Acute disseminated encephalomyelitis is also treated with high-dose anti-inflammatories (corticosteroids).

Central pontine myelinosis is not treatable: avoiding correction of hyponatraemia that is too rapid would prevent the majority of cases.

Peripheral nervous system

Guillain–Barré syndrome

(➔ *OHCM10* p.702.) The alternative name of this disease, acute inflammatory demyelinating polyradiculoneuropathy, describes the pattern of demyelination. The disease often occurs after an acute viral illness and presents as an ascending neuropathy which can progress from loss of plantar reflexes through to paralysis of respiratory muscles. The pathogenesis of the disease is still uncertain but an autoimmune reaction precipitated by the viral infection is currently favoured.

Box 4.2 summarizes the management of patients with Guillain–Barré syndrome.

Box 4.2 Management of Guillain–Barré syndrome

Patients may require many weeks of nursing in intensive care units when their breathing is affected, but go on to a complete recovery, as evidenced by some professional sports people returning to their sport after an episode of this illness.

Charcot–Marie–Tooth disease

(➔ *OHCM10* p.696.) A rare inherited peripheral neuropathy that is characterized by recurrent demyelination and remyelination of peripheral nerves.

Myasthenia gravis

(● *OHCM10* p.512.) Myasthenia gravis is an autoimmune disease in which autoantibodies inhibit neuromuscular transmission (● p.242).

The autoantibody is directed against the ACh receptor of the postsynaptic membrane of the NMJ. Binding of the antibody to this receptor causes an increased rate of degradation of this protein (rather than a simple blocking of the ACh binding site as originally thought) and damage to the postsynaptic membrane by complement binding.

Clinical effects

- Drooping eyelids (ptosis) and double vision due to weakness of the extraocular muscles
- Generalized muscular weakness—characteristically fluctuates
- Sensory and autonomic nervous systems are normal.

Box 4.3 summarizes the treatments for myasthenia gravis.

> **Box 4.3 Treatments and drug therapies for myasthenia gravis**
>
> - Anticholinesterase agents—which lead to an increased concentration of ACh at the postsynaptic membrane and thus maximal occupancy of the remaining ACh receptors. Administration of these is also a useful diagnostic test for the disease—reversing weakness, such as ptosis, almost immediately after intravenous injection
> - Thymectomy—a number of patients benefit from thymectomy, presumably as a result of modulation of the autoimmune activity
> - Supportive care including ventilation during acute exacerbations of the disease.

Lambert–Eaton myasthenic syndrome (LEMS) is caused by the presence of circulating antibodies against voltage-gated Ca^{2+} channels. It can be a paraneoplastic disorder, usually in association with lung cancer, or can arise without cancer as part of a generalized autoimmune state.

Motor neurone disease

(● OHCM10 p.506.) This is a rare disease but one with a high public profile in the UK, mainly due to sufferers of the disease seeking to legally end their lives by euthanasia.

Motor neurone disease is characterized by a loss of neurones in both the anterior horns of the spinal cord and the corticospinal tracts, producing a mixture of upper and lower motor neurone signs.

Clinical features

- Early—weakness of fine hand muscles, cramping and spasticity of arms and legs
- Later—gross motor weakness, weakness of respiratory muscles
- Course of disease—progressive, with death within a year or two usually due to recurrent respiratory infections.

Pathogenesis

- Inherited—about 10% of cases are familial and, in some of these, there is a gain of function mutation in the copper-zinc SOD gene on chromosome 21
- Sporadic—there is no clear mechanism for the disease. The motor neurones appear to die individually by apoptotic processes but the factors which precipitate this have not been identified.

Box 4.4 summarizes the treatments for motor neurone disease.

Box 4.4 Treatments and drug therapies for motor neurone disease

- There is no cure for motor neurone disease
- Riluzole is the only pharmaceutical that is used in this arena—its beneficial effects are limited in terms of prognosis and symptomatic relief. The mechanism of action of riluzole is controversial, but it seems to inhibit the actions of the neurotransmitter glutamate, perhaps through activation of uptake
- Treatment of patients with motor neurone disease revolves around the activity of a range of healthcare professionals to help with symptomatic relief (physiotherapists, speech therapists, occupational therapists among others).

Spinal cord damage

Neurones in the central and peripheral nervous systems do not regenerate if the cell body dies. The axons of peripheral nerves do regrow if the surrounding myelin sheath is still intact (at a rate of ~1mm a day). This lack of regenerative capacity has profound consequences for people who experience injury to their spinal cord since such injuries usually cause disruption of the whole cord, so there is no regrowth of neurones to connect across the deficit. This leaves these people with complete paralysis of any muscles supplied by neurones below the level of the injury.

Causes of spinal cord damage

- Trauma (➔ *OHCS11* pp.558–66)—is the most common cause and is usually sustained in car/motorbike accidents or in a sporting context, especially in the front row of the scrum in rugby union and falling from a height (horse riding, diving). There is usually transection of the cord by a displaced vertebra. The usual features of damage in the nervous system, such as initial inflammation followed by gliosis, are seen in the damaged cord
- Syringomyelia (➔ *OHCM10* p.516)—is characterized by development of a fluid-filled cavity in the centre of the spinal cord which causes pressure atrophy of neurones. Since the first neurones to be affected by this atrophy are the crossing anterior spinal commissural fibres, this produces the unusual neurological pattern of dissociated sensory loss of pain and temperature senses in the upper limbs. Syringomyelia may be associated with a congenital Arnold–Chiari malformation but some occur after spinal cord trauma.

Box 4.5 summarizes the management of patients with spinal cord damage.

Box 4.5 Management of spinal cord damage

At present, there are no therapies which can be used to induce neuronal regeneration across the damaged segment of cord and all treatment is directed to supporting the patients physically and emotionally, and trying to improve and utilize residual function. The most likely source of therapy is the use of pluripotent stem cells to produce neurones which can reconnect across the deficit.

Musculoskeletal system

Connective tissue

- Connective tissues are broadly responsible for providing mechanical support, protection, and definition to other specialized tissue types. They include adipose tissue, tendons, aponeuroses, fascia, ligaments, and other fibrous and elastic tissues, including parts of the haematopoietic system and the highly specialized tissues of cartilage and bone
- Significant differences exist between these tissue types, but each comprises a cellular component that is responsible for the production of an extracellular matrix of fibrous proteins and ground substance. In most connective tissue types, the extracellular matrix determines the mechanical properties of the tissue which are the result of its precise composition and arrangement
- In general, the fibrous components of connective tissue provide tensile strength, while the highly hydrated ground substance gives resistance to compression.

Connective tissue fibres

There are several different forms of connective tissue fibre. The type, concentration, and arrangement of these fibres are each important in determining the functional properties of different connective tissues.

- The most common fibrous protein is collagen, which is found in most types of connective tissue. This is a thick, strong, unbranched fibrous protein. There are many different types of collagen, each with different structural properties. The most abundant forms (e.g. types I and II) form fibrils with great tensile strength. The fibrils are formed from tropocollagen molecules, which consist of three peptide chains intertwined in a right-handed helix
- When stretched, elastic fibres return to their original size. These fibres are small, thin, and branching. In comparison to collagen, their tensile strength is low. They are abundant in the lungs, bladder, skin, and aorta, where they allow stretching without distortion or breaking of the tissue. They are formed in a three-stage process. Initially, a fibre comprised of fibrillin (a large molecule) and glycoproteins (called oxytalan) is formed. A protein called elastin is then deposited, through the oxytalan, in a disordered fashion to form elaunin. The third stage in this process involves the organization of elastin into the centre of fibril bundles to form elastic fibres
- Reticular fibres are thin fibres which are visible when stained with silver. They form a fragile network in the haematopoietic organs, spleen, liver, and lymph nodes. They filter blood and lymph and provide support for capillaries, muscle, and nerve cells.

Ground substance

Ground substance is the major 'packing' component of connective tissue and surrounds its cells and fibres.

- It consists of glycoproteins, glycosaminoglycans, and hyaluronic acid and forms an amorphous colourless, gel-like substance which is highly hydrated
- In cartilage and developing bone, the ground substance is mineralized by the deposition of calcium-containing crystals which leads to a marked alteration in its properties—particularly an increase in its strength.

Connective tissue cells

- Fibroblasts, or similar specialized cells (e.g. osteoblasts and chondroblasts) are responsible for the synthesis and turnover of extracellular matrix and are usually the most abundant cell type found in connective tissue. They are flattened cells which, when active, are specialized for the production and secretion of proteins (e.g. collagen) and other macromolecules
- Fibrocytes are quiescent fibroblasts and are usually smaller with less prominent nuclei
- Adipose cells, which share some of the features of fibroblasts, are specialized for the storage of fat in connective tissue. Where adipose cells account for the majority of cells present, the tissue is referred to as adipose tissue, which frequently has an important cushioning role as well as providing thermal insulation. As histological sections, adipose cells appear empty because the fat is dissolved during the process of staining the slides. The nuclei are usually at the periphery of the cell with a tiny band of cytoplasm around the edge of the cell
- Macrophages and leucocytes are also present in connective tissue, as well as plasma cells which are produced from the bone marrow by lymphocytes. They migrate into the connective tissue, particularly in lymphatic or respiratory tissue where they play an important role in immune defence by secreting antibodies
- Mast cells are found in connective tissue, often associated with blood vessels. They are dark staining because they contain lots of histamine granules.

Connective tissue types

In addition to more specialized connective tissue types (e.g. cartilage and bone), there are two main types of connective tissue: loose and dense.

- Loose connective tissue comprises fibroblasts, collagen fibres, and macrophages. It is the more common of the two tissue types and is characterized by an abundance of ground substance and an irregular, loose structure of collagen fibres. Loose connective tissue is found around groups of muscle cells, blood vessels, and supporting epithelia
- Dense connective tissue has less ground substance and a more limited number of cell types, as it is densely packed with collagen fibres. These can be arranged regularly or irregularly and define the dense connective tissue as regular or irregular respectively, on the basis of the collagen fibre arrangement:
  - Dense regular connective tissue is found in tendons and ligaments
  - Dense irregular connective tissue is found in the capsules of organs the dermis of the skin, and where strong support is required.
  In tendons, collagen fibres are arranged in parallel bundles with fibroblasts arranged in rows between the densely packed bundles of fibres. This provides great tensile strength against forces pulling in a straight line.

Cartilage

- Cartilage covers and protects the articulating surfaces of bone as well as forming part of the septum of the nose, the external ear, and the embryological skeleton
- It can bear mechanical stress without permanent distortion and is an efficient shock absorber. It is made up of an extracellular matrix which is produced and maintained by specialized cells called chondrocytes and chondroblasts
- Chondroblasts are mesenchymal in origin and synthesize the extracellular matrix of cartilage. When these cells become surrounded by cartilage in lacunae they are called chondrocytes (mature cartilage cells), which maintain cartilage
- Cartilage is an avascular tissue which is surrounded by vascular connective tissue. Diffusion is the only means by which nutrients can enter and waste products leave.

There are three main types of cartilage, and the difference between them is the result of differences in composition of the extracellular matrix.

- Fibrocartilage has an irregular dense collagen fibre structure. It is formed of layers of thick collagen fibres and layers of cartilage matrix. It is found in the pubic symphysis and intervertebral discs
- Hyaline cartilage is the most common type of cartilage and contains abundant type II collagen fibres and large aggregating proteoglycan (aggrecan). It forms the basic structure of bones which then undergo endochondral ossification in the embryo. In adults, it forms part of the nose, larynx, trachea, and bronchi, and covers the articular surfaces of bones. Its smooth surface allows virtually friction-free movement at articulating joints
- Elastic cartilage contains more elastic fibres in its matrix and lines the walls of the auditory canal and forms part of the ear, the larynx, and the epiglottis.

Bone

- Bone is a rigid, tough structure which is important for support, load bearing, and protection. It provides for the attachment of muscles, storage of calcium and phosphate, and contains bone marrow which is important for haematopoiesis
- Bone consists of a cellular component as well as an extracellular matrix mineralized by the deposition of hydroxyapatite (basic calcium phosphate) and is a specialized type of connective tissue
- Osteogenic cells are derived from mesenchymal tissue and are undifferentiated, pluripotent stem cells with the capacity to produce bone-forming cells. The cells remain in the periosteum and endosteum in adults and are important for the remodelling, growth, and repair of bones by the provision of new osteoblasts when required:
  - Osteoblasts secrete osteoid—the organic component of bone. This is initially uncalcified, but rapidly becomes mineralized by the deposition of hydroxyapatite
  - Osteoclasts are large, multinucleated cells which are important in bone remodelling because they resorb bone. They form Howship's lacunae which are depressions of the bone surface where resorption has occurred
  - Osteocytes are important in preserving the bone matrix. They sit in lacunae and are surrounded by the bony matrix
- Bone formation occurs through two different processes—intramembranous ossification and endochondral ossification. Intramembranous ossification occurs in bones such as the mandible, maxilla, clavicles, and most of the flat skull bones. It involves bone formation from a connective tissue matrix rather than from cartilage. The bone matrix is produced by osteoblasts. Endochondral ossification occurs in most of the rest of the bones of the body: this involves an initial hyaline cartilage structure that continues to grow. Gradually, cartilage is calcified by chondrocytes
- Osteoblasts are formed by osteoprogenitor cells which form an ossification centre within the bone. This is known as the primary ossification centre in the diaphysis (centre of the bone). Later, a secondary ossification centre forms at the epiphysis (swellings at the distal end of the cartilage framework). The epiphyseal plate region does not undergo mineralization until maturity, to allow long bone growth until adulthood. At the end of each long bone, a layer of articular cartilage covers the bone as part of the joint and comprises hyaline cartilage.

The structure of bone

- Cancellous (spongy) bone comprises many interconnecting cavities. It forms the interior and epiphyses (the growth plate ends) of long bones
- This spongy bone is surrounded by an outer layer of compact bone which is much denser (Fig. 5.1). This is much stronger and forms most of the outer layer of the bone shaft (the diaphysis)
- The bone marrow core of long bones is covered with a small amount of cancellous bone. In flat bones like those of the skull, bone forms with only a sliver of cancellous bone (diploë) sandwiched between the two plates of compact bone

- There are two types of bone—primary and secondary:
  - Primary bone comprises collagen fibres in an irregular arrangement, with a high number of osteocytes and little mineralization. This bone is usually replaced by secondary bone tissue in adults. In adults, it is only found in a few places (e.g. tooth sockets)
  - Secondary bone is formed as the Haversian system (or osteon). This structure is found in most adult bone. Collagen fibres are arranged in lamellae (rows). These are often arranged around a central canal in a concentric fashion. The central canal contains the neurovascular bundle and loose connective tissue. This canal is known as the Haversian canal. Osteocytes are found between the concentric layers of lamellae in lacunae
- Mineralization of bone prevents the diffusion of nutrients. Instead, canaliculi (channels within the bone which contain extracellular fluid) allow efficient exchange of nutrients and waste products. Osteocytes in lacunae extend their cytoplasmic processes into the canaliculi, thereby allowing communication between osteocytes and nutrient uptake from the blood
- Bone remodelling is a continuous process. This characteristic is particularly important during growth and fracture healing. Remodelling enables the shape of the bone to be maintained as it grows.

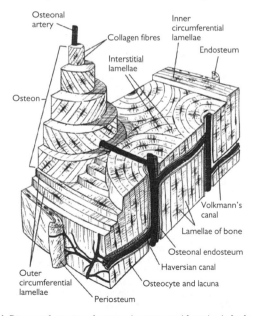

Fig. 5.1 Diagram of a section of compact bone removed from the shaft of a long bone.

Reproduced with permission from Ross MH, Kaye GI, Pawlina W (2003). *Histology: A Text and Atlas*, 4th edn. Baltimore: Lippincott, Williams and Wilkins.

Skin

- Skin is a large organ that almost completely covers the exterior surfaces of the body. It comprises a specialized epithelium which includes various glands (e.g. sweat glands, sebaceous glands, mammary glands) and associated supporting tissues
- Skin has a number of important functions including protection of the body from damaging external agents (water, infection, sunlight), detection of sensory stimuli, thermoregulation, and prevention of dehydration
- Skin is arranged as a number of layers including the epidermis (outermost layer), dermis, and subcutaneous tissue (hypodermis). The relative thickness and structure of each of these layers is dependent on the area of the body and relates to specific functional specializations.

Epidermis

- The epidermis is the surface layer of skin cells in contact with the outside world. Sweat glands and hair follicles are downgrowths of this layer
- It is a stratified epithelium with a tough keratinized upper layer. It is formed from keratin-producing cells (keratinocytes) which die, forming keratin plates (squames). The outermost layer of keratin is constantly being shed and replaced by new keratinocytes from deeper layers
- The epidermis is divided into distinct layers:
  - Stratum basale (basal layer): this layer sits on the basement membrane which separates the supporting dermis from the epidermis and is responsible for keratinocyte production. The cells are cuboidal or columnar and are attached to each other and to the underlying basement membrane. These cells are rich in ribosomes and mitochondria enabling rapid cell turnover and protein synthesis. Melanin granules are also present in pigmented skin
  - Stratum spinosum (prickle layer): this layer, which sits on the basal layer, is formed from polyhedral cells with round central nuclei. A system of intercellular bridges made of cytoplasmic projections connect these prickle cells. These projections terminate as desmosomal junctions on the cell surface
  - Stratum granulosum (granular layer): in this layer, the cells contain abundant keratohyaline granules. In the upper layers, cells become flattened and densely packed, with little cytosol
  - Stratum lucidum: this layer merges with the upper layers of the stratum granulosum and consists of extremely flattened cells in which organelles and nuclei are not readily apparent
  - Stratum corneum: the uppermost layers of the epidermis in which keratin is formed from dead cells. The waterproofing effect of this layer is a result of the hydrophobic properties of glyco phospholipids, which have a glue-like effect, sticking the dead flakes of cells together. This can be washed away as shown by the wrinkling of the skin of the hands after extended exposure to water
- There is a constant need to replenish the epidermis, as the outer surface is constantly being sloughed off. Turnover of cells from basal cells to desquamated keratin varies from site to site in the body, with traumatized sites having a faster turnover (e.g. soles of the feet: ~25 days; the back: ~45 days)

- Skin can be classified according to whether the epidermis is thin or thick. Thick skin has a much deeper stratum corneum, stratum granulosum, and stratum lucidum. It is found in areas which are most exposed to abrasive forces (e.g. palms of the hands, soles of the feet).

Other cell types within the epidermis
- Melanocytes produce the skin pigment melanin, which is responsible for skin colouring and reduces damage caused by ultra-violet radiation. They are located in the basal layer. They synthesize melanin in cytoplasmic membrane-bound granules. These granules advance along cytoplasmic processes into the cytoplasm of basal and prickle layer keratinocytes
- Langerhans cells present phagocytosed antigenic material to lymphocytes. These cells possess cytoplasmic processes which increase the surface area of the cell membrane. Langerhans cells are found in all skin layers, but are most frequently seen in the prickle layer of the skin. In inflamed skin, the number of the cytoplasmic processes increases, especially during autoimmune or allergy-related skin disorders
- Merkel's cells are sensory receptors in the skin and synapse with peripheral nerve endings. They are found as solitary cells or grouped together to form touch receptors where they are related to hair discs. They contain neuroendocrine granules.

The basement membrane
- The basement membrane comprises three layers and separates the epidermal layer from the underlying dermis
- Basal cells are attached to the outermost layer—the lamina densa—and from here, anchoring proteins cross to the lamina lucida
- Fibronectin is abundant in the zone below the lamina densa (fibroreticular lamina)
- The lower surface of the lamina densa is attached to collagen fibres in the papillary dermis by fibrils made of type VII collagen.

Dermo-epidermal junction
- The dermo-epidermal junction binds the dermis and the epidermis together. Tethering fibres pass between the two layers and the intervening basement membrane
- The rete system is a series of downgrowths of the epidermis into the dermis which increases the area of attachment between the two layers. This is minimal over the back and more extensive over the fingertips and soles where shearing forces are increased.

Dermis
- The middle layer of skin, the dermis, includes blood vessels, lymphatics, nerves and nerve endings, and epidermal appendages. These structures are embedded in a connective tissue stroma that is produced by fibroblasts
- The dermis consists of two layers:
  - The superficial loose papillary dermis, a thin layer of loosely arranged collagen and elastin fibres, containing small capillary-sized blood vessels, fine nerves, and nerve endings

- The dense reticular dermis, which contains mainly collagen and elastic fibres. The reticular dermis is thicker, forming the bulk of the dermis. It contains blood vessels, nerves, and lymphatics supplying the skin and comprises dense bands of collagen and long, thick fibres of elastin. These generally run parallel with the skin
- The dermis contains two vascular plexuses—a deep plexus in the lower reticular dermis close to the border with the subcutis and a superficial plexus in the upper reticular dermis close to the junction with the papillary dermis
- Variations in blood flow within the dermis allow the skin to participate in thermoregulation. This is controlled by the many arteriovenous anastomoses within the dermis, including the glomus bodies. Loops from the superficial plexus extend upwards to the papillary dermis to form capillaries near the basement membrane of the epidermis.

The nerve supply of the skin

There are four major specialized nerve endings detecting cutaneous sensation in skin:

- Free nerve endings detect pain, itch, and temperature and can be myelinated or unmyelinated
- Meissner's corpuscles detect touch, are found mainly on the hands and feet, and have ordered nerve endings which are confined to dermal papillae
- Merkel's cells are slow adapting touch receptors
- Pacinian corpuscles detect pressure and vibration and are encapsulated nerve endings with a characteristic structure. They are found mainly in deep dermis and subcutaneous fat on palms and soles.

The nerve supply to the skin consists of a sympathetic supply of unmyelinated nerves which control skin appendages (e.g. sweat glands) and vascular flow.

Subcutaneous tissue

The subcutis is the deepest layer of skin and varies in size over the body. Adipose tissue forms the majority of this layer. It contains a network of arteries and veins which extend upwards into the dermo-subcutaneous junction, forming a cutaneous plexus.

Sweat glands

The autonomic nervous system (⊃ 'Synaptic transmission' topics in Chapter 4) controls the secretion of sweat from eccrine sweat glands. The secretory component is found in the subcutis or deep dermis and ducts communicate with the exterior. These ducts are coiled in the deep dermis and epidermal layers with straight connections between them. Apocrine glands are stimulated during the fight-or-flight response. They are downgrowths of the epidermis with unknown function. They are concentrated in the genito-anal region and axilla.

Hair and nails

- Hair grows from hair follicles, which are invaginations of epidermal tissue. A bulbous expansion at the lower end of the hair follicle (Fig. 5.2) is called the hair bulb where specialized dermis forms the hair papilla
- Hair follicles are well supplied with small blood vessels and nerve endings
- Hair is formed of organized keratin. The hair follicle and shaft are controlled by the erector pili muscle which allows hair to stand on end
- Sebaceous glands around the follicle secrete sebum—a mixture of lipids
- Fingernails and toenails are hard plates of keratinized epithelium.

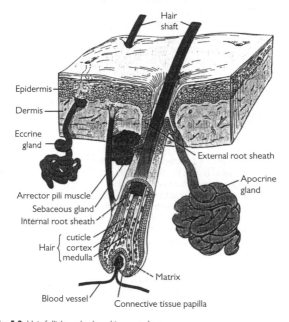

Fig. 5.2 Hair follicle and other skin appendages.

Reproduced with permission from Ross MH, Kaye GI, Pawlina W (2003). *Histology: A Text and Atlas*, 4th edn. Baltimore: Lippincott, Williams and Wilkins.

Upper limb bones

Shoulder girdle

Comprises the collar bone (or clavicle) and the shoulder blade (or scapula) (Figs 5.3, 5.4). Together, they provide a mobile base to attach the upper limbs to the trunk.

The clavicle:
- Acts as a strut to distance the scapula and upper limb from the chest wall
- Forms articulations with the sternum and with the acromion of the scapula
- Transmits forces to the axial skeleton (fractures frequently result from falls on the outstretched arm)
- Lacks marrow and ossifies in membrane.

The triangular scapula:
- Articulates with the humerus at the glenoid fossa
- Possesses a prominent ridge on the posterior surface: the spine
- Provides large, flat surfaces and roughened processes for muscle attachments
- Maintains a strong purchase on the chest wall.

The bones of the arms

Comprise the bone of the upper arm (the humerus), the bones of the forearm (the radius), and the ulna (Figs 5.3, 5.4).

The humerus:
- Acts as a mobile lever to direct forearm movement in any direction
- Consists of a head (in the shape of a half-sphere which articulates with the scapula) together with a neck and a shaft
- Articulates with the radius at the rounded capitulum
- Articulates with the ulna at the trochlea (the trochlea projects further at its medial border than at its lateral border, angling the forearm laterally with respect to the upper arm—this accounts for the carrying angle)
- Is often fractured at the 'surgical neck', below the anatomical head at the start of the shaft.

The radius:
- Possesses a head (proximal end), neck, and shaft
- Head is a thick, disc-like shape which articulates with the humerus
- Articulates with the scaphoid and lunate carpal bones at its distal end
- Is expanded at its distal end with an ulnar notch
- Is often fractured by a fall on an outstretched hand (a 'Colles fracture').

The ulna:
- Is not directly a part of the wrist joint
- Possesses a head (distal end), neck, and shaft
- Articulates with the humerus at the crescent-shaped trochlear notch
- Possesses a beak-shaped process (the olecranon) at its proximal extremity which, in combination with the humerus, locks the extended elbow joint to prevent overextension
- Articulates with the radius head at the radial notch.

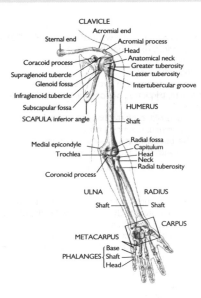

Fig. 5.3 Bones of shoulder girdle and upper limb; anterior view.
Reproduced from MacKinnon, Pamela and Morris, John, *Oxford Textbook of Functional Anatomy*, vol 1, p46 (Oxford, 2005). With permission of OUP.

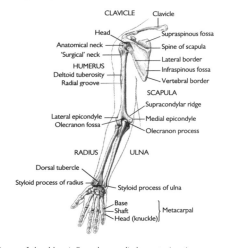

Fig. 5.4 Bones of shoulder girdle and upper limb; posterior view.
Reproduced from MacKinnon, Pamela and Morris, John, *Oxford Textbook of Functional Anatomy*, vol 1, p46 (Oxford, 2005). With permission of OUP.

The bones of the hand

Comprise eight small bones of the wrist (the carpus, Fig. 5.5) and 19 bones of the fingers (the metacarpals and phalanges).

The carpus bones:
- Are distal to the wrist joint
- Are pebble-like and arranged in two rows of four:
  - Scaphoid, lunate, triquetral, pisiform (a sesamoid bone) in proximal row
  - Trapezium, trapezoid, capitate, hamate in distal row
- Are transversely arched, creating a hollow
- Provide a flexible but firm basis on which muscles can exert their action
- Articulate distally with the metacarpal bones.

The scaphoid is prone to fracture through a fall on the outstretched hand.

The metacarpal bones:
- Are five in number:
  - The first is the bony support for the base of the thumb
  - The remaining four provide the framework for the palm of the hand
- Comprise a head, neck, and shaft
- Articulate distally with the first row of phalanges
- Are relatively immobile; only that of the thumb is truly mobile.

The phalanges:
- Are the bones of the fingers and thumb—there are three in each finger, two only in the thumb
- Comprise a base, shaft, and head
- Of the fingers are designated proximal, middle, distal
- Function as a unit, rather than as individual bones.

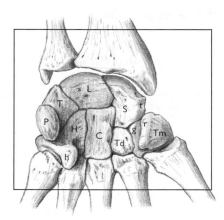

Fig. 5.5 Carpal bones; anterior view. Scaphoid (S), with its tubercle (t), waist, and proximal pole, lunate (L), triquetral (T), pisiform (a sesamoid bone) (P); distal row: trapezium (Tm), with its groove (g) and ridge (r), trapezoid (Td), capitate (C), hamate (H), and its hook (h).

Joints of the upper limb

Shoulder girdle

- Sternoclavicular joint (Fig. 5.6):
  - Between medial end of clavicle and upper lateral edge of manubrium of sternum
  - Atypical synovial joint—surfaces covered by fibrocartilage
  - Synovial capsule encloses joint—reinforced by anterior and posterior sternoclavicular ligaments, costoclavicular ligament
  - Fibrocartilaginous disc attaches to capsule; divides the joint into two cavities
- Acromioclavicular joint (Fig. 5.6):
  - Between lateral end of clavicle and medial edge of acromion of scapula
  - Atypical synovial joint
  - Weakly encapsulated
  - Stabilized mostly by coracoclavicular and acromioclavicular ligaments.

These joints transmit little of the force associated with upper limb movements, so dislocation is uncommon.

Shoulder joint

- Glenohumeral joint (Fig. 5.7)—between glenoid fossa of scapula and head of humerus:
  - Freely mobile, ball-and-socket synovial joint
  - Encapsulated—synovial capsule is thin and loose inferiorly to permit wide range of movement
  - Contains the glenoid labrum—rim of fibrocartilage attached to cavity margins
  - Capsule strengthened by tendons of the rotator cuff
  - Three glenohumeral ligaments, together with coracohumeral ligament, add stability.

Elbow joint

- Synovial hinge joint (Fig. 5.8)
- Articulations between:
  - Trochlea of humerus and trochlear notch of ulna
  - Capitulum of humerus and head of radius
  - Head of radius and radial notch (the superior or proximal radio-ulnar joint)
- Capsule envelops all these articulations
- Capsule reinforced by lateral and medial ligaments (the radial and ulnar collateral ligaments)
- Superior radio-ulnar joint held together by annular ligament—permits rotation of radial head on ulna
- Posterior dislocation by a fall on outstretched hand.

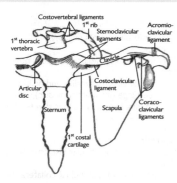

Fig. 5.6 Bones and ligaments of the shoulder girdle.

Reproduced from MacKinnon, Pamela and Morris, John, *Oxford Textbook of Functional Anatomy*, vol 1, p53 (Oxford, 2005). With permission of OUP.

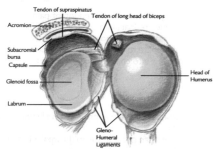

Fig. 5.7 Interior of shoulder joint.

Reproduced from MacKinnon, Pamela and Morris, John, *Oxford Textbook of Functional Anatomy*, vol 1, p60 (Oxford, 2005). With permission of OUP.

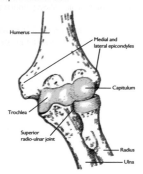

Fig. 5.8 Articular surfaces of elbow joint (anterior aspect).

Reproduced from MacKinnon, Pamela and Morris, John, *Oxford Textbook of Functional Anatomy*, vol 1, p70 (Oxford, 2005). With permission of OUP.

Joints of the forearm and the wrist
- Proximal and distal radio-ulnar joints (Fig. 5.9):
  - Synovial pivot joints:
    - Proximal: between head of radius with radial notch of ulna (see 'Elbow joint')
    - Distal: between ulnar notch of radius and head of ulna
  - A triangle of fibrocartilage (intra-articular disc) unites radius and ulna
  - Shafts of radius and ulna joined by fibrous sheet—the interosseous membrane
- Radio-carpal joint:
  - Synovial ellipsoid joint
  - Distal end of radius and intra-articular disc articulate with proximal row of carpal bones
  - Capsule strengthened by radial and ulnar collateral ligaments and by dorsal and palmar ligaments
  - Flexor retinaculum bridges the concavity of the carpal bones, retains flexor tendons; runs from scaphoid, trapezium to pisiform and hook of hamate
  - Protective palmar aponeurosis from tendon of palmaris longus merges with the flexor retinaculum
  - Thinner extensor retinaculum across the back of the wrist from pisiform, hook of hamate to radius; it bridges grooves on dorsal aspect of lower radius into channels for extensor tendons.

Joints of the hand
See Fig. 5.10.
- Intercarpal joints:
  - Adjacent carpal bones articulate at plane synovial joints to effect small sliding movements
  - Significant movement can occur at the mid-carpal joints between carpal bones in the proximal and distal rows
- Carpometacarpal joints:
  - The thumb joint (1st carpometacarpal joint) is the most important
  - Articulation between trapezium and 1st metacarpal
  - This is a synovial saddle joint, which allows opposition of the thumb
- Metacarpophalangeal joints:
  - Condylar synovial joints allowing side-to-side movements
  - Collateral ligaments between metacarpal and proximal phalanx
- Interphalangeal joints
- Synovial hinge joints.

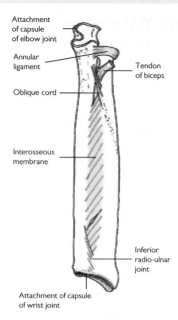

Fig. 5.9 Capsule and ligaments of the radio-ulnar joints.

Reproduced from MacKinnon, Pamela and Morris, John, *Oxford Textbook of Functional Anatomy*, vol 1, p78 (Oxford, 2005). With permission of OUP.

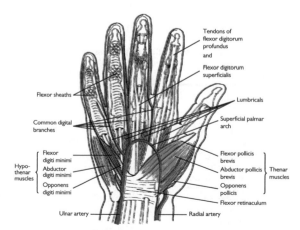

Fig. 5.10 Superficial aspect of palm.

Reproduced from MacKinnon, Pamela and Morris, John, *Oxford Textbook of Functional Anatomy*, vol 1, p91 (Oxford, 2005). With permission of OUP.

Movements of the upper limb

Nerve supply is indicated by[superscript]. See Figs 5.11 and 5.12 for terms denoting position and movement of the body.

Shoulder girdle

- Protraction (forward rotation):
  - Serratus anterior[long thoracic nerve]
  - Pectoralis minor[medial pectoral nerve]
- Retraction (backward rotation or bracing):
  - Rhomboideus major and minor[nerve to rhomboids]
- Elevation (shrugging):
  - Trapezius[spinal accessory nerve]
  - Levator scapulae[dorsal scapular nerve, cervical roots]
- Depression:
  - Trapezius (lower fibres)
  - Pectoralis minor
- Rotation:
  - Trapezius (lower fibres).

Shoulder joint

- Flexion (forward elevation):
  - Pectoralis major (clavicular head)[pectoral nerves]
  - Deltoid (anterior)[axillary nerve]
  - Coracobrachialis[musculocutaneous nerve]
  - Biceps[musculocutaneous nerve]
- Extension (backward swinging):
  - Latissimus dorsi[thoracodorsal nerve]
  - Triceps (long head)[radial nerve]
  - Deltoid (posterior)
- Abduction (sideways movement away from the body):
  - Supraspinatus (<15°)[suprascapular nerve]
  - Serratus anterior and trapezius (>90°)
  - Deltoid (middle)
- Adduction (movement towards the body, across the chest):
  - Pectoralis major
  - Latissimus dorsi
  - Teres major[lower subscapular nerve]
  - Triceps (long head)
- Medial (internal) rotation:
  - Pectoralis major
  - Subscapularis
  - Teres major
  - Latissimus dorsi
  - Deltoid (anterior)
- Lateral (external) rotation:
  - Infraspinatus[suprascapular nerve]
  - Teres minor[axillary nerve]
  - Deltoid (posterior)
- Circumduction (circular):
  - A combination of all of the above-listed muscles.

Elbow joint

- flexion:
  - biceps
  - brachioradialis[radial nerve]
  - brachialis[musculocutaneous nerve]
  - coracobrachialis
- extension:
  - triceps.

Radio-ulnar joints

- pronation (palm turned down):
  - pronator teres and quadratus[median nerve]
  - biceps
  - supination (palm turned up)
  - supinator[radial nerve].

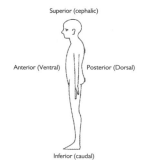

Fig. 5.11 Anatomical terms denoting position.

Reproduced from MacKinnon, Pamela and Morris, John, *Oxford Textbook of Functional Anatomy*, vol 1, p3 (Oxford, 2005). With permission of OUP.

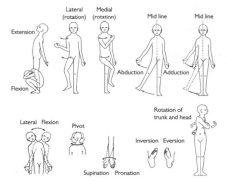

Fig. 5.12 Movement of the body.

Reproduced from MacKinnon, Pamela and Morris, John, *Oxford Textbook of Functional Anatomy*, vol 1, p4 (Oxford, 2005). With permission of OUP.

Wrist joint
- palmar flexion (wrist dropped):
  - flexor carpi radialis[median nerve]
  - flexor carpi ulnaris[ulnar nerve]
  - palmaris longus[median nerve]
  - long flexors of digits
- extension (or dorsiflexion—wrist bent backwards):
  - extensor carpi radialis[radial nerve]
  - extensor carpi ulnaris[radial nerve]
  - long extensors of digits
- abduction (radial deviation):
  - flexor carpi radialis
  - extensor carpi radialis
  - long flexor and extensor of the thumb
- adduction (ulnar deviation):
  - flexor carpi ulnaris
  - extensor carpi ulnaris
- circumduction and fixation:
  - A combination of all the above-listed muscles.

Movements of the digits

See Fig. 5.13.
- flexion:
  - flexor digitorum superficialis[median nerve]
  - flexor digitorum profundus[median, ulnar nerves]
  - lumbrical muscles[median, ulnar nerves]
- extension:
  - extensor digitorum, indicis and minimi[radial nerve]
- abduction:
  - dorsal interossei[ulnar nerve]
- adduction:
  - palmar interossei[ulnar nerve]
- little finger movements:
  - hypothenar eminence[ulnar nerve]:
    - abductor, flexor, and opponens digiti minimi
- thumb movements:
  - thenar eminence[median nerve]:
    - abductor, flexor pollicis brevis, and opponens pollicis
  - abductor and extensor pollicis longus[radial nerve].

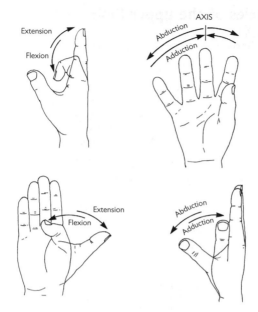

Fig. 5.13 Movements of the fingers and thumb.

Reproduced from MacKinnon, Pamela and Morris, John, *Oxford Textbook of Functional Anatomy*, vol 1, p89 (Oxford, 2005). With permission of OUP.

Muscles of the upper limb

See Figs 5.14–5.18.

- muscles connect vertebrae to the upper limb:

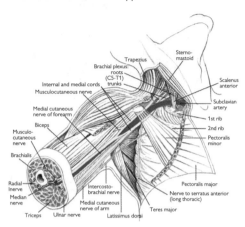

Fig. 5.14 Axilla, to expose the neurovascular bundle; the axillary vein and most of pectoralis major has been removed and pectoralis minor has been divided. To show the origins of the plexus from between the scalene muscles, the clavicle has been divided and sternomastoid reflected.

Reproduced from MacKinnon, Pamela, and Morris, John, *Oxford Textbook of Functional Anatomy* Vol 1, p112 (Oxford, 2005). With permission from OUP.

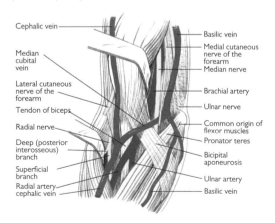

Fig. 5.15 Anterior aspect of elbow; 'cubital fossa'. Note the retraction of biceps and brachioradialis.

Reproduced from MacKinnon, Pamela, and Morris, John, *Oxford Textbook of Functional Anatomy* Vol 1, p202 (Oxford, 2005). With permission from OUP.

- To move the shoulder girdle:
  - trapezius
  - rhomboideus major, minor
  - levator scapulae
- To move the upper arm:
  - latissimus dorsi
- muscles connect the thoracic wall to the upper limb:
  - To draw the upper limb forward, inward:
    - pectoralis major and minor
  - To rotate the scapula:
    - serratus anterior

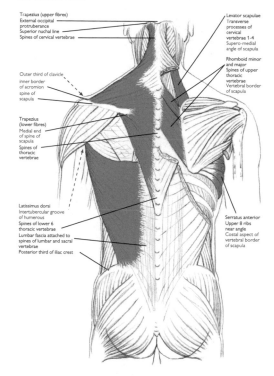

Fig. 5.16 Superficial muscles of the shoulder girdle and back.

Reproduced from MacKinnon, Pamela and Morris, John, *Oxford Textbook of Functional Anatomy*, vol 1, p55 (Oxford, 2005). With permission of OUP.

- muscles within the shoulder girdle:
  - Move the arm and control movements at the shoulder joint:
    - deltoid
    - supraspinatus
    - infraspinatus
    - teres major and minor
    - subscapularis

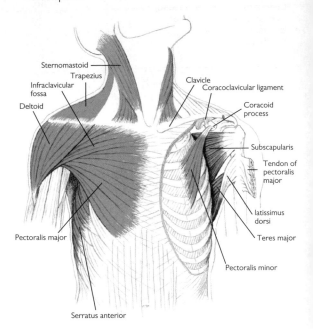

Fig. 5.17 Muscles of the axilla and shoulder (anterior view).

Reproduced from MacKinnon, Pamela and Morris, John, *Oxford Textbook of Functional Anatomy*, vol 1, p62 (Oxford, 2005). With permission of OUP.

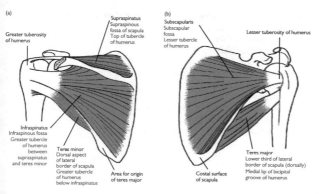

Fig. 5.18 (a) Supraspinatus, infraspinatus, teres minor; acromion removed (posterior view). (b) Subscapularis and teres major (anterior view).

Reproduced from MacKinnon, Pamela and Morris, John, *Oxford Textbook of Functional Anatomy*, vol 1, p62 (Oxford, 2005). With permission of OUP.

- muscles in the arm in the anterior compartment (Fig. 5.19):
  - Flex the elbow joint:
    - biceps brachialis
    - coracobrachialis
    - brachialis
- Muscles in the arm in the posterior compartment:
  - Extend the elbow joint:
    - triceps
- muscles in the forearm in the anterior compartment (Fig. 5.20):
  - Flex the wrist and digits:
    - flexor digitorum superficialis and profundus

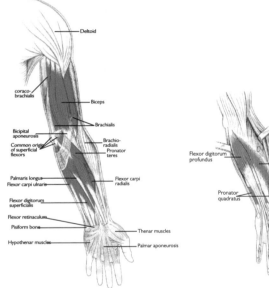

Fig. 5.19 Superficial flexor muscles of arm and forearm.

Reproduced from MacKinnon, Pamela and Morris, John, *Oxford Textbook of Functional Anatomy*, vol 1, p81 (Oxford, 2005). With permission of OUP.

Fig. 5.20 Deep flexor muscles of the forearm.

Reproduced from MacKinnon, Pamela and Morris, John, *Oxford Textbook of Functional Anatomy*, vol 1, p81 (Oxford, 2005). With permission of OUP.

- Muscles in the forearm in the posterior compartment (Figs 5.21, 5.22):
  - Extend the wrist and digits:
    - extensor digitorum, indicis, and digiti minimi
  - Rotate the forearm:
    - pronator teres and quadratus
    - supinator
    - biceps

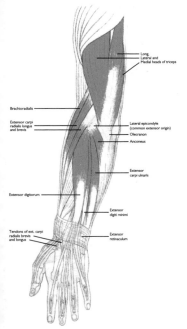

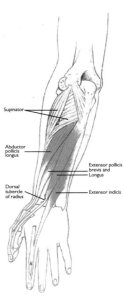

Fig. 5.21 Superficial extensor muscles of arm and forearm.

Reproduced from MacKinnon, Pamela and Morris, John, *Oxford Textbook of Functional Anatomy*, vol 1, p83 (Oxford, 2005). With permission of OUP.

Fig. 5.22 Deep extensor muscles of forearm.

Reproduced from MacKinnon, Pamela and Morris, John, *Oxford Textbook of Functional Anatomy*, vol 1, p83 (Oxford, 2005). With permission of OUP.

- muscles of the hand (Fig. 5.23):
  - Move the thumb, little finger:
    - thenar and hypothenar eminences
  - Move fingers from side to side:
    - interossei.

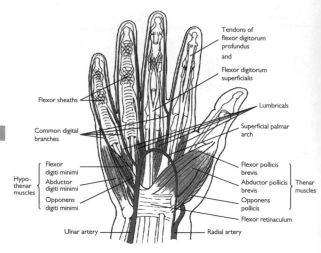

Fig. 5.23 Superficial aspect of palm.

Reproduced from MacKinnon, Pamela and Morris, John, *Oxford Textbook of Functional Anatomy*, vol 1, p91 (Oxford, 2005). With permission of OUP.

Upper limb innervation

Motor and sensory fibres are distributed to particular nerves that supply muscles groups, joints, and skin via the brachial plexus (Figs 5.24, 5.25). Spinal nerve root supply is indicated by [superscript].

Roots arise from the anterior primary rami of cervical spinal nerves C5–C8 in the neck and thoracic nerve T1 in the thorax. The spinal nerves are formed from the unification of ventral (motor) and dorsal (sensory) nerve roots from the spinal cord.

- The long thoracic nerve arises from C5, 6, and 7
- The dorsal scapular nerve arises from C5.

Trunks are formed in the posterior triangle of the neck. C5 and C6 unite to form the upper, C7 becomes the middle, C8 and T1 unite to form the lower. Trunks pass down and laterally over the first rib.

- The suprascapular nerve[C5,6] and the nerve to subclavius[C5,6] arise from the upper trunk.

Beneath the clavicle, fibres from each trunk redistribute into anterior and posterior divisions. Anterior supply flexor compartments and the overlying skin, posterior supply extensor equivalents.

Divisions recombine to form three cords: lateral, medial, and posterior to the axillary artery.

- The lateral pectoral nerve[C5,6,7] arises from the lateral cord
- The medial pectoral nerve[C8, T1] and the medial cutaneous nerves[C8, T1] of arm and forearm arise from the medial cord
- The upper subscapular nerve[C5,6], thoracodorsal nerve[C6,7,8], and lower subscapular nerve[C5,6] arise from the posterior cord.

Terminal nerves arise from the cords.

- The lateral cord gives the musculocutaneous nerve[C5,6,7] and contributes to the median nerve[C5,6,7, T1]
- The medial cord gives the ulnar nerve[C8, T1] and contributes to the median nerve
- The posterior cord gives the axillary nerve[C5,6] and the radial nerve[C5,6,7,8, T1].

Flexor (anterior) compartments are supplied by nerves originating from the lateral and medial cords.

- The musculocutaneous nerve (Fig. 5.26) supplies coracobrachialis, biceps, brachialis. It then continues as the lateral cutaneous nerve of the forearm, which supplies skin of the lateral forearm
- Median nerve branches (Fig. 5.27) supply:
  - flexor muscles (except flexor carpi ulnaris and medial half of flexor carpi profundus), thenar eminence, lumbricals
  - skin of the palm; skin of palmar surface and nail bed of lateral 3½ digits
- Ulnar nerve branches (Fig. 5.28) supply:
  - flexor carpi ulnaris and medial half of flexor carpi profundus, muscles of the hand except thenar eminence, lumbricals
  - skin of palmar and dorsal aspects of medial 1½ digits
  - Extensor (posterior) compartments are supplied by nerves originating from the posterior cord (Fig. 5.29)

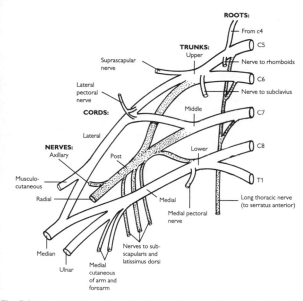

Fig. 5.24 Branchial plexus: schematic diagram of roots, trunks, cords, and branches.
Reproduced from MacKinnon, Pamela and Morris, John, *Oxford Textbook of Functional Anatomy*, vol 1, p112 (Oxford, 2005). With permission of OUP.

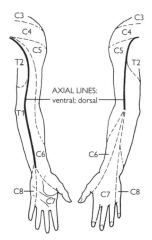

Fig. 5.25 Dermatomes of upper limb: note the axial lines.
Reproduced from MacKinnon, Pamela and Morris, John, *Oxford Textbook of Functional Anatomy*, vol 1, p114 (Oxford, 2005). With permission of OUP.

- Axillary nerve branches supply:
  - teres major and minor and deltoid
  - the badge of skin overlying the deltoid
- Radial nerve branches supply:
  - nearly all extensor muscles
  - Skin of the posterior aspect of the upper limb (as the posterior cutaneous nerves of the arm and forearm)
  - the dorsal aspect of the lateral 3½ digits.

Tendon reflexes (➲ *OHCM10* pp.67, 68, 466–7) are fundamental to the examination of the PNS. Diminished or brisk reflexes suggests lesion of the upper or lower motor neurone (upper motor neurones are first-order neurones which do not leave the CNS; lower motor neurones are the second-order neurones—the cranial and spinal nerves).

By tapping specific tendons, a monosynaptic reflex arc is triggered. The muscle spindle is stretched, which stimulates afferent fibres, and these synapse with efferent motor neurones of the anterior horn. This results in the muscle contracting and a subsequent 'jerk'.

Important 'jerk' reflexes in the upper limb and their nerve roots are:
- Biceps jerk: C5
- Brachioradialis jerk: C6
- Triceps jerk: C7.

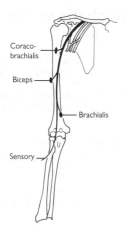

Fig. 5.26 Musculocutaneous nerve: supply to muscles.

Reproduced from MacKinnon, Pamela and Morris, John, *Oxford Textbook of Functional Anatomy*, vol 1, p116 (Oxford, 2005). With permission of OUP.

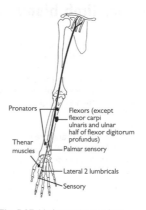

Fig. 5.27 Median nerve: supply to muscles.

Reproduced from MacKinnon, Pamela and Morris, John, *Oxford Textbook of Functional Anatomy*, vol 1, p116 (Oxford, 2005). With permission of OUP.

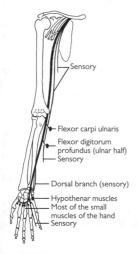

Fig. 5.28 Ulnar nerve: supply to muscles.

Reproduced from MacKinnon, Pamela and Morris, John, *Oxford Textbook of Functional Anatomy*, vol 1, p117 (Oxford, 2005). With permission of OUP.

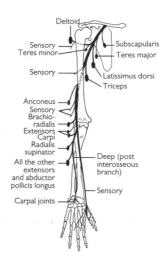

Fig. 5.29 Posterior cord, axillary and radial nerves: supply to muscles.

Reproduced from MacKinnon, Pamela and Morris, John, *Oxford Textbook of Functional Anatomy*, vol 1, p120 (Oxford, 2005). With permission of OUP.

Upper limb blood and lymphatic vessels

Arterial supply

(See Fig. 5.30.) The upper limb is supplied by the subclavian artery which arises on the left from the aorta and on the right from the brachiocephalic trunk. The subclavian artery passes beneath the clavicle and enters the axilla, giving branches to the anterior and lateral chest wall.

At the outer border of the first rib, it becomes the axillary artery and the artery accompanies the brachial plexus, located medially within the axillary sheath.

Branches:

- to the chest wall and breast (superior and lateral thoracic arteries)
- to medial superficial shoulder tissue (thoracoacromial artery)
- around the upper shaft of the humerus (circumflex humeral artery)
- along the lateral scapula (subscapular artery)

On leaving the axilla, the artery becomes the brachial artery. It runs along the medial aspect of the arm; its superficial lie allows it to be felt at the front of the elbow.

Branches:

- near its origin to supply posterior compartment muscles and elbow joint (profunda brachii artery); this follows the course of the radial nerve
- to supply the humerus (nutrient artery) and anterior compartment muscles.

As the artery enters the forearm, it divides into the radial artery and the ulnar artery, which pass towards the wrist beneath the superficial muscles in the anterior compartment.

The ulnar artery branches soon after formation to establish the common interosseous artery, which in turn branches into the anterior (to the flexor compartment) and the posterior (to the extensor compartment) interosseous arteries.

The ulnar artery and radial arteries pass through the wrist into the palm and divide into:

- superficial branches which anastomose to form the superficial palmar arch
- Deep branches which anastomose to form the deep palmar arch.

The superficial arch gives rise to four digital arteries; the deep arch gives rise to three metacarpal arteries. These unite to supply the digits. Princeps pollicis and radialis indicis arise from the radial artery to supply the thumb and index finger.

The radial pulse can be felt at the wrist where the artery lies on the distal radius.

In general, muscles and joints are supplied by adjacent arteries. Anastomoses between adjacent arteries are present at joints and provide a collateral circulation to sustain perfusion should the primary supply be compromised by joint movement.

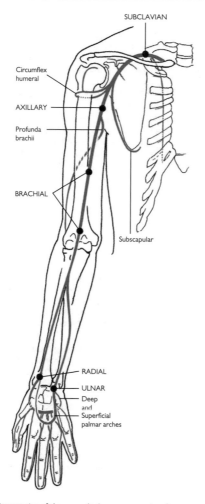

Fig. 5.30 Major arteries of the upper limb: pressure points for arrest of haemorrhage.

Venous drainage

Venous drainage can be divided into superficial and deep systems (Fig. 5.31).

Digital veins drain along with veins in the palm into the dorsal venous arch on the back of the hand. This network is drained by superficial veins:

- the ulnar side of the arch drains to the basilic vein, which ascends along the medial forearm
- the radial side is drained by the cephalic vein, which passes laterally up the forearm.

Tributaries draining superficial tissues join these two vessels as they ascend.

The two veins are connected by the median cubital vein which crosses the front of the elbow in the cubital fossa, and is often used for venepuncture.

Deep veins (the venae comitantes) usually run in pairs alongside arteries. Above the elbow, the basilic vein unites with veins draining deeper structures of the forearm and upper arm (the venae comitantes of the brachial artery). The united vessels form the axillary vein, into which other venae comitantes in turn drain. The cephalic vein drains into the axillary vein beneath the clavicle.

The axillary veins progress into the base of the neck where they become the subclavian veins. These unite with veins draining the head to form the superior vena cava.

Lymphatic drainage

Lymphatic drainage parallels venous drainage (Fig. 5.32).

There are superficial and deep lymphatic vessels, with few interconnections. Superficial drainage:

- of the radial side is through vessels accompanying the cephalic vein
- of the ulnar side occurs through vessels running alongside the basilic vein.

Deep tissue drains through vessels accompanying the deep blood vessels.

Most superficial and deep vessels ultimately drain into the axillary lymph nodes which lie along the axillary artery. There are up to 50 such nodes:

- the lateral nodes, medial to the axillary vein, drain most of the lymph from the arm
- in turn, the lateral nodes drain into the central nodes
- The central nodes feed into the apical nodes.

Vessels accompanying the cephalic vein drain directly into the apical nodes.

Lymph passes from the nodes to the subclavian lymph trunk alongside the subclavian artery. The trunk joins others, including the thoracic duct on the left side, to drain into the venous system at the unification of the subclavian and internal jugular veins.

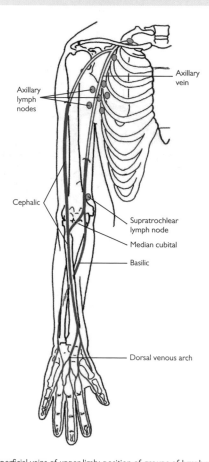

Fig. 5.31 Superficial veins of upper limb: position of groups of lymph nodes.
Reproduced from MacKinnon, Pamela and Morris, John, *Oxford Textbook of Functional Anatomy*, vol
, p102 (Oxford, 2005). With permission of OUP.

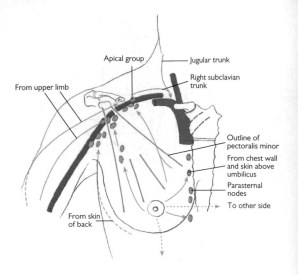

Fig. 5.32 Lymph nodes of the axilla and the internal thoracic chain.

Reproduced from MacKinnon, Pamela and Morris, John, *Oxford Textbook of Functional Anatomy*, vol 1, p105 (Oxford, 2005). With permission of OUP.

Important anatomical features of the upper limb

The axilla:
- lies between the upper end of the arm and the chest wall
- is continuous above with the space between the upper ribs and the shoulder girdle
- inferior limit is the armpit
- is a three-sided pyramid
- anterior wall is formed by pectoralis major and minor, together with the clavicle
- medial wall is formed by the upper ribs and serratus anterior
- posterior wall is formed by the scapula and muscles covering its anterior surface.

The axilla contains:
- the cords and branches of the brachial plexus
- the axillary artery
- the axillary vein
- lymph vessels and nodes

which are surrounded and protected by fat.

The cubital fossa:
- is a triangular region anterior to the elbow joint
- lateral boundary is brachioradialis
- medial boundary is pronator teres
- base is the line between the epicondyles of the humerus
- floor is supinator
- roof is fascia and skin.

The biceps tendon can be palpated within the cubital fossa. The brachial artery and median nerve are found medial to the tendon within the fossa. The radial and ulnar nerves are outside the fossa.

The carpal tunnel (Fig. 5.33) is formed by the concavity of the palmar surface of the carpal bones and its overlying flexor retinaculum, a fibrous band attached to the pisiform, hook of hamate, scaphoid, and trapezium. The tunnel contains:
- the long flexor tendons of the thumb and fingers (except that of flexor carpi radialis)
- the median nerve

but lacks veins and arteries.

Compression of the median nerve results in carpal tunnel syndrome (➔ *OHCM10* p.503), with motor and sensory impairment.

The anatomical snuffbox is a depression on the dorsal aspect of the hand. It is defined by:
- the scaphoid and trapezium, which form its base
- the tendons of abductor pollicis longus and extensor pollicis brevis on the anterior aspect
- the tendon of extensor pollicis longus posteriorly.

The radial artery passes through the snuffbox. Scaphoid fractures result in acute tenderness in this region.

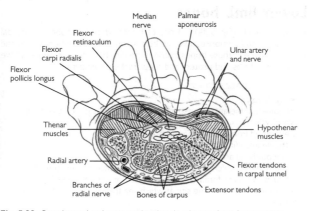

Fig. 5.33 Carpal tunnel and contents (tendon sheaths not shown); transverse section.

Reproduced from MacKinnon, Pamela and Morris, John, *Oxford Textbook of Functional Anatomy*, vol 1, p93 (Oxford, 2005). With permission of OUP.

Lower limb bones

The pelvic girdle (Fig. 5.34) is a ring of bones which surrounds the caudal body cavity and comprises two hip bones (or innominate bones) and the sacrum. Together, these bones transmit the weight of the structures above to the lower limbs and protect and support the pelvic organs.

The innominate bones

(See Fig. 5.35.)

* are joined at the anterior aspect by a cartilaginous joint—the pubic symphysis
* are formed from three separate bones:
  * the ilium, a flat blade which is superior to
  * the pubis, an arch of bone, which is anterior to
  * the ischium, which forms the bony prominence of the buttock
* articulate with the femur at a socket (the horseshoe-shaped acetabulum), which is the site at which the three constituent bones are fused
* articulate with the sacrum at sacroiliac joints.

The sacrum

* comprises five fused vertebrae
* It articulates:
  * at its superior aspect with the 5th lumbar vertebra
  * inferiorly with the coccyx, formed from three to five rudimentary fused vertebrae
  * laterally with the ilium on each side.

The bones of the legs comprise the thigh bone—the femur; the knee cap—the patella; the shin bone—the tibia; the fibula.

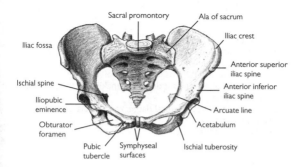

Fig. 5.34 Anterior aspect of adult (female) pelvis.

Reproduced from MacKinnon, Pamela and Morris, John, *Oxford Textbook of Functional Anatomy*, vol 1, p129 (Oxford, 2005). With permission of OUP.

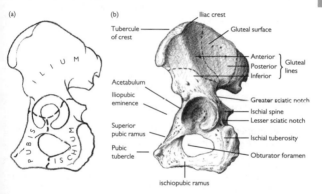

Fig. 5.35 (a) Diagram of pelvis (lateral view) showing fusion lines where the three constituent bones (ilium, ischium, pubis) meet. (b) Lateral aspect of adult pelvis.

Reproduced from MacKinnon, Pamela and Morris, John, *Oxford Textbook of Functional Anatomy*, vol 1, p130 (Oxford, 2005). With permission of OUP.

The femur

(See Fig. 5.36.)
- is the longest bone in the body
- possesses a smooth, rounded head, which articulates with the acetabulum and permits movement in any direction
- has a prominent neck which makes an angle of 125° with the shaft
- slopes in a medial direction as it descends, making an angle of 170° with the shaft of the tibia
- distal end is flared, with the lateral (larger) and medial condyles, which articulate with the tibia
- is often fractured at the neck.

The patella

(See Fig. 5.36.)
- is a roughly circular sesamoid bone in the tendon of quadriceps
- is located anterior to the knee joint:
  - posterior surface articulates with the condyles of the femur, with the lateral condyle providing a stop to resist lateral dislocation
- glides in the groove between the condyles, and so acts as a pulley to improve leverage.

The tibia

(See Fig. 5.36.)
- transfers weight from the femur to the talus
- possesses a flattened plateau with lateral and medial condyles which articulate with those of the femur
- shaft is straight, triangular in cross-section, and tapered towards its distal end
- The anterior border and medial surface of the shaft lie subcutaneously and constitute a common site for open fractures
- articulates with the superior bones of the talus
- projects inferiorly as the medial malleolus, which articulates with the medial talus forming one half of the mortice to stabilize the ankle joint
- derives its name from the similarity of its shape to a Roman flute.

The fibula

(See Fig. 5.36.)
- is a long, thin ('brooch') bone, with head, neck, shaft, and inferior end
- is not part of the knee joint
- does not transmit weight
- provides a framework for muscle attachments, protects blood vessels, participates in the ankle joint
- head articulates with the lateral condyle of the tibia
- at its inferior end articulates with the fibular notch on the lateral aspect of the inferior tibia
- at its most inferior, constitutes the lateral malleolus, the other half of the mortice.

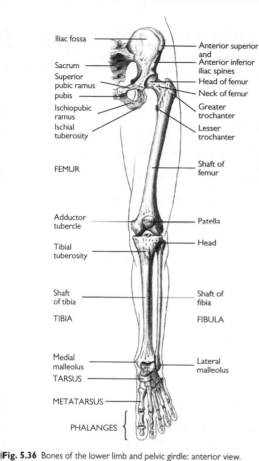

Fig. 5.36 Bones of the lower limb and pelvic girdle: anterior view.

Reproduced from MacKinnon, Pamela and Morris, John, *Oxford Textbook of Functional Anatomy*, vol 1, p130 (Oxford, 2005). With permission of OUP.

The foot

The bones of the foot comprise the seven bones of the main part of the foot—the tarsus—together with the five bones of the forefoot—the metatarsals—and 14 bones of the toes—the phalanges.

The tarsus bones (Fig. 5.37), located distal to the ankle joint, comprise the:
- talus
- calcaneus
- cuboid
- navicular
- three cuneiforms.

The talus bone
- body articulates superiorly with the tibia and, to the sides, with the lateral malleolus and medial malleolus
- possesses a neck connecting the body to a head
- head is rounded for articulation distally with the navicular bone
- makes three small articulations inferiorly with the calcaneus bone.

The calcaneus bone
- possesses a long body
- articulates with the talus bone superiorly
- makes an articulation at its anterior surface with the cuboid bone
- receives the insertion of the tendocalcaneus (Achilles' tendon) of superficial flexor muscles on its posterior aspect
- has two pronounced tubercles on its undersurface.

The three cuneiform bones
- articulate at their posterior aspect with the navicular bone
- articulate anteriorly with the first three metatarsals
- are wedge-shaped so that together they help to maintain a transverse arch.

The cuboid bone
- Articulates with the fourth and fifth metatarsals.

The metatarsal bones
- are equivalent to the metacarpals of the hand
- comprise a base, shaft, and head
- articulate anteriorly with the first row of phalanges.

The first metatarsal is not as free-moving as the first metacarpal. The second metatarsal forms a mortice joint between the lateral and medial cuneiforms to limit its movements. The heads of the metatarsals, together with the tubercles of the calcaneus form the longitudinal arch through which the weight of the body is distributed.

The phalanges
- are shorter and thicker than those in the hand
- possess a base, shaft, and head
- are three in number, except in the great toe, where there are only two.

C   = Calcaneus
T   = Talus
N   = Navicular
Cu = Cuboid
M   = Medial cuneiform
I   = Intermediate cuneiform
I   = Lateral cuneiform

Fig. 5.37 Tarsal bones: dorsal view.

Reproduced from MacKinnon, Pamela and Morris, John, *Oxford Textbook of Functional Anatomy*, vol 1, p132 (Oxford, 2005). With permission of OUP.

Joints of the lower limb

Pelvic girdle and hip

See Figs 5.38 and 5.39.

Pubic symphysis

- midline fibrocartilaginous joint between the two pubic bones
- strengthened by:
  - superior pubic ligament above
  - arcuate ligament spanning the pubic arch
- movements are usually limited, but mobility increases during pregnancy.

Sacroiliac joints

- synovial joints between the anterior surface of the sacrum and articular surface of the ilium (of the innominate bone)
- irregular surfaces within the joint interlock
- joint is strengthened by posterior, anterior, and interosseous sacroiliac ligaments
- sacrotuberous ligaments (between lateral sacrum, coccyx, and ischial tuberosity) and sacrospinous ligaments (between lateral sacrum, coccyx, and ischial spine) resist forward rotatory thrust from weight of spine on sacrum.

Hip joint

- ball-and-socket joint, capable of movements in any plane
- between head of femur and cartilage lining the horseshoe outline of the acetabulum
- for stability, the socket is far deeper than the equivalent in the shoulder joint
- possesses a stabilizing ring of fibrocartilage (the labrum) around the margin
- the transverse ligament bridges the notch of the acetabulum to complete the socket
- round ligamentum teres runs from the transverse ligament to fovea at centre of head of femur
- capsule lined by synovial membrane and strengthened by:
  - anterior Y-shaped iliofemoral ligament
  - pubofemoral ligament
  - ischiofemoral ligament.

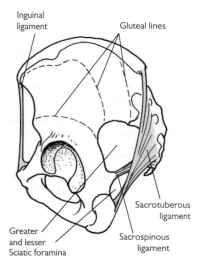

Fig. 5.38 Pelvis showing sacrotuberous and sacrospinous ligaments (and also inguinal ligament).

Reproduced from MacKinnon, Pamela and Morris, John, *Oxford Textbook of Functional Anatomy*, vol 1, p140 (Oxford, 2005). With permission of OUP.

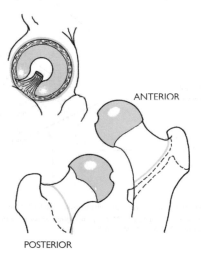

Fig. 5.39 Attachments of capsule of hip joint.

Reproduced from MacKinnon, Pamela and Morris, John, *Oxford Textbook of Functional Anatomy*, vol 1, p140 (Oxford, 2005). With permission of OUP.

Knee joint

See Figs 5.40–5.42.

- compound hinge synovial joint with hyaline cartilage
- primary movements are flexion and extension; some rotation possible
- between medial, lateral condyles of femur and tibial plateau, posterior patella
- femoral and tibial articulating surfaces are partially replaced by two crescent-shaped fibrocartilaginous menisci
- fibula plays no part
- encapsulated, with patella contributing the anterior portion
- patellar retinacula (fascial expansions of tendons) confer further support
- medial (tibial) and lateral (fibular) collateral ligaments strengthen the capsule
- posterior (oblique popliteal) ligament resists torsional stresses
- anterior and posterior cruciate ligaments lie outside the capsule; these keep the articulating surfaces apposed
- synovial membrane extends superiorly to form suprapatellar bursa, which facilitates movements of quadriceps over femur
- prepatellar bursa lies in front of the patella; subcutaneous and deep infrapatellar bursae sandwich the patellar tendon (termination of quadriceps).

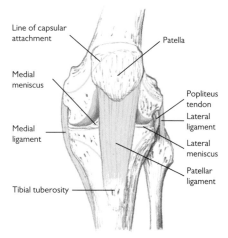

Fig. 5.40 Knee joint: anterior view showing capsule attachments and ligaments.
Reproduced from MacKinnon, Pamela and Morris, John, *Oxford Textbook of Functional Anatomy*, vol 1, p150 (Oxford, 2005). With permission of OUP.

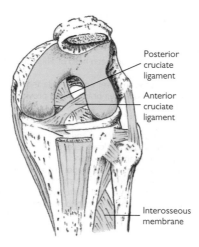

Fig. 5.41 Cruciate ligaments, anterior view.
Reproduced from MacKinnon, Pamela and Morris, John, *Oxford Textbook of Functional Anatomy*, vol 1, p151 (Oxford, 2005). With permission of OUP.

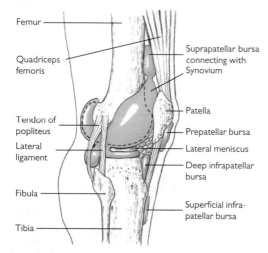

Fig. 5.42 Synovial membrane and bursae of knee.
Reproduced from MacKinnon, Pamela and Morris, John, *Oxford Textbook of Functional Anatomy*, vol 1, p152 (Oxford, 2005). With permission of OUP.

Joints of the lower leg

See Fig. 5.43.

Tibiofibular joint
- superior plane synovial joint between head of fibula and lateral condyle of tibia
- inferior fibrous joint between lower end of fibula and tibia
- shafts united by interosseous membrane, which provides a framework for muscle attachments.

Ankle joint
- synovial hinge joint, permitting flexion and extension
- between upper surface of talus and lower ends and malleoli of tibia and fibula
- malleoli restrict abduction and adduction
- anterior flaring of talus forms wedge shape; in dorsiflexion the wedge is driven between the malleoli to maximize stability
- encapsulated, with reinforcing medial (deltoid) and lateral (anterior talofibular, calcaneofibular, posterior talofibular) ligaments
- lateral ligament is weaker than the medial one—most commonly damaged in a sprain
- extensor retinacula retain extensor tendons to dorsum of foot:
  - superficial between anterior borders of lower tibia and fibula
  - inferior Y-shaped between calcaneus and medial malleolus, plantar fascia
- flexor retinaculum retains flexor tendons:
  - runs from medial malleolus to calcaneus and plantar fascia
- the protective plantar aponeurosis extends from calcaneus to the flexor sheaths at the base of the toes.

Joints of the foot

See Fig. 5.44.

Subtalar joint
- a compound joint comprising:
  - A posterior synovial plane joint (the talocalcaneal joint) strengthened by the intraosseous ligament
  - An anterior synovial ball-and-socket joint (the talocalcaneonavicular joint) strengthened by the calcaneonavicular ('spring') ligament
- The curvatures of the two articulating portions of the talus permit inversion and eversion.

Mid-tarsal joint
- a compound joint comprising:
  - Calcaneocuboid joint (a synovial joint) strengthened by the bifurcated ligament
  - The talonavicular articulation of the talocalcaneonavicular joint
- These also permit a degree of inversion and eversion
- Other tarsal joints—between the three cuneiform bones; cuneiform bones and the navicular bone; lateral cuneiform and cuboid bones; cuboid and navicular bones—are strengthened by interosseous, plantar and dorsal ligaments. These joints permit very little movement
- Tarsometatarsal joints—synovial plane
- Intermetatarsal joints—synovial plane
- Metatarsophalangeal joints—synovial condylar
- Interphalangeal joints—synovial hinge.

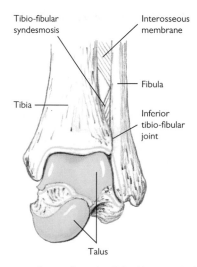

Fig. 5.43 Anterior attachments of capsule of left ankle joint, and inferior tibio-fibular joint.

Reproduced from MacKinnon, Pamela and Morris, John, *Oxford Textbook of Functional Anatomy*, vol 1, p162 (Oxford, 2005). With permission of OUP.

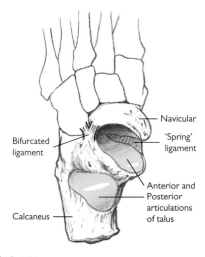

Fig. 5.44 Subtalar joint.

Reproduced from MacKinnon, Pamela and Morris, John, *Oxford Textbook of Functional Anatomy*, vol 1, p171 (Oxford, 2005). With permission of OUP.

Movements of the lower limb

Nerve supply is indicated by ^{superscript}.

Pelvic girdle and hip

- flexion:
  - Psoas major^{lumbar nerves L1–4}
  - Pectineus^{femoral nerve}
  - Iliacus^{pectoral nerves}
  - Rectus femoris^{femoral nerve}
- extension:
  - Gluteus maximus^{inferior gluteal nerve}
  - The hamstring muscles^{sciatic nerve}:
    - ○ semimembranosus
    - ○ biceps femoris
    - ○ semitendinosus
- abduction:
  - gluteus medius and minimus^{superior gluteal nerve}
  - tensor fasciae latae^{superior gluteal nerve}
- adduction:
  - adductor longus^{obturator nerve}
  - adductor magnus^{obturator nerve}
  - adductor brevis^{obturator nerve}
  - gracilis^{obturator nerve}
- medial rotation:
  - adductor longus
  - tensor fasciae latae
  - gluteus medius
- lateral rotation:
  - gluteus maximus
  - piriformis^{sacral nerves S1–2}
  - obturator internus^{nerve to obturator internus}
  - quadratus femoris^{nerve to quadratus femoris}.

Knee joint

- flexion:
  - semimembranosus
  - biceps femoris
  - semitendinosus
  - gastrocnemius^{tibial nerve}
- extension:
  - quadriceps^{femoral nerve}
    - ○ rectus femoris
    - ○ vastus medialis
    - ○ vastus intermedius
    - ○ vastus lateralis.

Joints of the foot

- subtalar eversion:
  - peroneus longus and brevis[deep peroneal nerve]
- subtalar inversion:
  - tibialis anterior[deep peroneal nerve]
  - extensor hallucis longus[deep peroneal nerve]
- plantar flexion:
  - Flexor hallucis longus[tibial nerve] and brevis[medial plantar nerve]
  - flexor digitorum longus[tibial nerve] and brevis[medial plantar nerve]
  - flexor digiti minimi brevis[medial plantar nerve]
  - flexor accessorius[medial plantar nerve]
- extension (dorsiflexion):
  - extensor hallucis longus[deep peroneal nerve]
  - extensor digitorum longus and brevis[deep peroneal nerve]
- lateral rotation ('unlocking'):
  - popliteus[tibial nerve]
- Medial rotation ('locking'):
  - passive process
- sitting cross-legged (the 'tailor's position'):
  - sartorius[femoral nerve].

Ankle joint

- dorsiflexion (foot points up):
  - tibialis anterior
  - extensor digitorum longus
  - extensor hallucis longus
- plantar flexion (foot points down):
  - gastrocnemius
  - flexor digitorum longus
  - tibialis posterior
  - soleus[tibial nerve]
  - flexor hallucis longus.

Muscles of the lower limb

See Figs 5.45–5.51.
- muscles of the pelvis:
  - Flex and rotate the thigh:
    - ○ psoas
    - ○ iliacus
- muscles of the buttocks (gluteal region):
  - Extend and rotate the thigh:
    - ○ gluteus maximus
    - ○ piriformis
    - ○ quadratus femoris
    - ○ gluteus medius, minimus
    - ○ obturator internus
    - ○ tensor fascia latae

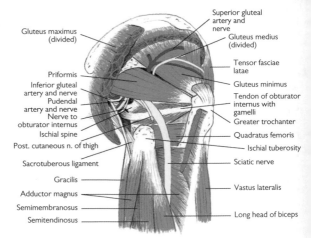

Fig. 5.45 Deep dissection of gluteal region; gluteus maximus and medius largely removed.

Reproduced from MacKinnon, Pamela, and Morris, John, *Oxford Textbook of Functional Anatomy* Vol 1, p142 (Oxford: 2005) With permission from OUP

- muscles of the thigh in the anterior compartment:
  - Flex and rotate the hip; extend and flex the knee:
    - quadriceps
    - pectineus
    - sartorius
- Muscles of the thigh in the posterior compartment:
  - Flex the knee; extend the hip:
    - biceps femoris
    - semitendinosus
    - semimembranosus
- Muscles of the thigh in the medial (adductor) group:
  - Adduct the hip:
    - adductor longus
    - adductor brevis
    - adductor magnus

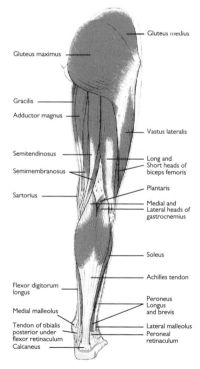

Fig. 5.46 Muscles of the back of the lower limb.

Reproduced from MacKinnon, Pamela and Morris, John, *Oxford Textbook of Functional Anatomy*, vol 1, p154 (Oxford, 2005). With permission of OUP.

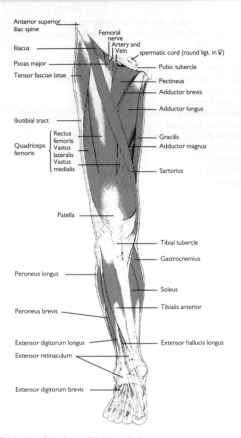

Fig. 5.47 Muscles of the front of the lower limb.

Reproduced from MacKinnon, Pamela and Morris, John, *Oxford Textbook of Functional Anatomy*, vol 1, p155 (Oxford, 2005). With permission of OUP.

- muscles of the lower leg in the anterior compartment:
  - Extend the toes; dorsiflex, invert the foot:
    - extensor digitorum longus
    - tibialis anterior
    - extensor hallucis longus
- Muscles of the lower leg in the posterior compartment:
  - Flex the toes; plantar flex the foot:
    - flexor digitorum longus
    - tibialis posterior
    - soleus
    - flexor hallucis longus
    - gastrocnemius
    - plantaris
- Muscles of the lower leg in the lateral compartment:
  - Evert the foot:
    - peroneus longus and brevis

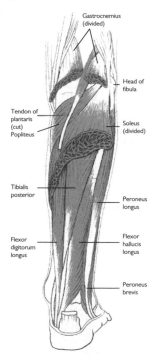

Fig. 5.48 Deep muscles of the calf.

Reproduced from MacKinnon, Pamela and Morris, John, *Oxford Textbook of Functional Anatomy*, vol 1, p164 (Oxford, 2005). With permission of OUP.

- muscles in the foot:
  - Flex and extend the toes:
    - ○ flexor digitorum brevis
    - ○ extensor digitorum brevis
    - ○ flexor digiti minimi brevis
  - move the toes from side to side:
    - ○ abductor and adductor hallucis
    - ○ Metatarsal interossei.

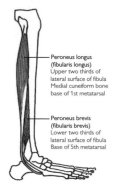

Fig. 5.49 Peroneus longus, peroneus brevis.

Reproduced from MacKinnon, Pamela and Morris, John, *Oxford Textbook of Functional Anatomy*, vol 1, p171 (Oxford, 2005). With permission of OUP.

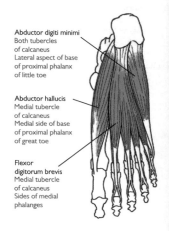

Fig. 5.50 Sole of foot, muscle layer 1.

Reproduced from MacKinnon, Pamela and Morris, John, *Oxford Textbook of Functional Anatomy*, vol 1, p172 (Oxford, 2005). With permission of OUP.

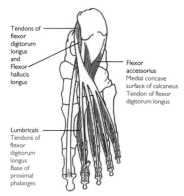

Fig. 5.51 Sole of foot, muscle layer 2.

Reproduced from MacKinnon, Pamela and Morris, John, *Oxford Textbook of Functional Anatomy*, vol 1, p172 (Oxford, 2005). With permission of OUP.

Lower limb innervation

See Figs 5.52–5.58.

Spinal nerve root supply indicated by [superscript].

A similar arrangement to that in the upper limb applies. Nerve supplies originate from the lumbosacral plexus which comprises:

- a lumbar part, formed from lumbar spinal nerve roots L2–5 located within psoas major
- a sacral part, formed from sacral spinal nerve roots S1–3 lying on the posterior pelvic wall.

Lumbar spinal nerve roots provide nerves to psoas major[L1,2,3,4], the ilioinguinal nerve[L1], the genitofemoral nerve[L1,2], and the lateral cutaneous nerve of the thigh[L2,3], and give rise to two major branches:

- Posterior divisions are arranged as the femoral nerve[L2,3,4] which:
  - emerges from the lateral side of psoas major
  - passes under the inguinal ligament
  - gives off the intermediate and medial cutaneous nerves of the thigh to supply overlying skin and the saphenous nerve, which supplies medial aspects of knee and lower leg
  - provides terminal branches supplying anterior (extensor) muscles
- Anterior divisions are arranged as the obturator nerve[L2,3,4] which:
  - emerges from the medial side of psoas major at pelvic brim
  - passes through obturator foramen of innominate bone
  - gives off a nerve to obturator externus[L3,4]
  - enters the adductor compartment and supplies:
    - Adductor muscles
    - Skin of medial thigh.

A branch of lumbar spinal nerve L4 unites with spinal nerve L5 to form the lumbosacral trunk. This combines with sacral spinal nerves S1–3 to form the sciatic nerve.

Superior to the input from S3, the superior[L4,5,S1,2] and inferior gluteal nerves[L5,S1,2] arise to supply the gluteal muscles and tensor fascia latae. Other branches include the nerve to quadratus femoris[L4,5,S1], the nerve to obturator internus[L5, S1,2], the posterior cutaneous nerve to the thigh[S1,2,3], and the pudendal nerve[S2,3,4].

The sciatic nerve:

- descends medial to psoas major, over the sacroiliac joint, then through the greater sciatic foramen, under gluteus maximus into the posterior compartment
- gives branches to semimembranosus, semitendinosus, biceps femoris, adductor magnus.

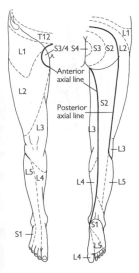

Fig. 5.52 Dermatomes of lower limb; note the axial lines.

Reproduced from MacKinnon, Pamela and Morris, John, *Oxford Textbook of Functional Anatomy*, vol 1, p188 (Oxford, 2005). With permission of OUP.

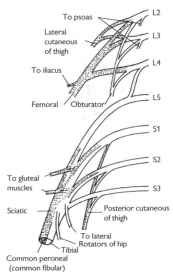

Fig. 5.53 Lumbosacral plexus.

Reproduced from MacKinnon, Pamela and Morris, John, *Oxford Textbook of Functional Anatomy*, vol 1, p188 (Oxford, 2005). With permission of OUP.

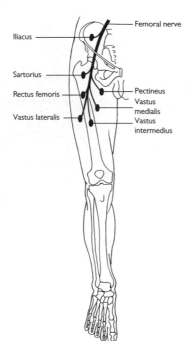

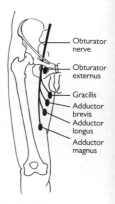

Fig. 5.55 Obturator nerve; supply to muscles.

Reproduced from MacKinnon, Pamela and Morris, John, *Oxford Textbook of Functional Anatomy*, vol 1, p189 (Oxford, 2005). With permission of OUP.

Fig. 5.54 Femoral nerve; supply to muscles.

Reproduced from MacKinnon, Pamela and Morris, John, *Oxford Textbook of Functional Anatomy*, vol 1, p189 (Oxford, 2005). With permission of OUP.

- Divides at the mid-length of the thigh into:
  - The tibial nerve[L4,5,S1,2,3] which:
    ○ provides branches to flexor muscles of the leg, muscles of sole of foot
    ○ gives rise to the medial and lateral plantar nerves; contributes to the sural nerve and calcaneal nerve
    ○ supplies skin of calf, heel, and sole through these sensory nerves
  - The common peroneal nerve[L4,5,S1,2] which:
    ○ provides branches to extensor muscles of leg and foot—the peroneal muscles
    ○ gives rise to the superficial and deep peroneal nerves; contributes to the sural nerve and calcaneal nerve
    ○ supplies skin on lateral aspect of leg and dorsal surface of foot.

Tendon reflexes (◑ OHCM10 pp. 67, 68, 466–7) are fundamental to the examination of the PNS. Diminished or brisk reflexes suggest lesion of the lower or upper motor neurone.

Important 'jerk' reflexes in the lower limb and their nerve roots are:
- knee jerk of patellar tendon: L3, L4
- ankle jerk of Achilles tendon: L5, S1.

In addition, a plantar reflex can be elicited. On stroking the sole of the foot (in a specific manner), the toes normally point downwards (plantar flex). If the toes point upwards (dorsiflex) and spread outwards, this is a positive test result (or Babinski response) and indicates an upper motor neurone lesion.

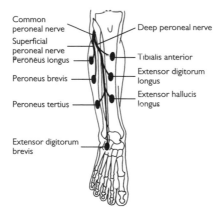

Fig. 5.56 Common peroneal nerve: supply to muscles.
Reproduced from MacKinnon, Pamela and Morris, John, *Oxford Textbook of Functional Anatomy*, vol 1, p192 (Oxford, 2005). With permission of OUP.

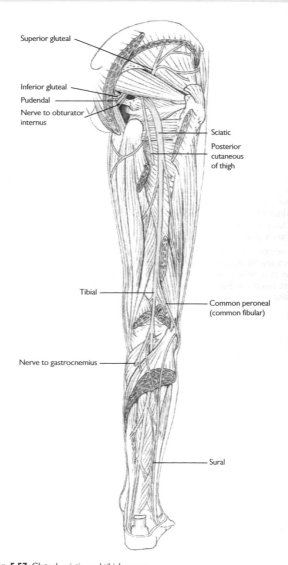

Fig. 5.57 Gluteal, sciatic, and tibial nerves.

Reproduced from MacKinnon, Pamela and Morris, John, *Oxford Textbook of Functional Anatomy*, vol 1, p190 (Oxford, 2005). With permission of OUP.

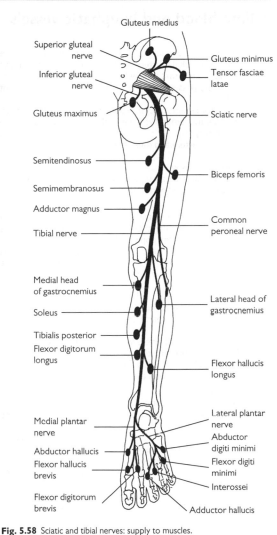

Fig. 5.58 Sciatic and tibial nerves: supply to muscles.

Reproduced from MacKinnon, Pamela and Morris, John, *Oxford Textbook of Functional Anatomy*, vol 1, p191 (Oxford, 2005). With permission of OUP.

Lower limb blood and lymphatic vessels

Arterial supply

(See Fig. 5.59.) The lower limb is supplied by the external iliac artery. The superior gluteal and inferior gluteal artery which arise from the internal iliac artery pass either side of piriformis to supply the gluteal region.

The external iliac artery passes under the inguinal ligament at its midpoint and enters the thigh.

At this point, it lies in the femoral triangle, and becomes the femoral artery which lies alongside the femoral vein within the femoral sheath.

The deep femoral artery:
- arises from the femoral artery and runs between adductors longus and magnus
- provides circumflex arteries which anastomose with descending branches of the gluteal arteries around the upper femur (trochanteric and cruciate anastomoses) and perforating arteries which supply muscles of the thigh.

The (superficial) femoral artery then passes under sartorius, runs medially down the thigh, passes between the adductor and hamstring components of adductor magnus, and enters the popliteal fossa.

From this point, the artery becomes the popliteal artery that provides genicular branches which anastomose around the knee joint. It exits the fossa and divides into anterior and posterior tibial arteries.

The anterior tibial artery:
- runs over the upper border and then descends in front of the interosseous membrane as far as the front of the ankle joint, with branches supplying anterior compartment (extensor) muscles of the lower leg
- then becomes the dorsalis pedis artery which runs over the dorsal aspect of the foot and supplies the toes through:
  - dorsal first metacarpal and arcuate branches
  - a branch which anastomoses with the deep plantar arch in the sole.

The posterior tibial artery:
- descends the posterior (flexor) compartment
- gives rise to the peroneal artery which supplies the lateral compartment of the leg
- passes behind the medial malleolus, after which it divides into medial and lateral plantar arteries under the flexor retinaculum:
  - the lateral plantar artery is larger and runs to the lateral aspect of the sole; its deep branch forms the deep plantar arch from which plantar metacarpal branches supply the toes
  - the medial plantar artery runs on the medial aspect of the sole; branches join plantar metacarpal branches from the lateral artery.

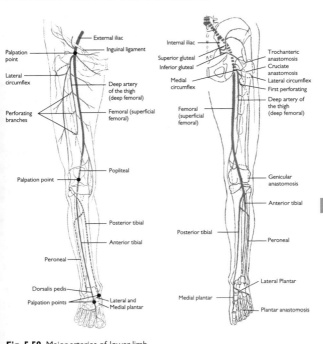

Fig. 5.59 Major arteries of lower limb.

Reproduced from MacKinnon, Pamela and Morris, John, *Oxford Textbook of Functional Anatomy*, vol 1, p178 (Oxford, 2005). With permission of OUP.

Venous drainage

(See Fig. 5.60.) A similar arrangement exists to that in the upper limb. Drainage can be divided into superficial and deep systems, connected by communicating veins.

The superficial system drains skin and superficial tissue. Veins from the toes pass in a dorsal direction to the dorsal venous arch which drains:
- medially to the small saphenous vein which passes behind the lateral malleolus, over the back of the calf, and drains into the popliteal vein
- laterally to the great saphenous vein which passes in front of the medial malleolus, along the medial aspect of the calf and thigh, enters the femoral triangle, and drains into the femoral vein.

The deep system comprises the venae comitantes accompanying the tibial arteries, which become, first, the popliteal vein and, subsequently, the femoral vein. The deep veins
- form an extensive network—the soleal plexus
- are connected by perforating veins at multiple levels to the great saphenous vein
- are emptied by the contraction of surrounding muscles which force blood upwards—the 'muscle pump'.

Veins in the lower limb have many valves to minimize hydrostatic pressure associated with upright posture and thereby facilitate venous return.

Lymphatic drainage

(See Fig. 5.60.) A similar pattern exists to that in the upper limb. Superficial lymphatic vessels draining skin and subcutaneous structures accompany superficial veins; deep tissue drains though deep lymphatic vessels which run with deep blood vessels.

Lymph nodes in the groin are arranged in superficial and deep groups. Superficial nodes are arranged as:
- a longitudinal chain along the great saphenous vein
- a horizontal chain along the line of the inguinal ligament

which drain the superficial tissues of the lower limb and the trunk, below the umbilicus.

Deep nodes receive the drainage from the superficial nodes. They are located medial to the femoral vein in the femoral canal and drain deeper tissues.

Lymphatic vessels leave the deep nodes and convey lymph to nodes along the external iliac artery, which in turn drain into paraaortic nodes. Lymph passes through the cisterna chyli and thoracic duct, and is returned to the circulation at the unification of the subclavian and internal jugular veins.

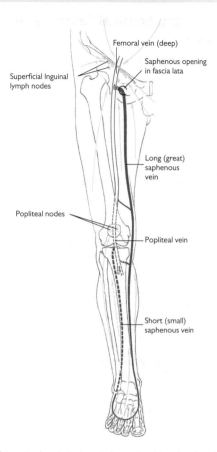

Fig. 5.60 Major veins, lymphatic channels (arrows), and lymph nodes of the lower limb.

Reproduced from MacKinnon, Pamela and Morris, John, *Oxford Textbook of Functional Anatomy*, vol 1, p182 (Oxford, 2005). With permission of OUP.

Important anatomical features of the lower limb

The femoral triangle (Fig. 5.61) boundaries are:
- base—the inguinal ligament
- lateral—the medial border of sartorius
- medial—the medial border of adductor longus.

The floor is contributed by:
- psoas
- pectineus
- iliacus.

The roof is defined by fascia and is punctured by the saphenous opening.
  The triangle contains:
- The femoral sheath enveloping the femoral artery and vein and the femoral canal (containing lymphatic vessels)
- the femoral nerve.

The femoral vessels and branches of the femoral nerve pass through the adductor canal—a channel which starts at the apex of the triangle and runs beneath sartorius towards the medial aspect of the thigh. It is defined by:
- adductor longus and, inferiorly, adductor magnus (posterior wall)
- vastus medialis (lateral wall)
- fascia-enveloping sartorius (roof).

The femoral artery and vein pass into the popliteal region (Fig. 5.62) through an opening in adductor magnus (the adductor hiatus).
  The diamond-shaped popliteal fossa boundaries are:
- superior—biceps femoris (lateral), semitendinosus and semimembranosus (medial)
- inferior—gastrocnemius
- floor—posterior surface of lower femur, oblique popliteal ligament, posterior surface of tibia, popliteus
- roof—deep fascia.

The fossa contains:
- popliteal artery (the continuation of the femoral artery)
- popliteal vein
- tibial nerve
- common peroneal nerve.

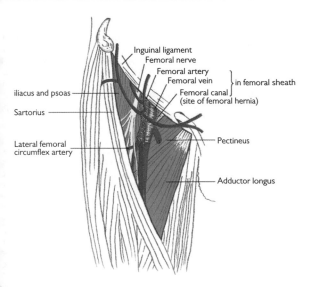

Fig. 5.61 Femoral triangle.

Reproduced from MacKinnon, Pamela, and Morris, John, *Oxford Textbook of Functional Anatomy* Vol 1, p179 (Oxford, 2005) With permission from OUP

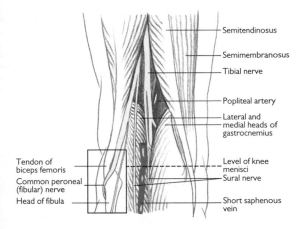

Fig. 5.62 Popliteal fossa; popliteal vein removed above entrance of short saphenous vein. In the boxed area, the muscles have been removed to show the relationship of the common peroneal nerve to the neck of the fibula.

Reproduced from MacKinnon, Pamela, and Morris, John, *Oxford Textbook of Functional Anatomy* Vol 1, p180 (Oxford, 2005) With permission from OUP

Vertebrae and discs

The vertebral column is the main bony supporting structure of the trunk and is critical for locomotion, support of the upper trunk and head, stabilization of the pelvis, posture, and protection of the delicate structures of the spinal cord. It consists of seven cervical vertebrae, 12 thoracic vertebrae, five lumbar vertebrae, five fused sacral vertebrae forming the sacrum and three to five fused coccygeal vertebrae forming the coccyx. In between the cervical, thoracic, and lumbar vertebrae, the intervertebral discs separate the bony elements (Figs 5.63, 5.64).

Curvatures of the spine

In utero, the spinal column develops with a single mild curvature in the median plane which is convex when viewed from a posterior position. During infancy, extra curvatures develop. Four curvatures are apparent in the adult. When viewed from a posterior angle, cervical and lumbar curvatures are convex, while thoracic and sacral curvatures are concave. Excessive curvature of the spine can result in disability. Kyphosis (➔ *OHCM10* p.55), which follows erosion of the anterior areas of the vertebral bodies, is an increase in the thoracic curvature, while lordosis is an increase in the lumbar curvature (this occurs naturally during pregnancy to accommodate the foetus). Scoliosis (➔ *OHCM10* p.55) is an abnormal curvature of the spine in the coronal plane. It can result from leg length discrepancies, unilateral weakness of spinal muscles, or vertebral abnormalities.

Structure of the vertebrae

Vertebrae from different parts of the spinal column share certain key features. The vertebral body is the most anterior part of a vertebra and is critical for weight-bearing and support. From the posterior aspect of each vertebral body arises a vertebral arch which completely encloses the vertebral foramen containing the spinal cord. The arch comprises two pedicles—short processes arising from either side of the vertebral body, connected by a single lamina. On either side of the vertebral arch there is a bony spinous projection (transverse process) and a spinous process arises from the peak of the two laminae. It projects inferiorly and overlaps the vertebral bone below. Superior and inferior articular processes are found at each join between the laminae and pedicles and these articulate with the corresponding processes on the vertebral bodies immediately above and below.

The atlas and the axis

- The first two cervical vertebrae are important in attaching the skull to the vertebral column (atlanto-occipital joint) and are atypical
- The atlas (C1) has no body and is a ring-like structure which articulates with the skull permitting nodding motions of the head
- The axis (C2) includes a peg (odontoid peg or dens) which sits in the ring of the atlas. The odontoid peg allows rotation of the head and is formed by the fusion of the body of C1 with C2 during development
- Alar ligaments run from the lateral sides of the foramen magnum to the side of the odontoid process, preventing excessive rotation of the head

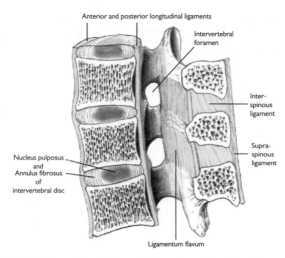

Fig. 5.63 Midline sagittal section through lumbar spine to show discs and ligaments.
Reproduced from MacKinnon, Pamela and Morris, John, *Oxford Textbook of Functional Anatomy*, vol 1, p204 (Oxford, 2005). With permission of OUP.

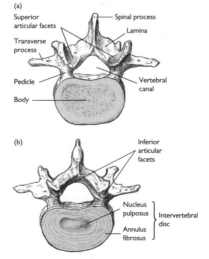

Fig. 5.64 Lumbar vertebrae viewed (a) from above, (b) from below; intervertebral disc.
Reproduced from MacKinnon, Pamela and Morris, John, *Oxford Textbook of Functional Anatomy*, vol 1, p205 (Oxford, 2005). With permission of OUP.

- The transverse ligament of the atlas extends between the lateral mass tubercles of C1, forming a sling around the odontoid process
- The cruciform ligament is a cross-shaped ligament between the occipital bone above and the body of C2 below
- The tectorial membrane is a continuation of the posterior longitudinal ligament, covering the alar and transverse ligaments. It runs from the body of C1 to the internal surface of the occipital bone.

Cervical vertebrae

Cervical vertebrae possess small bodies with transverse processes that have a transversarium foramen through which transmit the vertebral artery, vein, and sympathetic nerve fibres. The spinous processes the of cervical vertebrae are bifid (two-pronged).

Thoracic vertebrae

Thoracic vertebrae have heart-shaped bodies and, with the seventh cervical vertebra, articulate with the ribs at three different sites:

- Each vertebra possesses a superior and an inferior vertebral notch. These structures from adjacent vertebrae form an intervertebral foramen which articulates with the head of the rib. These foramina transmit the segmental spinal nerves
- The transverse processes of each vertebra articulate with the tubercle of the rib.

Lumbar vertebrae

The lumbar vertebrae have large bodies and transverse processes. The spinous processes are long and angled down and out. The vertebral canal at this level is triangular-shaped.

The sacrum

The sacrum is formed by the fusion of five vertebrae. It is a wide, large convex bone with four bilateral anterior and posterior sacral foramina through which the sacral nerves exit. The median sacral crest is formed by the remains of the spinous processes (of the fused vertebrae). The sacrum is important in the transmission of the weight of the body through the pelvis to the limbs, via the sacroiliac joints and the pelvic girdle. The upper surface of the sacrum articulates with the lower surface of the 5th lumbar vertebra.

The coccyx

The coccyx is a remnant of the embryonic tail and is the fusion of three to five vertebrae. It articulates with the sacrum. It has no spinous processes, no lamina, and no pedicles.

Intervertebral discs

Intervertebral discs are dense connective tissue structures which absorb loads distributed through the spine. Intervertebral discs form part of the secondary cartilaginous joints (symphyses) between adjacent vertebrae. Each disc comprises two parts—the central, gelatinous nucleus pulposus and the outer annulus fibrosus. Above and below each disc is a cartilage plate centrally, and outer epiphyseal rings, which the annuli insert into. There is no intervertebral disc between C1 and C2 and the most inferior disc is between L5 and S1.

Discs can degenerate (⊃ *OHCM10* p.542) and become thin (as a result of dehydration). The annulus of a disc may rupture or tear, allowing the nucleus pulposus to herniate (protrude) out of the intervertebral joint. The disc may impinge (⊃ *OHCM10* p.509) or compress spinal nerves, producing symptoms of pain and neurological deficits.

Ligaments of the spinal column

Long ligaments

* The anterior longitudinal ligament attaches to the anterior part of the vertebral bodies and the intervertebral discs. It extends from the inner surface of the sacrum to the anterior tubercle of the atlas (C1) and the occipital bone
* The posterior longitudinal ligament runs within the vertebral canal, attaching to the posterior surface of the vertebrae and the intervertebral discs. It extends from the sacrum to the axis and helps prevent disc herniation
* The supraspinous ligament extends between the tips of the spinous processes of the vertebrae. At cervical levels it becomes the ligamentum nuchae.

Short ligaments

* The interspinous ligaments extend between adjacent spinous processes
* The intertransverse ligaments extend between adjacent transverse processes but are only important in the lumbar region
* The ligamentum flava extends between adjacent lamina as a broad elastic band.

Musculature of the back

The muscles of the back (Fig. 5.65) support the vertebral column and, therefore, the rest of the torso and head. They are divided into three layers—superficial, intermediate, and deep. The intrinsic (deep) muscles cause movements of the vertebral column. The extrinsic muscles (superficial and intermediate) control movements of the limbs and respiration respectively. The musculature of the back also includes muscles which move the head (suboccipital muscles, longissimus capitis).

The intrinsic muscles

- The intrinsic muscles of the back are further subdivided into superficial, intermediate, and deep layers. They are important in:
  - Maintenance of posture
  - Control of movements of the vertebral column and the head
- The intrinsic muscles of the back are enclosed by a fascia. In the lower area, it is called the thoracolumbar fascia. The fascia attaches the back muscles to:
  - Cervical and lumbar transverse processes
  - Spinous processes
  - Ligamentum nuchae
  - Median crest of the sacrum
  - Supraspinous ligaments
- The superficial layer consists of splenius capitis and cervicis. They play a role in rotation of the head, lateral bending, and extension of the head and neck
- The intermediate layer is formed of erector spinae—the main extensors of the vertebral column. Erector spinae form a prominence either side of the spinous processes
- The deep layer is formed by the transversospinal muscles. From superficial to deep, they are semispinalis, multifidus, and rotators.

The extrinsic muscles

Superficial:
- Latissimus dorsi
- Trapezius
- Levator scapulae
- Rhomboids.

These are important in the control of the movement of the upper limbs and connect them to the trunk.

Intermediate:
- Serratus posterior.

Important as an accessory muscle of respiration.

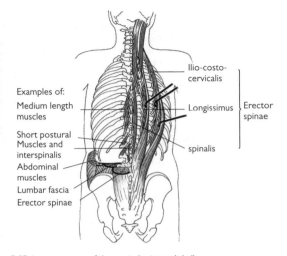

Fig. 5.65 Long extensors of the cervical spine and skull.
Reproduced from MacKinnon, Pamela and Morris, John, *Oxford Textbook of Functional Anatomy*, vol , p207 (Oxford, 2005). With permission of OUP.

The suboccipital muscles

These muscles are involved in maintenance of posture, as well as extension and rotation of the head on C1. They are found in the suboccipital triangle (the area between the occipital bone and the posterior part of C1 and C2 (Fig. 5.66). Four muscles are involved:
- Rectus capitis posterior major
- Rectus capitis posterior minor
- Inferior oblique
- Superior oblique.

Blood supply of the back musculature

The back is supplied according to the region:
- Cervical:
  - Occipital artery (from the external carotid artery)
  - Ascending cervical artery (from inferior thyroid artery)
  - Deep cervical artery (from the costocervical trunk)
  - Vertebral artery (from the subclavian artery)
- Thoracic:
  - Posterior intercostal arteries
- Lumbar:
  - lumbar and subcostal arteries
- Sacral:
  - Iliolumbar and lateral sacral arteries (from the internal iliac arteries).

Venous drainage

The venous drainage of the back is divided into two different systems:
- External vertebral venous plexus—lies outside the vertebral canal
- Internal vertebral venous plexus—lies inside the vertebral canal, but external to the dura mater of the spinal cord.

The two plexuses communicate freely with the veins in the neck, thorax abdomen, and pelvis. The basilar and occipital venous sinuses (in the head) communicate with the plexuses via the foramen magnum.

Lymphatic drainage of the back

Lymphatic drainage of the back is divided into superficial and deep layers.
- Superficial drainage:
  - Skin of neck to the cervical lymph nodes
  - Trunk (above the iliac crests) to the axillary lymph nodes
  - Trunk (below the iliac crests) to the superficial inguinal lymph nodes
- Deep drainage:
  - Deep cervical lymph nodes
  - Posterior mediastinal lymph nodes
  - Lateral aortic lymph nodes
  - Sacral lymph nodes.

Nerve supply of the back

Posterior rami of the spinal nerves supply the muscles and skin of the back.

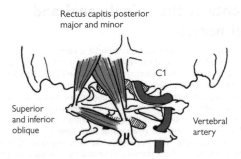

Fig. 5.66 Suboccipital muscles, posterior view.

Reproduced from MacKinnon, Pamela and Morris, John, *Oxford Textbook of Functional Anatomy*, vol , p208 (Oxford, 2005). With permission of OUP.

Contents of the spinal canal and spinal nerves

The spinal cord is the main pathway for the transmission of nervous impulses between the brain and the rest of the body. It lies in the spinal canal surrounded by three meningeal coverings and the CSF. The spinal canal is formed from interlocking vertebrae, which protect it.

The spinal cord emerges from the medulla oblongata at the base of the brainstem. It extends (in adults) from the foramen magnum to the level of the intervertebral disc between L1 and L2, where it becomes the cauda equina. Nerves enter and are given off at each level of the spinal cord as dorsal and ventral rami. The nerve root at each segment of the cord is formed by the rootlets of the dorsal and ventral rami as they merge. Dorsal roots contain afferent (sensory) fibres and ventral roots contain efferent (motor and preganglionic autonomic) fibres. The spinal ganglia, located outside the spinal cord at each vertebral level, contain the cell bodies of dorsal root fibres. The ventral grey horns, within the spinal cord, contain the cell bodies of the ventral nerve fibres.

The spinal nerves (eight cervical, 12 thoracic, five lumbar, five sacral, one coccygeal) are formed as the ventral and dorsal roots join together as they exit the vertebral canal via the intervertebral foramina. As a result of greater growth in the vertebral column than the spinal canal during childhood, the nerve roots are increasingly elongated distally and form the cauda equina below L2 in adults. The conus medullaris is a thinning extension of the spinal cord at its most distal end. The filum terminale extends distally from the conus medullaris, attaching to the posterior surface of the coccyx as a strand-like extension of the pia mater.

The spinal nerve divides into the anterior and posterior primary rami almost as soon as it leaves the vertebral canal. These are mixed, containing motor and sensory fibres. The skin and muscles of the back are supplied by the posterior primary rami. The limbs and trunk are supplied by the anterior primary rami.

The meninges and cerebrospinal fluid

The spinal cord is covered in three meningeal layers:
- Dura mater is the outer covering of the spinal cord and is a tough fibrous layer. It forms the dural sac and hangs freely in the vertebral canal. It is attached superiorly to the foramen magnum and is continuous with the cranial dura mater. The extradural or epidural space contains connective tissue and separates the dura mater from the periosteum of the vertebral canal. Dural root sleeves are extensions of this covering and extend along the nerve roots out of the intervertebral foramina
- Arachnoid mater forms the intermediate layer and is an avascular fragile membrane which lines the dural sac. The subdural space is an extremely thin layer of serous fluid, separating the dura mater and arachnoid mater. The pia mater is separated from the arachnoid mater by a layer of CSF in the subarachnoid space

Pia mater is the innermost covering of the spinal cord. It adheres very closely to the cord. It covers the anterior spinal artery and also forms part of the filum terminale. The dentate ligament on either side of the pia suspends it in the dural sac. There are 21 of these ligaments, extending on either side of the cord, halfway between the anterior and posterior nerve roots

CSF (⊕ *OHCM10* pp.768, 822) is a clear fluid produced by the choroid plexuses in the 3rd, 4th, and lateral ventricles, and flows into subarachnoid space. The lumbar cistern is a CSF-filled space, continuous with the subarachnoid space, which surrounds the cauda equina and filum terminale below L2.

Blood supply of the spinal cord

The spinal cord is supplied by three main arteries which run along its entire length:

The anterior spinal artery supplies the anterior two-thirds of the cord and lies in the anterior median sulcus of the spinal cord

Two posterior spinal arteries supply the posterior third and run down the lateral sides of the spinal cord, near the posterior spinal nerve roots

The spinal cord also has a rich anastomotic blood supply as radicular branches. These originate as branches from the vertebral, intercostal, inferior thyroid and lumbar arteries, and enter the spinal cord via the intervertebral foramina.

The internal vertebral venous plexus (formed by three anterior and three posterior spinal veins) drains venous blood from the spinal cord. The venous sinuses of the cranial dura mater superiorly, and the azygous venous system inferiorly, drain the internal vertebral venous plexus.

Arthritic joint disease

The term arthritis describes multiple joint conditions which are characterized by joint inflammation. Affected joints share certain symptoms, including pain, stiffness, and deformity. Either one or multiple joints can be affected depending on the underlying cause.

Osteoarthritis

Osteoarthritis is the most prevalent joint condition and is about three times more common in women than in men. It is primarily a disease of the elderly, although younger patients can also be affected. Osteoarthritis results from an imbalance between synthetic and degradative processes in articular cartilage (excess degradation, mediated by the cellular component, chondrocytes) although the mechanisms through which this occurs are unclear. In general, osteoarthritis is primary, meaning that the cause is unknown, although it can also be secondary to obesity, joint trauma or instability, or another known cause. There is evidence of genetic predisposition to osteoarthritis in some cases. Joint inflammation characterizes osteoarthritic joints although it is not the primary event.

Osteoarthritis is identified radiographically by a narrowing of the joint space, damage to subchondral bone, and the formation of osteophytes (bony swellings). The primary symptoms are pain, particularly during movement, stiffness, and deformity of affected joints. Osteoarthritis can affect one or many joints but the most common are the joints of the fingers, hips, knees, neck, and lower back.

On a microscopic level, cartilage of osteoarthritic joints is first softened, then fibrillated and fragmented, and underlying subchondral bone is stiffened. It may also contain calcium crystals (calcium deposition disease) and the zone of calcified cartilage adjacent to subchondral bone (tidemark) may be thickened. Eventually, total loss of articular cartilage can result.

Box 5.1 summarizes the treatments for osteoarthritis.

> **Box 5.1 Treatment and drug therapies for osteoarthritis**
>
> - Treatment is largely palliative and includes analgesic and anti-inflammatory therapies
> - Surgical insertion of prosthetic joints, particularly for hips and knees, is common and effective.

Rheumatoid arthritis

Like osteoarthritis, rheumatoid arthritis is more common in women than in men. However, unlike osteoarthritis, it usually affects those aged 20–50. Certain genes are linked to an increased incidence of the disease (e.g. HLA DR4). Unlike osteoarthritis, patients with rheumatoid arthritis can experience systemic effects such as pericarditis, pleurisy, and cachexia.

Typically, an autoimmune response is triggered within a joint which results in damage to articular tissues. Inflammatory cells, including monocytes, infiltrate affected joints, and in 80% of affected individuals antibodies (rheumatoid factors) are produced, particularly once the disease has progressed beyond its early stages. These can be detected in the blood. Inflammation results in the release and activation of cartilage degradative enzymes resulting in tissue damage. Rheumatoid arthritis affects many joints, particularly those of the hands and feet. Joint deformation occurs secondary to cartilage erosion.

Box 5.2 summarizes the treatments for rheumatoid arthritis.

Box 5.2 Treatment and drug therapies for rheumatoid arthritis

- Treatment of rheumatoid arthritis generally involves multiple anti-inflammatory and immunosuppressive drug therapy, combined with analgesia to help with the pain associated with the condition
- Recently, anti-cytokine therapy (e.g. anti-tumour necrosis factor α), has been shown to be highly effective in this disease
- Surgical joint replacement can also be used for the most affected joints.

Gout and pseudo-gout

(➡ OHCM10 p.548.)

- Gout is the presence of urate crystals in articular tissues, secondary to raised uric acid levels. These can initiate joint inflammation and tissue damage
- Pseudo-gout is a related disorder involving calcium pyrophosphate crystals, rather than urate.

Box 5.3 summarizes the treatments for gout.

Box 5.3 Treatment and drug therapies for gout

- Changes in lifestyle can have a significant impact on the occurrence of gout (e.g. weight loss, increased vitamin C intake, reduced alcohol intake)
- Non-steroidal and steroidal anti-inflammatory drugs are often prescribed to control the symptoms (pain) associated with the condition
- Allopurinol (an inhibitor of xanthine oxidase, an enzyme associated with the uric acid-forming pathway) may also have some benefit.

Infectious arthritis

Arthritis can also be induced by infectious agents.

- Septic arthritis (⊙ *OHCM10* p.544) involves direct infection of a (usually one) joint. Gonococci (secondary to gonorrhoea), staphylococci, and streptococci infections are most common
- Rheumatic fever occurs following streptococcal infection of the throat and causes generalized inflammation in a number of locations including the joints which are not themselves infected
- Viral infections of the joints can also occur and most frequently involve hepatitis B or C, parvovirus, rubella, Epstein–Barr virus, HIV, or mumps
- Lyme disease occurs secondary to a tick bite and is an infection caused by *Borrelia burgdorferi*. It involves direct infection of a joint and arthritic symptoms.

Box 5.4 summarizes the treatments for infectious arthritis.

Box 5.4 Treatment and drug therapies for infectious arthritis

- Antibiotics and analgesics are used extensively to treat the infection and pain, respectively
- The joint is also often washed out with saline to help remove bacteria
- Joint replacement is usually postponed until the infection is cleared from the joint.

Muscular dystrophies

The muscular dystrophies are diseases due to primary dysfunction of the muscle (Fig. 5.67), contrasting with neuropathies where the muscular weakness is due to lack of signal for muscle contraction. Although the muscular dystrophies are rare, they provide an interesting model of disease because the defects in muscle function have been found to be due to dysfunction of specific proteins secondary to mutations in the genes coding for those proteins.

Duchenne muscular dystrophy

(→ OHCM10 p.510.)
- Due to a mutation in the gene coding for the protein dystrophin on the X chromosome
- Mutations are usually deletions and dystrophin is usually absent
- Most common form of muscular dystrophy
- Affected males develop normally but, from age of 5 years, develop progressive muscle weakness with use of wheelchair by age of 10 and death due to respiratory failure in 20s.

Becker muscular dystrophy

- Due to a mutation in dystrophin but produces abnormal form with some function rather than absence of protein as in Duchenne type
- Later onset than Duchenne type with less muscle weakness
- Many affected individuals have normal lifespan.

All other types of muscular dystrophy are much rarer but are again due to mutations in proteins involved in muscle function (e.g. mutation of gene coding for β-sarcoglycan in type 2E limb girdle muscular dystrophy).

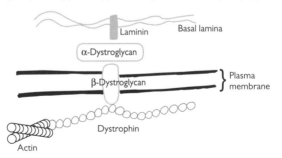

Fig. 5.67 Schematic diagram of muscle proteins which are affected in different types of muscular dystrophy.

Disuse atrophy

The volume of muscle tissue is regulated by the amount and range of use that is made of each particular muscle. Muscle cells do not divide in adults, so there is no increase in the number of cells in response to increased use. However, the volume of the existing cells can increase, as is illustrated, in exaggerated form, by bodybuilders (although there is often some pharmacological assistance in this as well as simple physical training). Similarly, when a muscle is not used, its volume will decrease.

Causes of disuse atrophy

- Physical immobilization of a limb, e.g. in plaster to treat a fracture
- Lack of nerve supply to a muscle, e.g. following trauma to the nerve supply.

Is atrophy reversible?

Muscle atrophy is completely reversible if the cause is removed within a short time (e.g. a few weeks immobilized in plaster) but if it occurs for longer, then irreversible changes occur (e.g. fatty infiltration of the muscle) and, although some function may return after removal of the cause, the muscle will not return to the same initial level of function.

For prevention and treatment, see Box 5.5.

Box 5.5 Prevention and treatment of atrophy

- Reversal of the cause of atrophy as soon as possible, e.g. restoration of nerve supply by repair of nerve damage
- Stimulation of muscle by other means during denervation or immobilization, e.g. cutaneous electrical stimulation
- Intensive physiotherapy, once the cause of disuse has been removed.

Respiratory and cardiovascular systems

Anatomy of the thorax and lungs

The thoracic cavity extends from the neck to the diaphragm. It is bounded by an osteocartilaginous framework and contains the lungs and associated structures, as well as the great vessels, the heart, and part of the oesophagus.

The lungs are spongy elastic tissue made up of alveoli, which are fed air via the bronchial tree of branching bronchi. The thoracic cage contains two lungs, one on either side of the mediastinum. The hilum is the root of each lung and the point where all the important vessels (including nerves, pulmonary vessels, bronchial vessels, lymph vessels) enter and leave. The lung is divided into lobes by the visceral fissures.

Right lung

The right lung comprises the superior, middle, and inferior lobes. The middle and inferior lobes are separated by the oblique fissure. This runs from the spinous process of the 2nd thoracic vertebra posteriorly to the 6th costal cartilage anteriorly. The horizontal fissure separates the superior and middle lobes. It runs from the point where the 4th rib crosses the oblique fissure and around to the 4th costal cartilage anteriorly.

Left lung

The left lung comprises the superior and inferior lobes. The two lobes are divided by the oblique fissure, which follows the corresponding surface markings to those of the right oblique fissure.

Borders of the lungs

- Anteriorly: costal and mediastinal surfaces of the lungs
- Inferiorly: diaphragmatic surface of the lungs
- Posteriorly: costal and mediastinal surfaces of the lungs.

The cardiac notch is found on the medial surface of the left lung, where the lateral heart border indents on the lung surface.

Lung surface markings

Posteriorly, starts at T12, moving laterally along the 12th rib to the mid-scapular point, and to the 10th rib at the mid-axillary line. It reaches the 8th costal cartilage at the mid-clavicular line. On the right, it moves medially to cross the 6th costal cartilage and end at the 4th costal cartilage. On the left, the heart border causes a steeper drop from the 4th costal cartilage to the 6th costal cartilage at the cardiac notch.

Lymphatic drainage of the lungs

The lymph drains inwards, from the pleura to the hilum of the lung to the bronchopulmonary lymph nodes. From there, it drains into the tracheobronchial nodes which are found at the bifurcation of the trachea; then to the paratracheal and mediastinal lymph trunks, and, finally, into the brachiocephalic veins.

Nerve supply of the lungs

At the root of the lung is a pulmonary nerve plexus comprising parasympathetic fibres from the vagus nerve and sympathetic fibres from the sympathetic trunk. Parasympathetic ganglia are located in the pulmonary plexuses and bronchial tree. Sympathetic ganglia (paravertebral ganglia) are located along the sympathetic trunk.

Parasympathetic fibres innervate smooth muscle found in the bronchial tree and pulmonary vessels, and the secretory glands of the bronchial tree. Parasympathetic fibres also carry sensation from stretch receptors in bronchial muscles, interalveolar connective tissue, baroreceptors in pulmonary arteries, and chemoreceptors in pulmonary veins.

Sympathetic fibres provide innervation to the smooth muscle of the bronchial tree and pulmonary vessels, and the secretory glands of the bronchial tree.

The parietal pleura of the lungs is innervated by the phrenic and intercostal nerves.

Blood supply of the lungs

The pulmonary arteries carry deoxygenated blood to the lungs for gas exchange in the alveoli. The reoxygenated blood returns to the heart via the pulmonary veins. The lungs themselves receive oxygenated blood from the bronchial arteries (branches of the descending aorta), which perfuse lung tissue.

The mediastinum

The mediastinum is located between the two lungs. It contains:
- The heart in its pericardial sac
- The great vessels entering and leaving the heart
- The thymus gland.

It is divided into superior and inferior areas by the lower border of the T4 vertebra (the angle of Louis). The inferior area is subdivided into the anterior, middle (contains the heart and great vessels), and posterior areas adjacent to the thoracic vertebrae T5–T12.

The thoracic duct

Lymph from the lower limbs and abdomen drains towards the cisterna chyli, which lies within the abdomen at the level of L1 and L2. This becomes the thoracic duct as it pierces the diaphragm at the aortic opening. It runs behind the oesophagus, moving to the left of the oesophagus at the level of T5. It drains into the start of the left brachiocephalic vein after ascending behind the carotid sheath, then turning back downwards over the subclavian artery.

Lymph drains from the head and neck, upper limbs, and thorax via the left jugular, subclavian, and mediastinal trunks either into the thoracic duct or directly back into large veins at the base of the neck. This is similar to the right side where the right jugular, subclavian, and mediastinal veins may drain into the large veins on the right side of the neck draining into the right brachiocephalic trunk.

Skeletal and soft tissue framework of the thorax

The thorax is bounded by a variety of structures (12 pairs of ribs, costal cartilages, thoracic vertebrae, intercostal muscles, sternum) which participate in ventilation as well as protecting the thoracic organs. Together they form the thoracic cage. The diaphragm is attached to the inferior margins of the thoracic cage and separates the thoracic cavity from the abdomen.

The ribs

A typical rib (Fig. 6.1) comprises several distinct parts:
* Head: articulates with the corresponding vertebra and the one above
* Neck: separates the head and the tubercle
* Tubercle: articulates with the transverse process of the corresponding vertebra
* Angle of the rib: divides the rib into two halves and is the weakest point of the rib
* Shaft: forms the flattened main portion of the rib
* Ribs 1–7 are true ribs (vertebrosternal ribs; the rib is fused anteriorly to the sternum via costal cartilages)
* Ribs 8–10 are floating ribs (vertebrochondral ribs; the rib is fused to the costal cartilage of the above rib)
* Ribs 11 and 12 are false ribs (the rib is not fused to the sternum by any means).

Costal cartilage increases the elasticity of the thoracic cage, making it less fragile and liable to fracture following trivial trauma.

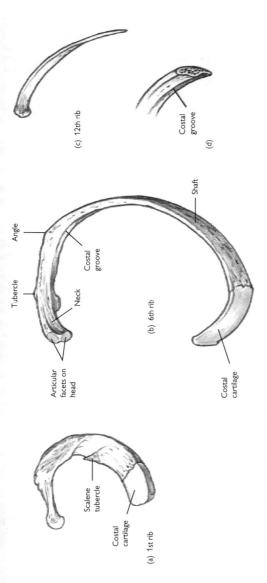

Fig. 6.1 (a) 1st rib; (b) 6th rib; (c) 12th rib.
Reproduced from MacKinnon, Pamela and Morris, John, *Oxford Textbook of Functional Anatomy*, vol 2, p40 (Oxford, 2005). With permission of OUP.

The 1st rib

Owing to its superior location, the first rib is associated with other structures and muscle attachments. It is also the most curved rib. Scalenus medius attaches to its upper surface posteriorly. Scalenus anterior attaches to its tubercle on the upper medial side of the rib. The subclavian vein runs in front of this attachment, and the subclavian artery and lowest branch of the brachial plexus run behind the attachment of scalenus anterior.

The 10th rib

The 10th rib only articulates with T10 and, hence, only has one articular facet.

The 11th and 12th ribs

These two ribs are short, have no necks or tubercles, and only have one large facet that articulates with the corresponding vertebra.

Cervical and lumbar ribs

An extra rib is found in 0.5% of the population, known as a cervical rib (since it articulates with the 7th cervical vertebra). Sometimes this can cause lower brachial plexus compression.

The sternum

This is a flat, long bone which articulates anteriorly with the ribs via costal cartilages. The sternum comprises three fused bones (manubrium, body, xiphisternum).

The manubrium is a flat, wide, and triangular-shaped bone, found at the level of T3 and T4 vertebrae. It articulates with the clavicle and the first two ribs. At the superior end is found the suprasternal notch as well as two notches either side of this where the head of the clavicle articulates. The sternal angle (or angle of Louis) defines the point where the manubrium fuses with the body of the sternum. The body is the largest part of the sternum and comprises four fused segments. It extends from T5 to T9 and articulates with the cartilage of ribs 2–7. The xiphisternum is found inferiorly and is the smallest sternal bone.

Musculature of the thorax

The intercostal muscles are located between neighbouring ribs. They comprise three layers of muscle, from inside to outside:

- Innermost intercostal muscles are separated incompletely from the middle intercostals by the neurovascular bundle. This layer may cross more than one rib, forming an incomplete fibrous layer
- Middle or internal intercostal muscles run obliquely away from the sternum
- External intercostal muscles form a membrane anteriorly (the external intercostal membrane). Fibres pass down and inwards, towards the sternum.

Several other muscles attach to the thoracic cage, including the accessory muscles of respiration such as sternocleidomastoid and some of the scalene muscles.

Nerve supply of the thoracic cage

Each thoracic spinal nerve (associated with the corresponding rib) gives off two nerve branches on leaving the intervertebral foramina. The dorsal roots supply the muscles, bones, joints, and skin of the back. The anterior roots form the intercostal nerves (T1–T11) and the subcostal nerve (T12). The intercostal nerves give off muscular branches and two cutaneous branches (lateral and anterior branches) and supply the muscular walls of the thorax and abdomen and the corresponding cutaneous area of skin (dermatome). The intercostal muscles are supplied by branches of the intercostal nerves called collateral nerves.

The neurovascular bundle (intercostal bundle) consists of the anterior and posterior vessels (arteries and veins) and the intercostal nerves. The intercostal bundle runs below the corresponding rib and includes, from top to bottom: vein, artery, and nerve. They enter the intercostal space between the pleura and the internal intercostal membrane posteriorly. Initially, the neurovascular bundle runs in the middle of the intercostal space, along the internal surface of the internal intercostal membrane. This neurovascular bundle runs under the rib in the costal groove at the angle of the rib where it gives off the collateral branch. The lateral cutaneous branch supplies the intercostal muscles. The anterior cutaneous branch runs on the inner surface of the internal intercostal muscle and ends anteriorly near the sternum as anterior cutaneous branches.

The blood supply of the thoracic cage

The thoracic cage is supplied by the anterior and posterior intercostal arteries. The 1st–6th anterior intercostal arteries are branches from the internal thoracic artery (from the subclavian artery). The 7th–9th anterior intercostal arteries are supplied by the musculophrenic artery. The 10th and 11th intercostal muscles only have a posterior supply.

The 1st and 2nd superior intercostal arteries are supplied by the costocervical trunk (a branch from the second part of the subclavian artery). The 3rd–11th posterior intercostal arteries are supplied by the thoracic aorta. Branches of the posterior intercostal arteries supply the skin, muscles, and the spinal cord. All posterior arteries run forwards and anastomose with the corresponding anterior intercostal artery. The anterior abdominal wall is supplied by the subcostal artery (a branch of the thoracic aorta).

Venous drainage of the thoracic cage

Veins follow the course of the corresponding artery. There are 11 intercostal veins and one subcostal vein on each side of the thorax. Generally, anterior and posterior intercostal veins anastomose to drain into the internal thoracic and then azygos vein, which drains into the superior vena cava, back to the heart. The 1st posterior intercostal vein drains into the left brachiocephalic or vertebral vein. The left 2nd and 3rd intercostal veins drain into the superior intercostal vein, which crosses the aorta to drain into the left brachiocephalic vein.

Ventilatory movements

During inspiration, the ribs move upwards and outwards like bucket handles lifting upwards. This increases the thoracic diameter in all directions. During expiration, the thoracic cage shrinks as the bucket handle-like motion of the ribs sinking reduces the volume of the thoracic cage. The diaphragm moves downwards with inspiration by contraction at the central tendon, and upwards with expiration.

The diaphragm

The diaphragm is a muscular, dome-like structure which separates the abdominal contents from the thorax (Fig. 6.2). It should be considered in two parts: a peripheral muscular part and a central aponeurosis. It is involved in ventilation of the lungs.

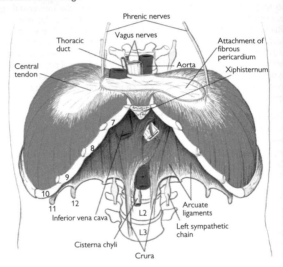

Fig. 6.2 The diaphragm and structures passing through it.

Reproduced from MacKinnon, Pamela, and Morris, John, *Oxford Textbook of Functional Anatomy* Vol 1, p98 (Oxford, 2005) With permission from OUP.

Attachments of the diaphragm:
- Costally to the inner sides of the lower six costal cartilages and ribs
- Inside surface of xiphisternum
- Front of upper three lumbar vertebrae and intervertebral discs form the attachment of the right crus
- 1st and 2nd lumbar vertebrae form the attachment of the left crus
- The central tendon formed by the insertions of the muscular attachments of the diaphragm (a trefoil-shaped area which partially fuses with the base of the pericardium).

The diaphragm is supplied by the phrenic nerve from the cervical roots 3, 4, and 5. Since sensory innervation to the diaphragm is via the phrenic nerve, when the diaphragm is inflamed, pain is referred to the shoulder tip which is the cutaneous portion of the phrenic nerve.

Openings of the diaphragm

Several important structures pass through the diaphragm between the thorax and the abdomen. The inferior vena cava and the right phrenic nerve pierce the diaphragm at the level of the 8th thoracic vertebra. The oesophagus, the left and right vagus nerves, and the left gastric artery and vein pierce the diaphragm at the level of the 10th thoracic vertebra. The aorta, azygous vein, and thoracic duct pierce the diaphragm at the level of the 12th thoracic vertebra. The diaphragm is also pierced by the sympathetic chain and greater and lesser splanchnic nerves.

Pleura and pleural cavities

The lungs are each surrounded by membranous pleural sacs, which are entirely separate from each other.

- Each pleural sac comprises an inner layer of visceral pleura (pulmonary pleura), closely adhered to the lung's surface, and an outer layer of parietal pleura, closely adhered to the inside of the thoracic cavity
- In different regions of the thorax, the parietal pleura is further subdivided into:
  - Costal pleura (inner thoracic wall)
  - Mediastinal pleura (mediastinum)
  - Diaphragmatic pleura (diaphragm)
  - Cervical pleura (apex of lung in the neck)
- Pleural reflections are junctions between different parts of pleura where a marked change in direction occurs (e.g. costal pleura merging with mediastinal pleura)
- Visceral and parietal pleura are continuous with each other at the hilum of the lung and, at this point, a double layer of parietal pleura extends inferiorly forming the pulmonary ligament which provides space for the pulmonary vessels to move during ventilation
- Between the two layers of pleura exists a potential space, the pleural cavity, which is normally filled with pleural fluid. This aids the movement of the pleural layers against each other during inspiration and expiration:
  - At certain points, when the volume of the thoracic cavity is submaximal (during expiration), larger spaces known as recesses are formed (Fig. 6.3). These are pronounced where costal pleura is in contact with diaphragmatic pleura (costodiaphragmatic recess) and where costal pleura is in contact with mediastinal pleura (costomediastinal recess). These cavities can become filled with pus or blood during pulmonary infection (➔ OHCM10 p.170) or injury. This is visible on a chest X-ray as a blunting of the angles of, for example, the costodiaphragmatic recess
  - When fluid needs to be drained from the pleural cavity, it is usually done by inserting a needle into the 5th intercostal space, anterior to the mid-axillary line, over the border of the lower rib to avoid damage to nerves or blood vessels (➔ OHCM10 p.766)
- The visceral pleura receives innervation from the anterior and posterior pulmonary plexuses. Parietal pleura is innervated by intercostal and phrenic nerves and inflammation of this layer frequently results in referred pain to other areas supplied by the same spinal segments. For example, inflammation of diaphragmatic pleura can result in abdominal wall pain, and inflammation of mediastinal pleura can be referred to the neck and shoulder. In contrast, visceral pleura receives no sensory innervation.

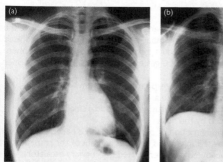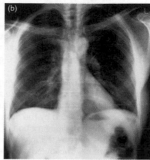

Fig. 6.3 Radiograph of the chest: (a) in inspiration; (b) in full expiration.

Reproduced from MacKinnon, Pamela and Morris, John, *Oxford Textbook of Functional Anatomy*, vol 2, p49 (Oxford, 2005). With permission of OUP.

Upper airways

The upper airways comprise those parts of the respiratory tract above the trachea. However, it must be remembered that the same term is also used to refer to all airways which conduct inspired gases from the atmosphere to the terminal bronchioles, where gas exchange starts. Here, the former definition will be used.

- The upper airways are lined by respiratory epithelium which is characteristically pseudostratified and ciliated
- Frequent goblet cells secrete mucus, which absorbs smaller inhaled particles not excluded by the nose
- The continuous beating motion of cilia prevents these particles from entering the lungs by shifting mucus upwards and out of the respiratory tract where it is swallowed or expectorated (mucociliary escalator). This is an important defence against the entry of foreign, potentially pathogenic, particles.

The nose

See Fig. 6.4.

As well as playing an important role in the sense of smell, the nose moistens and warms inhaled air while preventing particulate matter from entering the airways.

- Air enters the nose through the anterior nares (nostrils), passing the anterior nasal hairs (vibrissae). These trap and prevent inhalation of larger foreign particles
- The epithelial lining changes shortly after entering the nose from keratinized to respiratory epithelium
- The nasal septum, which is formed from part of the ethmoid bone of the skull, the septal cartilage, and the vomer, separates the nasal airway into left and right halves
- Conchae, swirl-like bony structures found on the lateral aspect of each side of the nasal airway, moisten and warm air passing by increasing the surface area of the nasal passage. There are three conchae on each side: an inferior, a middle, and a superior concha
- Olfactory epithelium is found in the upper regions of the nasal airway above the superior conchae and is specialized for the detection of smell. Olfactory nerves, hair-like projections that line the roof and lateral walls of the nose where olfactory epithelium is found, possess receptors that bind specific odorants as air circulates past them
- Inhaled air exits the nose through its posterior openings—the right and left choanae (posterior nares)—to enter the nasopharynx (the area lying behind the nasal passage and above the soft palate).

The paranasal sinuses

See Fig. 6.5.

- Hollow, air-filled bony cavities that surround the nose. There are four pairs: the maxillary, frontal, ethmoidal, and sphenoidal sinuses
- Lined with respiratory epithelium and produce mucus that drains into the nasal cavity via ostia (cavities or holes below each concha, also called meati). There is a meatus associated with each concha, as well as a spheno-ethmoidal recess above the superior concha:

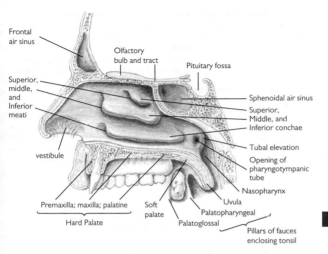

Fig. 6.4 Lateral wall of nasal cavity.

Reproduced from MacKinnon, Pamela and Morris, John, *Oxford Textbook of Functional Anatomy*, vol 3, p70 (Oxford, 2005). With permission of OUP.

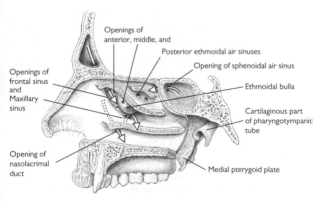

Fig. 6.5 Lateral wall of nose with conchae removed to show drainage of paranasal sinuses.

Reproduced from MacKinnon, Pamela and Morris, John, *Oxford Textbook of Functional Anatomy*, vol 3, p71 (Oxford, 2005). With permission of OUP.

- The spheno-ethmoidal recess drains the sphenoidal sinuses
- The superior meatus drains the posterior ethmoidal sinuses
- The middle meatus drains the rest of the ethmoidal sinuses, and all of the maxillary and frontal sinuses
- The inferior meatus receives drainage from the naso-lacrimal duct. This duct drains tears from the medial angle of the eye into the nose.

Blood supply to the nose

See Fig. 6.6.

The nose is supplied by several different arteries which anastomose at Little's area in the anterior part of the nasal septum. The roof and anterior and lateral walls are supplied by the anterior and posterior ethmoidal arteries, while the meati, septum, and conchae are supplied by the sphenopalatine arteries, superior labial artery, and a branch of the greater palatine artery.

The pharynx

See Fig. 6.7.

- A muscular tube that extends from the oesophagus to the base of the skull. Anteriorly, the pharynx opens into the back of the nose, mouth, and larynx
- The pathway of food and air to the oesophagus and trachea respectively
- Comprises three muscles:
  - Superior, middle, and inferior pharyngeal constrictor muscles
  - Fan-like structures that stack one inside the other and interdigitate
  - Attached to the side walls of the three orifices into which the pharynx opens anteriorly. All three muscles attach to the median raphe (fusion of the muscles) as they fan out and attach to the posterior wall of the pharynx
- The nasopharynx is the area behind the nose and above the soft palate, and plays an important role in respiration
  - It is protected from the regurgitation of food during swallowing by the soft palate rising upwards and closing it off from the rest of the pharynx. The pharyngeal tonsil (a collection of lymphoid tissue commonly known as the adenoids) is found in the posterior wall and roof of the nasopharynx
  - The eustachian tube, a conduit with the middle ear, enters at the level of the floor of the nose on the lateral walls. This explains the common concurrence of throat and middle ear infections
- The oropharynx is the area behind the mouth, between the soft palate and the hyoid bone, and is important in digestion and as part of the immune response:
  - It receives food boluses during deglutition (swallowing) and is part of the conduit between the mouth and the oesophagus
  - Involuntarily contracts on receiving food, squeezing the bolus into the laryngopharynx and into the oesophagus
  - Contains the palatine tonsils, between the palatoglossal and palatopharyngeal arches at the back of the throat
- The laryngopharynx is the area behind the larynx, from the epiglottis to C5, terminating at the start of the oesophagus.

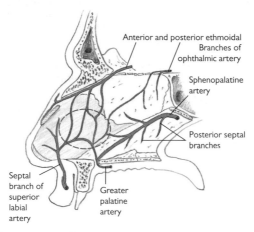

Fig. 6.6 Arterial supply of nose. Area of anastomosis (ringed).

Reproduced from MacKinnon, Pamela and Morris, John, *Oxford Textbook of Functional Anatomy*, vol 3, p72 (Oxford, 2005). With permission of OUP.

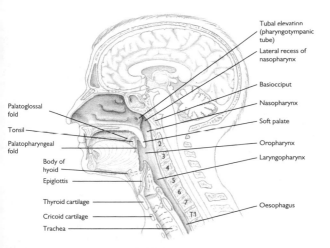

Fig. 6.7 Sagittal section of head and neck to show pharynx.

Reproduced from MacKinnon, Pamela and Morris, John, *Oxford Textbook of Functional Anatomy*, vol 3, p94 (Oxford, 2005). With permission of OUP.

A continuous lymphoid ring is formed by the palatine tonsils, Waldeyer's ring (lymphoid tissue on the dorsum of the tongue), and the adenoids (pharyngeal tonsil). Together, they act as one of the first lines of defence in the immune system.

Blood supply and innervation of the pharynx

The pharynx is supplied by branches from the external carotid and the superior thyroid arteries (Fig. 6.8). The pharyngeal venous plexus drains into the internal jugular vein.

Sensory innervation of the pharynx is via cranial nerve IX (via pharyngeal branches) and cranial nerve V (via the maxillary division), which supplies the nasopharynx. Motor innervation is by cranial nerve X (via pharyngeal branches).

The larynx

See Fig. 6.9.

- The larynx is a tube that conveys air to the lungs from the pharynx
- It plays an important role in producing speech and sound, allows for ventilation, and protects the trachea and bronchial tree during swallowing
- It is comprised of a framework of nine cartilages, bound together by ligaments and muscles, and contains the vocal cords which are responsible for vocalization
- The U-shaped hyoid bone within the neck is the framework by which the larynx is attached to other structures within the neck, including the pharynx, mandible, and the tongue. The hyoid bone lies at the level of cervical vertebrate 3 and 4. The larynx is attached to the hyoid bone by the thyrohyoid muscle and membrane
- The epiglottis is an elastic flap of cartilage, which lies behind the tongue and forms the entrance to the larynx. It attaches to the hyoid bone (in front) and posteriorly to the back of the thyroid cartilage. Laterally, the epiglottis is attached to the pyramid-shaped arytenoid cartilages by aryepiglottic folds which form the opening of the larynx
- The thyroid cartilage is V-shaped and, in men, forms the prominence in the neck called the 'Adam's apple'. The thyroid cartilage is attached to the hyoid bone by the thyrohyoid membrane
- The cricoid cartilage is the only complete ring of cartilage in the respiratory system and is signet ring-shaped. The widest part of the ring faces posteriorly and, either side of it, sit the arytenoid cartilages. The corniculate and cuneiform cartilages are small, paired cartilages which support the aryepiglottic folds and are found within them
- The cricothyroid membrane (cricovocal membrane) runs on the posterior surface of the thyroid cartilage, behind the vocal processes of the arytenoids, connecting the thyroid, cricoid, and arytenoid cartilages. This membrane is thickened between the thyroid and the cricoid and, anteriorly, it becomes the cricothyroid ligament. This is easily palpable since it is subcutaneous and, in an emergency, can be pierced to provide an airway during laryngeal obstruction (⊘ *OHCM10* p.772).

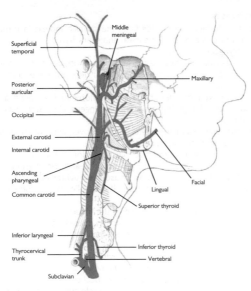

Fig. 6.8 Arterial supply to pharynx.

Reproduced from MacKinnon, Pamela and Morris, John, *Oxford Textbook of Functional Anatomy*, vol 3, p98 (Oxford, 2005). With permission of OUP.

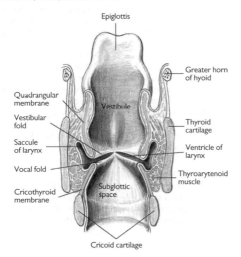

Fig. 6.9 Interior of larynx, coronal section viewed from posterior.

Reproduced from MacKinnon, Pamela and Morris, John, *Oxford Textbook of Functional Anatomy*, vol 3, p107 (Oxford, 2005). With permission of OUP.

Laryngeal muscles
- Muscles of the larynx are divided into the intrinsic and extrinsic muscles
- Extrinsic muscles consist of the infra- and supra-hyoid muscles and stylopharyngeus:
  - The infra-hyoid muscles are sternohyoid, omohyoid, thyrohyoid, and sternothyroid and are responsible for depressing the larynx and hyoid bone
  - The supra-hyoid muscles are digastric, stylohyoid, mylohyoid, and geniohyoid and, together with stylopharyngeus, elevate the larynx and hyoid bone
- The intrinsic muscles of the larynx control movements within the larynx, such as tension on the vocal cords:
  - The muscles include thyroarytenoid, posterior and lateral cricoarytenoid, interarytenoid, aryepiglottic, and cricothyroid
  - Cricothyroid is the only exterior muscle and tightens the vocal cords by tilting the cricoid cartilage. It is supplied by the superior laryngeal nerve
- All the intrinsic muscles are supplied by the recurrent laryngeal nerve and have a common sphincter action, since they form an encircling sheet. They have different attachments which are evident in their names:
  - Thyroarytenoid relaxes the vocal cords
  - The posterior cricoarytenoid abducts the vocal cords
  - The lateral cricoarytenoids adduct the vocal cords
  - The interarytenoids and aryepiglottic muscle close off the larynx during swallowing by forming a sphincter.

The vocal cords
- The vocal cords are formed by two different folds of mucosa to form a triangular-shaped membrane either side of the opening between them, called the rima glottidis. The shape of this area is constantly changing with vocalization
- They have a pearly white avascular appearance, as there is no submucosa between them, and only consist of tightly fused mucosa
- The superior vestibular fold forms the false vocal cord; the inferior vestibular fold forms the true vocal cord. The true cords are important for vocalization, while the false cords have a purely protective role
- The larynx is divided into three areas by these folds of mucosa:
  - Supraglottic compartment (above the vocal cords)
  - Glottic compartment (between the two types of vocal cords)
  - Subglottic compartment (below the true cords and terminating at the start of the trachea).

Nerve, blood, and lymphatic supply of the larynx
Sensory innervation, blood supply, and lymphatic drainage are different above and below the vocal cords (Figs 6.10, 6.11).
- The superior laryngeal nerve provides sensory innervation for laryngeal structures above the vocal cords and the recurrent laryngeal nerve below

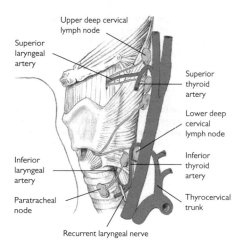

Fig. 6.10 Arterial supply and lymphatic drainage of larynx.

Reproduced from MacKinnon, Pamela and Morris, John, *Oxford Textbook of Functional Anatomy,* vol 3, p111 (Oxford, 2005). With permission of OUP.

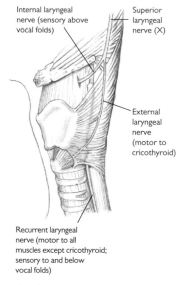

Fig. 6.11 Nerve supply to larynx.

Reproduced from MacKinnon, Pamela and Morris, John, *Oxford Textbook of Functional Anatomy,* vol 3, p111 (Oxford, 2005). With permission of OUP.

- The superior laryngeal branch from the superior thyroid artery supplies structures above the cords, while the inferior laryngeal branch from the inferior thyroid artery supplies structures below the cords
- Lymphatic drainage below the cords is to the lower group of deep cervical nodes; the upper group of deep cervical nodes drain structures above the cords.

The trachea

The trachea starts just below the cricoid cartilage, at the level of C6. It has C-shaped cartilaginous rings, with a fibrous muscular band (trachealis) over the cartilage-deficient area posteriorly. It is lined with respiratory epithelium, which acts as an escalator, wafting particulate matter in the mucus upwards, away from the lower airways.

Nerve, blood, and lymphatic supply of the trachea

Parasympathetic innervation is from the vagus nerve and recurrent laryngeal nerve, while sympathetic innervation is from the sympathetic trunk. The inferior thyroid artery supplies the trachea. The posteroinferior deep cervical nodes drain the trachea.

Lower airways

See Fig. 6.12.

- The distal segments of the pulmonary tree conduct gases between the upper airways and those areas of the lung which are highly specialized for gas exchange
- The airways divide 20–25 times before reaching the alveoli (where most gas exchange occurs) and, between each division, become smaller in length and diameter than more proximal segments
- In healthy subjects, the upper airways contribute most to total pulmonary resistance because their total cross-sectional area is markedly less than for more distal segments
- The angle of Louis lies at the level of thoracic vertebrae 4 and 5 and marks the bifurcation of the trachea into right and left main bronchi:
  - The right main bronchus is wider, more vertical, and shorter, making foreign bodies more likely to lodge in this tract. The right upper lobe bronchus is given off before the right main bronchus enters the hilum of the lung, below and anterior to the pulmonary artery
  - The left main bronchus has an inferolateral pathway to the root of the lung. It is inferior to the arch of the aorta, but anterior to the thoracic aorta and oesophagus. It gives off no branches before entering the lung at the level of T6
- These main bronchi enter each lung at the hilum which demarcates extrapulmonary and intrapulmonary bronchi
- Inside the lungs, bronchi divide into branches (lobar bronchi), each of which supplies a pulmonary lobe
- The right lung possesses three lobes: superior, middle, inferior. On the left, there are two main lobes—the superior and inferior lobes—and the lingular lobe, a tiny attachment which is the remnant of the middle lobe found on the right side
- The lobar bronchi continue to divide and, after approximately four divisions, form bronchioles which each supply a single lobule. Each bronchiole divides into 5–7 terminal bronchioles which then form 2–5 respiratory bronchioles (characterized by the presence of sporadic alveoli). Distally, respiratory bronchioles form 2–11 alveolar ducts, from which most alveoli lead via alveolar sacs
- Bronchopulmonary segments (Fig. 6.13) are wedge-shaped areas within the lung (smaller than a lobe) which are discrete anatomical and functional units supplied by an individual bronchus, artery, and vein. Different lobes exhibit different numbers of segments
- On the right:
  - Superior: apical, posterior, and anterior segments
  - Middle: lateral and medial segments
  - Lower: superior, anterior basal, medial basal, lateral basal, and posterior basal segments
- On the left:
  - Superior: the apical, posterior, superior, and inferior bronchopulmonary segments of the superior lobe
  - Inferior: anterior basal, medial basal, lateral basal, posterior basal segments.

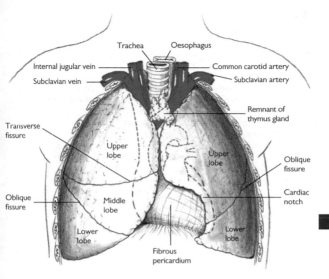

Fig. 6.12 Lungs and great vessels after removal of the anterior thoracic wall.

Reproduced from MacKinnon, Pamela and Morris, John, *Oxford Textbook of Functional Anatomy*, vol 2, p56 (Oxford, 2005). With permission of OUP.

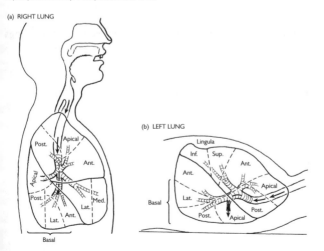

Fig. 6.13 Bronchopulmonary segments in (a) the right and (b) the left lung. Arrows represent likely course of inhaled material: (a) in upright position and (b) in recumbent position.

Reproduced from MacKinnon, Pamela and Morris, John, *Oxford Textbook of Functional Anatomy*, vol 2, p59 (Oxford, 2005). With permission of OUP.

Alveoli

- The alveoli form the major compartment specialized for gas exchange between blood and air
- There are ~300 million alveoli in the two lungs, with a combined surface area of 80–140m^2. This, coupled with their proximity to pulmonary capillaries, enhances the rapid exchange of gases between blood and air
- Adjacent alveoli are separated by a thin wall, the interalveolar septum, which comprises epithelial layers from each of the alveoli sandwiching a connective tissue matrix and dense network of pulmonary capillaries
- The alveoli and capillary blood are separated by the blood–air barrier. This comprises the thin, single-layered alveolar epithelial cells, the fused basal laminae of the epithelial layer and capillary endothelial cells, and the endothelial cells themselves. Together, these three layers are only about 1.5μm thick and greatly facilitate diffusional gas exchange
- Alveoli from adjacent alveolar ducts are linked by interalveolar pores of Kohn: 10μm openings in the interalveolar septum which are an alternative route for movement of gases between alveoli during local bronchiolar obstruction
- Type I alveolar cells comprise about 95% of the alveolar epithelial surface and are characteristically squamous
- Type II alveolar cells are more rounded and account for the remaining 5% of alveolar surface area. They secrete pulmonary surfactant which lines alveoli, reducing surface tension, maintaining their stability and minimizing work required to inflate the lungs
- Type I cells are involved in the absorption of surfactant, promoting its turnover. Type II cells are also a stem cell population, and can divide and differentiate into type I cells in the event that the alveolar epithelium is damaged.

Blood and nerve supply, and lymphatic drainage of the lungs

- The airways and parenchymal tissue of the lungs are supplied by bronchial arteries which arise from the descending thoracic aorta and enter the lungs at each hilum
- Corresponding bronchial veins drain blood from the lungs
- Branches of the pulmonary artery deliver deoxygenated blood to the alveolar capillaries, and the oxygenated blood is returned to the left side of the heart by the pulmonary veins
- Autonomic nerve plexuses (containing sympathetic and parasympathetic fibres) exist at the root of the right and left lungs. The lungs receive both efferent and afferent innervation
- Lymph drains from the lungs to the hilar tracheobronchial nodes.

Respiratory mechanics—static

Lung volume

Respiratory mechanics—static

Respiratory mechanics is the study of the forces, pressure, and work involved in ventilating alveoli. Air flow into and out of the lungs can only occur down pressure gradients and it is the function of the respiratory muscles to create these gradients and permit respiration.

Lung volumes

- Lung volumes vary between individuals and are influenced by age, gender, size, and posture
- However, standard values are available and variations from these can be useful in the diagnosis of lung pathology: restrictive and obstructive lung diseases (⊕ *OHCM10* p.162) affect these volumes differently. The most important examples are shown in Table 6.1
- Certain lung volumes can be measured by spirometry (⊕ *OHCM10* pp.162, 163) (Fig. 6.14), which involves the subject breathing into a sealed container and, in so doing, measuring the volume of inhaled and exhaled gases under variable conditions. These volumes can be recorded by a pen recording apparatus
- Of the lung volumes described in Table 6.1, residual capacity (and any other lung volumes including it) cannot be directly measured, since this is the volume that cannot be exhaled (e.g. into a spirometer). In order to assess these volumes, an alternative technique (e.g. nitrogen washout or helium dilution) is required:
  - Nitrogen washout involves the subject breathing 100% O_2 and collecting expired gases until expired N_2 is zero. At this point, the N_2 content of all of the expired gas is measured and, since N_2 content of air in the lungs is about 80%, total lung volume can be determined
  - Helium dilution involves allowing a known amount of helium (which is not absorbed by the pulmonary circulation) to equilibrate in the lungs. Since the concentration of helium in expired air can then be measured, functional residual capacity can be determined from the relationship:

$$\text{Volume} = \frac{\text{Concentration}}{\text{Amount}}$$

Intrapleural pressure

- Intrapleural pressure is the pressure of fluid within the pleural cavity
- At the end of a normal expiration, intrapleural pressure is negative relative to atmospheric pressure ($-5cm\ H_2O$). This is because of the inherent mechanical tendency of the lungs to collapse inwards and the chest wall to recoil outwards:
  - A collapsed lung, or pneumothorax (⊕ *OHCM10* pp.190, 812), where air is introduced into the pleural space results in the lungs collapsing inwards and the chest wall outwards

Table 6.1 Lung volumes—descriptions

Name (abbreviation)	Description
Residual volume (RV)	Volume of gas in the lungs after a maximal expiration
Functional residual capacity (FRC)	Volume of gas in the lungs after a normal expiration
Inspiratory reserve volume (IRV)	Volume of extra gas that can be inhaled at the end of a normal inspiration by a maximal inspiratory effort
Expiratory reserve volume (ERV)	Volume of extra gas that can be exhaled at the end of a normal expiration by a maximal expiratory effort
Inspiratory capacity (IC)	Volume of gas that can be inhaled following a normal expiration by a maximal inspiratory effort
Tidal volume (TV)	Amount of gas inhaled or exhaled during one normal breath
Vital capacity (VC)	Amount of gas that can be inhaled by a maximal inspiratory effort following a maximal expiration
Total lung capacity (TLC)	The total volume of the lungs at the end of a maximal inspiratory effort

(🔗 OHCM10 p.165.)

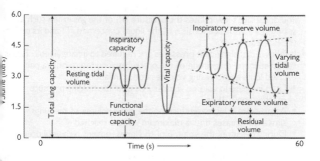

Fig. 6.14 The subdivisions of the lung volumes. Idealized spirometry record of the changes in lung volume during normal breathing at rest, followed by a large inspiration to total lung capacity, followed by a full expiration to the residual volume.

Reproduced with permission from Pocock G, Richards CD (2006). *Human Physiology: The Basis of Medicine*, 3rd edn, p318. Oxford: Oxford University Press.

- During inspiration, the muscles of chest wall and diaphragm expand the chest and increase intrathoracic volume, thus reducing intrapleural pressure further. In this way, alveolar pressure is reduced (by up to 5cm H_2O in normal subjects) and inspiration is initiated
- Conversely, during normal expiration, the muscles of the chest wall and diaphragm relax, decreasing intrathoracic volume, increasing alveolar pressure above atmospheric pressure, and causing expiration
- A forced expiration, during which contraction of certain chest wall muscles results in an even higher increase in intrapleural pressure, obviously increases expiratory rate further
- Intra-oesophageal pressure is approximately equal to intrapleural pressure and can be recorded by introducing a pressure transducer into the oesophagus.

Compliance

- Compliance is a measure of the pressure required to inflate the lungs by a certain incremental volume and is therefore expressed in units of, for example, L kPa^{-1}
- This pressure is acting against forces working to deflate the lung which include the inherent elasticity of the lung as well as forces which arise as a result of the surfactant which lines the alveoli
- During normal breathing, airways resistance and other frictional forces reduce the compliance of the lungs
- Total compliance is a combination of lung compliance and chest wall compliance.

Surfactant and surface tension

- Alveoli are small (~100μm in diameter), gas-filled spheres lined with liquid. Laplace's law relates the pressure (P) of the gas inside such a sphere to the surface tension of the liquid concerned (T) and its radius (r): $P = 2T/r$
- It follows that significant pressures are required to inflate alveoli. This is reflected in the magnitude of the pressures required to hold a pair of excised lungs at increasing volumes during inflation:
  - Little increase in volume is measured until the holding pressure reaches about 1kPa
  - As pressure is increased further, the static volume of the lungs similarly increases until a maximum value is reached
  - During deflation, the volume of the lungs remains high until pressure has dropped significantly
  - The difference in holding pressures required for any given lung volume during inflation and deflation is referred to as hysteresis (Fig. 6.15). Hysteresis is only marked for excised lungs which are initially collapsed
- A saline-filled lung is readily inflated, whereas surface tension in one with an air–water interface shows reduced compliance
- Surfactant reduces surface tension compared with water and renders the lung more compliant. On alveolar inflation, as the surface area of the surfactant film increases, surface tension is increased, thereby preventing smaller alveoli from emptying into larger ones. The static surface tension of a film of surfactant is greater when measured during

expansion than during contraction (this accounts for the observed hysteresis)
• Surfactant is secreted by type II alveolar cells from about 30 weeks of gestation (humans). Insufficient surfactant is believed to account for infant respiratory distress syndrome in premature babies.

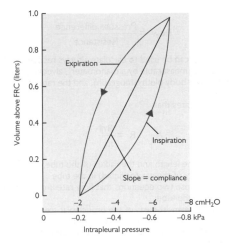

Fig. 6.15 The pressure–volume relationship for a single respiratory cycle.
Reproduced with permission from Pocock G, Richards CD (2006). *Human Physiology: The Basis of Medicine*, 3rd edn, p321. Oxford: Oxford University Press.

Respiratory mechanics—dynamic

Airways resistance

- The resistance of the airways to airflow determines the rate of laminar flow in the face of an applied pressure gradient and is given by the equation:

$$\text{Flow} = \frac{\text{Pressure difference}}{\text{Resistance}}$$

- This relationship can be used to calculate airways resistance when mouth pressure (measurable by a manometer), alveolar pressure (measurable by a body plethysmograph), and the rate of airflow are known
- Poiseuille's law states that:

$$R = \frac{8l\eta}{\pi r^4}$$

where l and r are the length and the radius of the tube, R is the resistance and η is the viscosity of the gas (or liquid) in the tube

- It is clear from these two equations that flow rate is critically dependent on the radius of the tube:
  - *In vivo*, the uppermost parts of the bronchiolar tree contribute most to the total resistance because, although individually their radius is large, small bronchioles and terminal bronchioles are much greater in number
  - Bronchiolar contraction (e.g. during anaphylaxis; ➔ *OHCM10* p.794) increases airways resistance and, similarly, decreases in lung volume compress the airways and increase resistance
- About one-third of total airway resistance arises from the nose, pharynx, and larynx. Mouth breathing (e.g. during exercise) significantly reduces this value. Of the lower bronchial tree, the greatest resistance to airflow occurs in medium-sized bronchi
- During a forced expiration, intrapleural pressure rises to positive levels. While this increases the driving force for air to exit the lungs, it also causes compression of the airways, which increases their resistance and decreases airflow:
  - Since the effect on airways resistance is greater, there is a certain peak expiratory flow rate (➔ *OHCM10* p.162), above which increases in expiratory effort do not result in increases in expiratory rate
  - Peak flow rate decreases as lung volume decreases, since airway resistance increases
  - Obstructive airway disease (e.g. asthma ➔ *OHCM10* pp.178–83, 810–11) reduces peak flow rate, whereas it is unchanged in restrictive airways disease (e.g. pulmonary fibrosis ➔ *OHCM10* p.201)

- Turbulent flow in the airways, which can be heard as wheezing under certain pathological conditions, is governed by different pressure–flow relationships. Whether airflow in a given tube is laminar or turbulent is determined by the Reynolds number. In healthy subjects, the airflow in the bronchial tree exhibits a combination of laminar and turbulent characteristics
- In addition to airway resistance, the tissues of the lung also provide some resistance to breathing as they move. Normally, this value is a relatively low contributor to total pulmonary resistance compared with airway resistance (~20%).

Work done by breathing

- Mechanical work done by breathing is given by the product of the total change in volume and the total change in pressure
- Normally, inspiration is an active process requiring muscular contraction, and expiration is passive (the required energy is obtained from the stretched elastic tissues of the lung and chest wall)
- Usually this value is low, although it is increased when airways resistance is increased (obstructive lung disease) or during forced inspiration and expiration.

Diffusion

- Under normal conditions, the process of ventilation continuously fills the alveoli with atmospheric air, while mixed venous blood enters the pulmonary circulation
- The gases in these two compartments are brought into close contact with each other and O_2 and CO_2 move in opposite directions, across the blood–gas barrier, by simple diffusion (Fig. 6.16). The blood–gas barrier is formed of the alveolar epithelium, the capillary endothelium, and their fused basement membranes and associated structures
- The rate at which gas moves from a region of high partial pressure to a region of low partial pressure is:
  - Proportional to the partial pressure difference and solubility of the gas concerned and the surface area of the barrier to be traversed
  - Inversely proportional to the thickness of the barrier and the square root of the molecular weight of the gas under consideration (Fick's law)
- The total surface area of the alveoli taking part in gas exchange in the lungs is large (80–140m²) and the thickness of this barrier is only 0.3μm. The structure of the blood–gas barrier is therefore optimized for rapid gas exchange:
  - At 37°C, CO_2 is some 20 times more soluble in water than O_2, and, since they are of similar molecular weight, the rate of diffusion of CO_2 is much greater, even though the partial pressure gradient for CO_2 is not as great
  - It must be remembered that the mechanisms that exist in red blood cells for increasing the transfer of O_2 and CO_2 do not exist in the blood–gas barrier and do not speed up the rate of diffusion
- On average, it takes ~0.75s for blood to traverse the length of an alveolar capillary:
  - This is sufficient time for rapidly diffusing gases (e.g. CO_2, normally O_2) to equilibrate across the blood–gas barrier, and they are said to be 'perfusion limited'. That is, the level of perfusion in the capillary limits the amount of the gas that can cross the blood–gas barrier and the alveoli are in equilibrium with the blood at the end of the capillary
  - More slowly diffusing gases (CO, O_2 under certain pathological conditions) are said to be 'diffusion limited'. That is, the alveoli are not in equilibrium with the capillary blood at the point at which the capillary ends
  - Whether a gas is perfusion or diffusion limited depends on the relative solubility of that gas in the blood and the alveolar wall
  - Owing to its high solubility, the transport of CO_2 is rarely diffusion limited
  - At altitude (when the partial pressure gradient for O_2 is reduced) or in diseases which lead to thickening of the alveolar wall, the transport of O_2 can become diffusion limited
  - Exercise significantly reduces the length of time taken for blood to traverse the length of a pulmonary capillary, although in healthy subjects, the rate of diffusion of O_2 is still sufficiently high to prevent its transport from becoming diffusion limited

- The transfer of CO across the blood–gas interface is always diffusion limited and is, therefore, used to measure the properties of the blood–gas interface. The amount of CO disappearing from an inhaled sample over 10s is measured and, assuming that the amount of starting CO in the blood is negligible, reflects the area and thickness of the blood–gas barrier
- Certain pathological conditions (e.g. pulmonary oedema; ➔ *OHCM10* pp.134–5, 800) significantly alter the rates at which gases can diffuse across the blood–gas interface.

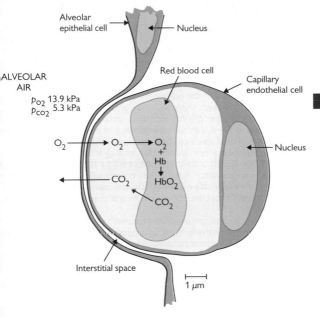

Fig. 6.16 Diagrammatic representation of the layers separating the alveolar air space from the blood in the pulmonary capillaries.

Reproduced with permission from Pocock G, Richards CD (2006). *Human Physiology: The Basis of Medicine*, 3rd edn, p316. Oxford: Oxford University Press.

Ventilation

- Ventilation is the process by which inspired gases reach the alveoli and the blood–gas barrier, and the process by which expired gases are removed from the alveoli
- Alterations in the rate and depth of ventilation affect the composition of alveolar gas and, therefore, the composition of the gases entering the blood
- The airways comprise:
  - Conducting airways (which deliver gas to the alveoli)
  - Exchange zones (where transfer of gases to and from the pulmonary circulation occurs)
- In a healthy individual, the volume of the conducting airways is ~150mL and this volume is frequently referred to as dead space (i.e. the volume of each breath which does not ventilate the exchange zones):
  - Anatomical dead space refers to the volume of the lung that is not alveoli
  - Physiological dead space is the volume of the lung that does not exchange gases with the pulmonary circulation
  - In healthy individuals, these values are approximately equal. However, during lung disease, physiological dead space may be significantly increased as a result of reduced efficiency of pulmonary gas exchange
  - Anatomical dead space can be measured by the nitrogen content in the expired gas from an individual inhaling 100% pure oxygen for a single breath. Initially, N_2 content will be zero as gas from the non-alveolar regions is exhaled (last in, first out) but, after a certain volume is exhaled, N_2 will rise to 80% (the same as N_2 content of alveolar gas). From this, the volume of anatomical dead space can be measured
  - Physiological dead space can be measured from the Bohr equation, which relies upon the fact that, in expired air, all CO_2 comes from the alveoli (the CO_2 content of atmospheric air and, therefore, the dead space is zero):

$$V_D = V_E \frac{\left(1 - F_E\right)}{\left(F_A\right)}$$

where V_D is the dead space volume; V_E is the volume of expired gas in one breath; and F_E and F_A are the fractional CO_2 content of expired air and alveolar air, respectively.

Composition of alveolar gases

- The composition of alveolar gas is determined by:
  - The rate at which O_2 is added to the alveoli by ventilation and removed by the blood
  - The rate at which CO_2 is removed from the alveoli by ventilation and added by the blood
- It follows from this that an increase in ventilation will be accompanied by an increase in alveolar O_2 and a decrease in alveolar CO_2

• The composition of alveolar gases can be calculated from the alveolar gas equation which, if the concentration of CO_2 in inspired air is close to zero, is given by:

$$P_AO_2 = P_IO_2 - \frac{P_ACO_2}{R}$$

• P_AO_2 and P_IO_2, are the partial pressures of oxygen in alveolar and inspired air respectively
• P_ACO_2 is the partial pressure of carbon dioxide in alveolar air
• R is the respiratory quotient, which is the ratio of CO_2 production to O_2 utilization. R is dependent upon dietary status and other factors but is normally about 0.8.

Regional differences in ventilation

• In an upright subject, the apices of the lungs are ventilated less efficiently than the bases. This can be detected by analysing the distribution of inhaled radioactive gas (e.g. ^{133}Xe)
• These variations in ventilation arise primarily because the base of the lung undergoes a larger volume change during ventilation than the apex: at the beginning of an inspiration the lung is relatively less inflated and thus more compliant. In contrast, the apex of the lung must support the weight of the lung and is, therefore, more distended and less compliant
• For a given pressure difference, the base of the lung is therefore ventilated to a greater extent than the apex. This effect is gravity dependent: if the subject lies on their back, posterior regions of the lung become the best ventilated.

Pulmonary perfusion

- Blood enters the pulmonary circulation from the right ventricle of the heart. The right and left pulmonary arteries progressively divide and accompany the airways as far as the terminal bronchioles, and from here they form the capillary network supplying the alveoli
- The very large surface area of pulmonary capillaries and alveoli, as well as the close apposition of these structures, make the lungs a very efficient device for mediating gas exchange
- The capillary network ultimately drains into pulmonary veins and into the left atria of the heart
- The mean pressure within the pulmonary system is 15mmHg, with systolic and diastolic pressures of 25mmHg and 8mmHg, respectively. This is lower than the systemic circulation, largely because the height through which blood is required to ascend is significantly less
- In addition to pulmonary blood pressure, the volume of alveolar blood vessels is determined by alveolar pressure, while the volume of extra-alveolar vessels is determined by lung volume and consequent pull of lung parenchyma on their walls. This distinction is critical to understanding the regulation of pulmonary blood flow
- The total pressure drop across the pulmonary circulation is only 10mmHg, compared with 100mmHg for the systemic circulation. It follows, therefore, that the total resistance of the system is much lower. This is largely because the muscular arterioles which are the major resistance vessel of the systemic circulation are not present in the pulmonary circulation
- Increased pressure within the pulmonary circulation (e.g. from increases in either pulmonary arterial or venous pressure) causes resistance to fall even further. This is the result of recruitment of normally closed vessels, as well as distension of already open vessels. Recruitment and distension normally occur together
- High lung volumes also reduce vascular resistance by opening vessels (pull of parenchymal tissue on capillary wall). If vessels are collapsed, the pressure in pulmonary arterial vessels must reach a critical opening pressure before blood flow can occur
- During a deep inspiration, alveolar pressure rises with respect to capillary pressure such that capillaries tend to be squashed
- The amount of O_2 taken up by lungs can be measured (spirometry), as can the concentration of O_2 in arterial (brachial or radial artery) and mixed venous blood (catheter in pulmonary artery). Therefore, the total amount of blood perfusing the lungs can be determined (Fick's principle).

Distribution of blood flow

- Considerable inequality of perfusion exists within the human lung. In the upright human, lung blood flow decreases linearly from base to apex. This is affected by changes in posture and activity:
  - Gravity is the major determinant of perfusion. The lower regions of the lung are perfused to a greater extent since hydrostatic pressure here is higher, allowing greater recruitment and distension of blood

vessels. The relative changes are much greater for the pulmonary circulation than for the systemic circulation, since the ambient pressure in the pulmonary circulation is much lower

The lung is divided into zones according to the perfusion pattern seen in the upright lung:

- Zone 1: alveolar pressure > than pulmonary arterial pressure—no flow (i.e. dead space)
- Zone 2: pulmonary arterial pressure > than alveolar pressure. Blood flow is therefore determined by arterial–alveolar difference (not arterial–venous difference, because venous pressure is so low and much lower than alveolar pressure)
- Zone 3: venous pressure > alveolar pressure (flow determined by arterial venous pressure difference in usual way). As one moves down this zone, perfusion increases due to distension of blood vessels.

Hypoxic vasoconstriction

- Hypoxic regions of the lungs undergo vascular vasoconstriction
- PO_2 of alveolar gas is chiefly responsible for determining this response. Above 100mmHg PO_2, little vasoconstriction is seen. Below this level, rapid vasoconstriction occurs
- The response persists in excised lungs and is, therefore, presumably an effect which is intrinsic to the lung
- The precise mechanism underlying this is unknown (although it is believed to involve an increase in or sensitization to Ca^{2+} in vascular smooth muscle cells). K^+ channels, and/or nitric oxide (NO) may be involved.
- Hypoxic vasoconstriction directs blood flow towards better-ventilated regions of the lung
- At high altitude, hypoxic vasoconstriction increases pulmonary blood pressure and can cause oedema. It is also responsible for the vascular changes seen at birth. Low blood pH also causes pulmonary vasoconstriction.

Water balance in lung

- The distribution of fluid between pulmonary capillaries and the pulmonary extracellular space is determined by Starling's forces
- There is low capillary, hydrostatic pressure: net fluid outflow is estimated to be ~20mL h^{-1}
- Increases in pulmonary capillary pressure (e.g. during hypoxic vasoconstriction or left ventricular failure ⊃ OHCM10 pp.134–6) increase fluid outflow. Pulmonary oedema (⊃ OHCM10 p.800) in which fluid crosses the alveolar epithelium and enters the alveolar spaces, can ensue, compromising gas exchange.

Other functions of the pulmonary circulation

The pulmonary circulation is also a reservoir for blood and has several metabolic functions: it is well suited to modifying blood-borne substances since it receives the entire circulation.

Ventilation–perfusion relationships

- The matching of ventilation and perfusion in all regions of the lung is a critical determinant of healthy gas exchange. The ventilation–perfusion (V/Q) ratio is a useful measure of this
- Ventilation and perfusion are both higher at the bottom of the lung, but perfusion varies to a greater extent than ventilation
- Gravity is for the most part responsible for these differences. Perfusion varies to a greater extent because the density of blood is greater than that of inspired air
- The V/Q ratio is therefore greatest at the apex of the lung and least at its base, such that the bottom of the lung is relatively under-ventilated while the top is relatively over-ventilated (Fig. 6.17):
  - In well-ventilated, well-perfused alveoli (V/Q ~1), blood equilibrates with alveolar air
  - In poorly ventilated but well-perfused alveoli (V/Q < 1, lower regions), alveolar PO_2 and PCO_2 tend towards the levels found in mixed venous blood
  - In well-ventilated but poorly perfused alveoli (V/Q > 1, higher regions), alveolar PO_2 and PCO_2 tend towards the levels found in inspired gas
- The overall effect of a regional mismatch in ventilation and perfusion on whole lung function is to reduce its efficiency as a gas exchanger (that is, arterial PO_2 is not as high and arterial CO_2 is not as low as it would otherwise be if uniform ventilation and perfusion existed):
  - The blood exiting the lungs comes to a greater extent from the lower regions, where perfusion exceeds ventilation and alveolar PO_2 is low and PCO_2 high. The O_2 content of the blood leaving these regions is therefore considerably reduced, while CO_2 is elevated
  - Furthermore, over-ventilated regions cannot compensate for under-ventilated ones. Blood from over-ventilated regions does not have appreciably higher oxygen content than that from well-matched regions: the O_2 dissociation curve is flat at high PO_2 values so haemoglobin is fully saturated by PO_2 levels lower than those found in the over-ventilated regions
- The effect of alterations in V/Q ratio can be plotted on an O_2–CO_2 diagram. This demonstrates all of the possible compositions of alveolar gas (or end-capillary blood) in a particular lung unit as the V/Q ratio is altered, assuming that the composition of inspired air and mixed venous blood remains constant
- In healthy, upright subjects, the depression of PO_2 in blood leaving the lung by V/Q mismatch is trivial. It can become more significant in disease where the degree of mismatch can be far greater:
  - A lowered V/Q ratio may be observed with asthma or acute pulmonary oedema
  - An elevated V/Q ratio can arise from pulmonary embolism
- Rises in arterial PCO_2 as a result of a lowered V/Q ratio are usually self-limiting: chemosensors act to increase ventilatory drive and return PCO_2 to normal levels.

Shunt

• Shunt is an extreme form of V/Q mismatch caused by the passage of blood through the pulmonary circulation which is not fully ventilated (e.g. bronchial artery blood which supplies the lung parenchyma and drains into the pulmonary veins, as well as coronary venous blood which drains into the left ventricle)

• An abnormal connection between the pulmonary artery and pulmonary vein (pulmonary arteriovenous fistula), as well as defects that allow blood to pass from the right- to the left-hand side of the heart, can also produce a shunt

• The shape of the O_2–Hb dissociation curve is such that the addition of a small amount of under-oxygenated shunted blood greatly reduces PO_2. Hypoxaemia in a shunted patient cannot be abolished by breathing 100% O_2 since the ventilated and perfused regions of the lung are already oxygenating the blood to the greatest possible extent

• Again, a shunt does not normally affect PCO_2 in arterial blood since the chemoreceptors sense increased arterial PCO_2 and make the appropriate adjustments (increase ventilatory rate).

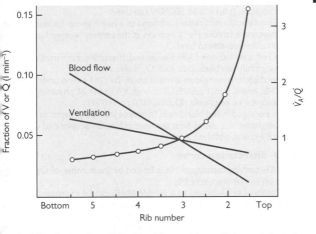

Fig. 6.17 Distribution of ventilation, blood flow, and the ventilation–perfusion ratio in the normal upright lung.

Reproduced with permission from Pocock G, Richards CD (2006). *Human Physiology: The Basis of Medicine*, 3rd edn, p330. Oxford: Oxford University Press.

Oxygen transport in the blood

- Oxygen is transported in the blood from the alveolar capillaries of the lungs (where blood is loaded with O_2) to the peripheral capillaries in th tissues (where O_2 is off-loaded)
- O_2 is transported in blood in two distinct ways: complexed to haemoglobin (Hb) or dissolved in solution in intracellular and extracellular fluids:
  - The amount of oxygen dissolved in the blood is proportional to its partial pressure (Henry's law). At 37°C, 0.003mL O_2 is dissolved in each 100mL blood per mmHg. Resting O_2 consumption is ~300L O_2 min^{-1} and, even if all of the O_2 in arterial blood (100mmHg) was extracted by the tissues, cardiac output would have to be ~100L min$^-$ to support the body's O_2 requirement. This is clearly not possible and the amount of O_2 transported in the blood in this way is normally negligible. However, dissolved O_2 represents the major pathway for transport of O_2 across capillary walls to respiring cells, and the only pathway from the alveoli to red blood cells.

Haemoglobin

- Haemoglobin (Hb) is a 64,500MW tetramer
- Each protein subunit (globin) is bound to a haem group containing four pyrroles and a ferrous (Fe^{2+}) iron ion at the centre. Neither haem nor globin alone are able to bind O_2
- Each Fe^{2+} can bind one O_2 molecule and, therefore, each molecule of haemoglobin may bind up to four O_2 molecules
- Normal adult haemoglobin is made up of 2α (141 amino acids) and 2β (146 amino acids) subunits, although a number of physiological and pathological variants exist (➔ *OHCM10* pp.340–43)
- Hb is packaged in red blood cells (erythrocytes) to prevent its filtration by the glomerulus and to limit the rises in blood viscosity that would arise if Hb was dissolved in plasma.

Hb–O_2 dissociation curve

- The O_2-carrying capacity of Hb is limited by the number of O_2-binding sites on each molecule of Hb
- The amount of O_2 bound to a sample of Hb can be expressed either as a *concentration* (normally mL O_2/100mL blood) or, alternatively, as a *percentage saturation* of maximal O_2 capacity
- The reaction between Hb and O_2 is both rapid and reversible. Once Hb is 100% saturated, the amount of O_2 bound to it cannot be increased by increasing the partial pressure of O_2
- Binding of O_2 to Hb is cooperative such that the binding of each O_2 molecule to the Hb tetramer facilitates the binding of the next (Figs 6.18, 6.19). This positive cooperativity is a particular property of tetrameric Hb and is not exhibited by the monomer
- The Hb–O_2 dissociation (or association) curve is therefore sigmoidal rather than hyperbolic, which facilitates O_2 loading at the lungs and O_2 unloading at the tissues.

Factors affecting O₂ binding to Hb

Increases in H^+, CO_2, and temperature each shift the Hb–O_2 dissociation curve to the right and favour the unloading of O_2:

- This is clearly of physiological benefit in, for example, a metabolically active muscle which will have a high demand for O_2 and where pH will be decreased, CO_2 production raised, and temperature will be raised
- The effects of pH and CO_2 on the Hb–O_2 dissociation curve are known collectively as the Bohr effect

The 2, 3-diphosphoglycerate (2,3-DPG) produced by erythrocytes during glycolysis, binds to Hb and reduces its affinity for O_2. The production of 2,3-DPG is raised during hypoxic conditions, favouring the delivery of O_2 to tissues.

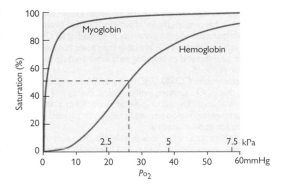

Fig. 6.18 Comparison between the oxygen dissociation curves for myoglobin and haemoglobin.

Reproduced with permission from Pocock G, Richards CD (2006). *Human Physiology: The Basis of Medicine*, 3rd edn, p235. Oxford: Oxford University Press.

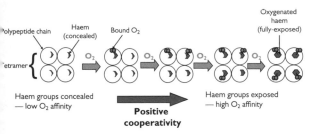

Fig. 6.19 Positive cooperativity in binding of O_2 to haemoglobin.

Variant forms of Hb

Many variant and modified forms of haemoglobin have been described only a few of which will be detailed here:

• Sickle haemoglobin (HbS) (➲ *OHCM10* pp.340–1) results from a mutation of the globin polypeptide. HbS polymerizes, especially under conditions where O_2 is low or acidity is high (e.g. in respiring tissues). The polymerized protein distorts the shape of the erythrocyte, making it sickle-shaped, and causes it to obstruct small capillaries, triggering sickling crises

• Foetal haemoglobin (HbF) (➲ *OHCM10* p.343) has a raised affinity for O_2 compared with adult haemoglobin. This facilitates delivery of O_2 to the foetus from maternal uterine blood which is at a lower partial pressure than normal arterial blood. HbF tends to disappear from foetal red blood cells a few months after birth

• Myoglobin is a monomeric form of haemoglobin expressed in striated muscle fibres. It has a much higher affinity for O_2 than haemoglobin and does not demonstrate cooperativity in its binding of O_2 (➲ p.412). Myoglobin acts as a store of O_2 available in hypoxic conditions, and also allows O_2 to be delivered to respiring cells when muscle is contracted and perfusion reduced

• Carboxyhaemoglobin: CO (➲ *OHCM10* p.842) has an affinity for Hb ~200 times that of O_2. Consequently, inhaling even a low concentration of CO causes anaemia by reducing the amount of Hb available to bind O_2. Carboxyhaemoglobin is red in colour, so patients with CO poisoning do not appear anaemic

• Methaemoglobin contains Fe^{3+} ions in its haem groups, rather than Fe^{2+}. Oxidizing agents such as nitrites and sulphonamides can cause this to occur. Methaemoglobin does not carry O_2 efficiently. Erythrocytes contain an enzyme methaemoglobin reductase, which catalyses the reduction of Fe^{3+} ion back to its Fe^{2+} form.

Carbon dioxide transport in the blood

- Carbon dioxide is transported in the blood from the tissues to the lung where it is excreted from the body (Fig. 6.20)
- CO_2 in transported by the blood in three ways:
  - As dissolved CO_2
  - In the form of HCO_3^-,
  - Complexed to the terminal amine groups of blood proteins as carbamino CO_2
- In arterial blood, transport in the form of HCO_3^-, makes up 90% of total CO_2 carriage; 5% is transported as dissolved CO_2, and 5% is transported as carbamino CO_2, 5%
- In venous blood, the equivalent proportions are 60%, 10%, and 30%.

Dissolved CO_2

- Dissolved CO_2 is carried in the blood in both intracellular and extracellular compartments. The solubility of CO_2 is 20 times greater than that of O_2 so, obeying Henry's law, transport as dissolved CO_2 constitutes a significant proportion of total CO_2 carriage in blood.

HCO_3^-

- At the tissues, CO_2 diffuses into erythrocytes and is hydrated to form carbonic acid, which subsequently dissociates into H^+ and HCO_3^-:

$$CO_2 + H_2O \leftrightarrow H_2CO_3 \leftrightarrow H^+ + HCO_3^-$$

- The hydration reaction is accelerated 13,000-fold by the intracellular enzyme carbonic anhydrase
- H^+ binds to intracellular buffers, primarily Hb
- HCO_3^- exits the erythrocyte in exchange for extracellular Cl^- on AE1 (anion-exchanger isoform 1), thereby pulling the equilibrium of the above reaction to the right
- At the lungs, the reverse sequence of events occurs, with the reformed CO_2 diffusing across the blood–gas interface into alveoli
- Intracellular $[Cl^-]$ is therefore higher for venous erythrocytes than for arterial erythrocytes (chloride shift).

Carbamino CO_2

CO_2 can bind to the terminal amine groups of proteins in blood cells and plasma. Hb is the most significant protein for carrying CO_2 in this way: deoxygenated Hb binds CO_2 more readily than oxygenated CO_2.

The CO_2 dissociation curve

- The CO_2 dissociation curve (Fig. 6.21) is almost linear in the working range
- CO_2 dissociates more readily from oxygenated Hb (i.e. the CO_2 dissociation curve is shifted to the right): this is termed the Haldane effect
- The Haldane effect occurs because deoxygenated Hb is a weaker acid than oxygenated Hb, so more readily binds either H^+ (promoting dissociation of carbonic acid) or the weak acid CO_2 (allowing the formation of carbamino CO_2). Similarly, under acid conditions, the off-loading of O_2 from oxyhaemoglobin is promoted (the Bohr effect).

(a) CO_2 uptake by red cells as the blood perfuses active tissues

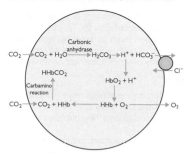

(b) O_2 uptake by red cells as the blood passes through the lungs

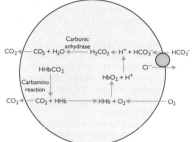

Fig. 6.20 A schematic representation of: (a) the exchange of CO_2 and O_2 that occurs between the blood and tissues; (b) the exchange that occurs in the lungs between the blood and the alveolar air.

Reproduced with permission from Pocock G, Richards CD (2006). *Human Physiology: The Basis of Medicine*, 3rd edn, p236. Oxford: Oxford University Press.

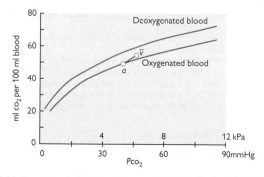

Fig. 6.21 The carbon dioxide dissociation curve for whole blood and the Haldane effect.

Reproduced with permission from Pocock G, Richards CD (2006). *Human Physiology: The Basis of Medicine*, 3rd edn, p236. Oxford: Oxford University Press.

Regulation of breathing

In healthy individuals, the act of breathing is largely automatic and is regulated to meet the body's requirements for O_2 uptake and CO_2 excretion. In addition, certain activities (such as sneezing, coughing, swallowing, or speech) also require short-term adjustments to breathing pattern. Of course, one can voluntarily override automatic breathing for short periods.

Central control

- The respiratory rhythm is generated in respiratory centres in the medulla. Sectioning of the brainstem above these areas abolishes voluntary (cortical) control of breathing but leaves its normal rhythmicity intact
- Two groups of upper motor neurones are important: the dorsal respiratory group initiate inspiration, while the ventral respiratory group are responsible for inspiration and expiration. Reciprocal inhibition is evident between inspiratory and expiratory cells. Both of these groups of neurones exhibit action potentials with a frequency that corresponds to the ventilatory cycle
- Higher inputs from the pons and the cortex can modify the rhythm of the respiratory group neurones. Furthermore, afferent fibres (largely from chemoreceptors) can also regulate their activity
- Certain reflex responses from the lungs modify breathing behaviour. The Herring–Breuer reflex describes the inhibition of inspiration as the lungs are stretched. This reflex pathway limits the depth of inspiration, particularly during heavy breathing
- Irritant receptors in the nasal mucosa and other airways result in a reflex sneeze or cough. This response helps to clear the airways of the original irritant. J-receptors, located in the lung interstitium, respond if the lungs become congested, and limit the rate and depth of breathing.

Chemical control of breathing

- Sensing of PO_2 and PCO_2 is performed by central and peripheral chemoreceptors that are located in the ventral surface of the medulla (central) and the aortic arch and carotid artery (peripheral). These organs respond to changes in blood gases and initiate rapid changes in respiratory rate in response
- Central chemoreceptors are responsive to changes in the pH of the extracellular fluid of the brain (Fig. 6.22):
  - In turn, the pH of this compartment is determined by the pH of blood and the CSF. The blood–brain barrier and the CSF–brain barrier are relatively impermeable to charged proton equivalents (e.g. $H+$ or HCO_3^-) and, therefore, it is PCO_2 which is the major determinant of pH of the brain interstitium. Increases in CO_2 shift the equilibrium:

$$CO_2 + H_2O \leftrightarrow H^+ + HCO_3^-$$

  to the right and thus decrease pH

- Central chemoreceptors respond to the fall in pH by increasing the frequency of action potentials in their afferent nerve, resulting in an increase in ventilatory rate and reduction of PCO_2
- The pH of the CSF is more sensitive to changes in PCO_2 than the pH of the brain interstitium, because the proton-buffering capacity of the CSF is lower (few proteins). Consequently, ventilatory rate is most sensitive to the composition of this compartment

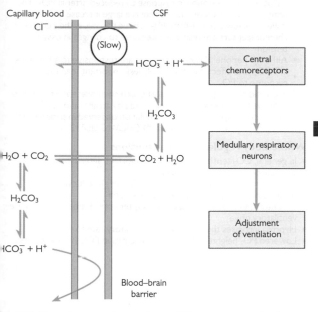

Fig. 6.22 Schematic diagram illustrating how the PCO_2 of the capillary blood in the brain stimulates the central chemoreceptors.

Reproduced with permission from Pocock G, Richards CD (2006). *Human Physiology: The Basis of Medicine*, 3rd edn, p336. Oxford: Oxford University Press.

- Peripheral chemoreceptors are small (7mm × 5mm), encapsulated organs that receive a large blood supply:
  - In contrast to central chemoreceptors, these receptors sample PO_2, pH, and PCO_2 in arterial blood. The afferent nerve fibres of peripheral chemoreceptors fire more frequently in response to lowered PaO_2 or arterial pH, or raised $PaCO_2$
  - They are the only part of the respiratory system that is able to elicit an increase in ventilation in response to reduced arterial PO_2. The firing rate of the afferent nerve shows a large increase as PO_2 is lowered below about 100mmHg—above this value, the peripheral chemoreceptors are relatively insensitive to changes in oxygen tension
  - Although peripheral chemoreceptors are responsive to alterations in PCO_2, this effect is quantitatively less important than their response to PO_2
- The carotid bodies (but not the aortic arch chemoreceptors) are also responsive to decreases in pH not elicited by an alteration to PCO_2. This response explains the hyperventilation observed in patients with, for example, diabetic ketoacidosis (→ *OHCM10* pp.832–3).

Whole-body regulation of gas tensions

- In general, CO_2 tensions are a more important determinant of ventilatory rate than O_2 tensions
- Until a threshold value is reached (~100mmHg), reducing oxygen tensions has only a minor effect on ventilation
- In contrast, even small increases in CO_2 tensions dramatically increase ventilatory drive
- Hypoxia increases the sensitivity of ventilatory rate to PCO_2
- Lowered PO_2 heightens the sensitivity to raised PCO_2 and vice versa.

Respiratory disorders

There are a number of disorders that, at best, are an irritant, but can be debilitating or even fatal in serious cases.

Cough

(→ *OHCM10* p.48.)
- Coughing is a reflex that is activated to remove mucus and extraneous material (e.g. dirt) in order to prevent occlusion of the airway
- It is sometimes overstimulated in response to local inflammation caused by viral or bacterial infection, or a persistent irritant such as cigarette tar.

Box 6.1 summarizes the treatments for cough.

Box 6.1 Treatments and drug therapies

- The mechanism underlying cough is very poorly understood but non-addictive analogues of morphine reduce the symptoms of coughing
- Codeine (methylmorphine) is the most commonly used opiate for symptomatic treatment of cough but it causes sputum to thicken, preventing its use in conditions such as chronic bronchitis and asthma, where thick sputum might exacerbate breathlessness.

Asthma

An inflammatory disorder resulting in episodic constriction of the bronchioles (bronchoconstriction), → acute breathlessness, wheezing, and cough (→ *OHCM10* pp.178–83, 810–11).
- Affects up to 10% of the population in Western countries and is increasing both in prevalence and severity
- Although there is likely to be a genetic and/or early childhood environmental influence on predisposition to asthma, an attack is generally prompted by an environmental stimulus (e.g. allergens, atmospheric pollutants, or even cold air)
- Episodes can be broadly divided into two phases, although the severity of each varies enormously between individual cases: the first phase is characterized by rapid bronchoconstriction in response to inflammatory cell-derived release of histamine, platelet-activating factor (PAF), prostanoids (PGD_2), and leukotrienes (e.g. LTC_4, LTD_4) caused by the irritant stimulus (Fig. 6.23). Activation of the inflammatory cells also releases chemokines that attract specific T-lymphocytes (Th2 cells) together with other inflammatory cells (eosinophils are particularly important). These cells generate increasing amounts of the inflammatory mediators, resulting in lung epithelial cell damage, bronchial hyper-reactivity, and inflammation (the late phase), culminating in an asthma attack (Fig. 6.23).

Box 6.2 summarizes the treatments for asthma.

Box 6.2 Treatments and drug therapies for asthma

- There is no cure for asthma. Bronchodilators (e.g. β_2-adrenoceptor agonists such as salbutamol and terbutaline and, to a lesser extent, caffeine-related compounds such as theophylline, or muscarinic receptor antagonists such as ipratropium) are an effective means of alleviating the symptoms of bronchospasm associated with asthma
- Chronic inflammatory asthma is best treated with glucocorticoids (e.g. prednisolone, hydrocortisone), which inhibit production of many of the cytokines and inflammatory mediators that are important in the recruitment and activation of inflammatory cells
- Severe allergic asthma can be treated with inhibitors of the interleukin IL-5 or its receptor, IL-5R. Mepolizumab and reslizumab are monoclonal antibodies that bind to IL-5 and reduce the number of eosinophils by reducing their rate of production. IL5R monoclonal antibodies are under development—these reduce eosinophil numbers by inducing antibody-dependent cell death.

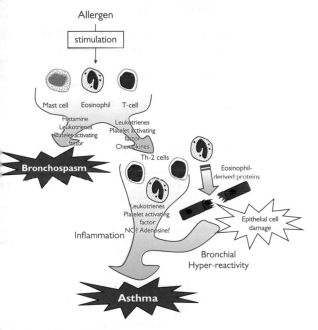

Fig. 6.23 Cellular events in asthma.

Chronic obstructive pulmonary disorder (COPD; chronic bronchitis and emphysema)

COPD is a term that is often attributed to conditions that cause constan[t] breathlessness and wheezing, such as chronic bronchitis and emphysem[a] (➔ *OHCM10* pp.184–5, 812–13).

Chronic bronchitis
- Inflammation of the bronchi in response to the permanent damage caused to the lungs by the effects of smoking, pollution, or infection
- Symptoms include constant breathlessness and a persistent, chesty cough. Both symptoms are brought about by inflammation-induced increases in mucus secretions, reducing the accessibility of air to the alveoli. The condition deteriorates with infections such as colds and influenza.

Emphysema
- In reality, can only be fully diagnosed histologically post-mortem
- Involves the destruction of supporting structures in the lung → the collapse of airways. Trapped air in the alveoli results in over-inflation of the lungs. Smoking is again the primary cause of the disease.

Box 6.3 summarizes the treatments for COPD.

Box 6.3 Treatments and drug therapies for COPD

Like asthma, COPD is not curable but the symptoms can be eased by:
- Stopping smoking
- Administration of bronchodilators (e.g. salbutamol)
- Administration of muscarinic antagonists to reduce sputum production
- Long-term oxygen therapy.

Adult (or acute) respiratory distress syndrome (ARDS)

ARDS is not a specific disease but is the result of lung injury caused by direct insult to the lung (e.g. inhalation of smoke or corrosive gas) or as [a] result of severe systemic inflammation in response to infection (e.g. sepsi[s,] severe viral infections like flu and COVID-19) that leads to multiorgan dys[-] function (➔ *OHCM10* p.186).

ARDS is attributed to massive capillary leak as a result of the recruit[-] ment and activation of large numbers of inflammatory cells at the site o[f] injury. This over-response leads to endothelial damage and increases th[e] permeability of alveolar capillaries, resulting in the flooding of alveoli. Th[e] syndrome is associated with the following symptoms:
- Alveolar collapse
- Poor lung compliance
- Gas exchange problems → hypoxaemia (lack of oxygen in the blood)
- Pulmonary hypertension (exacerbating the capillary leak and the resulting pulmonary oedema).

In the long term, inflammation gives way to fibrotic remodelling—the layin[g] down of the fibrous structures associated with scar tissue. Lung function i[s] yet more compromised because the scar tissue leads to yet more destruc[-] tion of the microcirculation and reduced compliance.

Box 6.4 summarizes the treatments for ARDS.

Box 6.4 Treatment and drug therapies for ARDS

The prognosis for ARDS is very poor (mortality rate of 50–75%). Therapeutic interventions are limited, but current management practice is as follows:

- Provide respiratory support (oxygen)
- Use inhaled NO to reduce pulmonary hypertension
- Treat underlying infection with antibiotics, if applicable.

Pneumonia

(🠊 OHCM10 pp.166–7, 816.)

Pneumonia is the collective term to describe infection of lung parenchymal tissue. The source of the infection is most often bacterial, but can also be viral or fungal. Before the advent of antibiotics, pneumonia was often fatal. Even today, mortality is ~10%, but most patients can be treated in the community and do not require hospitalization. Symptoms include:

- Breathlessness
- Fever
- Chest pain (especially on deep breaths)
- Cough (mucus can contain blood in severe cases)
- Anorexia
- Fatigue.

Pneumonia is classified as follows:

- Community-acquired pneumonia: most commonly caused by *Streptococcus pneumoniae* bacteria, but can be the result of a secondary infection in patients with COPD (*Pseudomonas* infection) or through viral infection (influenza, COVID-19)
- Nosocomial pneumonia: acquired >48h after hospital admission. Often Gram-negative enterobacterial infection or *Staphylococcus aureus*
- Aspiration pneumonia: aspiration is defined as the entry of a foreign substance into the respiratory tract and the associated pneumonia can refer to the resulting physical damage or to a subsequent infection of damaged parenchymal tissue. This condition most often affects people in unconscious states (e.g. alcohol-induced) but can also be related to oesophageal disorders such as reflux, or even poor dental hygiene. *Streptococcus pneumoniae* is the primary infective agent in this form of pneumonia
- Immunocompromised pneumonia: pneumonia is a common cause of death in patients with compromised immune systems (e.g. AIDS). A wide range of infective agents, including *S. pneumoniae*, Gram-negative bacteria, and influenza virus proliferate rapidly in the absence of a weakened immune response and can ultimately result in respiratory collapse.

Box 6.5 summarizes the treatments for pneumonia.

Box 6.5 Treatment and drug therapies for pneumonia

- Management and treatment involve provision of oxygen and treatment with appropriate antibiotics (🠊 OHCM10 p.167) (e.g. amoxicillin for *Streptococcus pneumoniae* (also known as pneumococcus)) and fluids, in severe cases
- Preventative measures include vaccination against pneumococcus and influenza in vulnerable groups.

Cystic fibrosis

(➔ *OHCM10* p.173.)

Genetics

Cystic fibrosis is transmitted by an autosomal recessive gene and this gene is the most frequent seriously deleterious recessive gene in the Northern European gene pool with a frequency of 1 in 25 in the UK and a disease incidence of 1 in 2500 live births.

Molecular mechanisms

The gene which is defective in cystic fibrosis codes for a protein which controls a chloride channel in the apical membrane of epithelial cells. The defective protein does not transport chloride into the epithelial cells from the surface and this has different effects in different epithelia. In the airways, there is reduced secretion of chloride into the surface mucus with subsequent sodium and water resorption into the cells causing dehydration of the surface mucus, which is then much more viscid and is not cleared by the mucociliary escalator, → recurrent respiratory infections.

Different mutations in the cystic fibrosis gene produce proteins with a range of function from none through to moderately reduced, so the phenotype of the disease varies with the extent of the protein dysfunction.

Complications of cystic fibrosis

- Bronchiectasis and recurrent respiratory infection: the viscid secretions in the bronchi cause obstruction → permanent dilation (bronchiectasis) and recurrent infection by organisms such as *Pseudomonas aeruginosa*. Respiratory failure is the cause of death in cystic fibrosis
- Malabsorption: viscid secretions block the pancreatic ducts with subsequent atrophy of the exocrine glands. The resultant lack of digestive enzymes leads to malabsorption, especially of fats and fat-soluble vitamins. Oral supplements of enzymes can improve this
- Hepatic cirrhosis: again secondary to viscid secretions obstructing the biliary tree
- Male infertility.

Box 6.6 summarizes the treatments for cystic fibrosis.

Box 6.6 Treatment and drug therapies for cystic fibrosis

- Management of cystic fibrosis centres on actions to help dislodge mucus in the lungs
- This can involve daily chest physiotherapy or use of mechanical devices and use of various medications to dislodge (hypertonic saline) or reduce the viscosity (deoxyribonuclease, N-acetylcysteine) of sputum and to cause bronchodilatation (β_2 receptor agonists or muscarinic acetylcholine receptor antagonists).

Potential of gene therapy

Since the airways are easily accessible to fluids dispersed as aerosols, it seems that a potential therapy for cystic fibrosis would be to inhale aerosols containing DNA fragments with a normal cystic fibrosis gene. If this could enter the epithelial cells lining the respiratory tract and encode mRNA, then these cells could produce a normal protein and the disease could be alleviated at this site. While there has been some progress, there are still problems with finding a vector that will mediate large-scale take-up of the DNA. A more promising, though less sophisticated, molecular therapy is inhalation of recombinant human DNAase, which breaks down the polymerized DNA from neutrophils. This makes a large contribution to the increased viscosity of bronchial secretions in cystic fibrosis.

Pulmonary embolism

(→ *OHCM10* pp.190, 818.)

Another cause of respiratory disorder is pulmonary embolism due to the occlusion of one or more pulmonary arteries by an embolus (usually a displaced thrombus) originating from the large veins in the pelvis or legs (deep vein thrombosis; Fig. 6.24). When the thrombus becomes dislodged, it travels relatively unimpeded through the veins (which increase in diameter as they approach the heart) and through the right side of the heart. Once the thrombus passes out of the right ventricle into the pulmonary artery, the arteries that it encounters are of progressively reducing diameter. Ultimately, the thrombus is unable to progress and lodges in a vessel, preventing blood flow to areas of the lung downstream of the occlusion.

The severity of the effect is determined by the size of the area served by the occluded artery: a large thrombus will block a large vessel, causing severe lung tissue damage and is usually fatal; smaller emboli (microemboli) may not inflict sufficient damage to induce symptoms. On occasion, the embolus may not be thrombotic in origin: fat or air can be similarly transported to the lung through the venous system, although this is usually only associated with surgery or scuba diving, respectively.

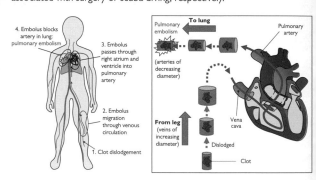

Fig. 6.24 Sequence of events → pulmonary embolism in an individual with deep vein thrombosis. Inset: the importance of variance in blood vessel diameter in determining embolic migration and final destination.

Symptoms

Diminished perfusion of the lungs results in reduced blood available for gas exchange, resulting in breathlessness and hypoxaemia.

Box 6.7 summarizes the treatments for pulmonary embolism.

Box 6.7 Treatments and drug therapies for pulmonary embolism

Treatments include immediate oxygen therapy and anticoagulant therapy (e.g. warfarin and low-molecular-weight heparin (LMWH)) to prevent further events.

Altitude and pulmonary oedema

With increasing altitude, air pressure falls, resulting in a fall in the amount of oxygen present in a given volume. Therefore, less oxygen is inhaled with each breath and the breathing rate has to increase to compensate. This short-term solution to the problem of reduced oxygen levels is accompanied by slower adaptive changes to help improve blood oxygenation, including an increase in red blood cell generation in response to kidney-derived erythropoietin. These adaptive changes are termed 'acclimatization' and can take several days; high-altitude climbers will progress slowly to give sufficient time for their bodies to acclimatize. Nevertheless, many climbers experience 'altitude sickness' (◐ *OHCM10* p.802) above altitudes of ~3000m (10,000ft), complaining of headaches, nausea, and breathlessness. In more serious cases, severe pulmonary or cranial oedema can ensue as a result of increased permeability of capillaries in the lungs or brain respectively—the mechanism remains largely unknown.

Pulmonary oedema results in severe breathlessness that is rapidly alleviated by taking the patient back to low altitude. Cranial oedema is more serious because the patient is often unaware of their condition, even if their delirious behaviour is obvious to others. Failure to return to low altitude very rapidly can result in death. Susceptibility is not predictable and is not related to fitness; susceptible individuals are probably genetically predisposed to altitude sickness.

Blood—plasma

Blood consists of a suspension of a number of different types of cells in an aqueous medium called plasma (Table 6.2). The primary function of blood is the transport of oxygen in red blood cells and essential nutrients in the plasma to tissues and organs for use in cellular respiration. Conversely, the waste products of cellular metabolism are also carried away from tissues by the blood. However, the transport function of blood extends well beyond the requirements for cellular metabolism: many hormones and other signalling molecules are transported from the glands where they are synthesized to their target tissues by the blood, while white blood cells and platelets are carried in the blood to help protect against infection and to repair tissue damage.

Plasma is an iso-osmotic aqueous medium, the basic constituents of which are sodium and potassium salts, glucose, and plasma proteins (Table 6.2). Plasma is isolated from the cellular fractions of blood by centrifugation in the presence of an anticoagulant to prevent the blood from clotting. Serum is similar to plasma but is isolated by allowing the blood to clot; as a result, serum does not include clotting proteins.

The concentration of the plasma constituents is regulated primarily by the kidney, where water, sodium, and urea are excreted. Plasma glucose is tightly regulated by the actions of the pancreatic hormones—insulin and glucagon—but can more than double immediately following a meal containing carbohydrates. The plasma proteins, of which albumin is most abundant, are mainly synthesized in the liver and they perform a number of functions, including the maintenance of osmotic balance in the plasma and removal of potentially toxic chemical entities that might enter the bloodstream (e.g. transition metal ions and some drugs and toxins). Plasma proteins do not normally pass through blood vessel walls into tissues.

Table 6.2 Blood composition and constituent function

Blood constituent	Content/cell count	Function
Plasma	Water Electrolytes Glucose Proteins Hormones	Transport, electrolytic, and osmolarity balance
Red blood cells	5×10^9/mL	O_2, CO_2 transport
White blood cells	9×10^6/mL	Defence
Platelets	3×10^8/mL	Haemostasis

Red blood cells

Red blood cells (also known as erythrocytes) are the most abundant cell type in the blood (Table 6.2). These cells do not have nuclei (they are anucleate) and account for ~40% of total blood volume (haematocrit) in healthy adults. They develop in the bone marrow from large, nucleated normoblastic cells that differentiate into mature red cells in response to the kidney-derived hormone, erythropoietin. Healthy red blood cells conform to a biconcave discoid shape (~8μm in diameter) that is sufficiently flexible to allow them to pass through capillaries that can be as narrow as 3μm.

Red blood cells are packed with the specialized, pigmented protein, haemoglobin, the primary function of which is to bind molecular oxygen (O_2) in vascular beds where O_2 is abundant (the lungs) and deliver it to regions where the oxygen is needed to maintain cellular respiration. Haemoglobin is a complex protein comprising four subunits (2α and 2β in adults), each of which contains a haem group. The haem groups are synthesized in the mitochondria of maturing red blood cells in a process requiring vitamin B_6 and circulating iron bound to transferrin. In order to bind O_2 efficiently, haem iron is kept in its reduced (Fe^{2+}) form as opposed to the met (Fe^{3+}) form by NADPH-fuelled methaemoglobin reductase and the abundant endogenous antioxidant GSH.

The interaction of haemoglobin with O_2 is covered in detail on ● p.412.

Blood groups

Red blood cells, along with many other cells in the body, express antigens on their cell membranes. These antigens are glycoproteins (proteins linked to carbohydrate chains), the precise structures of which are determined at a genetic level. In humans, two types of antigen can be expressed, each with a different sugar residue at a specific locus in its carbohydrate chain: type A has an acetylgalactosamine, whereas type B has galactose (Fig 1.26). Blood groups are assigned according to which of these antigens is expressed in an individual, hence the blood groups A, B, AB (where both are expressed), and O (where neither is expressed).

Blood groups are inherited according to Mendelian principles, with A and B co-dominant, while O is recessive. The blood group of an individual also determines the antibodies that they will produce against red blood cell antigens—the overriding determining factor for compatibility of blood used in transplants. Should plasma containing anti-A antibody come into contact with red blood cells expressing the A antigen, they congeal or agglutinate, making them functionally inactive. This provides the basis for the simple test that is routinely conducted to determine blood group to enable cross-matching before transfusions (Figure 1.27).

The rhesus (Rh) system

A and B antigens do not constitute the only types of antigen that are expressed on red blood cells. Indeed, there are a great number of antigens on red blood cells, the combination of which are probably unique to any given individual. Many of these antigens have not yet been fully characterized, but those associated with the Rh system are recognized, alongside the ABO antigens, as important identification tags.

Rh system antigens constitute a group of glycoproteins that are expressed in most people (85% of Caucasians and 99% of Asians are Rh positive (Rh⁺)).

The Rh system only becomes an issue when a Rh⁻ mother gives birth to a Rh⁺ child who is of the same ABO blood group as the mother. In the event of some red blood cells from the foetus crossing into the mother's circulation during childbirth, the cells will not be destroyed by the mother's immune system because they are of the same ABO group. However, the surviving cells will induce the mother to produce antibodies against Rh antigens and, while the first child is unaffected, should the mother have a second Rh⁺ child, her immune system will destroy the red blood cells of the child, → haemolytic disease. The issue is overcome in modern medicine by screening pregnant women for Rh and injecting them with Rh⁺ antibodies during pregnancy and immediately after birth to destroy any foetal red blood cells before the immune system has time to generate antibodies of its own.

White blood cells

The comparatively large nucleated cells found in the blood are collectively known as the white blood cells or leucocytes, the primary function of which is defence against infection. The white blood cell population can broadly be divided into two classes of cell:

- The lymphocytes that act in concert with immunoglobulins and the complement system to instigate immunity
- The phagocytes that contribute to inflammatory processes and actively ingest invading pathogens, diseased host cells, and cellular debris.

The function of white cells is covered in detail on p.430. The origins and localization of white blood cells are outlined in Fig. 6.25.

Lymphocytes and natural killer cells

Like all blood cells, lymphocytes are ultimately derived from pluripotent stem cells in the bone marrow, which differentiate into lymphoid stem cells in response to specific inflammatory mediators (e.g. interleukin 3 (IL-3)) and growth factors (e.g. granulocyte-macrophage colony stimulating factor (GM-CSF)), release of which is stimulated by infection. In this way, the body responds rapidly to infection by generating more white blood cells to help deal with the crisis.

The next stage of development is differentiation of lymphoid stem cells into B lymphocytes (in the bone marrow) and T lymphocytes (in the thymus). B lymphocytes are responsible for generating antibodies for immunity and are characterized by expression of specific surface markers (CD19, -20, and -22). T lymphocytes express CD4 or CD8 surface markers (among others) and have the ability to distinguish between healthy cells that belong to the host individual ('self') and foreign or diseased cells ('non-self'). CD4-expressing lymphocytes help to produce antibodies ('helper cells'), while CD8-expressing lymphocytes initiate cell-mediated immunity against intracellular organisms. Mature B and T cells circulate in the bloodstream but are able to migrate into tissues to fight infection and also gravitate towards the lymph nodes and spleen, via the lymphatic system, resulting in localized swelling of the nodes during infection.

Lymphocytes flow back into the bloodstream, via the thoracic duct, into the superior vena cava. Natural killer (NK) cells are also derived from lymphoid stem cells in the bone marrow. NK cells are designed to search and destroy cells that are infected by viruses or are cancerous.

The phagocytes

Phagocytes are also derived from pluripotent stem cells that differentiate into myeloid stem cells, via haemopoietic stem cells in the bone marrow. The fate of myeloid stem cells is determined by the relative abundance of specific monocyte, granulocyte, or eosinophil-derived growth factors and interleukins. This system results, therefore, in sophisticated self-regulation whereby generation of cells of a particular type is only initiated if the existing cells of that type are highly activated and in need of reinforcements.

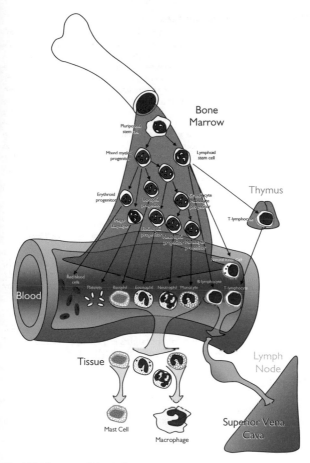

Fig. 6.25 The origin of blood cells.

Monocytes and macrophages

Monocytes are derived directly from myeloid cells and pass into the bloodstream, where they circulate in a quiescent form for up to 2 days before they migrate into the tissue and differentiate into mature phagocytic macrophages. Macrophages are the scavengers of the immune system and are particularly prevalent in the liver (where they are sometimes called Kupffer cells) and the lungs. However, they accumulate in specific sites of infection to help destroy invading pathogens and to clear cellular debris from the site.

Neutrophils

These granulocytes have a very distinctive nuclear arrangement consisting of densely packed chromosomal material in two to five distinct lobes. Like monocytes, mature neutrophils also differentiate from myeloid progenitor cells via granulocyte stem cells. The cytoplasm of neutrophils is packed with granules of different types:

- Primary granules contain a range of enzymes (e.g. myeloperoxidase) that generate highly toxic, oxygen-related species, including superoxide and hydrogen peroxide
- Secondary granules contain enzymes (e.g. lysozyme, collagenase) that act to lyse cells and digest their contents, or deprive them of essential iron (lactoferrin).

The primary function of neutrophils is the identification, phagocytosis, and killing of invading pathogens. Mature neutrophils only circulate for about 10h before they undergo programmed cell death and are cleared by macrophages.

Eosinophils

Eosinophils are very similar in structure, function, and origin to neutrophils. The distinguishing feature of eosinophils is their two- to three-lobed dense nucleus. They are often associated with allergic reactions and defence against parasites.

Basophils

Basophils are rarely found in peripheral blood and when they enter tissues, they become mast cells that are involved in the recruitment of other inflammatory cells to sites of infection or damage. These cells are packed with histamine and heparin-containing granules that can obscure the nucleus.

Platelets

Platelets are very small (1–2μm in diameter), discoid, subcellular fragments that do not contain nuclei but have most of the other features associated with cells, including mitochondria, ER, and a crude microfilament and actin-based cytoskeleton. Platelets are derived from megakaryocytes in the bone marrow in response to thrombopoietin synthesized in the kidneys and liver. Their generation is autoregulated on account of clearance of circulating thrombopoietin by platelets.

The function of platelets is to stop blood loss after injury by forming a plug in damaged blood vessels and releasing agents (e.g. thrombin) that contribute to rapid clot formation. They also release signals to recruit and activate further platelets (e.g. ADP, thromboxane A$_2$, 5-HT) and to attract inflammatory cells (e.g. platelet-derived growth factor (PDGF)) to the site of injury in order to ward off any potential infection. This process of haemostasis is covered in detail on p.488.

Heart morphology

The heart is a four-chambered, muscular pump responsible for perfusing the vascular network with blood. The left and right sides of the heart effectively operate as two pumps, respectively, sending blood through the systemic and pulmonary circulations that are arranged in series.

Gross anatomy

See Fig. 6.26.

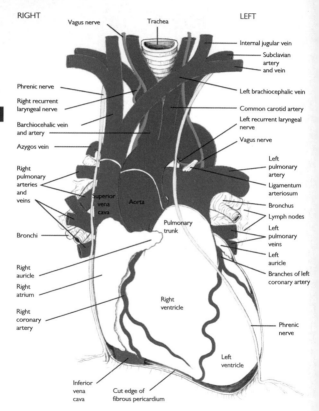

Fig. 6.26 The heart and great vessels viewed from the front.

Reproduced from MacKinnon, Pamela, and Morris, John, *Oxford Textbook of Functional Anatomy* Vol 2, p73 (Oxford, 2005) With permission from OUP.

- The outline of the heart, which is roughly conical in shape, can be consistently defined on the surface of the chest according to the following guidelines:
  - The superior border of the heart is defined as a line following the 2nd intercostal space ~4cm either side of the sternum
  - The right border runs from the 3rd right costal cartilage to the 6th right costal cartilage
  - The apex of the heart is located in the 5th intercostal space in the mid-clavicular line. Its inferior border runs from this point to the right border
  - The left border of the heart connects the apex to the superior border
  - These relationships are of importance to clinicians during auscultation and palpation (➲ OHCM10 pp.38–41)
- The heart is located in the middle mediastinum within the pericardium: a double-layered sac which completely surrounds the heart apart from the points where the great vessels enter and leave
  - The external pericardial layer is referred to as fibrous pericardium: a structure which prevents excessive distension of the heart
  - On its internal face, the heart is lined by a parietal layer of serous pericardium. This is continuous with the visceral layer of serous pericardium (epicardium) forming a potential space (pericardial cavity), which contains a thin layer of fluid permitting the heart to move within the pericardial sac
  - Pericardial sinuses are formed by the reflection of the pericardium around the heart. These sinuses are small, blind-ending spaces between the heart and the great vessels (oblique sinus) and around the aorta and pulmonary trunk posterior to the heart (transverse sinus)
- The heart is fist-sized and lies obliquely within the thorax such that its anterior surface is formed largely of the right atrium and the right ventricle, while the left atrium and the left ventricle are orientated posteriorly
- The wall of the heart is made up of three layers: the epicardium is the most superficial of these and is lined by the muscular myocardium, which in turn is separated from the chambers by a layer of endocardium (monolayer of endothelial cells)
- The fibrous skeleton of the heart is formed by a cartilaginous ring at the level of the membranous ventricular septum, separating the atria from the ventricles. It contains the AV, aortic, and pulmonary orifices. It provides electroinsulation, so that electrical impulses cannot pass directly from the atria to the ventricles except via the AVN. This fibrous skeleton also supports the cardiac valves at the base of the cusps, preventing stretching and incompetence of the valves.

Blood flow through the heart

- The systemic circulation drains into the right atrium via the superior and inferior venae cavae, while the cardiac veins enter the right atrium via the coronary sinus
- The right atrium is separated from the left atrium by the interatrial septum within which can be found the fossa ovalis—a vestige of the foramen ovale which permits the shunting of blood from the right to the left atrium in the foetus

- Right atrial contraction forces blood into the right ventricle through the right atrioventricular (AV) orifice. This structure is bounded by the right AV (tricuspid) valve which prevents any back-flux of blood during ventricular contraction. The papillary muscles which arise from the ventricular wall are attached to the loose edges of the cusps of the right AV valve (usually one per cusp) via chordae tendineae which maintain the direction of the cusps. The three cusps of this valve are attached to a fibrous ring surrounding the AV orifice
- Right ventricular contraction forces blood into the pulmonary trunk via the infundibulum. The interventricular septum separates the right and left ventricles and bulges into the right ventricle because of higher pressure in the left ventricle. Back-flux of blood from the pulmonary circulation into the right ventricle is prevented by the presence of the pulmonary valve; this comprises three semi-lunar cusps. Stenosis of the pulmonary valve, frequently occurring alongside infundibular pulmonary stenosis, narrows the outflow from the right ventricle and causes right ventricular hypertrophy
- The left atrium receives the four pulmonary veins (two inferior and two superior). Its wall is slightly thicker than that of the right atrium owing to the higher pressures within the systemic circulation. The fossa ovalis is a visible part of the interatrial septum. Blood is expelled from the left atrium past the mitral valve, via the AV orifice, to the left ventricle. The mitral valve is analogous in structure and function to the tricuspid valve, although it comprises only two cusps and its papillary muscles are larger than their counterparts in the right side of the heart
- The wall of the left ventricle is approximately twice as thick as the right ventricle. It pumps blood into the aorta via the aortic orifice. The three semi-lunar cusps of the aortic valve guard this opening
- Aortic sinuses are formed behind the cusp of each valve as a bulge in the aortic wall. The posterior sinus supplies the origin of the left coronary artery, and the anterior cusp provides the origin of the right coronary artery, which supplies the heart.

Blood and nerve supply to the heart

See Fig. 6.26, ➔ p.438.

- The coronary circulation must provide the myocardium with sufficient blood to meet its high basal oxygen consumption, and have the capacity to increase during exercise
- The high pressure that develops in the ventricular wall during systole transiently shuts off the coronary circulation. This effect means that 80% of coronary blood flow occurs during diastole. In order to cope with these challenges, myocardium contains a high density of capillaries, increasing the efficiency with which nutrients and waste products can be exchanged. Furthermore, the total oxygen extraction from the coronary circulation is high
- The branches of the coronary arteries are particularly sensitive to obstruction (e.g. during atherosclerosis; ➔ OHCM10 pp.116–17), since anastomoses are infrequent and inefficient (functional end arteries). Such events underlie myocardial infarction (MI) and angina. Sites of obstruction can be localized by coronary angiography (➔ OHCM10 pp.112–13).

Arterial supply

See Fig. 6.27 (➔ OHCM10 p.113.)

- The right and left coronary arteries arise from the aorta just distal to the aortic valve at the coronary sinus. These vessels supply the myocardium and the epicardium:
  - The right coronary artery passes from the anterior aortic sinus anteriorly, past the pulmonary trunk in the right AV groove, in which it passes under the inferior border of the heart; ultimately it anastomoses with the circumflex branch of the left coronary artery at the posterior interventricular groove. A branch of the right coronary artery (the posterior interventricular artery) runs in the inferior interventricular groove and anastomoses with the anterior interventricular artery near the apex of the heart, which arises from the left coronary artery. The right posterior interventricular artery supplies the AVN in 90% of people, while in the remaining 10%, the AVN is supplied by a branch of the left coronary artery. Clearly, this is of significance following myocardial ischaemia. A marginal branch also arises from the right coronary artery and passes along the inferior border of the heart
  - The left coronary artery passes from the posterior aortic sinus and runs in the left AV groove to anastomose with the right coronary artery. It gives off several key branches analogous to those from the right coronary artery—the anterior interventricular artery (which runs in the anterior interventricular groove) and the circumflex artery (which anastomoses with the right coronary artery). The left marginal artery follows the left margin of the heart
- In general, the right and left ventricles are supplied by the right and left coronary arteries, respectively, while the atria and interventricular septum can be supplied by both. There can be considerable variations from this scheme, however.

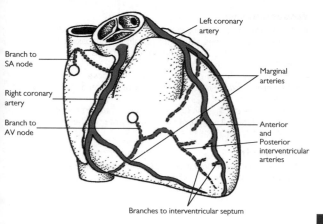

Fig. 6.27 The arterial supply of the heart.

Reproduced from MacKinnon, Pamela and Morris, John, *Oxford Textbook of Functional Anatomy*, vol 2, p75 (Oxford, 2005). With permission of OUP.

Venous drainage

- The majority of the venous drainage of the heart (Fig. 6.28) empties into the coronary sinus, although a certain amount passes directly into the right atrium (largely from anterior cardiac veins which drain the anterior aspect of the heart)
- The coronary sinus runs in the posterior AV groove and receives:
  - The great cardiac vein (from the anterior interventricular groove) at its left end
  - The middle cardiac vein (from the inferior interventricular groove)
  - The small cardiac vein (from the lower border of the heart) at its right end.

Nerve supply

- The heart receives innervation from parasympathetic, sympathetic, and sensory fibres (Table 6.3), which together form the superficial and deep cardiac plexuses below the aortic arch
- Sensory fibres innervating the heart run in close proximity to cervical and thoracic spinal nerves. This explains the phenomenon of referred cardiac pain to the chest, arms, and neck during myocardial ischaemia (➔ OHCM10 pp.94–5).

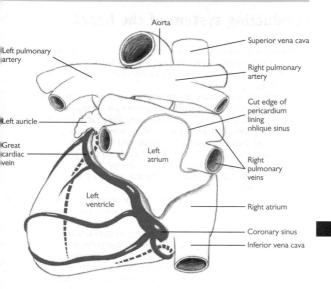

Fig. 6.28 The venous drainage of the heart.

Reproduced from MacKinnon, Pamela and Morris, John, *Oxford Textbook of Functional Anatomy*, vol 2, p76 (Oxford, 2005). With permission of OUP.

Table 6.3 Nerve supply to the heart

Type	Nerve	Innervated site	Function
Parasympathetic (efferent)	Vagus nerve	• SAN • AVN • Coronary arteries	• Decreases heart rate • Constriction of coronary arteries • No appreciable effect on contractility
Sympathetic (efferent)	Cervical and upper thoracic spinal nerves via sympathetic ganglia	• SAN • AVN • Cardiac muscle fibres • Coronary arteries	• Increases heart rate • Increases force of contraction • Dilation of coronary arteries
Sensory (afferent)	Follows sympathetic ganglia via white rami communicantes and spinal nerves	• Myocardium	• Pain from myocardium ischaemic

Conducting system of the heart

See Fig. 6.29.

- Membrane depolarization is the stimulus for contraction of cardiac myocytes to occur, and the coordinated and regulated spread of electrical excitation from the atria to the ventricles is essential for efficient pumping activity
- The cells of the heart are arranged as a branching syncytium in which depolarization passes from one cell to the next via gap junctions. Dome cardiac muscle cells contain few myofibrils and are specialized conducting fibres that allow a wave of depolarization to spread throughout the heart in a rapid, coordinated manner
- Unlike skeletal muscle, electrical impulses within the heart are generated intrinsically and are not dependent on external nervous input, although cardiac function can be modulated by the activity of the autonomic nervous system:
  - The SAN—an area of specialized cardiac tissue on the posterior wall of the right atrium—is the normal pacemaker region that initiates electrical excitation
  - The SAN exhibits the highest frequency of spontaneous activity and overrides other potential pacemaker regions (ectopic pacemakers)
  - SAN cells contain few myofibrils and are not specialized for contraction. Instead, they spontaneously depolarize in a rhythmic manner, triggering action potentials which are conducted to the surrounding atrial tissue
  - Following an action potential, the resting membrane potential of a SAN cell is −55 to −60mV. A slow inward leak of Na^+ ions (I_f, 'funny' current) then depolarizes SAN cells until an action potential is fired at about −40mV. The upstroke of the action potential in a SA node cell is a result of the influx of Ca^{2+} ions through voltage-gated Ca^{2+} channels
- The atria are almost completely electrically insulated from the ventricles by the annulus fibrosus and electrical impulses can only pass between them via the AVN:
  - The AVN is a specialized area of conducting tissue within the atrial septum that slows the conduction of the electrical impulse, allowing the atria to contract before the ventricles
  - Electrical impulses arrive at the AVN via conducting pathways from the SAN. From here, the bundle of His transmits depolarization across the annulus fibrosus and along the interventricular septum. The bundle of His divides into anterior and posterior, left and right bundle branches that pass down the left or right side of the interventricular septum and transmit impulses initially to the endocardial regions of the left and right ventricles, respectively
  - Fibres from the left and right bundle branches transmit impulses to Purkinje fibres, comprising large-diameter cells that conduct electrical impulses very rapidly. From the endocardium, contractile cells transmit impulses to each other. This network of rapid conducting fibres therefore allows all parts of the ventricles to contract virtually simultaneously

- The function of the pacemaker and conducting system of the heart can be modulated by autonomic nervous system activity
  - Sympathetic nerve activity to the SAN increases heart rate (positive chronotropic effect; tachycardia). This is mediated by catecholamines binding to β_1 adrenoceptors, → an increase in intracellular cAMP:
    - The open-state probability of channels that conduct the funny current increases, thereby increasing the magnitude of this current and decreasing the time taken for depolarization to occur.
    - The time taken for conduction through the AVN is reduced
  - Parasympathetic (vagal) nervous activity slows heart rate (negative chronotropic effect; bradycardia). This is mediated by acetylcholine binding to muscarinic (M_2) receptors:
    - The rate of depolarization of SAN cells is reduced, an effect mediated by reduced cAMP
    - Activation of K^+ channels hyperpolarizes the pacemaker cells, increasing the time required to reach threshold for an action potential.

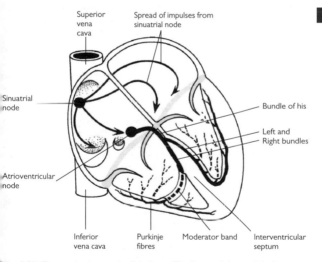

Fig. 6.29 The conducting system of the heart. The fibrous skeleton of the heart separates the muscle of the atria from that of the ventricles, which are connected only by the bundles of His.

Reproduced from MacKinnon, Pamela and Morris, John, *Oxford Textbook of Functional Anatomy*, vol 2, p77 (Oxford, 2005). With permission of OUP.

The electrocardiogram (ECG)

- The ECG (➔ *OHCM10* pp.96–107) is a record of the electrical events associated with depolarization and repolarization of the myocardium, measured as changes in the surface potential of the skin
- During the cardiac cycle, as the atria and then the ventricles undergo sequential depolarization followed by repolarization, the extracellular myocardial compartment can be treated as two moving dipoles of opposite charge
- Each dipole is essentially an aggregation of the depolarized (negative) and hyperpolarized (positive) regions of the heart. Charge flows between these two dipoles, and it is the potential arising from these minute currents that can be picked up as small (up to 1mV) potential differences at the skin
- Normally, only potentials arising from myocardial depolarization and repolarization can be recorded: the conducting system is too small to produce potential changes of sufficiently large magnitude to be measurable by ECG
- The ECG is routinely used clinically and allows abnormalities in the electrical activity of the myocardium to be detected and diagnosed. It should be remembered that the ECG is a record of electrical, and not contractile, events.

Properties of the ECG

The nature of the ECG trace varies according to the 'limb leads' used to record it (➔ see next section), although certain features are invariant in healthy subjects (Fig. 6.30):

- The P wave is the first event of the cardiac cycle observable by ECG. It arises from depolarization of the atria and lasts ~0.08s
- The PR interval is the period from the start of the P wave to the start of the QRS complex (it should more logically be known as the PQ interval). A large proportion of the PR interval is flat (after the P wave) and this represents the time taken for conduction through the AVN—the heart is essentially isoelectric during this period. The PR interval lasts ~0.2s
- The QRS complex is a record of ventricular depolarization and, as such, is analogous to the P wave for the atria. It lasts for only a short time, ~0.1s, demonstrating the almost synchronous depolarization of the ventricular myocardium
- The ST segment corresponds to the plateau phase of the ventricular action potential and, like the major part of the PR interval, represents the heart in an isoelectric state
- The T wave is a record of ventricular repolarization. It is normally in the same orientation as the QRS complex (large upward deflection), since repolarization occurs in the opposite direction to depolarization.

Standard limb leads

- The ECG is traditionally recorded using three electrodes: one on each arm and one on the left leg. Together, these three points make up Einthoven's triangle

- An ECG trace is generated by resolving the electrical vector arising from the electrical dipole onto one of the three leads connecting the above three electrodes
- During the cardiac cycle, this electrical dipole changes in magnitude and direction and, therefore, its resolution onto each of the three leads is continually changing
- Each of the three leads is in a different orientation and emphasizes different features of the ECG:
  - Lead I: right arm (−) to left arm (+)—horizontal
  - Lead II: right arm (−) to left leg (+)—60° below horizontal
  - Lead III: left arm (−) to left leg (+)—120° below horizontal
- The standard leads are therefore designed such that a positive deflection results when a positive dipole points towards the left arm (lead I) or the left leg (lead II or III).

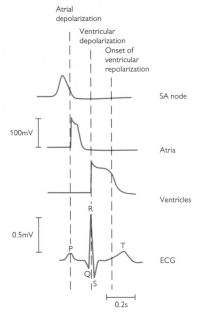

Fig. 6.30 The relationship between the onset and duration of the action potentials of cardiac cells during a single cardiac cycle and the ECG trace.

Reproduced with permission from Pocock G, Richards CD (2006). *Human Physiology: The Basis of Medicine*, 3rd edn, p271. Oxford: Oxford University Press.

Normal cardiac rhythm and arrhythmias

The action potential in the heart

There are a number of features that make the action potentials of the heart different from those seen in other excitable tissue. Most importantly, there are special pacemaker cells (found primarily in the SAN) that spontaneously generate action potentials in a cyclical fashion, thanks to their unusually high Na^+ permeability. The key events in the cardiac action potential are summarized in Fig. 6.31 and described on p.270.

The pacemaker potential is transmitted to surrounding cells in the atrium because the cells are electrically coupled. These cells do not usually have pacemaker activity and action potentials are only triggered by depolarization of adjacent cells.

Depolarization of atrial cells is of shorter duration (200ms) than that in Purkinje fibres and ventricles, where Ca^{2+} channel activation is prolonged, → a longer plateau phase and action potentials that last for 300–400ms (Fig. 6.31).

Refractory period

The refractory period is the time taken between action potentials for the cell to prepare itself for another depolarization (e.g. reactivation of Na^+ channels that are inactivated during depolarization). During the refractory period, cells cannot be activated, even if a stimulus arrives—a crucial factor in ensuring that the action potential is propagated in one direction only and cannot double-back on itself to create chaotic contractile patterns.

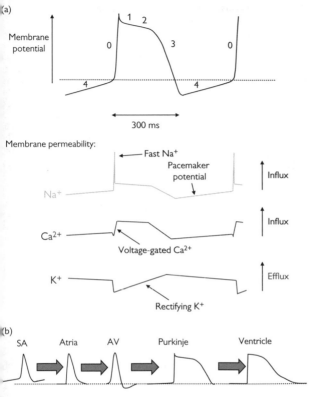

Fig. 6.31 (a) Cardiac action potential and membrane permeability; (b) change in action potential response profile across cardiac tissue—longer depolarizations correspond to protracted Ca²⁺ channel opening.

Arrhythmias (also known as dysrhythmias)

(➲ *OHCM10* pp.124–31, 804–9.)

Maintenance of a consistent rhythm is central to efficient cardiac func
tion. The electrical conductance system described is cleverly designed to
only allow electrical waves to travel in one direction and to prevent their
arriving too close together, making for inefficient contraction of cardiac
muscle. However, an injury to the heart muscle (e.g. caused by a heart at
tack) results in the death of a discrete area of the muscle, as defined by the
location of the thrombus of the coronary artery that led to the heart attack.
The infarcted area is dominated by scar tissue, which alters the electrical
conduction properties of that region. The site of the infarct has serious
implications for the efficient transmission of the electrical activity across the
heart and can lead to potentially fatal changes in heart rhythm—so-called
arrhythmias.

- Abnormal pacemaker activity: the heart rate is normally determined
  by the pacemaker cells in the SAN, but other cells in the heart can
  undertake pacemaker activity during or after ischaemic damage. The
  cellular mechanisms have not been fully elucidated but may involve a
  decrease in Na^+/K^+-ATPase pump activity, → membrane depolarization,
  and pain-mediated adrenaline release might play a part during a
  heart attack
- Heart block: if an infarct encompasses the AVN, the wave of activity
  may not be properly transmitted from the atria to the ventricles. The
  atria will continue to beat at the rate set by the SAN; the ventricles
  will beat independently, at a rate set by ventricular pacemaker cells.
  Sequential contraction of the atria, followed by the ventricles, is crucial
  for efficient pumping of blood and 'heart block' is best treated by
  implantation of an artificial pacemaker (➲ *OHCM10* pp.132–3)
- Re-entry: the refractory nature of cardiac muscle immediately after
  depolarization normally ensures that the impulse wave only travels in
  one direction. Re-entry applies to a 'ring' of cardiac tissue, which can be
  anatomically distinct from the surrounding tissue, but is more commonly
  only functionally distinct. The concept dictates that an impulse arising
  from any point in the ring will propagate in both directions, until the
  waves of depolarization meet and both impulses are cancelled out
  (Fig. 6.32). However, if part of the ring is damaged, such that the
  impulse is not transmitted in the normal (anterograde) direction but
  can still be conducted in the retrograde direction, the impulse can
  cycle around the ring continuously, if the time taken for each cycle
  exceeds the refractory period. Although the re-entrant circuit may
  only occupy a small area of cardiac tissue, its effects are transmitted
  to the surrounding cardiac muscle and the impact on heart rhythm
  can be dramatic. Drugs that are useful in treating re-entry prolong the
  refractory period (Table 6.4)
- Delayed after-depolarization: normally myocardial cells do not contract
  when they are stimulated by the wave of activity that originates in the
  SAN. In between times, the cells are first refractory and then quiescent,
  until the arrival of the next wave of depolarization. However, if
  intracellular Ca^{2+} levels increase excessively during depolarization, an

after-depolarization can result from a net influx of Na^+ ions in exchange for the Ca^{2+} (in the ratio 3 Na^+ in: 1 Ca^{2+} out) and the opening of Ca^{2+}-sensitive non-selective cation channels. If the heart rate is slow, the after-depolarization may not be sufficient to elicit an action potential, and it gradually subsides. However, as the heart rate increases, the after-depolarization increases until it is sufficiently high to elicit an action potential: the effect is self-perpetuating, → an indefinite chain of action potentials in quick succession (tachycardia).

Normal tissue:

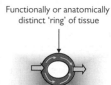

Functionally or anatomically distinct 'ring' of tissue

- Free transmission in both directions around the ring from the point of stimulation—circus movement not possible
- Depolarization wave transmitted to downstream cardiac tissue

Damaged tissue:

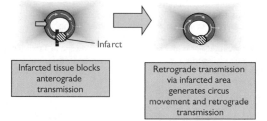

Infarct

| Infarcted tissue blocks anterograde transmission | Retrograde transmission via infarcted area generates circus movement and retrograde transmission |

Fig. 6.32 Circus movement caused by re-entry in cardiac tissue.

Bradycardia and atrial fibrillation (AF)

Bradycardia (◔ OHCM10 p.124) is an unusually slow heart rate that can be brought on by sinus dysfunction (sick sinus syndrome) or hypothyroidism and exacerbated by heart-slowing drugs (β-blockers and cardiac glyco-sides). Severe cases can lead to cardiac arrest, AF, or thromboembolism. Mild cases can be effectively treated with the muscarinic ACh receptor antagonist, atropine, which prevents the slowing effect of parasympathetic (vagal) stimulation of the heart. More serious cases may require temporary or permanent implantation of pacemaker devices (◔ OHCM10 pp.132–3)

AF (◔ OHCM10 p.130) is defined as irregular and extremely rapid (300–600/min) contractions of the atria. The AVN is only intermittently activated by this chaotic activity, giving rise to irregular ventricular function. The cause is often MI, but heart failure, hypertension, bronchitis, and hyperthyroidism can also result in AF. The main risk is thromboembolism (◔ OHCM10 p.470), which is prevented by the anticoagulant, warfarin (◔ OHCM10 p.350). The key to treatment of AF is to deal with the underlying cause, if possible, and to use drugs to slow the atrial contractions (digoxin, β-blockers, verapamil, amiodarone). If patients fail to respond to drug treatment or in emergency, cardioversion (electrical shock treatment; ◔ OHCM10 p.770) might help to restore normal rhythm (Box 6.8).

Box 6.8 Treatments and drug therapies for arrhythmias

- Some arrhythmias are best treated by implantation of an artificial pacemaker (abnormal pacemaker activity), by surgical intervention to ablate re-entry circuits (e.g. accessory pathways between the atria and ventricles in Wolff–Parkinson–White syndrome), or by cardioversion in serious acute cases
- Drug therapies, however, are primarily aimed at the electrophysiological events of cardiac contractility and crudely fit into the Vaughan Williams' classification. As can be seen in Table 6.4, the overall effect is generally a reduction in cardiac contractility, but the underlying cause of this effect varies with the different classes of drug
- Venous thromboembolism is a risk of AF on account of blood pooling due to inefficient heart function. The anticoagulant, warfarin or, increasingly, new oral anticoagulants are prescribed to reduce risk by inhibiting the coagulation cascade at a variety of points (◔ p.490).

Table 6.4 Vaughan Williams' classification of anti-arrhythmic drugs

Class	Example	Mechanism	Rate of depolarization	Action potential duration	Refractory period	Atrioventricular conduction	Cardiac contractility
Ia	Procainamide	Use-dependent inhibition of Na^+ channels	Reduced	Increased	Increased	Reduced	Reduced
Ib	Lidocaine	Inhibition of fast-dissociation Na^+ channels	Reduced	Reduced	Increased	No effect	No effect
Ic	Flecainide	Inhibition of slow-dissociation Na^+ channels	Reduced	No effect	No effect	Reduced	Reduced
II	Propranolol	β block	No effect	No effect	No effect	Reduced	Reduced
III	Amiodarone	K^+ channel block	No effect	Increased	Increased	Reduced	No effect
IV	Verapamil	Ca^{2+} channel block	No effect	Reduced	No effect	Reduced	Reduced

The heart as a pump

The role of the heart is to supply sufficient blood to the tissues to satisf
their O_2 and nutrient requirements and to remove waste products, includin
urea and CO_2. This role is fulfilled due to the synchronized contraction o
the cardiac myocytes that constitute the walls of the heart chambers in re
sponse to the wave of electrical activity that is conducted by the myocyte
themselves. What follows (and in Figs 6.33, 6.34) is a summary of the mech
anical events that contribute to the cardiac cycle; we join the cycle durin
the relaxation phase (diastole), just before the next wave of excitation i
initiated in the SA node. Under resting conditions in humans, the whol
cycle is complete in ~1s:

1. The atria and ventricles are relaxed and the pressure in the heart
   chambers is low. Blood from the large systemic veins (superior and
   inferior vena cavae) and that returning from the lungs (pulmonary vein
   flows into the right and left atria respectively. The AV valves are open
   at this stage of the cycle, so the majority of blood passes passively
   from the atria to the ventricles.
2. The wave of depolarization emitted from the SAN in the right atrium
   sweeps across the atria and causes the cardiac myocytes in their walls
   to contract, forcing most of the remaining blood from the atria into
   the ventricles via the open AV valves. The volume of the ventricles
   increases as they fill (to a maximum of ~130mL under resting
   conditions).
3. The conduction wave has now passed through the AVN and been
   conducted, via the fast Purkinje fibres in the bundle of His, to the apex
   of the ventricles, whereupon it sweeps across the ventricles from
   bottom to top, initiating ventricular contraction (systolic phase).
4. As soon as ventricular contraction starts, the AV valves snap shut,
   trapping the blood in the ventricles and causing ventricular pressure to
   rise without any significant change in ventricular volume.
5. Pressure in the ventricles continues to rise until it exceeds that in the
   outflow arteries (pulmonary artery and the aorta; ~80mmHg). Now
   the valves at the openings to these arteries are forced open by the
   pressure, and blood flows down its pressure gradient into the arteries.
   Ventricular volume falls as the blood is forced out. Ultimately, the
   pressure in the ventricles will fall below that in the arteries and the
   pulmonary and aortic valves will shut. Under resting conditions, only
   about half of the total volume of blood in the ventricles is ejected (i.e.
   the ejection fraction is ~50%).
6. The wave of contraction is followed by a relaxation phase. The
   refractory nature of the cardiac myocytes at this stage prevents
   another contraction occurring too soon after the first; it is essential
   that sufficient time is given between heartbeats to allow the chambers
   of the heart to fill properly. The relaxation (diastolic) phase of the
   cycle therefore allows blood to flow back into the heart, via the atria,
   before the start of another cycle.

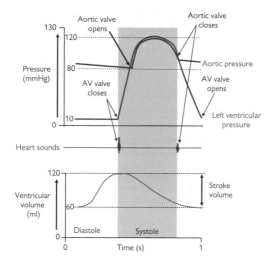

Fig. 6.33 Cardiac cycle at rest: pressure, volume, and heart sounds.

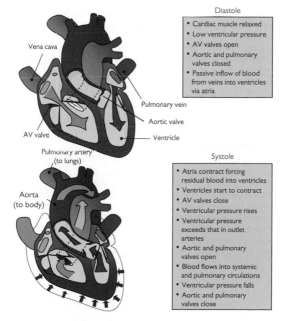

Diastole
- Cardiac muscle relaxed
- Low ventricular pressure
- AV valves open
- Aortic and pulmonary valves closed
- Passive inflow of blood from veins into ventricles via atria

Systole
- Atria contract forcing residual blood into ventricles
- Ventricles start to contract
- AV valves close
- Ventricular pressure rises
- Ventricular pressure exceeds that in outlet arteries
- Aortic and pulmonary valves open
- Blood flows into systemic and pulmonary circulations
- Ventricular pressure falls
- Aortic and pulmonary valves close

Fig. 6.34 The cardiac cycle.

Each stroke of a healthy adult human heart under resting conditions ejects ~70mL into the systemic circulation, via the aorta. This volume is known as the 'stroke volume'. The heart rate under the same conditions is usually ~70 beats/min. Knowing these two parameters, we are able to calculate the amount of blood that is pumped out of the heart every minute (the cardiac output):

$$\text{Stroke volume} \times \text{heart rate} = \text{cardiac output}$$

$$70\text{mL} \times 70 \text{ beats/min} = 4900\text{mL/min} \ (4.9\text{L/min})$$

Preload and the Frank–Starling law

The force generated by contraction of cardiac myocytes is dependent on their length—just as it is for skeletal muscle fibres. The 'length–tension' relationship for cardiac tissue therefore bears a close similarity to that for skeletal muscle, but the length of the muscle is uniquely determined by the amount of blood in the ventricle when the heart is fully relaxed (the end-diastolic volume; Fig. 6.35).

The Frank–Starling law establishes that the force of contraction of the heart is related to the end-diastolic volume (Fig. 6.35). Sympathetic nervous stimulation of the heart increases the efficiency of contractility—more force is generated for a given end-diastolic volume.

The end-diastolic volume is dependent on the amount of blood returning to the heart, also known as preload. In a non-compliant, fixed-volume, closed system, this would necessarily equal the amount of blood leaving the heart, meaning that increased cardiac output would be reflected in increased preload and increased force of contraction. Cardiac output is indeed a determinant of preload, but the issue is complicated by the fact that contraction of veins alters the volume of the vascular system and influences preload (Fig. 6.35): venous contraction reduces venous volume and increases preload and vice versa.

Therefore, while arterial tone is regarded as the primary determinant of blood pressure, venous tone has an impact on cardiac work and output, hence the ability of veno-selective dilators (e.g. nitrates) (➡ p.464) to reduce cardiac O_2 consumption by reducing cardiac work.

Heart valve disease

(➡ *OHCM10* pp.146–51.)

Effective valves in the heart are essential for optimal pumping conditions because they prevent backflow of blood against the desired direction of flow. Valves are found between the atria and ventricles (the AV valves; mitral—left, tricuspid—right) and between the ventricles and the major arteries (aortic valve—left; pulmonary valve—right).

There are a number of potential causes of valve malfunction, which can be crudely divided into those that cause narrowing of the valve opening (stenosis) and those that cause the valve to leak, → regurgitation.

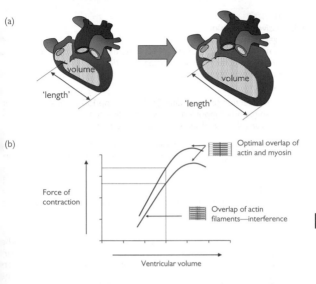

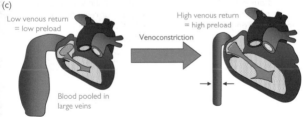

Fig. 6.35 Preload and the Frank–Starling law. (a) Ventricular volume determines cardiac muscle strength; (b) the Frank–Starling relationship between ventricular volume (muscle length) and force of contraction; (c) venoconstriction is a determinant of ventricular volume (preload).

Stenosis
- Excessive calcification
- Congenital malformation
- Rheumatic fever (⊃ *OHCM10* p.142) (autoimmune damage to valve tissue—more common in developing countries)
- Atherosclerotic degeneration.

Regurgitation
- Bacterial infection (endocarditis; ⊃ *OHCM10* pp.150–1) or inflammation
- Prolapse (poorly supported or weak valve leaflets)
- Ventricular (AV valve disease) or aortic dilatation.

The physiological impact of heart valve disease is a loss of effective pumping in the heart resulting in symptoms consistent with reduced cardiac output:
- Fatigue
- Breathlessness
- Angina (left-side valve disease)
- Oedema (pulmonary oedema for right-side valve disease, systemic oedema for left-side).

Diagnosis
Dysfunctional valves are often first diagnosed by general practitioners, who detect abnormal heart noises (or murmurs). The normal 'click' of heart valves closing becomes a prolonged flutter on account of the valve leaflets fluttering or blood regurgitating through an insufficiently closed valve. The time and duration of the murmur is indicative of the valve that is damaged and the type of valve dysfunction (e.g. stenosis or regurgitation). However, diagnosis can only be confirmed with an ECG.

Box 6.9 summarizes the treatments for valvular diseases.

**Box 6.9 Treatments and drug therapies
of valvular diseases**

- Advanced deterioration of valve function usually requires cardiological
  or surgical intervention, depending on the nature of the disease
- Valvuloplasty (⊃ *OHCM10* p.148) is a procedure conducted by a
  cardiologist in a patient with stenosed pulmonary or mitral valves; it
  does not require general anaesthesia. A balloon-tipped catheter is
  inserted into the femoral artery in the leg and manipulated remotely
  using X-ray imaging until the tip is across the stenosed valve, where it
  is inflated to increase the size of the opening and improve blood flow
  through it
- Valve replacement (⊃ *OHCM10* p.148) involves open-heart surgery.
  Damaged valves can be replaced by artificial valves or valves taken
  from cadavers (homografts) or from pigs (porcine xenografts).
  Artificial valves have the advantage of durability but patients with
  artificial grafts have to be maintained on antithrombotic drugs, while
  grafts from natural sources require replacement after ~10 years but
  do not require antithrombotics.

Heart failure

The inability of the heart to meet the supply needs of the body is known as heart failure (→ *OHCM10* pp.134–7). Heart failure is often precipitated by a heart attack, which leads to the death of an area of ventricular myocardium and a reduction in the efficiency of ventricular contraction.

In patients with heart failure, the ejection fraction at rest falls considerably, with critical effects on cardiac output. For example, if the ejection fraction falls to 25%, only ~35mL is ejected with every heartbeat and cardiac output falls to ~2.5L/min (from the usual ~5L/min) → the following chain of events (see also Fig. 6.36):

- Blood pressure falls and is sensed by the baroreceptors and through a fall in renal blood flow
- Signals from the baroreceptors result in the stimulation of the sympathetic nervous system, which restores blood pressure to normal levels by increasing heart rate (via cardiac β_1 adrenoceptors), blood volume (β adrenoceptors in the kidney), and vascular resistance (α_1 receptors in arterioles)
- The shortfall in cardiac output is therefore compensated for at the expense of increased heart rate, peripheral vascular resistance (afterload), and blood volume (preload).

The vicious cycle

Unfortunately, this short-term solution has long-term consequences. The increased work rate of the heart, coupled with the increased resistance against which it has to pump blood, leads to the thickening of the ventricular walls (remodelling). While cardiac remodelling might be conceived to be advantageous on account of the strengthening of the muscle, it also leads to a further reduction in the volume of the ventricular chamber, reducing the stroke volume further.

Simultaneously, the reduction in renal perfusion activates the renin–angiotensin–aldosterone pathway (Fig. 6.36) → Na^+ and water retention in the tissue (oedema) and pooling of blood in the central veins. The increase in central venous pressure constitutes an increase in cardiac preload, which might be predicted to improve cardiac contractility according to the Frank–Starling law (Fig. 6.35). However, the sympathetic compensation that has already taken place means that increasing the preload fails to increase contractility—instead, the heart becomes overloaded with blood (dilated).

This vicious cycle (Fig. 6.36) means that the initial reduction in stroke volume is compensated for by mechanisms that, ultimately, lead to a further reduction in stroke volume and a gradual deterioration in the condition. Heart failure is classified according to symptoms (New York Heart Association (NYHA) classification of heart failure; → *OHCM10* p.135) (Table 6.5).

Table 6.5 NYHA classification of heart failure

Class I: heart disease diagnosed but no breathlessness during ordinary activity	Class II: comfortable at rest but breathless with ordinary activity (e.g. walking)
Class III: breathlessness apparent with very mild activity; moderately debilitating	Class IV: breathless at rest; highly uncomfortable and debilitating

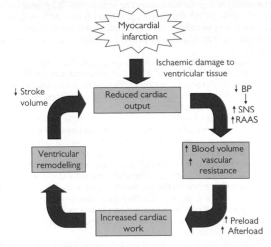

Fig. 6.36 The vicious cycle of heart failure is triggered by ischaemic damage to the ventricle. RAAS, renin–angiotensin–aldosterone system; SNS, sympathetic nervous system.

For treatment of heart failure, see Box 6.10.

Box 6.10 Treatment and drug therapies of heart failure

(➜ OHCM10 pp.134–37; ➜ OHPDT2 Ch. 2.)

The aim of treatment of heart failure is to break the vicious cycle to slow progression of the disease and prolong the life of sufferers:

- Reduce volume overload: as with hypertension, the kidney is a major target for therapeutic intervention in heart failure. Diuretics (e.g. furosemide) or angiotensin-converting enzyme (ACE) inhibitors (e.g. lisinopril) are the first-line drugs in heart failure. ACE inhibition has a secondary benefit of causing vasodilatation and reducing afterload, and may also slow or prevent processes involved in cardiac remodelling

- Veno-/vasodilatation: organic nitrates (e.g. isosorbide mononitrate) are veno-selective NO donor drugs that reduce cardiac workload primarily through reducing preload (although they may have some impact on afterload through vasodilatation as well). These drugs have been shown to reduce mortality

- Increase the force of ventricular contraction: cardiac glycosides (e.g. digoxin) can be prescribed if none of these treatments are showing benefit. These drugs are especially effective in patients with a dilated heart and work primarily by inhibiting the Na^+/K^+-ATPase, → an increase in intracellular Na^+, which exchanges with calcium through the Na^+/Ca^{2+} exchange pump. Ultimately, the extra Ca^{2+} swells the intracellular Ca^{2+} stores in the SR, meaning that more Ca^{2+} is released when the cells are stimulated. Cardiac glycosides can also slow the heart and improve rhythm. In acute heart failure, where a rapid increase in contractility is imperative, β_1 adrenoceptor agonists (e.g. dobutamine) can be used

- Inhibit sympathetic activity: increased sympathetic activity contributes to the vicious cycle that exacerbates heart failure. Inhibition of β adrenoceptors in the heart has long been avoided on account of the perceived danger of reducing the force of contraction. However, it is now recognized that low doses of the β-blocker, carvedilol, in conjunction with a cardiac glycoside, ACE inhibitor, and diuretic can reduce mortality, although the precise mechanism of this action is still unknown. Physicians are advised to proceed cautiously with β-blockers in heart failure

- Neprilysin inhibitors (e.g. sacubitrilat—derived from the oral pro-drug, sacubitril) are the latest drugs to be approved in heart failure. These drugs inhibit a neutral endopeptidase (neprilysin) that ordinarily deactivates several vasodilator peptides, including bradykinin, substance P, natriuretic peptides (A, B, and C) and adrenomedullin. Some of these peptides also increase sodium retention. The effect of inhibiting neprilysin is to increase vasodilatation and sodium (and therefore water) excretion. Typically, sacubitril is prescribed as an adjunctive therapy with the AT_1 receptor antagonist, valsartan, in heart failure.

The vascular system

Blood is transported between the heart and the tissues by blood vessels (Fig. 6.37).

- Arteries carry blood from the heart to the tissues. They have thick, muscular walls to cope with the high pressures that they are exposed to, and to facilitate their constriction and dilatation to modulate blood pressure and flow distribution. Arteries become progressively smaller but more numerous with distance from the heart; the smallest arteries are called arterioles and are the primary determinant of resistance to flow, often termed peripheral vascular resistance or afterload
- Capillaries are very fine vessels ($<50\mu m$) that distribute blood from the arterioles throughout tissues. The walls of capillaries are only one endothelial cell in thickness and do not contract. The thin walls facilitate easy diffusion of O_2 and glucose necessary for cellular respiration down the concentration gradient from the incoming blood, into the tissues. Waste metabolites (CO_2, urea) diffuse in the opposite direction. An exception to this basic rule applies in the lungs (pulmonary circulation), where capillaries come into close contact with alveolar air to facilitate gaseous exchange, with the loss of CO_2 to the atmosphere and the uptake of O_2 into the red blood cells, where it is transported bound to haemoglobin
- Veins carry blood away from tissues and back to the heart. They have some vascular smooth muscle, but not as much as arteries; as a result, they can contract and relax, but the changes in diameter are far less dramatic than in arteries. Blood leaving the capillaries enters small veins (venules), which progressively converge, pooling blood into increasingly large vessels. There is little pressure difference across the venous circulation, meaning that unaided flow of blood would be very slow. As a result, veins contain valves to prevent retrograde flow (backflow) and the venous return of blood to the heart is aided by contraction of the surrounding skeletal muscles. This is particularly important in humans, where our upright stance means that the effects of gravity have to be overcome to ensure the return of blood to the heart from our feet. The amount of blood returning to the atria of the heart determines preload.

Modulators of vascular tone

An important feature of our blood vessels is that they contract and dilate in response to numerous signalling molecules in the body. The mechanism by which these effects occur are summarized and the cellular mechanisms are illustrated in Fig. 6.38.

Systemic vasoconstriction

- The primary stimulus for vascular smooth muscle contraction is activation of the sympathetic nervous system that innervates blood vessels. Increased sympathetic drive results in release of noradrenaline from sympathetic nerve terminals, which activates α_1 and β_2 adrenoceptors on the smooth muscle cells. Most arteries and arterioles have α_1 receptors and contract in response to noradrenaline; α_2 adrenoceptors are also found in these arteries, but they are probably stimulated by circulating adrenaline rather than by sympathetic

nerve-derived noradrenaline. Arteries that supply skeletal muscle and
some veins have a predominance of β_2 adrenoceptors, which causes
them to dilate in response to noradrenaline and circulating adrenaline.
The net effect of increased sympathetic nervous system activity is to
redistribute blood flow away from the internal organs to the skeletal
muscles to prepare for 'fight or flight'

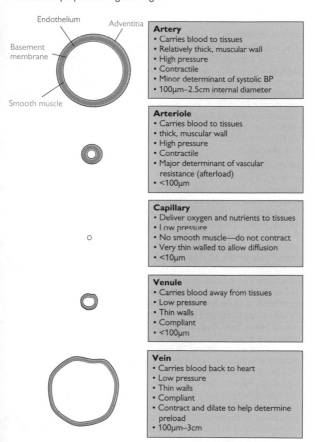

Artery
- Carries blood to tissues
- Relatively thick, muscular wall
- High pressure
- Contractile
- Minor determinant of systolic BP
- 100µm–2.5cm internal diameter

Arteriole
- Carries blood to tissues
- thick, muscular wall
- High pressure
- Contractile
- Major determinant of vascular resistance (afterload)
- <100µm

Capillary
- Deliver oxygen and nutrients to tissues
- Low pressure
- No smooth muscle—do not contract
- Very thin walled to allow diffusion
- <10µm

Venule
- Carries blood away from tissues
- Low pressure
- Thin walls
- Compliant
- <100µm

Vein
- Carries blood back to heart
- Low pressure
- Thin walls
- Compliant
- Contract and dilate to help determine preload
- 100µm–3cm

Endothelium, Adventitia, Basement membrane, Smooth muscle

Fig. 6.37 Characteristics of different blood vessel types, showing the relative proportions of smooth muscle.

- ATP and neuropeptide Y are co-transmitters that are often released, with noradrenaline, to cause rapid or long-lasting vasoconstrictor effects respectively. Stimulation of β-adrenoceptors in the kidney also increases the amount of renin available to catalyse the first step in the renin–angiotensin–aldosterone system. One of the products of this system is angiotensin II, which is a powerful vasoconstrictor through stimulation of angiotensin (AT) receptors on the smooth muscle (primarily AT_1 receptors).

Local vasoconstrictors

- The endothelins (ET-1, ET-2, ET-3—of which ET-1 is the most abundant) are endothelium-derived vasoconstrictor peptides, acting through ET_A and ET_B receptors on the smooth muscle. However, the action of ET-1 is modulated through stimulation of ET_B receptors on the endothelium, which leads to the release of an endothelium-derived vasodilator, NO (Fig. 6.39)
- Thromboxane A_2 (TXA_2) is a prostanoid synthesized in platelets in response to vascular injury. As well as stimulating platelet activation, TXA_2 is a powerful local vasoconstrictor, which helps to reduce blood loss after injury.

Local vasodilators

See Figs 6.38 and 6.39.

- Adenosine is primarily produced as a by-product of ATP breakdown, and can either be seen as a local or systemic vasodilator through stimulation of A_2 receptors on the smooth muscle (except in the kidney, where stimulation of A_1 receptors causes vasoconstriction). Adenosine is important in the heart, where it blocks AV conduction and reduces the force of contraction; adenosine release might be partly responsible for the pain associated with heart attacks. Adenosine is also a neuromodulator (A_1 receptors), a bronchoconstrictor (A_1), and a pro-inflammatory mediator (A_3)
- NO is one of a number of endothelium-derived vasodilators that are generated to cause local vasodilatation. Stimuli for NO generation include shear stress (the lateral stress experienced by endothelial cells due to blood flow), hypoxia, and circulating neurohormonal factors (e.g. substance P, bradykinin) that act to increase endothelial intracellular Ca^{2+}. Endothelium-derived NO also acts as a powerful inhibitor of platelet activation and inflammatory cell adhesion. Dysfunction in NO production has been implicated in many cardiovascular diseases, including atherosclerosis. NO is the most important endothelium-derived relaxing factor in large arteries
- Prostacyclin (PGI_2) is a product of arachidonic acid metabolism, stimulated in response to many of the same mediators as NO. PGI_2 acts synergistically with NO (the effect of combined release is greater than the sum of the two parts)
- Endothelium-derived hyperpolarizing factor (EDHF) is the dominant endothelium-derived factor in small (resistance) arteries. Its identity is still an unresolved issue, but K^+ ions appear to play a prominent role.

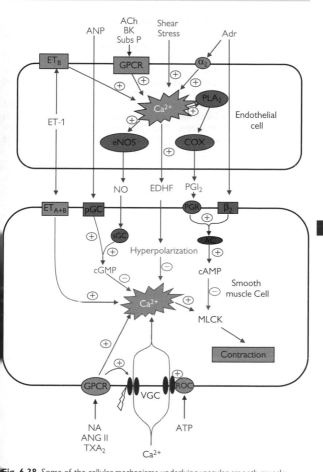

Fig. 6.38 Some of the cellular mechanisms underlying vascular smooth muscle contraction and dilation in response to endogenous signals. AC, adenylate cyclase; ACh, acetylcholine; Adr, adrenaline; α_2, α_2 adrenoceptor; Ang II, angiotensin II; ANP, atrial natriuretic peptide; BK, bradykinin; β_2, β_2 adrenoceptor; COX, cyclo-oxygenase; EDHF, endothelium-derived hyperpolarizing factor; eNOS, endothelial nitric oxide synthase; ET$_A$ and ET$_B$, endothelin receptors A and B; ET-1, endothelin-1; GPCR, G-protein-coupled receptor; MLCK, myosin light chain kinase; NA, noradrenaline; PER, prostaglandin receptor; PGI$_2$, prostacyclin; PLA$_2$, phospholipase A$_2$; ROC, receptor-operated channel; sGC/pGC, soluble and particulate guanylate cyclase; Subs P, substance P; TXA$_2$, thromboxane A$_2$; VGC, voltage-gated channel.

Signal integration and intracellular contractile processes

The extent of constriction of any artery depends on the balance of vaso-constrictor and vasodilator stimuli. In the interests of efficiency, the sig-nals from all of the different mediators are almost exclusively channelled through a single intracellular entity—the concentration of cytoplasmic cal-cium (Ca^{2+}_i): vasoconstrictors stimulate an increase in intracellular Ca^{2+}_i and vasodilators reduce Ca^{2+}_i, thus avoiding the potentially inefficient stimula-tion of two or more competing pathways (Fig. 6.38). Ca^{2+}_i derived from intracellular stores in the SR and through voltage-gated Ca^{2+} channels binds to calmodulin, which stimulates myosin light chain kinase (MLCK) to phos-phorylate myosin—an essential step in the interaction of smooth muscle myosin with actin.

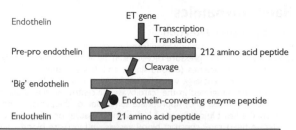

Endothelin

ET gene

Transcription
Translation

Pre-pro endothelin ⬛⬛⬛⬛⬛⬛⬛⬛ 212 amino acid peptide

Cleavage

'Big' endothelin ⬛⬛⬛⬛⬛

Endothelin-converting enzyme peptide

Endothelin ⬛⬛ 21 amino acid peptide

- Endothelin is produced continuously by the endothelium, contributing to vascular tone through stimulation of ET_A and ET_B receptors on smooth muscle
- Stimulation of ET_B receptors on the endothelium causes release of the vasodilator NO
- ET_B receptors might also be involved in clearing ET-1 from the circulation
- Changes in ET-1 levels requires modification to gene expression and is therefore relatively slow (hours)

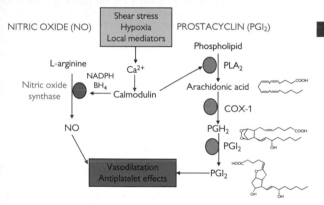

NITRIC OXIDE (NO)

Shear stress
Hypoxia
Local mediators

PROSTACYCLIN (PGI_2)

Phospholipid

L-arginine

Ca^{2+}

PLA₂

Nitric oxide
synthase

NADPH
BH_4

Calmodulin

Arachidonic acid

COX-1

NO

PGH_2

PGI_2

Vasodilatation
Antiplatelet effects

PGI_2

Fig. 6.39 Some of the endothelium-derived local regulators of vascular tone.

Haemodynamics

Blood pressure detection

Blood pressure is constantly monitored by special receptors (baroreceptors). There are 'high-pressure receptors' in the aortic arch, pulmonary artery, and carotid arteries (carotid sinus), and 'low-pressure receptors' in the atria and adjacent large veins. Signals from both high- and low-pressure receptors are integrated in the 'cardiovascular centres' in the upper medulla and responded to by appropriate stimulation of the parasympathetic (to slow the heart in response to high blood pressure) or sympathetic (to accelerate heart rate, constrict blood vessels, and increase blood volume in response to low blood pressure) branches of the autonomic nervous system (Fig. 6.40).

Blood pressure determination

The relationship between blood flow, resistance, and pressure bears close similarity to Ohm's law for electricity (Fig. 6.41). The cardiovascular system is a closed circuit and the pressure within the system is determined by a number of parameters, all of which can be controlled from the cardiovascular centres in the medulla.

- Blood volume: the volume of fluid within a system with fairly rigid walls is an important determinant of pressure. Just as pumping more air into a tyre increases the tyre pressure, so increasing the volume of blood increases blood pressure. Indeed, the effect is far more dramatic in the blood system because, unlike air, blood is virtually incompressible—the molecules cannot be forced to come closer together by external force. However, the impact of changes in blood volume are partially damped out by the fact that arteries (like tyres) are compliant—the walls are able to stretch to accommodate more fluid without a proportional increase in pressure. As we age, our arteries stiffen (become less compliant—arteriosclerosis; ➋ OHCM10 p.116)—a factor that contributes to the gradual increase in blood pressure that we experience with age. The kidney controls the blood volume by modulating the amount of salt and, consequently, water that is reabsorbed in the distal tubule. Blood volume-mediated changes are slow in onset and are not responsible for rapid compensatory changes in blood pressure, but are central to hypertension (➋ OHCM10 p.138)
- Cardiac output: as mentioned previously, cardiac output is determined both by heart rate and stroke volume. The relationship between cardiac output and blood pressure is proportionate: if all other parameters remain unchanged (and ignoring compliance), doubling cardiac output would be expected to double blood pressure
- Vascular resistance: the resistance against which the heart has to work to drive blood through the arteries (or afterload) is a crucial determinant of blood pressure. The most important blood vessels involved in blood pressure determination are the arterioles because they are the smallest vessels in the arterial tree and are responsible for the resistance to flow (hence the term 'resistance arteries'). The importance of arteriolar contraction is demonstrated in the simple

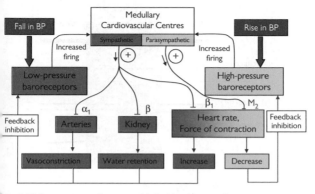

Fig. 6.40 Baroreceptor reflex responsible for blood pressure homeostasis.

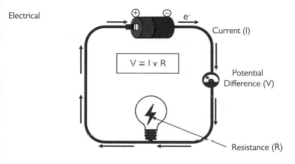

Fig. 6.41 Relevance of Ohm's law to fluid dynamics in the cardiovascular system. BF, blood flow; BP, blood pressure; R, resistance.

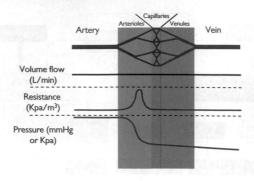

The importance of arteriolar radius (tone) on blood

$$\text{Vascular Resistance} = \frac{\text{Length} \times \text{fluid viscosity}}{\text{Radius}^4}$$

Blood Pressure = Flow × Vascular Resistance

If radius is reduced by a factor of 2 (i.e. halved), pressure increases by a factor of 2^4 (=**16-fold**).

Fig. 6.42 The importance of arteriolar radius (tone) on blood:

model shown in Fig. 6.42: in this case, contraction of an isolated arteriole, so that the diameter is halved (decreased by a factor of 2), causes the pressure in the system to rise by a factor of 2^4 (=16). Although the diameter of the large arteries is not the primary determinant on blood pressure, their contraction reduces their compliance, → a small increase in systolic blood pressure.

Reflex responses

The rapid and integrated response to changes in blood pressure in humans is best illustrated with the example of what happens when we stand up (Fig. 6.43):
1. The action of standing leads to a rapid pooling of venous blood in the legs due to gravity, leaving less blood in the large central veins for return to the heart. More than 0.5L of blood is redistributed in this way upon standing.
2. The reduction in preload leads to a reduction in stroke volume (Frank–Starling law; Fig. 6.35, ⊃ p.445) and, consequently, cardiac output. Arterial pressure falls momentarily. Baroreceptors in the large veins and atria detect the fall in central venous pressure.
3. Signals from the low-pressure baroreceptors are processed in the cardiovascular centres of the medulla and the sympathetic nervous system is stimulated (Fig. 6.43).
4. Heart rate increases, peripheral vascular resistance increases, central veins contract—all of which returns arterial and venous pressure to near-normal levels within a few seconds.

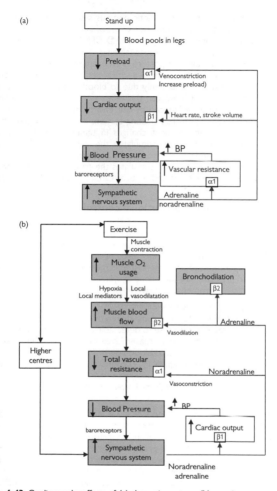

Fig. 6.43 Cardiovascular effects of (a) change in posture; (b) exercise.

The importance of the sympathetic nervous system in this process is highlighted by the fact that a major side effect of inhibitors of the synthesis of noradrenaline is postural hypotension (➲ OHCM10 p.41) (low blood pressure on standing), which can cause fainting.

The cardiovascular reflex to posture is an example of our response to a sudden change in blood pressure—a similar response would be experienced if blood volume fell suddenly due to blood loss through haemorrhage. For the most part, however, this reflex system responds to minor changes in blood pressure induced by our environment. It is comparable to a thermostatically controlled heating system, switching on and off in response to moment-by-moment changes in temperature. The effects are widespread—low pressure stimulates the entire sympathetic nervous system, which impacts all our arteries and veins as well as the heart and kidney. High blood pressure stimulates the parasympathetic nervous system, which slows the heart and reduces cardiac output, affecting blood flow to all parts of the body.

Regional blood flow is under local control

Although there is some modification of the pattern of blood flow upon sympathetic stimulation, determined by the distribution of α and β adrenoceptors in blood vessels of different tissues, there is little or no capability for responding to the specific needs of a particular organ or tissue. Therefore, superimposed on these systemic mechanisms that keep a tight grip on blood pressure, are local control systems that react to the immediate metabolic requirements of a given tissue.

Control of regional flow—the endothelium

There are several key features that are shared by local mediators:

- They are generated in response to detectable changes in the local environment (e.g. ischaemia, increased levels of metabolites, increased shear stress caused by elevated blood flow)
- They are generated very rapidly in order that they can execute a rapid response
- They are metabolized or inactivated rapidly so that their effects remain local.

The endothelium that lines all blood vessels is ideally situated to detect and respond to changes in the local environment, as it is the interface between the flowing blood and the vessel wall. It is not surprising, therefore, that the endothelium is a hotbed for the production of local modulators of blood vessel tone, particularly vasodilators. The importance of the endothelium is highlighted by the fact that so-called endothelial dysfunction has been implicated in a range of cardiovascular disease, including atherosclerosis.

NO is synthesized in response to an increase in Ca^{2+} within the endothelial cells, triggered by ischaemia, shear stress, or circulating modulators (including bradykinin and substance P). Ca^{2+} binds to calmodulin and stimulates the endothelial isoform of the enzyme, NO synthase (eNOS) → increased generation of the free radical NO (Fig. 6.38). This is a small molecule which diffuses rapidly into both the vessel wall and the lumen to cause vasodilatation and inhibition of platelet and monocyte function. Endothelial NO is usually generated in very low concentrations (low nM range), indicating its potency as a signalling molecule. Its nature as a free radical means that it is highly reactive, with a biological half-life of only a few seconds. Most of the effects of NO are cGMP mediated (Fig. 6.38), but there is evidence of cGMP-independent mechanisms, particularly when NO is generated in higher concentrations. Organic nitrates that are often used in angina undergo tissue-mediated metabolism to release NO.

NO is also known to be the non-adrenergic, non-cholinergic (NANC) neurotransmitter found in specific nerves and an inducible isoform of the enzyme (iNOS) is expressed in response to inflammatory stimuli; local concentrations of NO from iNOS are believed to be ~1000 times higher than those from eNOS, reflecting its change in function from a highly controllable local mediator to a cytotoxic agent for use by the immune system.

Prostacyclin (PGI_2) acts synergistically with NO and is generated in response to similar stimuli. It has a relatively short half-life (<5min), but its dilution in flowing blood reduces its activity as it is washed away from its site of production. PGI_2 is synthesized from arachidonic acid by a three-step process involving phospholipase A_2, COX-1, and prostacyclin synthase (Fig. 6.39, ➲ p.471). PGE_2 is a closely related prostanoid that also causes vasodilatation.

Exercise

Exercise is defined as an increase in skeletal muscle activity, which requires modifications to the cardiovascular system to accommodate the increased metabolic needs of muscular tissue (Fig. 6.43, ⊃ p.475). In the first instance this might be achieved simply by activation of local mediators in response to the hypoxia that rapidly develops during exercise. However, exercise is also associated with an increase in sympathetic drive, resulting in release of noradrenaline and adrenaline, with the following effects:

• Vasodilatation of arteries that supply skeletal muscle through β_2 adrenoceptors (partly mediated by the endothelium)
• Vasoconstriction of blood vessels supplying the major organs and the gut through α_1 adrenoceptors
• Increased heart rate and stroke volume through stimulation of β_1 adrenoceptors
• Bronchodilatation mediated by circulating adrenaline acting on β_2 adrenoceptors in the bronchi; the breathing rate will also increase in response to activation of oxygen and carbon dioxide chemoreceptors.

These processes combine to cause a considerable increase in cardiac output to account for the massive increase in oxygen consumption and a prioritization of blood distribution to favour muscles at the expense of other tissues. In a trained athlete, heart rate can more than treble (from ~50 to >180 beats/min), stroke volume can more than double (~80 to >160mL/min), resulting in an increase in cardiac output from ~4L/min to up to ~40L/min. If peripheral resistance were to remain constant, systolic blood pressure under these conditions would rise above 1000mmHg, which would clearly exceed the limits of blood vessel strength. In the event, blood pressure usually only reaches approximately double the normal values (~200mmHg), indicative of an overall decrease in peripheral resistance. Clearly, the signals from the higher centres that are driving the exercise overcome the signals from the high-pressure baroreceptors, which would normally act to return blood pressure to normal by reducing heart rate through vagal stimulation. The body relinquishes its tight grip on blood pressure in order to meet the immediate metabolic needs of the muscles.

Hypertension

(→ *OHCM10* pp.138–41.)

Hypertension is a condition that is characterized by chronically elevated blood pressure and is a risk factor for other cardiovascular diseases including coronary artery disease, MI, stroke, and heart failure.

Diagnosis

In reality, an artificial limit has to be defined to enable the distinction between hypertension and normal blood pressure. From the clinical perspective, two levels have been set:

- Patients with blood pressure >160/100mmHg should be treated
- Blood pressures of 140/90–159/99mmHg are in the 'grey area': the physician must decide whether the elevated blood pressure constitutes a significant risk of heart disease on a case-by-case basis. In order to make this judgement, other risk factors (e.g. diabetes, smoking, high LDL) will be taken into account. The more risk factors, the greater the need to treat.

Mild hypertension is asymptomatic and regular blood pressure measurements conducted by general practitioners is sufficient to ensure early diagnosis and improved prognosis in hypertension. Originally, it was assumed that diastolic pressure was the best indicator of hypertension but, more recently, it has become generally accepted that the risk of conditions such as coronary artery disease is more closely linked to systolic blood pressure, resulting in this measure being the prime consideration in diagnosis of hypertension.

Causes of hypertension

The specific cause of hypertension can only be clearly identified in a small percentage of cases (~5%)—so-called secondary hypertension, which can be due to:

- Renal disease, e.g. renal stenosis (blockage of the renal arteries), glomerular nephritis (inflammation of the glomerulus), diabetic nephropathy (damage to the nephron induced by diabetes)
- Endocrine diseases, e.g. tumour-related overproduction of aldosterone (Cushing's syndrome, → *OHCM10* p.224; Conn's syndrome, → *OHCM10* p.228) or adrenaline (phaeochromocytoma; → *OHCM10* p.228)
- Other clearly identifiable causes, e.g. monoamine oxidase inhibitors (amphetamines), pregnancy.

The remaining vast majority of hypertensive cases (~95%) are collectively diagnosed as essential hypertension, for which the cause is undefined. However, it is a broadly held view that sufferers are genetically predisposed to the condition through a renal disorder.

For treatment of hypertension, see Box 6.11.

Box 6.11 Treatment and drug therapies for hypertension

(● OHPDT2 Ch. 2.)

Irrespective of the cause of hypertension, the primary aim of treatment is to reduce blood pressure in order to lower the risk of accelerated atherosclerotic disease (Fig. 6.46, ● p.485) that can lead to heart attack or stroke (Fig. 6.45, ● p.484).

As with most cardiovascular disorders, the primary target for drug intervention in hypertension is not necessarily the heart. Instead, it has proved more profitable to target drugs at the kidney (as renal dysfunction is intrinsically linked to the pathogenesis of the condition) or the blood vessels.

Diuretics

Thiazides (e.g. bendroflumethiazide) can be used to treat hypertension: they inhibit the Na^+/Cl^- transporter in the distal convoluted tubule, resulting in an increase in sodium excretion, which is associated with increased water excretion and reduced blood volume. Side effects include K^+ loss (● OHCM10 p.674) (hypokalaemia) and postural hypotension (● OHCM10 p.41) (low blood pressure when upright → dizziness or fainting).

β-blockers

Although $β_1$ adrenoceptors in the heart are a logical target for β-blockers, these receptors are only significantly activated during stress, exercise, or heart failure, when the sympathetic nervous system is stimulated. β-blockers are, therefore, only likely to have an inhibitory effect on heart rate when there is increased sympathetic drive; they will have little impact on the heart at rest, when its rate and force of contraction is primarily determined by the parasympathetic nervous system (vagus nerve).

In reality, the primary effect of β-blockers is not mediated by inhibition of cardiac β-receptors but those in the kidney that ordinarily activate synthesis of the enzyme renin, which is required to convert angiotensinogen to angiotensin I in the renin–angiotensin–aldosterone system (Fig. 6.44). Angiotensin I is subsequently converted to angiotensin II, which increases blood pressure by the combined effect of increased systemic vasoconstriction and through salt and water retention. The immediate benefit of β-blockers in reducing blood pressure is likely to be coupled to long-term benefits through inhibition of angiotensin II-mediated vascular hypertrophy and hyperplasia, which exacerbates hypertension and plays a role in the progression of conditions such as arteriosclerosis and atherosclerosis.

ACE inhibitors, angiotensin receptor antagonists

The renin–angiotensin–aldosterone system is clearly an important therapeutic target in hypertension because of its dual impact on salt handling by the kidney and vasoconstriction of blood vessels (Fig. 6.44). While β-blockers act on the first step in this process, ACE inhibitors act on the enzyme that converts angiotensin I to angiotensin II—ACE. Like β-blockers, these drugs ultimately reduce the amount of circulating angiotensin II that can activate AT receptors in the kidney and in the blood vessels to cause salt retention and vasoconstriction respectively.

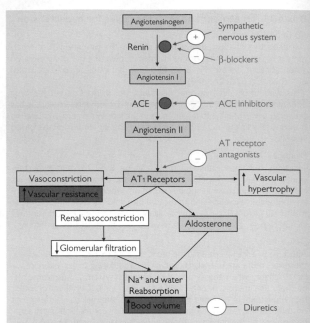

Fig. 6.44 The renin–angiotensin–aldosterone system.

ACE inhibitors will also share the long-term benefits of β-blockers, by inhibiting vascular remodelling that plays a role in the progression of hypertension and other vascular diseases.

The side effects of ACE inhibitors include increased K^+ retention and cough, which is because ACE also mediates the metabolism of the peptide, bradykinin, through its neutral endopeptidase activity. Increased bradykinin in the bronchial mucosa is responsible for stimulation of the cough reflex.

Some of the side effects of ACE inhibitors are avoided by inhibition of the renin–angiotensin–aldosterone system at the AT receptors. AT_1 receptors mediate most of the pro-hypertensive effects of angiotensin II (Fig. 6.44) (AT_2 receptor function is still largely unknown) and a number of AT_1-specific receptor antagonists (e.g. losartan) are increasingly popular in hypertensive therapy. An added benefit of AT_1 receptor antagonists over ACE inhibitors surrounds a possible alternative route for angiotensin II synthesis via the enzyme chymase. While it is unclear whether the chymase route of synthesis is clinically relevant, the possibility remains that ACE inhibition might be partially circumnavigated by this route.

Calcium antagonists

Ca^{2+} is central to contraction of both smooth and cardiac muscle. Inhibition of the mechanisms that cause the levels of cytoplasmic Ca^{2+} to rise and mediate contraction, are legitimate therapeutic targets in hypertension because they might cause vasodilatation (reduced afterload) and reduced cardiac output. A reduction in cytoplasmic Ca^{2+} can be effected by a number of means:

• Inhibition of release from ER
• Inhibition of activation of voltage-gated (e.g. L-type) Ca^{2+} channels in the plasma membrane
• Increased sequestration into the ER
• Increased extrusion via calcium pumps or exchange mechanisms.

The majority of therapeutic agents act to prevent stimulation of L-type voltage-gated Ca^{2+} channels. These drugs fall into three broad categories: dihydropyridines (e.g. nifedipine, amlodipine; vessel selective), benzothiazepines (e.g. diltiazem; fairly non-selective), and phenylalkylamine (e.g. verapamil; cardiac selective). Dihydropyridines are favoured in hypertension—particularly amlodipine, which has a long plasma half-life. Their action is primarily on resistance arteries that determine blood pressure (peripheral vascular resistance, afterload) but their use is often associated with reflex tachycardia (compensatory increase in heart rate), unless administered in conjunction with β-blockers.

Side effects of Ca^{2+} antagonists used in hypertension are mainly associated with their vasodilator action: headache, flushing, and oedema (swelling), particularly in the ankles.

Other drugs

Many of the drugs (other than β-blockers) that reduce blood pressure through actions on the sympathetic nervous system, have been used as antihypertensives in the past. However, these usually carry fairly severe side effects, particularly with respect to postural hypotension. Nevertheless, methyl-dopa, which is metabolized to the false transmitter, α-methylnoradrenaline, is still occasionally used in pregnancy; the $α_2$ agonist, clonidine (which enhances the negative feedback system in sympathetic nerve terminals), is sometimes used in refractory cases; and $α_1$ receptor antagonists (e.g. doxazosin) might have an added benefit on the plasma lipid profile (LDL/HDL). Hydralazine can be used as an alternative vasodilator to reduce afterload; its mechanism of action is poorly understood but might involve the inhibition of Ca^{2+} release from the smooth muscle SR. Like Ca^{2+} antagonists, its use is associated with reflex tachycardia and it should be given with a β-blocker. Mixed endothelin receptor antagonists (e.g. bosentan) are showing promise in the treatment of pulmonary hypertension, by preventing the powerful vasoconstrictor action of the endothelium-derived peptide, ET-1.

There are considerable differences in sensitivity of patients to these drugs and it is often a case of trial and error to find the most suitable drug, or combination of drugs, for a patient. This is an area of medicine that might realize considerable benefit from pharmacogenomics.

Atherosclerosis

Atherosclerosis is a complex disease process that results in the deposition of lipids in discrete lesions (plaques) found in the walls of large conduit arteries. Although atherosclerotic lesions are found in almost all of us from an early age, their prevalence is greatly increased by a number of risk factors, including genetic predisposition, sex (male), a high-lipid diet, smoking, hypertension, and diabetes (Fig. 6.45).

Plaque distribution is not random: plaques are absent from veins and the microvasculature and their distribution in large arteries coincides with bifurcations, bends, and branch points, where blood flow is disturbed (turbulent). Coronary arteries are particularly susceptible to plaque formation because, as well as their being tortuous and highly branched, the flow is particularly disturbed by the beating heart in which they are embedded.

The response to injury model shown in Fig. 6.46 is widely accepted to explain the initiation and progression of the disease.

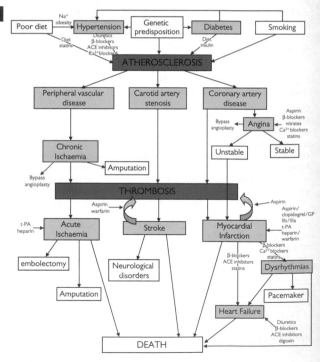

Fig. 6.45 Inter-relationships between risk factors and disease processes: interventional strategies.

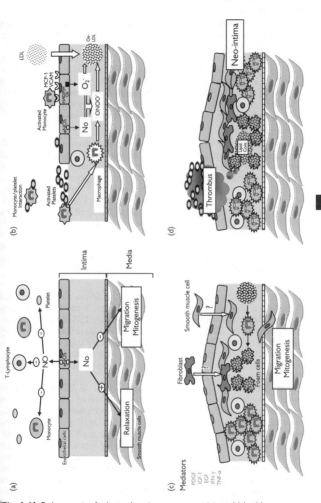

Fig. 6.46 Pathogenesis of atherosclerosis—response to injury: (a) healthy endothelium; (b) damaged/dysfunctional endothelium; (c) inflammatory phase; (d) unresolved inflammation; plaque rupture; thrombosis.

Response to injury

Endothelial injury

Disturbed flow leads to endothelial dysfunction or erosion, with the loss of the protective effects of NO in particular. The affected endothelium becomes 'activated', expressing a range of adhesion molecules (e.g. vascular cell adhesion molecule 1 (VCAM-1)), which 'capture' circulating monocytes. Endothelial erosion exposes the collagen-rich prothrombotic subendothelium, to which platelets adhere, forming microthrombi. There may also be increased release of proatherogenic endothelium-derived ET-1. A further consequence is the generation of the oxidizing free radical, superoxide, from NAD(P)H oxidases in the endothelial membrane.

Inflammation

Captured monocytes infiltrate through the endothelium, where they differentiate into macrophages in response to growth factors, cytokines, and chemoattractants generated by infiltrating T lymphocytes (e.g. granulocyte colony stimulating factor (G-CSF)), which go on to generate high concentrations of several pro-oxidant species (superoxide, NO, peroxynitrite) designed to kill pathogens. The inflammatory process is exacerbated by adherent platelets, which degranulate, releasing a number of pro-inflammatory mediators (e.g. PDGF). Neighbouring smooth muscle cells begin to proliferate and migrate to form the 'neointima' in response to growth factors and in the absence of anti-mitogenic NO. The smooth muscle cells of the neointima conform to a non-contractile, secretory phenotype, generating extracellular matrix to stabilize the developing plaque (fibrosis).

Lipid accumulation

Normally, circulating lipids, in the form of LDLs, diffuse readily in and out of the vessel wall without consequence. However, in the highly oxidizing environment of a developing atherosclerotic lesion, LDLs are rapidly oxidized (forming ox-LDLs), which are recognized by scavenger receptors on macrophages, prior to phagocytosis. The ox-LDLs are trapped in the vessel wall in macrophages (now called foam cells), which ultimately die, releasing their contents to form the lipid-rich core of the plaque. Calcification is also a feature of mature plaques in humans.

Most atherosclerotic plaques stabilize at this point, as inflammation is resolved. A stable plaque will partially occlude the lumen of the artery and the extent and site of the occlusion (or 'stenosis') will determine whether the individual has symptoms. Stable angina pectoris (➲ *OHCM10* p.116) is caused by restricted blood flow through a stenosed coronary artery. Patients with angina, therefore, experience severe chest pain caused by hypoxia associated with insufficient blood supply to part of the myocardium in response to increased demand (e.g. exercise). Symptoms can be successfully managed using organic nitrate drugs (glyceryl trinitrate spray or sublingual tablet) immediately before exercise. Patients are also recommended to take low-dose aspirin daily to reduce the risk of thrombosis, as well as β-blockers to reduce the work rate and oxygen demand of the heart. Severe cases may be treated with balloon angioplasty and stenting (➲ *OHCM10* p.116) or bypass surgery (➲ *OHCM10* p.116), although both procedures carry risks of re-occlusion (due to restenosis or thrombosis). Stenoses in the large conduit arteries of the leg (e.g. femoral arteries) can lead to ischaemia (lack of oxygen) to the affected limbs, causing severe pain and, in some cases, infection or gangrene. Treatments for this so-called peripheral vascular disease include angioplasty and stenting, or bypass grafting. Very severe cases require amputation to prevent sepsis and gangrene.

Plaque rupture

Plaques that remain inflamed can become unstable (prone to rupture). The mechanism that determines the stability of atherosclerotic plaques is not yet fully understood, but the consequences of plaque rupture can be devastating. Material from the core bursts through the weakened neo-intima, where it comes in contact with the blood. This material is highly thrombogenic, → rapid platelet adhesion and aggregation, with the associated activation of the coagulation cascade. The resulting thrombus may completely occlude the artery at the site of the plaque or become dislodged, forming an embolus that passes further down the arterial tree before occluding a smaller vessel. The result is an acute ischaemic event: In the heart (coronary arteries), this causes MI (● OHCM10 pp.118–22); in the brain (carotid arteries), a stroke; and in the leg, acute peripheral ischaemia (● OHCM10 p.657) (femoral arteries, although this may also be caused by an embolus from elsewhere). All are extremely serious and require emergency treatment, although the severity of the event is entirely dependent on the site of the thrombus, the size of the ischaemic area, and the speed at which the correct medical attention is provided (Box 6.12).

Box 6.12 Treatments and drug therapies of atherosclerosis

It is the lipid accumulation stage of the atherosclerotic process for which the most effective drug therapies have been targeted (● OHPDT2 p.56).

- Lowering plasma LDL levels is known to reduce mortality in patients with atherosclerosis-related conditions
- Moderate benefits can be seen with improved diets, but the recent introduction of the drug class known as statins (● OHCM10 p.115) (which inhibit *de novo* synthesis of cholesterol in the liver resulting in up-regulation of LDL receptors that effectively reduce circulating LDL) has led to dramatic improvements in lipid lowering
- Statins are routinely prescribed to 'at-risk' patients, and it has since transpired that statins also have a range of other benefits, particularly with respect to reparative effects on the endothelium, antiplatelet effects, and plaque stabilization
- Other primary prevention is aimed at reducing the prevalence of pro-oxidant species by stopping smoking and treating diabetes
- Other drug therapies are targeted at the thrombotic consequences of atherosclerosis—aspirin (MI), warfarin and new oral anticoagulants (embolic events), or heparin (peripheral ischaemia). Patients with MI should also be prescribed thrombolytic therapy (● OHCM10 p.796) (tissue plasminogen activator (t-PA), streptomycin) as soon as possible. ACE inhibitors and β-blockers have both been shown to reduce subsequent mortality
- Patients who are unable to tolerate statins or for whom maximum therapy is insufficient can now be treated using proprotein convertase sutilisin kexin 9 (PCSK9) inhibitors, blocking the enzyme that normally degrades LDL receptors in the liver. The impact of the inhibitors is, therefore, to increase the population of LDL receptors, resulting in a reduction in circulating LDL. The drugs (e.g. evolocumab and alirocumab) are 'humanized' antibodies raised in mice
- Patients with familial hypercholesterolaemia can also be treated with PCSK9 inhibitors.

Haemostasis

Platelets play a central role in haemostasis—the process that stops blood loss after blood vessels injury. There are three components of haemostasis:
- Platelet activation to form a loose plug as a stop-gap measure
- Local vasoconstriction to reduce blood flow to the affected area
- Activation of the coagulation cascade to convert soluble fibrinogen to fibrin strands that form a mesh around the platelet plug and trap other blood cells, to generate a more permanent repair to the damaged vessel.

These processes are closely interlinked. Platelet activation in response to exposed collagen at the wound results in their synthesizing the vasoconstrictor and platelet activator, TXA_2, and their release of granules containing the vasoconstrictors (adrenaline and 5-HT), as well as inflammatory mediators (e.g. platelet-activating factor (PAF), PDGF). Meanwhile, collagen and platelets stimulate the intrinsic pathway for blood coagulation, while the tissue damage stimulates the extrinsic pathway. These pathways converge to convert prothrombin (so-called factor II) to thrombin (factor II activated (IIa)), which acts to convert fibrinogen to fibrin and to further recruit platelets. The processes of platelet activation and coagulation will now be looked at in more detail.

Platelet activation

There are a range of different glycoprotein receptors on platelet membranes, which recognize and bind to a variety of ligands, including collagen and von Willebrand factor (vWF)—a factor secreted by the endothelium in response to injury. Stimulation of these receptors triggers the platelet activation pathway (Fig. 6.47), resulting in an increase in intracellular Ca^{2+}—the trigger responsible for the following cellular effects:
- Change of shape: pseudopodia emerge from the normal smooth discoid platelet surface, vastly increasing the surface area and, consequently, the adhesiveness of the platelets
- Degranulation: release of vasoconstrictor and platelet-activating factors to cause vasoconstriction and platelet recruitment respectively
- GPIIb/IIIa exposure: a conformational change in the membrane leads to the exposure of the otherwise hidden glycoprotein, GPIIb/IIIa, which binds to fibrinogen to help stabilize the platelet plug.

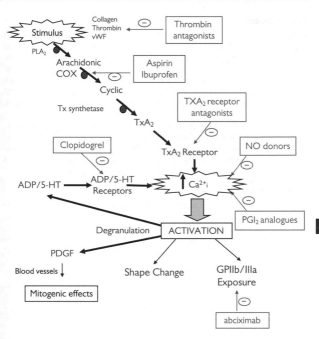

Fig. 6.47 Platelet activation pathways and therapeutic interventions.

Coagulation

Coagulation consists of two cascade pathways (Fig. 6.48a) that converge to generate the activated serine protease, thrombin (factor IIa), which is responsible for the polymerization of soluble fibrinogen into fibrin strands. The purpose of the cascade systems is to amplify the signal: activation of a small amount of one factor in the cascade generates large amounts of the next factor downstream, and so on. The result is rapid formation of large amounts of fibrin in response to what may have been a fairly weak initial signal.

Each step in the cascade involves the activation of normally inactive circulating enzymes (known as 'factors'), most of which are serine proteases. As each factor becomes activated, it catalyses a specific proteolytic event in the subsequent factor in the cascade to activate it. Fibrin formed from soluble fibrinogen is finally stabilized by the action of factor XIIIa.

Haemophilia

Haemophilia (Ⓓ *OHCM10* p.344) is a sex-linked genetic disorder that affects men only and constitutes an inability to synthesize factor VIII (classical haemophilia) or, more rarely, factor IX (haemophilia B or Christmas disease). Until recently, transfusions of whole plasma, or purified, concentrated preparations of missing factors (from the blood of healthy donors) has been the means of factor supplementation. However, with the risk of transfusion-transmitted infectious diseases (e.g. HIV and hepatitis), recombinant factor VIII and IX are now available, although their manufacture has proved difficult because of the need for post-translational modification of the proteins. Genetically modified animals offer hope in this area: they can be genetically manipulated to produce essential coagulation factors in their milk.

Thrombosis and endogenous anti-thrombotic mechanisms

Haemostasis is clearly a crucial process in the very rapid formation of a temporary patch in damaged blood vessels, in advance of the healing process effecting a permanent repair. However, it is essential that the haemostatic process is only stimulated in damaged vessels and that inappropriate clotting (known as thrombosis), that would prevent blood flowing to tissues and organs, is avoided.

- Antithrombin III (ATIII) is central to preventing thrombosis by binding to the active site of all of the factors involved in the coagulation cascade and inhibiting their action (Fig. 6.48b)
- The endothelial lining of blood vessels is now recognized to be central to preventing thrombosis in undamaged vessels by a number of mechanisms (Fig. 6.49):
  - Presenting a physical barrier that separates platelets and coagulation factors in the blood from stimulatory collagen in the subendothelial layers of the blood vessel
  - Secretion of heparan sulphate on the luminal surface to activate ATIII and prevent activation of the clotting factors
  - Synthesis of powerful inhibitors of platelet activation. Prostacyclin (PGI_2) and NO act synergistically to prevent the increase in intra-platelet Ca^{2+} that is essential for activation

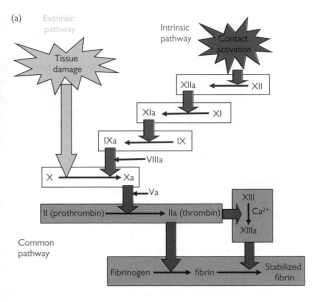

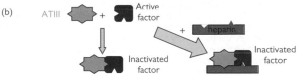

Fig. 6.48 (a) The coagulation cascade; (b) inhibition of coagulation by antithrombin III—effect of heparin.

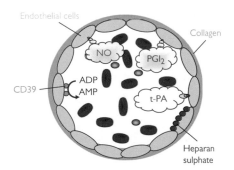

Fig. 6.49 Antiplatelet and anticoagulation properties of the endothelium.

Expression of the enzyme CD39 on the luminal surface to convert the
   platelet activator, ADP, to inactive AMP
• Release of t-PA to convert plasminogen to plasmin, which cleaves
  fibrin strands, earning t-PA the title 'clot-buster'.

Damage to, or dysfunction of, the endothelium compromises some or all
of these protective effects and also leads to the release of the platelet activator, vWF, and the t-PA inhibitor, plasminogen activator inhibitor-1 (PAI-
1). This is believed to be a vital step in a number of pathological conditions
that cause thrombosis, with potentially fatal consequences. The site (arterial or venous) of thrombus generation is important in determining the
morphology and cardiovascular impact of thrombosis:

• Arterial thrombi are predominantly composed of activated platelets
  (so-called white thrombus) with some fibrin. They can be caused by
  atherosclerotic plaque rupture or release of pro-thrombotic debris
  during vascular surgery or interventional cardiology procedures. They
  may remain at the initial site of thrombosis or be carried downstream
  as an embolus where they will block a smaller artery in the same
  vascular tree
• Venous thrombi are mainly fibrin with few platelets and a large
  proportion of red blood cells that become trapped in the mesh (red
  thrombus). Venous thrombi tend to form in conditions where venous
  flow has stopped (deep vein thrombosis caused by extended physical
  inactivity, such as long flights in cramped conditions) or AF, where blood
  pools in the atrial chambers of the heart. Venous thrombi are of little
  consequence while they remain at their site of origin, but at some point,
  they will become dislodged and will pass through the heart into the
  arterial circulation as an embolus, with potentially catastrophic effects.
  Those that originate in the systemic venous circulation will pass through
  the heart into the pulmonary artery, causing pulmonary thrombosis,
  those that originate in the left atrium will pass into the systemic arterial
  circulation, where will they cause thrombosis in whatever tissue or
  organ chance takes them (e.g. the brain to cause a stroke, the heart
  to cause MI or the limbs to cause acute ischaemia). A recognized
  side effect of some oral contraceptives is an increased risk of venous
  thromboembolism.

Box 6.13 summarizes the treatments for thrombosis.

Box 6.13 Treatments and drug therapies

Antiplatelet agents

Antiplatelet agents (⊃ *OHPDT2* Ch. 2) are a particularly useful therapeutic approach to prophylaxis against arterial thrombotic events associated with atherosclerotic disease. Indeed, low-dose aspirin is recognized to be a cheap and effective means of reducing the risk of thrombotic events and is routinely prescribed for patients with a history of cardiovascular disease.

Platelet activation pathways provide several points for drug intervention (Fig. 6.47):

- COX-1: conversion of arachidonic acid to the TXA_2 precursor, prostaglandin H_2, is an essential intermediate step in collagen-stimulated platelet activation. Aspirin and ibuprofen inhibit the enzyme by covalent modification (acetylation) of a critical serine residue in the enzyme. The inhibition is irreversible for aspirin and, uniquely in platelets, will persist for the lifetime of the platelet because, unlike other cells, they have no DNA from which to synthesize replacement COX. The lack of ability of platelets to replace inhibited COX may contribute to the apparent platelet selectivity of COX inhibitors. Their use is limited by gastric irritation, which can lead to gastric ulcers
- ADP receptors: clopidogrel is an irreversible ADP receptor ($P2Y_{12}$) antagonist that is an antiplatelet agent with clinical benefit in a number of settings, particularly in the prevention of thrombosis related to interventional cardiology (e.g. balloon angioplasty and stenting). Given that ADP is unlikely to be the initial trigger for activation, these agents probably block the recruitment phase of thrombosis that is in part mediated by ADP secreted in granules from activated platelets (Fig. 6.47). Few side effects have been reported although there is a risk of bleeding. Clopidogrel itself is a pro-drug and requires metabolism by specific cytochrome P450 enzymes; some individuals are resistant to clopidogrel treatment because they lack the necessary P450 enzymes to metabolize the pro-drug to the active agent. Ticagrelor is a more recent alternative
- GPIIb/IIIa receptors: these agents (e.g. abciximab) have been developed to block the receptors that bind to fibrinogen. They have been found to be particularly powerful and are associated with an increased risk of haemorrhage. ADP antagonists appear to have a preferable therapeutic profile
- Other possible targets: TXA_2 receptor antagonists are under investigation as antiplatelet agents. NO donor drugs that inhibit the increase in Ca^{2+} through cGMP and cAMP production respectively, might be highly effective inhibitors of platelet activation. Combined NO/aspirin drugs are under development, with the possible advantage of reduced gastric toxicity.

Anticoagulants

(⊃ *OHCM10* pp.350–1; ⊃ *OHPDT2* Ch. 2.)

Drug interventions that impact on the coagulation pathway focus on the following targets:

- Inhibition of the vitamin K-dependent post-translational modification of clotting factors: factors II, VII, IX, and X require post-translational γ-carboxylation before they are functional. This conversion involves the carboxylation of glutamic acid to γ-carboxyglutamic acid and requires the co-factor, vitamin K (Fig. 6.50), which is oxidized to an epoxide in the process. Reconstitution of vitamin K to the reduced form involves the enzyme, vitamin K reductase. Warfarin is an oral anticoagulant (first developed as a rat poison) that inhibits vitamin K reductase and reduces the amount of available vitamin K for post-translational modification of clotting factors. The effects of warfarin are slow in onset (~40h) and reversal by oral vitamin K supplementation is also slow. As a result, careful dose titration is critical to ensure that sufficient drug is available to inhibit synthesis of functional clotting factors without causing potentially fatal haemorrhage by preventing the clotting process altogether
- Potentiation of the inhibitory effects of the endogenous anticoagulant, ATIII: each step in the coagulation cascade is inhibited by ATIII, which acts to prevent inappropriate coagulation *in vivo*. Heparin has long been recognized to potentiate the effects of ATIII by simultaneously binding to it and the target activated clotting factor (Fig. 6.50). Factor Xa is an exception, where it appears that binding of heparin to ATIII alone is sufficient to modulate its activity. LMWHs have been developed to act primarily on factor Xa. Heparin and LMWHs carry the side effect of an increased risk of bleeding, but prolonged use of heparin can trigger an immune response which causes paradoxical thrombotic thrombocytopenia (a pro-thrombotic state)
- Thrombin inhibitors: a number of direct inhibitors of thrombin are under development. Hirudin is the anticoagulant used by the medical leech to prevent blood clotting and facilitate continuous feeding. Trials of hirudin have been disappointing to date, but several other thrombin inhibitors are being trialled, including hirudin and PPAK, both of which are able to inhibit thrombin even once it has bound to fibrinogen.

New oral anticoagulants (NOACs) have been developed to overcome the problems associated with warfarin. These have been particularly targeted at venous thromboembolism associated with:
- Deep vein thrombosis
- Pulmonary embolism
- AF
- Stroke in patients with non-valvular AF
- Thrombophilia
- Rivaroxaban, apixaban, and edoxaban are anti-factor Xa anticoagulants
- Dabigatran is an anti-factor IIa (thrombin) anticoagulant (hirudin analogue).

The advantages of NOACs over warfarin are that they are rapid in onset and offset and do not need constant monitoring for dosage. They are not yet approved in children or pregnancy, patients with mitral valve issues or malignant disease and should only be used with caution in patients with chronic kidney disease. They are contraindicated in patients with hepatic disease.

Fibrinolytic drugs (clot-busters)
(➲ *OHCM10* pp.345, 470, 796.)

In the unfortunate event of a thrombotic event, it is essential to re-store blood flow to the ischaemic area as soon as possible to minimize the area of infarct. The drugs of choice are recombinant t-PAs (alteplase, duteplase, reteplase) and streptokinase, which act to support endogenous fibrinolytic agents, t-PA, and urokinase-type plasminogen activator (u-PA) in converting inactive plasminogen in clots to the active fibrinolytic en-zyme, plasmin (Fig. 6.51). Plasmin induces the lysis of fibrin, fibrinogen, and many of the clotting factors. Fibrinolytic agents administered within 12h of MI and 3h of stroke have shown real clinical benefit in terms of the post-trauma impairment suffered by patients. However, fibrinolytics can cause severe bleeding and it is recommended that they are not used again in the same patient for 1 year. Bleeding complications can be reversed by tranexamic acid.

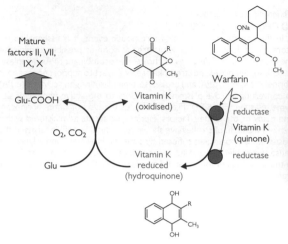

Fig. 6.50 Role of vitamin K in post-translational modification of clotting factors—effect of warfarin.

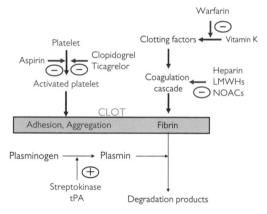

Fig. 6.51 Interventional strategies for preventing thrombosis.

Urinary system

The kidney

- The kidneys are found bilaterally either side of the vertebral column at T12–L3, outside the peritoneal cavity (Table 7.1)
- Their role is to filter and excrete waste products from blood and urine, as well as to regulate body fluid composition. In addition, the kidneys are important endocrine organs
- They each measure ~11cm long × 6cm wide × 4cm deep
- The right kidney sits slightly lower (about 12mm) than the left kidney, since it is displaced by the right lobe of the liver. The kidneys move up and down with respiration but demonstrate little side to side movemen
- The renal hilum (the point at which the renal vessels, ureters, and nerves enter and leave the kidney) is an opening on the medial aspect o each kidney. The left renal hilum is at the same level as the transpyloric plane, and the right is usually slightly lower
- Moving anteriorly to posteriorly, the hilum usually contains:
  - Renal vein
  - Renal artery
  - Renal pelvis
  - Subsidiary renal artery
- Lymph vessels, nerves, and fat occupy a more variable position within the hilum
- The renal pelvis drains urine from the two or three major calyces of the kidney, with two or three minor calyces draining into each major calyx. In turn, each minor calyx is fed by renal papillae tissue, and this represents the point at which the collecting ducts of the kidney transmi urine into the ureter. The renal pelvis can be intra or extra-renal, depending on whether it is completely enclosed by the kidney.

Coverings of the kidney

- The kidneys are surrounded by four distinct coverings:
  - The fibrous capsule of the kidney almost completely encloses it and is separated from the renal fascia by a layer of perirenal fat
  - Renal fascia is a fibrous tough tissue around the kidney and adrenal glands that projects bundles of collagen into the surrounding fat, helping to anchor the kidney in position
  - Together, the perirenal fat and pararenal fat (which lies outside the layer of renal fascia, particularly posteriorly) form a double protective layer of fat around the kidney.

Gross structure

- The kidney is made up of an outer cortex and an inner medulla (Fig. 7.1):
  - The medulla is lighter brown in colour and comprises up to a dozen renal pyramids that are oriented such that the point feeds into the minor calyces
  - The cortex comprises all of the outer lateral regions of the kidney, as well as renal columns between the renal pyramids. Medullary rays are striated areas which project from the bases of the renal pyramids, through the renal cortex.

Table 7.1 Anatomical relations

	Right kidney	Left kidney
Superiorly	Diaphragm separates from pleura and 12th rib	Adrenal glands cap superior pole of each kidney
Posteriorly	Quadratus lumborum, transversus abdominis, psoas	Iliohypogastric nerve, ilioinguinal nerve, subcostal nerve/vessels
Anteriorly	Liver, second part of duodenum, ascending colon	Spleen, jejunum, pancreas and blood vessels, stomach, descending colon

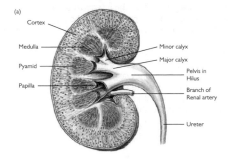

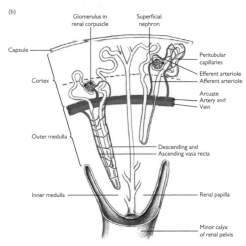

Fig. 7.1 (a) Diagram of hemisected kidney to show its component parts. (b) Arrangement of kidney microvasculature.

Reproduced from MacKinnon, Pamela, and Morris, John, *Oxford Textbook of Functional Anatomy* Vol 2, p172 (Oxford, 2005). With permission from OUP.

Blood supply and lymphatic drainage

- The renal arteries arise from the aorta at L1–2:
  - The right artery is longer and runs behind the inferior vena cava to cross to the right side
  - Close to the hilum of each kidney, the artery divides into five segmental end arteries. These segmental arteries supply the segments of the kidney
  - Each segmental artery gives rise to lobar arteries, which each supply an individual renal pyramid. Two or three interlobar arteries are given off by each lobar artery
  - Interlobar arteries then enter the renal cortex on either side of the renal pyramid
  - Arcuate arteries are given off by the interlobar arteries at the intersection of the cortex and medulla over the base of the pyramids
  - These arteries give off several interlobular arteries, which enter the cortex
  - From these arteries, the afferent glomerular arterioles arise to supply the glomerulus, and ultimately form the vasa recta of the medulla
- The venous drainage of the kidney largely parallels the arterial supply, with veins following a similar route to arteries. The renal vein runs anterior to the renal artery as it leaves the hilum. The left vein is longer, having to cross in front of the abdominal aorta
- Lymphatic drainage of the kidneys is to the para-aortic and lumbar lymph nodes.

Nerve supply

Thoracic splanchnic nerves form a renal plexus containing mainly sympathetic and parasympathetic vasomotor nerves.

The ureters

These are muscular tubes lined with transitional epithelium which, by peristalsis, carry urine produced by the kidneys to the bladder for storage and
excretion. They are ~25cm long and are commonly considered in four
parts, starting with the renal pelvis, which narrows to form the abdominal,
then pelvic, and, finally, intravesical (bladder wall) sections of the ureter.

Anatomical relations

The right ureter is covered by the second part of the duodenum anteriorly
at the renal pelvis, and lies behind the posterior peritoneum, lateral to the
inferior vena cava. It is crossed by three vessels (Table 7.2):
• Right colic artery
• Testicular/ovarian artery
• Ileocolic artery.

The left ureter passes along the medial border of psoas and behind the sig-
moid mesocolon and sigmoid colon. It is crossed by two vessels:
• Left colic artery
• Testicular/ovarian artery.

Both ureters cross the pelvic brim level with the division of the right and
left common iliac arteries into internal and external segments, and these
points mark the beginning of the pelvic sections of the right and left ureters.
    The final intravesical part of the ureters runs obliquely through the wall
of the bladder. The oblique course allows the opening of the ureters to act
like a valve, preventing the back flow of urine from the bladder.
    There are three narrowings of the ureter where kidney stones are most
likely to become lodged:
• The junction of the renal pelvis with the abdominal part of the ureter
• The pelvic brim, where the ureter enters the pelvis
• The pelviureteric junction, where the ureter enters the bladder wall.

The ureters are drained by the testicular/ovarian veins.
    Lymphatic drainage of the ureters is to the lumbar lymph nodes and to
the internal, external, and common iliac lymph nodes.

Table 7.2 Blood supply to the ureters

Section of ureter	Blood supply
Renal pelvis	Aorta and renal arteries
Abdominal	Aorta, renal, and testicular/ovarian arteries
Pelvic	Testicular/ovarian and internal iliac arteries
Intravesical	Internal iliac and inferior vesicle arteries

The bladder

The bladder is a balloon-like structure, lined with transitional epithelium. It expands and contracts as it stores and excretes urine, which is produced by the kidneys and drains into it via the ureters. In humans, urine is excreted via the urethra under the voluntary control of the internal meatus. Most adult subjects begin to feel the sensation of a full bladder once it contains ~250mL.

Relations of the bladder

- The bladder is a relatively mobile organ within the pelvic cavity, surrounded largely by extraperitoneal fat. It always contains some urine but, after emptying, it is pyramidal in shape, with the apex formed by the bladder wall behind the pubic symphysis. The base is formed by the posteroinferior aspect of the bladder (the fundus)
- The bladder is anchored at its neck by the pubovesical ligament (in females) or the puboprostatic ligament (in males):
  - In females, the bladder neck sits on the pelvic fascia, which surrounds the short urethra
  - In males, the bladder neck merges with the prostate and the urethra is much longer, extending along the length of the penile shaft to the external meatus
- In adults, the bladder is an extraperitoneal organ but, as it fills, it expands upwards into the abdomen, stripping the peritoneum upwards from the anterior abdominal wall. In children up to 3 years old, the bladder is an intra-abdominal, extraperitoneal organ, due to the relatively small pelvis of a child
- The trigone is an area of the bladder wall which is smooth, even when the bladder is empty, because the mucosa in this area of the bladder is adherent to the underlying smooth muscle. This area is bounded by the internal meatus and the two ureteric orifices, forming the triangular area
- The internal urethra sphincter (internal meatus) is formed by circular fibres from the smooth muscle of the trigone area. The external sphincter is made of striated (skeletal) muscle as part of the urogenital diaphragm muscle
- In the rest of the bladder, the mucosa is loosely adherent to the underlying detrusor muscle, causing folding or rugae of the mucosa when the bladder is empty. An interureteric ridge runs between the two ureters, formed by a band of muscle underlying the mucosa
- Anteriorly, the bladder is bounded by the pubic symphysis and, laterally, by the obturator internus and levator ani muscles
- Posteriorly: in the male, the bladder is surrounded by the rectum, seminal vesicles, and termination of the vas deferens. In the female, it is bounded, posteriorly, by the vagina and superior part of the cervix
- Superiorly, the bladder is covered by the peritoneum. Coils of small intestine and sigmoid colon lie above the layer of the peritoneum. The uterus of the female can lie against the posterosuperior aspect of the bladder.

Blood supply

The internal iliac arteries supply the bladder via the superior and inferior vesicle branches:

- The superior vesicle artery supplies the anterosuperior part of the bladder
- In males, the fundus of the bladder is supplied by the superior vesicle artery; in females, the fundus is supplied by the vaginal arteries

The vesicle venous plexus drains the bladder:

- In men, this plexus is formed by the vesicle veins and combines with the prostatic venous plexus. The vesicle plexus drains via the internal vesicle veins to the internal iliac veins
- In females, the vesicle venous plexus communicates with the vaginal venous plexus and also receives blood from the dorsal vein of the clitoris. As in males, blood is eventually drained to the internal vesicle and the internal iliac veins.

Lymphatic drainage

Lymph is drained from the bladder in parallel to the vesicular blood vessels and mainly to the iliac and para-aortic lymph nodes.

Innervation

The inferior hypogastric plexuses innervate the bladder and contain sympathetic postganglionic fibres (from L1 and L2):

- Preganglionic parasympathetic fibres form part of the plexus via the splanchnic nerves (S2–4)
- These preganglionic fibres synapse with the postganglionic fibres in the inferior hypogastric plexus

The pelvic splanchnic nerves also carry afferent sensory fibres to the CNS. Other afferent sensory fibres are transmitted, via the inferior hypogastric plexus, to the L1 and L2 segments.

Histology of the urinary tract

The functional unit of the kidney is the nephron (Fig. 7.2). This is divide
into a number of sections that have distinct structural features related
their function in urine production.

The renal corpuscle

- The endothelial cell wall of the glomerular capillary, the podocytes, and
  the fused basement membrane of these two cell types make up the
  filtration barrier of the kidney (Fig. 7.3):
  - Ions, water, and small solutes (below the size of albumin, ~60kDa)
    are filtered out of the blood and enter the proximal tubule
- The macula densa cells, extraglomerular cells, and juxtaglomerular (or
  granular) cells, with the last part of the loop of Henle/the initial part o
  the distal tubule, form the juxtaglomerular apparatus (Fig. 7.4):
  - This is responsible for controlling the Na^+ level of the plasma (➲
    OHCM10 p.668), which in turn contributes to regulation of blood
    pressure (➲ p.472)
  - The granular cells secrete renin in response to low distal tubule Na^+.

The proximal tubule

- Arises directly after the Bowman's capsule
- Region of nephron where the majority of filtered solutes are reabsorbe
- Formed from cuboidal epithelial cells:
  - Rich in mitochondria (needed for high levels of active transport)
  - Apical microvilli form brush-border—greatly increases surface area
    for absorption
  - Basolateral cell membranes have deep infoldings which increase their
    surface area.

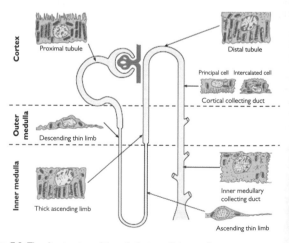

Fig. 7.2 The ultrastructure of the cells that constitute a nephron.

Reproduced with permission from Pocock G and Richards CD (2006), *Human Physiology: The Basis of Medicine*, 3rd edn, p350 Oxford University Press.

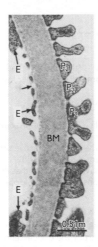

Fig. 7.3 Electron micrograph of the glomerular filter.

Reproduced with permission from Young B and Heath JW (ed) (2000), *Wheater's Functional Histology*, 4th edn, Churchill Livingstone.

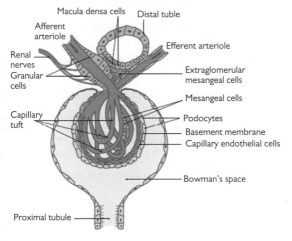

Fig. 7.4 The principal features of a renal glomerulus and the juxtaglomerular apparatus.

Reproduced with permission from Pocock G and Richards CD (2006), *Human Physiology: The Basis of Medicine*, 3rd edn, p349 Oxford University Press.

Loop of Henle

Can be subdivided into three regions based on histology (and indeed function; ➲ p.518):
- Thin descending limb:
  - Squamous epithelial cells (thin and flattened cells)
  - Few mitochondria, compatible with their low active transport activity
- Thin ascending limb:
  - Similar to thin descending limb
- Thick ascending limb:
  - Cuboidal epithelial cells, like those of proximal tubule but without a brush-border
  - Rich in mitochondria (high levels of active transport)
- Not all loops of Henle are the same length:
  - Superficial nephrons have short loops that do not go into the medulla
  - Juxtamedullary nephrons have longest loops which go deep into the medulla.

Distal tubule

- Cells very similar to those of thick ascending limb
- The cells from the last part of the thick ascending limb of the loop of Henle/initial part of the distal tubule contact the afferent glomerular arteriole to form the juxtaglomerular apparatus (Fig. 7.4).

Cortical collecting duct (CCD)

CCD epithelium is made up of two cell types:
- Principal (P) cells:
  - Important in regulation of sodium balance (➲ p.521; see also ➲ OHCM10 p.668)
- Intercalated (I) cells:
  - Important in acid–base regulation (➲ OHCM10 pp.670–2).

Renal blood supply

- The renal tubules are encased in a rich network of peritubular capillaries (Fig. 7.5):
  - Afferent arterioles (from the arcuate artery) give rise to the capillary tufts of the glomerulus, and then the outer cortex efferent arterioles give rise to the peritubular capillaries
  - Arterioles close to the medulla give rise to the vasa recta that supply the inner and outer medulla
  - The peritubular capillaries and vasa recta drain into the stellate and arcuate veins, respectively.

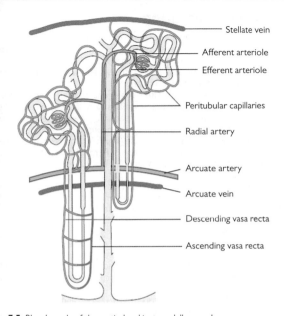

Fig. 7.5 Blood supply of the cortical and juxtamedullary nephrons.
Reproduced with permission from Pocock G and Richards CD (2006), *Human Physiology: The Basis Medicine*, 3rd edn, p351 Oxford University Press.

Glomerular filtration

- The glomerulus is a capillary knot at which blood is filtered into renal tubules. It provides a high-pressure, large-surface area arrangement to maximize filtration. There are ~1 million glomeruli
- Blood enters glomerular capillaries in afferent arterioles from the renal artery and leaves in efferent arterioles
- Each glomerulus is associated with a single renal tubule. The epithelial cells defining the tubule form the Bowman's capsule. They envelop the glomerulus and define the Bowman's space. Together, these structures form the functional unit of the kidney: the nephron (Fig. 7.6)
- Ultrafiltration occurs within the capsule (Fig. 7.7). Approximately one-fifth of cardiac output supplies the kidneys. Renal plasma flow (RPF) is 625mL min^{-1}; glomerular filtration rate (GFR) = 125mL min^{-1}; filtration fraction = GFR/RPF = 20%
- No volume of blood is entirely swept of a substance: most substances are incompletely filtered from a larger volume
- The filtrate lacks cells: its composition is essentially that of plasma minus the proteins. The filter comprises three barriers:
  - Fenestrated endothelial capillary cells
  - Negatively charged basement membrane of capillary endothelium
  - Interdigitating finger-like processes (pedicels) from specialized epithelial cells called podocytes, which encircle capillaries. Negatively charged nephrin molecules from the pedicels further interdigitate to define slit pores
- The filter discriminates on the basis of size, charge, and shape
- Cell passage is blocked by fenestrations; proteins are excluded by negative charges of basement membrane and slit pores. In nephrotic syndrome (→ OHCM10 p.312), damage to the basement membrane leads to filtration of proteins and proteinuria
- Filtration is driven by Starling forces:
  - Forces driving filtration:
    - Capillary hydrostatic pressure (HP$_{cap}$)
    - Osmotic pressure within Bowman's space (OP$_{BS}$)
  - Forces opposing filtration:
    - Hydrostatic pressure in Bowman's space (HP$_{BS}$)
    - Osmotic pressure in capillary (OP$_{cap}$)
- Since plasma and filtrate have identical composition (except for proteins), only the osmotic pressure exerted by proteins (oncotic pressure, π) differs. π_{BS} is essentially zero. Therefore:

$$\text{Net filtration pressure}\left(P_{uf}\right) = HP_{cap} - \left(HP_{BS} + \pi_{cap}\right)$$

and for any nephron, GFR = $K_f \times P_{uf}$

where K_f = surface area × permeability
  - As blood is filtered, HP$_{cap}$ falls only slightly, but π_{cap} rises as [protein] increases. HP$_{BS}$ is low and does not increase—the tubule acts a 'sink' for filtrate. As a result, filtration pressure equilibrium (where HP$_{cap}$ = (HP$_{BS}$ + π_{cap})) is achieved very late along capillary length, if at all

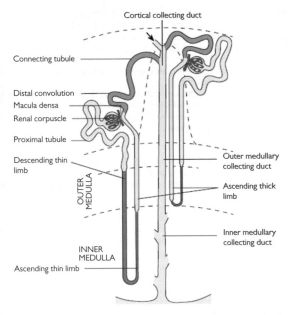

Fig. 7.6 Diagram of a short looped (cortical) and long-looped (juxtamedullary) nephron to show their basic organization.

Reproduced with permission from Pocock G and Richards CD (2006), *Human Physiology: The Basis of Medicine*, 3rd edn, p349 Oxford University Press.

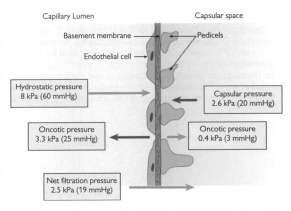

Fig. 7.7 Diagrammatic representation of the filtration barrier in the glomerulus and the hydrodynamic forces that determine the rate of ultrafiltration.

Reproduced with permission from Pocock G and Richards CD (2006), *Human Physiology: The Basis of Medicine*, 3rd edn, p355 Oxford University Press.

- HP_{cap} is adjusted to hold GFR steady. Afferent arteriole dilation increases HP_{cap}; efferent dilation reduces it. Mesangial cells in Bowman's capsule can contract to alter surface area and hence K_f
- There is autoregulation of GFR. If blood pressure rises, GFR is held steady by two mechanisms:
  - Myogenic mechanism: increased stretch of afferent arterioles causes smooth muscle contraction
  - Flow-dependent mechanism (tubuloglomerular feedback): increased GFR increases flow to distal nephron. The macula densa senses distal flow rate; increased flow causes release of mediator(s) which constricts afferent arteriole.

Measurement of GFR

- GFR can be measured using clearance (◐ *OHCM10* pp.668–9)
- The renal clearance of a substance, X, is the volume of plasma from which the substance has been completely removed and excreted to the urine per unit time
- Clearance is an idealized parameter: the minimum volume from which the kidneys could have obtained the excreted amount in a given time. Clearance can be calculated on the basis that the amount of X removed from the blood = the amount appearing in the urine.

Clearance is calculated by measuring $[X]_{plasma}$, $[X]_{urine}$, and urine flow rate V. If removal rate = excretion rate, then

$$[X]_{plasma} \times volume\ cleared = [X]_{urine} \times V$$

i.e. volume cleared = GFR = $([X]_{urine} \cdot V)/[X]_{plasma}$

- Clearance will equal glomerular filtration rate only if the marker is:
  - Freely filtered
  - Not reabsorbed from the nephron
  - Not secreted into the nephron
  - Not metabolized or synthesized by the kidney
- Inulin clearance obeys these rules: GFR = 125mL min⁻¹. If the marker is reabsorbed, $[X]_{urine}$ will be reduced, so clearance is lowered, and the apparent GFR is lower. Creatinine is commonly used for clinical measurements
- Similar methods can be used with substances (e.g. para-aminohippurate (PAH)) that are totally removed from the blood during passage through the kidney (i.e. secreted into the tubule) to obtain RPF.

Tubular transport

- The renal tubule modifies the composition of the primary filtrate to generate urine with appropriate composition for body fluid homeostasis
- Most of the filtered load is reabsorbed. Why filter and reabsorb so much?:
  - Only transporters for essential solutes that must be recovered are required
  - Water balance is facilitated by using a reabsorptive process for H_2O, rather than a secretory one
  - Energetically advantageous—many other solutes can be recovered in association with the reabsorption of Na^+
- The tubule follows the Ussing model (❷ p.62) of epithelial transport:
  - Apical membranes with specialized carriers for reabsorption—carrier-mediated processes saturate, so renal tubules exhibit transport maxima for solutes
  - Basolateral membranes with cellular homeostatic functions (especially Na^+-K^+-ATPase); solute efflux pathways (e.g. GLUT for glucose)
  - Tight junctions with varying degrees of 'leakiness': the tubule becomes increasingly 'tight' along its length—the early stages recover filtered essentials (e.g. ions, glucose), the later stages fine-tune urine composition to regulate body fluid composition
- The proximal tubule performs isotonic fluid reabsorption. It is a leaky bulk reabsorptive tissue, with high water permeability. Around two-thirds of filtered solutes (Na^+, Cl^-, HCO_3^-, glucose, amino acids, Ca^{2+}) and water are reabsorbed here. All absorption is ultimately dependent on transepithelial Na^+ absorption:
  - Apical Na^+ entry is coupled to movement of other solutes:
    - Na^+-glucose co-transport (SGLT)
    - Na^+-amino acid co-transporters (systems for cationic, basic, neutral amino acids) (Fig. 7.8)
    - Na^+-H^+ exchange (NHE) drives HCO_3^- reabsorption and Cl^- reabsorption (Fig. 7.9)
    - Na^+-Cl^- co-transport (NCC)
    - $3Na^+$-HPO_4^{2-} (or $2Na^+$-$H_2PO_4^-$) co-transport (NaPi)
  - Much Cl^- movement happens in later proximal tubule, through paracellular pathways, after electrical and chemical gradients have been established by Na^+ movement in preceding processes. Paracellular Mg^{2+} reabsorption can occur in a similar fashion
  - Ca^{2+} enters through apical epithelial Ca^{2+} channels (ECaC) and is extruded across the basolateral membrane by Ca^{2+}-ATPase, $Na^+ \times Ca^{2+}$ exchange (process stimulated by parathyroid hormone). Similar processes for Ca^{2+} operate in the ascending limb of the loop of Henle and distal tubule
  - There is also secretion of organic anions such as PAH using anion exchangers on each membrane. Basolateral accumulation of organic anions is achieved by exchange with a divalent cation (e.g. α-ketoglutarate). PAH then exits to the lumen using another apical anion exchanger.

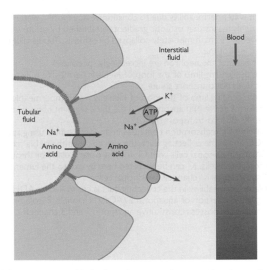

Fig. 7.8 Process responsible for the reabsorption of amino acids in the proximal tubule.

Reproduced with permission from Pocock G and Richards CD (2006), *Human Physiology: The Basis of Medicine*, 3rd edn, p360 Oxford University Press.

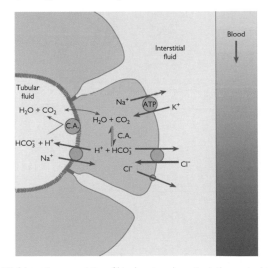

Fig. 7.9 Schematic representation of bicarbonate reabsorption in the proximal tubule.

Reproduced with permission from Pocock G and Richards CD (2006), *Human Physiology: The Basis of Medicine*, 3rd edn, p361 Oxford University Press.

- High water permeability due to constitutive aquaporin I water channels means that osmotic gradients established by solute movements are immediately collapsed by osmosis. Tubular fluid remains isotonic
- Other parts of the nephron are increasingly specialized:
  - The descending limb of the loop of Henle is permeable only to water, but does not transport solute
  - The ascending limb of the loop of Henle is water impermeable but performs active Na^+ absorption in the thick segment using apical Na^+-K^+-$2Cl^-$ co-transport (NKCC)
  - Na^+ and Cl^- reabsorption continues in the distal tubule using apical NCC and in the collecting duct, through apical epithelial Na^+ channels (ENaC) in principal cells with Cl^- moving paracellularly or through intercalated cells. K^+ can be secreted here by exit to the lumen through apical K^+ channels in principal cells
  - Water permeability in the collecting duct is controlled by ADH-regulated insertion of aquaporin 2 (AQP2) channels. Basolateral membranes possess constitutive aquaporin 1 and 3 channels.

Dilute and concentrated urines

- The kidney has the capacity to produce dilute or concentrated urines, according to water balance
- The default is to produce dilute urine, but a countercurrent multiplier system can extract large volumes of water to concentrate urine when water intake is inadequate (Fig. 7.10)
- The loop of Henle is responsible: comparison of different species shows that the longer the loop, the more concentrated the urine
- To understand the process, start in the thick ascending limb (TALH): here, active Na^+ absorption on NKCC is the key event. This explains the action of loop diuretics such as furosemide (→ p.481; see also → *OHPDT2* Ch. 2), which inhibit NKCC. This transport initiates a cascade of responses:
  - The interstitial fluid surrounding the tubule is made hypertonic
  - Water moves by osmosis from the water-permeable (but largely Na^+-impermeable) descending limb of the loop (DLH). $[Na^+]$ rises in the tubule fluid, making the fluid hypertonic. Some influx of Na^+ into the thin ascending limb (TALH) from interstitium may also occur
  - The ascending limb is permeable to Na^+ but not to water. In the TALH (which lacks Na^+-K^+ ATPases), the high $[Na^+]$ established in the tubule fluid encourages passive absorption of Na^+
  - The absorption of Na^+ continues on NKCC, so the tubular fluid is hypotonic when it reaches the distal tubule
  - Na^+ absorption also occurs in the distal tubule and collecting duct, further diluting the tubule fluid
- The apical membrane of the collecting duct lacks a constitutive permeability to water:
  - When water intake is sufficient, water is not reabsorbed from the collecting duct: large volumes of hypotonic urine are produced = diuresis
  - In hydropenia, circulating ADH raises water permeability of the collecting duct. The hypertonic interstitium draws water from the collecting duct. Small volumes of hypertonic urine are produced = antidiuresis.

ADH actions

- ADH promotes the production of a concentrated urine by:
  - Increased water permeability in collecting ducts: V2 receptor occupation; cAMP generation; PKA activation; insertion of vesicles containing AQP2
  - Increased urea permeability in inner medullary collecting ducts: V1 receptor-mediated PKA phosphorylation of apical urea carrier UT1
  - Increased Na^+ reabsorption in TALH: V1 receptor-mediated PKA phosphorylation of NKCC
- Diabetes insipidus (→ *OHCM10* p.240) results when ADH secretion fails (neurogenic) or when renal responses are absent (nephrogenic)
- Not all of the interstitial hypertonicity in antidiuresis is due to Na^+ reabsorption. Up to 50% can be contributed by urea in inner medulla. Urea, concentrated as tubular fluid is reabsorbed along the nephron, exits on UT1 to the interstitium down a steep concentration gradient.

Urea cycles—it diffuses back into TALH, minimizing loss to blood (this trapping establishes high [urea] in interstitial fluid)
- The system is a multiplier: from a 'standing start', NaCl extraction in TALH promotes water extraction in DLH, raises Na^+ in TALH, allows further NaCl reabsorption. The increase in medullary [NaCl] is hence amplified
- Wash-out of interstitial hypertonicity is minimized by low rates of blood flow in medulla and by vasa recta (hairpin capillaries which parallel the loop of Henle). Water leaves and the Na^+ enters as blood passes down the descending limb; water moves back in, Na^+ leaves as blood flows up the ascending limb. The vasa recta countercurrent exchange provides nutrients and O_2 to tubules and removes excess solutes and water from the medulla.

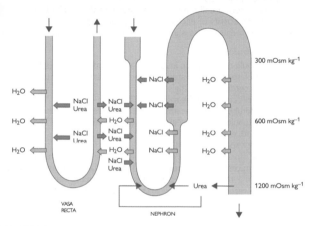

Fig. 7.10 Schematic representation of the countercurrent multiplier of the renal medulla.

Reproduced with permission from Pocock G and Richards CD (2006), *Human Physiology: The Basis of Medicine*, 3rd edn, p369 Oxford University Press.

Regulation of tubule function

- The capacity of the collecting duct to separate water and solute movements underlies body fluid homeostasis. Renal regulation of extracellular osmolarity and volume relies on maintaining a constant delivery to the tight, distal epithelium which is regulated by hormones
- GFR (the volume of plasma filtered by the kidneys) = 180L day^{-1}. GFR remains constant in the face of changes to blood pressure (although GFR does change when renal plasma flow changes). This is due to:
  - Myogenic autoregulation of renal arterioles
  - Tubuloglomerular feedback. The macula densa in the distal tubule acts as a flow sensor (probably by sensing Cl$^-$ absorption); if altered GFR is detected, a mediator (believed to be adenosine) is released that adjusts arteriole resistance and resets GFR
- As a result, a constant filtered load is delivered to the renal tubules. The functional activity of each nephron segment depends on events in preceding segments. There is intrinsic and extrinsic regulation.

Proximal tubule

- Proximal tubule reabsorption involves Na$^+$-dependent transepithelial movement of solutes and water, followed by uptake into peritubular capillaries according to Starling's forces
- Absorptive processes can be modulated by:
  - Glomerulotubular balance: absorption is load dependent: the proximal tubule normally reabsorbs a constant fraction (2/3) of filtered load, even when GFR changes:
    - ○ Increased solute availability stimulates carriers
    - ○ Starling's forces promote enhanced capillary uptake. Increased GFR reduces hydrostatic pressure in the capillary, increases oncotic pressure in the capillary—promotes greater uptake into capillaries
  - Hormone regulation of Na$^+$-H$^+$ exchange: angiotensin II stimulates, PTH inhibits
  - Renal sympathetic activity.

Loop of Henle

- The loop of Henle reabsorbs ~20% of the filtered load. Again, reabsorption is proportional to the load delivered
- This is the result of dependence on flow rate, which determines contact time and so extent of Na$^+$ transport
- If flow rate rises, NaCl stays higher, for longer, in the ascending limb. Increased NKCC activity 'mops up' some of the excess Na$^+$, but more NaCl is delivered to the macula densa. Tubuloglomerular feedback corrects flow rate by altering GFR.

Distal tubule

- The distal tubule reabsorbs ~7% of the load using NCC; regulatory mechanisms are unknown. The connecting tubule is a major site of Ca^{2+} reabsorption, stimulated by PTH.

Collecting tubule

- The collecting tubule is a tight epithelium, a heavily regulated site, for 'fine-tuning' urine composition
- Principal cells reabsorb ~5% Na^+ through apical channels, H_2O through ADH-regulated AQP2 channels. K^+ is secreted through apical K^+ channels. Reabsorption of Na^+ is:
  - Proportional to load
  - Increased by aldosterone
  - Inhibited by atrial natriuretic peptide (ANP)
- Aldosterone stimulates transcription of ENaC, the Na^+-K^+-ATPase, and metabolic enzymes. Increased exchange of Na^+ for K^+ on the ATPase means that K^+ secretion is proportional to Na^+ absorption. Mutations in ENaC lead to constitutive activation of reabsorption: pseudohyperaldosteronism (Liddle's disease)
- ANP antagonizes the renin–angiotensin–aldosterone cascade, and reduces ADH release
- Type A intercalated cells perform active H^+ secretion; type B cells perform HCO_3^- secretion and Cl^- reabsorption. Type A cells predominate, but numbers of type B cells are up-regulated in alkalosis.

Diuretics

(➔ OHCM10 p.136; ➔ OHPDT2 Ch. 2.)

Diuretics increase Na+ excretion by inhibiting the reabsorptive process in nephron. Water excretion increases, extracellular fluid volume falls. Used in cases of fluid retention, where effective circulating volume compromised by heart, liver, or renal dysfunction and to treat hypertension.

Diuretics are classified by their action. They can have disadvantageous effects on electrolyte balance.

Common diuretics

Loop diuretics
- Example: furosemide
- Uses: acute and chronic oedema (➔ OHCM10 pp.134–6), hypertension (➔ OHCM10 pp.138–40)
- Action: inhibits NKCC in TALH
- Consequence: compromises generation of medullary interstitial hypertonicity
- Side effects: excessive depletion of plasma volume; K+ depletion, alkalosis, uric acid retention.

Thiazides
- Example: bendroflumethiazide
- Uses: hypertension (➔ OHCM10 pp.138–40), oedema in chronic heart failure (➔ OHCM10 pp.134–6)
- Action: inhibits NCC in distal tubule
- Consequence: compromises generation of medullary interstitial hypertonicity
- Side effects: K+ depletion, alkalosis, Ca^{2+} retention, uric acid retention.

K+ sparing
- Example: amiloride, spironolactone
- Uses: adjunct to K+-wasting diuretic, in mild hypokalaemia
- Action: amiloride inhibits ENaC in collecting tubule duct; spironolactone antagonizes aldosterone receptor in collecting tubule duct
- Consequences: reduces regulated Na+ reabsorption through channels in late nephron
- Side effects: K+ retention with subsequent acidosis.

Osmotic diuretics
- Example: mannitol
- Uses: raised intracranial, intraocular pressure, forced diuresis in overdose
- Action: trapped in lumen, exerts osmotic potential pull, traps water in tubule
- Consequence: osmotic pull traps water in lumen
- Side effects: Na+ wash-out.

Carbonic anhydrase inhibitors (obsolete)
- Example: acetazolamide
- Uses: glaucoma (➔ OHCM10 pp.456, 561), mountain sickness
- Action: inhibits carbonic anhydrase in proximal tubule
- Consequence: prevents Na+-associated HCO_3^- reabsorption
- Side effects: acidosis, K+ depletion.

Regulation of extracellular fluid volume and osmolarity

- The processes regulating extracellular fluid volume and osmolarity are overlapping. Volume must be regulated to maintain appropriate perfusion pressure; osmolarity changes affect cell volume and function (➲ *OHCM10* pp.668–9)
- Volume is regulated by changing NaCl; osmolarity by changing water content
- The primary effector for body fluid homeostasis is the kidney:
  - Aldosterone and ANP (➲ *OHCM10* p.668) regulate Na^+ excretion
  - ADH (vasopressin) regulates water balance
- Changes in one parameter often cause knock-on effects in the other.

Volume

- Na^+ is the primary determinant of volume: along with accompanying anions it is the dominant osmolyte. Since osmolarity is held constant at ~290mOsm, changes in Na^+ alter osmolarity, which modifies water excretion. Gain or loss of water to restore osmolarity brings about the volume change
- Effective circulating volume (ECV) is detected by:
  - Cardiovascular sensors:
    - Baroreceptors in carotid sinus, aortic arch
    - Stretch receptors in renal afferent arterioles
    - Stretch receptors in atria, pulmonary circulation
  - Renal sensors: the macula densa senses flow of luminal fluid in the distal tubule, which reports GFR and hence ECV
- Afferent neurones from baroreceptors send signals to the medulla in the brainstem. Sympathetic nervous system discharge is altered. Extreme (>15%) falls in ECV also trigger ADH release from the posterior pituitary
- Reduced ECV increases sympathetic discharge and enhances processes that conserve Na^+ and thereby raise ECV by the accompanying H_2O conservation:
  - Renin release
  - Afferent arteriole constriction—reduces GFR
  - Proximal tubule Na^+ reabsorption
- Renin release is also induced by reduced distension of afferent arterioles and by reduced reabsorption across macula densa cells. Renin is an enzyme, and initiates a cascade which occurs in the systemic circulation:
  - Angiotensinogen from liver → angiotensin I
  - Angiotensin I → angiotensin II by angiotensin-converting enzyme in lungs, kidneys
  - Angiotensin II → angiotensin III by aminopeptidases
- Drugs can interfere with the cascade: renin inhibited by enalkiren; ACE by captopril; angiotensin II receptor antagonized by saralasin

- Angiotensin II has a number of actions that conserve Na^+:
  - Constriction of efferent > afferent arterioles. Increased GFR lowers hydrostatic pressure in the capillary and raises the oncotic pressure—promotes greater uptake into capillaries
  - Constriction of the vasa recta to conserve medullary hypertonicity
  - Stimulation of $Na^+ \times H^+$ exchange in the proximal tubule
  - Stimulation of thirst and ADH release
  - Release of aldosterone from adrenal cortex
- Aldosterone stimulates Na^+ reabsorption in collecting duct: steroid actions to increase transcription of ENaC, Na^+-K^+ ATPase, metabolic enzymes. Spironolactone antagonizes the aldosterone receptor. K^+ secretion is increased when Na^+ absorption is increased. Raised plasma $[K^+]$, ACTH also stimulate aldosterone release
- The 28-amino acid peptide, ANP, antagonizes the renin–angiotensin II–aldosterone axis. It is released from atrial myocytes in response to stretch (i.e. raised ECV). It has diverse actions, mediated through cGMP. Selected actions include:
  - Inhibition of angiotensin II stimulation of proximal tubule transport
  - Inhibition of renin release
  - Inhibition of NCC in distal tubule
  - Inhibition of aldosterone receptor
- Net effect is to raise distal tubule load, inhibit reabsorption, increase excretion.

Osmolarity

- Osmolarity is regulated by modulating water excretion (➲ p.517)
- Osmoreceptors in the hypothalamus monitor osmolarity. Raised osmolarity (1% change) triggers cell bodies in the supraoptic and paraventricular nuclei
- As a result, the nine-amino acid peptide ADH, is released from nerve endings in the posterior pituitary. ADH release also induced by larger (15%) falls in volume pressure
- ADH conserves water to reduce osmolarity:
  - V_2 receptor-mediated adenylate cyclase activation, cAMP generation, protein kinase A activation, insertion of vesicles containing AQP2 water channels
  - Stimulation of UT1 urea carriers
  - Stimulation of NKCC in TALH
  - Vasoconstriction.

Renal regulation of plasma pH

- The kidneys work in concert with the lungs to regulate plasma pH
- They have two roles:
  - Reabsorption of filtered HCO_3^- ions
  - Excretion of non-volatile acids (NVAs)
- NVAs include HCl, H_2SO_4 from amino acid metabolism, and H_3PO_4 from ingested phosphate
- Typically, 70mmol of NVA must be excreted. NVAs are initially buffered in extracellular fluids by HCO_3^-. The second function of the kidney is, therefore, in reality, the regeneration of HCO_3^- lost in buffering.

Reabsorption and regeneration of HCO_3^-

- The renal cellular mechanisms for reabsorption and regeneration of HCO_3^- are the same. The key components of these mechanisms are:
  - Carbonic anhydrase in tubule cells to catalyse hydration of CO_2 to yield H^+ and HCO_3^-
  - Secretion of H^+ into the lumen across the apical membrane
  - Efflux of HCO_3^- from the cell across the basolateral membrane
- It is the fate of secreted H^+ ions in the tubule lumen that differs
- For reabsorption:
  - H^+ secreted into lumen, where it reacts with HCO_3^- to form CO_2 and H_2O: catalysed by carbonic anhydrase on apical membrane
  - CO_2 diffuses into cell, where carbonic anhydrase catalysed hydration to re-form HCO_3^- and H^+ occurs
  - HCO_3^- exits at basolateral membrane, H^+ cycles again across apical membrane
  - Net result: HCO_3^- transfers from lumen to interstitial fluid
- For regeneration:
  - H^+ generated by CO_2 hydration is secreted into lumen where it titrates urinary buffers: HPO_4^{2-}, $H_2PO_4^{4-}$, or NH_3. The titrated species are lost in urine:
    - Phosphate species are dietary in origin. The NH_3 used to buffer secreted H^+ is synthesized from glutamine in proximal tubule cells, and diffuses into lumen across the apical membrane. Ionic NH_4^+ formed by titration is trapped in lumen. The increasingly acidic environment of the lumen means dissociation becomes increasingly unlikely along tubule length. Some NH_4^+ can be reabsorbed on NKCC into medullary interstitium, but NH_3 reforms there and diffuses back into collecting duct where NH_4^+ is once more generated
  - HCO_3^- generated by the CO_2 hydration cell exits at basolateral membrane
  - Net result: H^+ excreted, HCO_3^- returned to interstitial fluid. Balance restored. (The HCO_3^- recovered is equivalent to the one consumed by the original HCO_3^- buffering of NVAs.)

Acid–base transporters in the kidney

- Na^+-H^+ exchange (NHE) is responsible for the greatest portion of apical acid secretion. It is especially prominent in the proximal tubule, where bulk Na^+ reabsorption is occurring. H^+-ATPases are most prominent in type A intercalated cells in the collecting duct; this primary active transport system can establish large H^+ gradients
- Basolateral efflux of HCO_3^- is largely mediated by Na^+–$3HCO_3^-$ co-transport (NBC) in the proximal tubule, where there are large fluxes of Na^+. Later nephron segments use $Cl^- \times HCO_3^-$ exchange (AE)
- Acid is also secreted across apical membranes in the medullary collecting duct by an H^+-K^+-ATPase, although this pump may be more important as a K^+ absorber. The operation of this ATPase is one reason why disturbances of K^+ or H^+ homeostasis are often interlinked.

Regulation of renal acid–base transport

- Acid secretion across the apical membrane can be regulated:
  - NHE: protein kinases can stimulate (PKC-mediated activation by angiotensin II) or inhibit (PKA-mediated inhibition by parathyroid hormone)
  - ATPase: acidosis stimulates insertion of vesicles containing the ATPase into apical membrane
- Respiratory and metabolic acidosis (● OHCM10 pp.670–1) can up-regulate NHE, NBC, and ammoniagenesis
- Type B intercalated cells are normally few in number. Their transporter expression is reversed: ATPase is basolateral, AE is apical. As predicted from the Ussing model of epithelial function, this reverses the function of these cells: they secrete HCO_3^-, reabsorb H^+. In normal circumstances, there is a chronic acidosis, so this function is less important. Numbers of type B cells increase in alkalosis
- A reciprocity exists between plasma pH and $[K^+]$:
  - Hyperkalaemia (● OHCM10 p.674) → acidosis:
    ○ Hyperkalaemia inhibits H^+-K^+-ATPase, ammoniagenesis, so limiting H^+ excretion
  - Acidosis → hyperkalaemia:
    ○ Acidosis reduces K^+ secretion in the distal nephron: this occurs because raised H^+ inhibits (i) basolateral Na^+-K^+ ATPase, reducing cell K^+ available for secretion; (ii) apical membrane permeability to K^+
- Mutations in carbonic anhydrase or in the transporters that mediate H^+ and HCO_3^- transport lead to renal tubular acidosis (i.e. a systemic acidosis caused by impaired renal acid–base handling, not acidosis within the renal tubule).

Micturition

- Emptying the bladder is referred to as micturition, urination, or voiding. It involves the synchronous contraction of the bladder wall muscle (detrusor) and relaxation of the urethral sphincters. These processes are coordinated by a combination of autonomic spinal cord reflexes and voluntary control of the external urethral sphincter which is made of striated muscle (Table 7.3). Consequently, micturition, like breathing, is a mixture of reflex and voluntary actions
- The urinary bladder is progressively filled by inflow of urine from the ureters. At low volumes (<200–300mL) the bladder is a relatively compliant organ and filling is accompanied by only modest increases in tension in the bladder wall. At volumes >300mL, tension increases more markedly and this is detected by stretch receptors in the bladder wall smooth muscle. It is at this level of filling that a sensation of fullness is felt in the bladder
- Impulses from the stretched detrusor muscle are sent, via sensory nerves (S2–S4), to the pontine micturition centre in the brainstem where the micturition reflex is integrated. In response to these sensory inputs, parasympathetic efferent signals (S2–S4) initiate contraction of detrusor which, since it is a syncytium of smooth muscle cells, contracts. This results in increased pressure and tendency towards expulsion of urine and opening of the bladder neck (internal sphincter)
- Voluntary control of the bladder arises largely as a result of the external urethral sphincter. This structure is formed where the urethra passes through the urogenital diaphragm (located in the pelvic arch) which comprises skeletal muscles of the pelvic floor under voluntary, cortical control. Contraction of this sphincter prevents the flow of urine and voluntary relaxation has the reverse effect and initiates the flow of urine during micturition
- Further, voluntary control of micturition arises from inhibitory inputs from cortical and suprapontine centres to the pontine micturition centre, which inhibit the micturition reflex until it is socially acceptable to urinate or the sensation of bladder fullness becomes too intense. Sympathetic nerves from the hypogastric plexus also inhibit contraction of detrusor and aid the inhibition of micturition by higher cortical centres
- During bladder emptying, sensory receptors in the urethra sense the flow of urine and feedback to the micturition centre. This has the effect of enhancing the micturition reflex and increasing the flow of urine. Contraction of abdominal and pelvic muscles can also increase the pressure on the bladder wall, thus aiding micturition
- The bladder can be voluntarily emptied at any time, even when not full, indicating that higher areas of the brain can initiate the micturition reflex in the absence of afferent input from stretch receptors
- Voluntary control of micturition can be lost following damage to inhibitory descending pathways in the spinal cord. Such an insult results in urinary incontinence (➔ OHCM10 pp.648–9). Furthermore, lesion of the parasympathetic nerve supply to the bladder results in incomplete emptying of the bladder during micturition, which can lead to recurrent urinary tract infections (➔ OHCM10 pp. 296–7).

Table 7.3 Innervation of the micturition reflex

Type	Nerve	Spinal segment	Function
Parasympathetic (efferent)	*Preganglionic:* pelvic nerves *Postganglionic:* in bladder wall	S2–4	• Contraction of detrusor • Opening of internal sphincter
Sympathetic (efferent)	*Preganglionic:* splanchnic nerve/ paravertebral sympathetic chain *Postganglionic:* pelvic nerves/hypogastric plexus	T10–L2	• Inhibit contraction of detrusor • Closing of internal sphincter
Somatic (efferent)	Pudendal nerve	S2–3	Contraction of external urethral sphincter
Sensory (afferent)	Pelvic nerves/ pudendal nerve	S2–4	Senses bladder filling, pain, and urinary flow

Renal failure

(→ OHCM10 pp.298–305.)

Definition

Reduction in the function of the kidneys such that there is an accumulation of waste products that the kidneys usually excrete, most easily measured by urea and creatinine in the blood. There are about 1 million nephrons in each human kidney but these do not divide or regenerate after foetal development, so any reduction in their number will lead to a reduction in renal function. However, the kidneys have a large reserve capacity and actual renal failure only occurs when about 90% of nephrons are non-functional. Thus, chronic renal failure only becomes manifest after the disease process has been present for years, and it is often difficult to discover the original precipitating cause.

Classification

- Aetiological—but often unknown
- Site of problem—but unhelpful because, for example, a pre-renal cause such as low cardiac output acts at a glomerular and tubular level:
  - Pre-renal
  - Renal
  - Post-renal
- Time course—but does not give any indication of cause:
  - Acute—occurring suddenly over days or weeks
  - Chronic—insidious development over years.

Causes

- Glomerulonephritis (→ OHCM10 pp.310–11)—deposition of material on the glomerular basement membrane and/or inflammation. Mediated by many mechanisms including immune complex deposition, anti-glomerular basement membrane antibodies, sensitized lymphocytes, and neutrophils. The classification is complex (and confusing) and is really only useful for postgraduate nephrologists
- Tubule dysfunction (→ OHCM10 pp.318–19):
  - Acute ischaemia—low cardiac output, disseminated intravascular coagulation, polyarteritis nodosa, or malignant hypertension in intrarenal vessels
  - Toxins—heavy metals
  - Tubulointerstitial nephritis—hypersensitivity to drugs
- Outflow obstruction (→ p.534).

Box 7.1 summarizes the treatments for glomerulonephritis.

Box 7.1 Treatment and drug therapies for glomerulonephritis

(➔ *OHPDT2* Ch. 7.)
- Treatment of the cause
- Renal replacement therapy:
  - Dialysis
  - Transplantation
- Drug therapy for renal disease is particularly difficult because:
  - Reduced renal perfusion reduces drugs reaching the target organ
  - Impaired clearance of the drug through the kidney increases the risk of drug accumulation and nephrotoxicity
- There is currently no recognized effective drug therapy for acute renal failure.
- Drugs prescribed in chronic renal failure include ACE inhibitors and statins with a view to helping to treat potential causes and reducing risk of death from cardiovascular causes. Loop diuretics are also often prescribed to reduce oedema (➔ p.522), but are not effective in severe renal insufficiency.

Obstructive uropathy

(⊃ *OHCM10* pp.640–1.)

Definition

Obstruction of the urinary tract at some point from the renal pelvis to the end of the urethra (Fig. 7.11).

Causes (⊃ *OHCM10* **p.640**)

- Pelvis:
  - Staghorn calculus
  - Pelviureteric obstruction—congenital anomaly due to disarray of the smooth muscle in the wall
- Ureter:
  - Calculus
  - Tumour—transitional cell carcinoma (⊃ *OHCM10* p.644)
- Bladder:
  - Calculus
  - Tumour—transitional cell carcinoma
- Urethra:
  - Prostatic enlargement—benign prostatic hypertrophy (⊃ *OHCM10* p.642), prostatic carcinoma (⊃ *OHCM10* p.644) less commonly
  - Urethral valves—congenital anomaly, a neonatal emergency (recognized by no urine in nappy) which must be treated surgically
  - Phimosis—constriction of the end of the urethra by a tight foreskin in males; causes include scarring post-circumcision, balanitis xerotica obliterans (inflammatory condition analogous to lichen planus in general skin)
  - Tumour—transitional cell carcinoma.

Complications

- Compression atrophy of renal tissue—if there is complete obstruction of a kidney's outflow, then a large hydronephrosis (⊃ *OHCM10* p.641) will form as the urine fills up the collecting system to the level of the obstruction. If this is left for any length of time, there will be pressure atrophy of the renal tissue and the kidney will become rapidly non-functional. The end stage of hydronephrosis is a distended bag of urine (the dilated collecting system) with a tiny rim of renal tissue around the outside
- Urinary infection (⊃ *OHCM10* p.296)—the static urine provides a culture medium for bacterial organisms
- Hypertrophy of smooth muscle proximal to the obstruction, e.g. hypertrophy of the bladder muscle in cases of benign prostatic hypertrophy.

  Box 7.2 summarizes the management of obstructive uropathy.

Box 7.2 Management of obstructive uropathy

Rapid drainage of the urine from upstream of the obstruction is essential to avoid complications. This can be achieved by simple instrumentation, endoscopy or lithotripsy, depending on the cause.

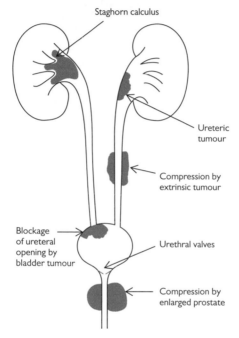

Staghorn calculus

Ureteric tumour

Compression by extrinsic tumour

Blockage of ureteral opening by bladder tumour

Urethral valves

Compression by enlarged prostate

Fig. 7.11 Diagram showing the different site in the urinary tract where obstruction can occur and the most common pathological causes at those sites.

Digestive system

Abdominal wall, peritoneal cavity, and the peritoneum

The structures of the abdominal wall are of obvious significance for surgical approaches to the abdomen. The abdominal wall consists of several distinct layers which are mostly consistent throughout. Skin is the outermost layer and overlies superficial fascia. Beneath this is a tough muscular layer superficial to another layer of fascia and extraperitoneal fat above the peritoneum. These layers protect and contain the abdominal viscera, thereby preventing herniation.

Superficial fascial layer

The subdermal superficial fascia comprises two distinct layers:
- Camper's fascia is a superficial fatty layer containing variable amounts of fat. It covers the anterior abdomen and blends inferiorly with the superficial fascia of the thigh
- Scarpa's fascia is fibrous tissue that forms the deeper of the two fascial layers. It is not present in the superior and lateral areas of the abdominal wall, but inferiorly merges with the fibrous fascial layer of the upper thigh (fascia lata, below the inguinal ligament) and with Colles' fascia, which extends into the genital area and perineum

The deep fascia of the abdomen is indistinct and is tightly bound to the superficial muscles.

Abdominal wall muscles

The muscles of the anterior abdominal wall are arranged in layers (Fig. 8.1). All the muscular layers thin to become fibrous, tendon-like sheets (aponeuroses) across the medial part of the anterior abdominal wall. The aponeuroses fuse in the midline to form the linea alba (white line).
- External oblique is the most superficial muscular layer with fibres running inferiorly from lateral to medial. It attaches laterally to the anterior surface of the 5th–12th ribs, the xiphoid process, linea alba, anterior surface of iliac crest, and pubis. It forms the inguinal ligament inferiorly as the aponeurosis folds back on itself, between the anterior superior iliac spine and the pubic tubercle. An opening in external oblique just medial to the pubic tubercle forms the superficial inguinal ring
- Internal oblique lies beneath external oblique, with muscular fibres running at 90° to external oblique fibres. It arises from the anterior two-thirds of the iliac crest and the lateral half of the inguinal ligament, in front of the deep ring, and attaches to the linea alba, pubic crest, as well as the lumbar fascia. This fascial layer splits medially to ensheath rectus abdominis as two aponeuroses
- Transversus abdominis is the deepest muscular layer with muscular fibres running transversely. It attaches to the 7th–12th costal cartilages, the lumbar fascia, the anterior two-thirds of the iliac crest, the inguinal ligament, linea alba, and pubis. Inferiorly, transversus abdominis forms part of the roof of the inguinal canal as it arches over to become the conjoint tendon with the aponeurosis of internal oblique.

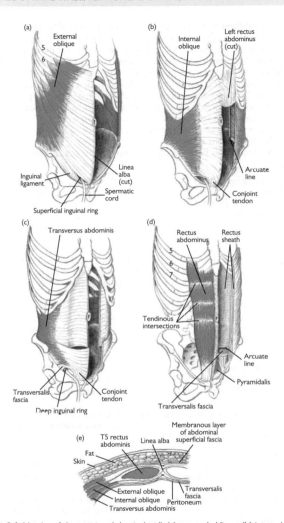

Fig. 8.1 Muscles of the anterior abdominal wall: (a) external oblique; (b) internal oblique; (c) transversus abdominis; (d) rectus abdominis; (e) section through anterior abdominal wall. The aponeuroses of the abdominal wall muscles which form the rectus sheath interdigitate to form the midline linea alba.

- Rectus abdominis is a wide band of muscle, enveloped by the rectus sheath (aponeurotic layers) and divided into two bands of muscle by the fusion of the aponeurotic layers of the abdomen at the linea alba. Rectus abdominis is attached superiorly to the costal cartilages of the 5th–7th ribs and inferiorly to the pubic bone. It is marked by three tendinous intersections which form the 'six pack' seen in muscular, lean people. These adhere to the anterior (but not the posterior) part of the rectus sheath, thereby allowing rectus abdominis to lie free posteriorly. The intersections have constant positions and occur at:
  - The tip of the xiphoid process
  - The level of umbilicus
  - A point halfway between the two intersections already described.

The rectus sheath

- The coverings of the rectus sheath vary, depending on abdominal level
- Above the costal margin, external oblique forms the anterior sheath and no posterior sheath is present. From the costal margin, down to a point halfway between the umbilicus and the pubis, the anterior sheath is formed by external oblique and the anterior aponeurotic slip of internal oblique, with the posterior part being formed by the other aponeurotic slip of internal oblique and the aponeurosis of transversus abdominis
- Below this halfway point, the sheath is formed by all three aponeurotic expansions lying anterior to the rectus muscle with only the transversalis fascia and the peritoneum lying between rectus abdominis and the viscera
- The arcuate line represents the inferior limit of the posterior rectus sheath. The inferior epigastric vessels pass upwards from the external iliac vessels and pierce the rectus sheath at the level of the arcuate line where they anastomose with the superior epigastric vessels.

Innervation

- The nerve supply to the abdominal wall is segmental and follows dermatomal distribution of skin and muscular innervation
- T7–L1 supply the abdomen, with the umbilicus supplied by T10, and the groin and scrotal area supplied by L1
- Above the arcuate line, a neurovascular plane runs between internal oblique and transversus abdominis, and contains nerves and arteries supplying the abdominal wall.

Blood supply

- The inferior epigastric artery and deep circumflex iliac arteries are branches of the external iliac artery and supply the inferior part of the anterior abdominal wall
- The superior part of the anterior abdominal wall is supplied by the superior epigastric artery, a terminal branch of the internal thoracic artery, and branches from the posterior intercostal arteries of the 10th and 11th ribs and the subcostal arteries. These anastomose with the blood vessels supplying the inferior abdominal wall
- The inferior part of the abdominal wall is drained by three superficial inguinal veins into the great saphenous vein of the lower limb
- The superior part of the abdominal wall is drained by the superficial epigastric vein and the lateral thoracic vein
- The para-umbilical veins form a porto-systemic anastomosis between the superficial veins and the deep (portal) venous system.

Peritoneum and the peritoneal cavity

See Figs 8.2 and 8.3.

- Peritoneum is a thin, single-celled layer of mesothelium that covers the internal surfaces of the abdominal wall (parietal peritoneum) and envelops abdominal viscera (visceral peritoneum)
- It originates from the endothelial lining of the primitive coelomic cavity of the embryo and forms a cavity (peritoneal cavity) which is punctured only by the uterine tubes in females
- The peritoneal cavity is filled with peritoneal fluid which lubricates parietal and visceral peritoneum, thereby permitting movement of abdominal viscera
- Peritoneum has several important functions, including support of viscera within the abdomen, fat storage, and sealing off infected bowel segments
- Peritoneal ligaments are formed from double layers of the peritoneal membrane connecting an organ with another organ or the abdominal wall. These include:
  - The falciform ligament (which extends from the anterior abdominal wall to the liver)
  - The gastrophrenic ligament (which extends from the stomach to the inferior surface of the diaphragm)
  - The gastrosplenic ligament (which extends from the stomach to the hilum of the spleen)
  - The gastrocolic ligament (which forms part of the greater omentum and extends from the stomach to the transverse colon).

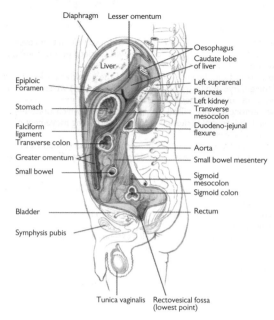

Fig. 8.2 Arrangement of peritoneum of the greater and lesser sacs in a parasagittal section of the abdominal cavity.

Reproduced from MacKinnon, Pamela, and Morris, John, *Oxford Textbook of Functional Anatomy*, Vol 2, p134 (Oxford, 2005). With permission from OUP.

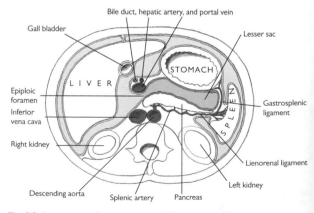

Fig. 8.3 Arrangement of the peritoneum of the greater and lesser sacs in a horizontal section of the abdomen through the stomach, viewed from below.

- Certain bowel segments and abdominal organs, in particular the small intestine, are attached to the posterior abdominal wall by a mesentery—a double-layered flap of peritoneum which is reflected from the abdominal wall and contains blood vessels, nerves and lymphatics, and fat stores. Certain abdominal viscera (e.g. the ascending colon) have no mesentery, are only covered by peritoneum anteriorly, and are retroperitoneal
- The omentum is a double-layered segment of peritoneum that connects the stomach to other organs (in contrast to a mesentery, which connects viscera to the abdominal wall):
  - The lesser omentum runs between the liver and the lesser curvature of the stomach and proximal part of the duodenum as two peritoneal ligaments—the hepatogastric ligament and the hepatoduodenal ligament
  - The greater omentum starts from the greater curvature of the stomach and proximal part of the duodenum as a large flap of tissue that hangs like an apron and passes back up to the transverse colon, where it merges with the transverse mesocolon. At the inferior fold of the greater omentum, its four layers are fused

Two continuous sacs are partially separated within the abdomen by the greater and lesser omentum:
- The lesser sac lies posterior to the lesser omentum and the stomach
- The greater sac forms the anterior cavity of the abdomen
- The foramen of Winslow (or epiploic foramen) connects these two cavities and is bounded anteriorly by the free border of the lesser omentum (which contains the common bile duct, hepatic artery, and the portal vein), posteriorly by the inferior vena cava, inferiorly by the first part of the duodenum, and superiorly by the caudate process of the liver.

Peritoneal recesses

Where peritoneum folds, pouch-like peritoneal recesses are formed
These are potential spaces that may become filled with pus or blood
They include:
- The right and left subphrenic spaces (which lie between the liver and the diaphragm and are divided by the falciform ligament)
- The right subhepatic space (which lies between the liver and the posterior abdominal wall)
- The left subhepatic space (which is the lesser sac)
- The right extraperitoneal space (which lies between the diaphragm and the bare area of the liver).

The inguinal canal

OHCM10 pp.614–15.)
The inguinal canal connects the abdomen to the scrotum
It is a weakness in the abdominal wall and is clinically important as a site prone to herniation of abdominal contents
The inguinal canal is ~6cm long and passes obliquely between the deep and superficial inguinal rings. It carries the spermatic cord in men or the round ligament in women, as well as the ilio-inguinal nerve

- The spermatic cord contains several important structures including the vas deferens, the artery to the vas, the testicular artery, the cremasteric artery, the genital branch of the genitofemoral nerve supplying the cremasteric muscle, the pampiniform plexus of veins, autonomic nerve fibres, and lymphatics which drain to the aortic nodes
- The spermatic cord is covered in three layers of fascia:
  - The outermost external spermatic (continuous with external oblique)
  - The cremasteric muscle and fascia (continuous with internal oblique)
  - The innermost internal spermatic fascia (continuous with transversalis fascia)
- The relations of the inguinal canal are important when considering surgical approaches to repairing inguinal hernias:
  - The base of the canal is formed by the inguinal ligament and the lacunar ligament (which forms the medial part of the floor as a continuation of the inguinal ligament)
  - The roof of the canal is formed by transversus abdominis, internal oblique, and the pectineal line of the pubic bone
  - The anterior wall of the canal is formed by the external oblique medially and the internal oblique laterally, while the posterior wall is formed by the transversalis fascia and the conjoint tendon.

The deep inguinal ring

- The deep inguinal ring is an opening in the transversalis fascia at the midpoint of the inguinal ligament (halfway between the anterior superior iliac spine and the pubic tubercle), 1.3cm above the ligament
- It is just lateral to the mid-inguinal point (halfway between the anterior superior iliac spine and the pubic symphysis)
- It is bounded by the transversalis fascia medially, the inferior epigastric vessels posteriorly, and, laterally, by the angle formed by the inguinal ligament and transversus abdominis.

The superficial inguinal ring

See Fig. 8.4.
- The superficial inguinal ring is an opening in the external oblique aponeurosis and lies above and medial to the pubic tubercle
- The medial crus of external oblique attaches to the pubic crest and the lateral crus attaches to the pubic tubercle. The pubic crest forms the base of the superficial inguinal ring
- The inferior epigastric artery runs around the medial edge of the superficial inguinal ring
- A direct hernia passes through a weakness in the transversalis fascia (i.e. the floor of the inguinal canal) and lies medial to the inferior epigastric vessels. An indirect hernia runs through the deep ring, the canal, and the superficial ring and lies lateral to the inferior epigastric vessels.

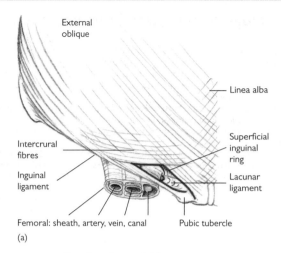

External oblique

Linea alba

Intercrural fibres

Superficial inguinal ring

Inguinal ligament

Lacunar ligament

Femoral: sheath, artery, vein, canal

Pubic tubercle

(a)

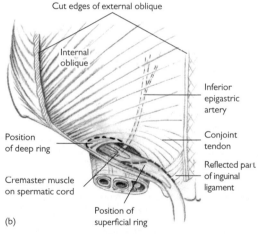

Cut edges of external oblique

Internal oblique

Inferior epigastric artery

Position of deep ring

Conjoint tendon

Reflected part of inguinal ligament

Cremaster muscle on spermatic cord

Position of superficial ring

(b)

Fig. 8.4 (a) Position of superficial inguinal ring relative to the femoral sheath and its contents. Note the sharp edge of the lacuna ligament in relation to the femoral canal, which is continuous with the abdominal (extraperitoneal) cavity. (b) External oblique has been largely removed to show the floor and roof of the inguinal canal and the spermatic cord.

Reproduced from MacKinnon, Pamela and Morris, John, *Oxford Textbook of Functional Anatomy*, vol p115 (Oxford, 2005). With permission of OUP.

Histology of the gastrointestinal tract

- The primary function of the GI tract is to digest and absorb food for energy, growth, and maintenance of the body. It is also an important immune organ for protecting against bacterial invasion
- Certain generalizations can be made about the histological structure of the digestive tract, specifically with regard to the layers of tissue surrounding the bowel lumen
- Differences in microstructure between different regions of the GI tract reflect location and functional specializations
- There are four main layers of the GI tract, moving from the lumen outwards:
  - The mucosa consists of an epithelial layer immediately adjacent to the lumen, underneath which is a connective tissue layer (lamina propria) comprising a small amount of smooth muscle, blood vessels, and sometimes glands and lymphoid tissue. The muscularis mucosa is part of the mucosal layer and the innermost layer of smooth muscle. The mucosa contains gut-associated lymphoid tissue (GALT) in the lamina propria
  - The submucosa is a layer of dense connective tissue containing a nervous plexus (Meissner's plexus) as well as the main blood and lymph vessels
  - Muscularis comprises two layers of smooth muscle—an inner circular layer (muscularis interna) and an outer longitudinal layer (muscularis externa)—sandwiching a myenteric (Auerbach's) plexus
  - The serosa comprises loose connective tissue, adipose tissue, and blood and lymph vessels, surrounded by the mesothelium (simple squamous epithelial layer).

The oesophagus

- Functions:
  - Transmits and propels food from mouth to stomach
  - Lubricates food to aid propulsion
- Lined with non-keratinized, stratified, squamous epithelium, which becomes columnar epithelium at the junction with the stomach
- Within the lamina propria, certain glands (e.g. cardiac glands) secrete mucus into the lumen
- The muscularis mucosae are only found in the lower section
- Mucous glands in the submucosa secrete mucus into the lumen of the oesophagus, via ducts, aiding the lubrication of food boluses and protecting the lining from abrasions
- The muscularis externa in the upper third is made of striated muscle; the middle third is mixed (striated and smooth muscle); the lower third near the junction with the stomach, consists of smooth muscle only. These muscles are responsible for propelling swallowed food to the stomach
- The adventitia forms the outer layer in the non-abdominal sections. This is a connective tissue layer which is only loosely restraining. In the abdominal sections, this layer is replaced by serosa.

The stomach

Functions:
* Digestion of food
* Secretion of hormones

The mucosa of the stomach is highly specialized for secretory functions. The single-layered, columnar epithelium has many invaginations (folds) that extend into the lamina propria, forming branched specialized glands (gastric pits) for the secretion of acid, mucus, and digestive enzymes

The lamina propria is thin and contains large numbers of capillaries, nerve fibres, and lymphatics

The muscularis is made up of three smooth muscle layers with fibres running in different directions—longitudinal, circular, and oblique when running from the external to the internal layer. This promotes mixing of the stomach contents and propulsion towards the pyloric sphincter to enable emptying into the duodenum. The stomach wall is arranged into numerous folds or rugae that allow the stomach to distend and increase in size

Different regions of the stomach have different functions and, thus, different lining specializations:
* The cardia: short simple or branched glands with mucus-secreting cells and a few parietal cells
* The fundus and body: branched tubular gastric glands with mucous neck cells (secrete mucus) and oxyntic (parietal) cells (secrete hydrochloric acid and intrinsic factor) are found at the neck of the glands. Chief cells (secrete pepsin and lipase) and endocrine cells (secrete 5-HT) are found at the base of these glands
* The pylorus: shorter branched glands (pyloric glands), which secrete mucus, gastrin (stimulates the production of acid secretion) and enteroendocrine cells, which secrete somatostatin (an inhibitor of gastrin).

The small intestine

Functions:
* Digestion of food
* Absorption of nutrients
* Secretion of hormones

Several structural adaptations increase its surface area, enhancing absorptive and secretory processes:
* Plicae circulares are permanent folds in the mucosa and submucosa which are particularly prominent in the jejunum
* Villi are formed by outgrowths of the mucosa (epithelia and lamina propria), with simple tubular glands (glands of Lieberkühn) found at their base
* Absorptive cells are specialized epithelial cells with an absorptive surface of densely packed microvilli extending into the intestinal lumen as an apical brush border.

Peyer's patches

Aggregates of lymphoid material (in the lamina propria and submucosa along the small intestine, covered by specialized epithelial cells (M (microfold) cells). M cells have pits on their basal membrane which contain antigen-presenting cells (APCs). M cells endocytose antigens and transport them to the underlying APCs which are able to trigger appropriate immune responses to foreign antigens.

Paneth's cells

Exocrine cells in the basal section of intestinal glands. They secrete lysozyme which digests bacterial walls of some types of bacteria and is, therefore, antimicrobial.

Brunner's glands

Deep coiled glands extending into the muscularis mucosa, with extensive branching. These glands secrete alkaline mucus and distinguish the duodenum from the rest of the small intestine.

The large bowel

- Functions:
  - Absorption of water
  - Production of mucus (lubricates lining of intestine)
  - Formation of faecal material
- The large intestine has no mucosal folds (except the rectum) or villi. It has deep glands, lined by specialized columnar epithelium, including extensive goblet and absorptive cells, and few enteroendocrine cells. The absorptive cells are columnar with irregular, short microvilli
- Large amounts of bacteria are present in the lumen (this aids in the breakdown of cellulose material) and there is extensive lymphoid tissue in the lamina propria and submucosa to prevent bacterial invasion
- The outer layer of the muscularis differs from the small intestine, with the longitudinal layer forming three longitudinal bands of muscle: the taeniae coli. Appendices epiploicae are small outpouchings of adipose tissue in the serous layer. This helps distinguish the large from the small bowel during surgery
- Epithelial cell turnover is high, as cells are sloughed off the walls by passing matter. Stem cells at the base of the glands (the proliferative zone) are constantly replacing epithelial cells by mitosis.

The rectum and anal canal

- Functions:
  - Storage of faecal material
  - Excretion of faecal material at socially acceptable times
- The rectum is lined with columnar epithelium, with lots of goblet and absorptive cells
- The mucus membrane has vertical folds (rectal columns of Morgagni) forming valve-like folds (valves of Ball). The upper anal canal is also lined by a columnar epithelium, which becomes stratified squamous epithelium at the dentate line. The mucocutaneous junction is the area between the two types, where the epithelium is transitional
- The pectineal line defines the boundary of embryological development, with mucosa above endodermal in origin, and that below ectoderm in origin. Blood, lymphatic, and nervous supply, and histology reflect this boundary.

The oesophagus

- The oesophagus is a long tube (~25cm), responsible for transporting food from the mouth to the stomach
- It is divided into three sections, named according to the vertebral level. This division is particularly useful when considering blood, nerve, and lymphatic supply:
  - The cervical oesophagus starts at the level of the cricoid cartilage and runs posterior to the trachea and anterior to the cervical vertebrae and prevertebral fascia. The common carotid arteries and recurrent laryngeal nerves run on either side of the cervical oesophagus
  - The thoracic oesophagus lies within the superior and posterior mediastinum, posterior to the trachea, the left main bronchus, and the pericardial cavity which encloses the heart. The thoracic vertebrae, thoracic duct, the azygous vein, and descending aorta run posterior to this section of the oesophagus. It becomes the abdominal oesophagus after it pierces the right crus of the diaphragm at the T10 level
  - The abdominal oesophagus is very short and terminates at the cardia of the stomach. It lies in the oesophageal groove of the posterior surface of the left liver lobe. Its anterior surface is covered with parietal peritoneum and, in common with the rest of the GI tract, the abdominal oesophagus lies outside the peritoneal cavity
- Narrowings of the oesophagus are clinically important (◑ OHCM10 pp.250–1) since they are likely points for foreign bodies to become lodged if swallowed and are also likely sites for stricture and carcinoma development. They delay the passage of food and liquid after it has been swallowed. Four such sites are of greatest importance:
  - The cricopharyngeal sphincter
  - Where the aortic arch crosses the oesophagus
  - Where the left main bronchus indents the oesophagus
  - The point at which the oesophagus pierces the diaphragm.

Blood supply

- Cervical: inferior thyroid artery
- Thoracic: oesophageal branches from aorta and bronchial arteries
- Abdominal: branches from the left inferior phrenic and left gastric artery.

Venous drainage

- Cervical: vertebral, brachiocephalic, and inferior thyroid veins
- Thoracic: azygous and hemiazygos veins
- Abdominal: left gastric vein, which drains to the portal system
- The oesophageal veins form an anastomosis between the left gastric (portal) and azygous (systemic) venous drainage systems. During portal hypertension, distension of the veins draining the oesophagus can occur → oesophageal varices. These varices are at risk of rupture and severe haemorrhage.

Lymphatic drainage

Cervical: deep cervical nodes
Thoracic: tracheobronchial and posterior mediastinal nodes
Abdominal: left gastric and coeliac nodes.

Innervation

Cervical: recurrent laryngeal nerve and middle cervical ganglion (sympathetic)
Thoracic: vagus, sympathetic trunk, and greater splanchnic nerve
Abdominal: vagus-forming plexuses (anterior and posterior).

The stomach

- The stomach is a J-shaped tube that receives food from the oesophagus at the cardia. It is responsible for initiating many digestive processes and for mixing and breaking up swallowed food before presenting it to the duodenum in a more digestible form (called chyme)
- Food is mixed with a number of secretions, including digestive enzymes (e.g. pepsin), hydrochloric acid (which aids with digestion and kills bacteria), and intrinsic factor (essential for vitamin B_{12} absorption)
- The stomach is also a reservoir for food, and regulates its release into the duodenum.

Gross anatomy

See Fig. 8.5.

- The stomach has two curves: the greater (which forms the left border of the stomach) and the lesser (which forms its right border)
- The cardia is the narrow neck of the stomach and, although it is not anatomically distinct, this whole region functions as a 'physiological' sphincter to prevent 'reflux' of stomach contents into the oesophagus (GORD, ⊃ *OHCM10* p.254). This is dependent on several separate factors:
  - The oesophagus is narrowed as it pierces the right crus of the diaphragm
  - A valve is formed at the lower oesophagus by circular muscle fibres
  - Pressure differences between the thorax and the abdomen compress the walls of the intra-abdominal oesophagus
  - The acute angle of entry of the abdominal oesophagus has a valve-like effect and the muscularis mucosa forms mucosal flaps, which also act as valves
- The fundus is the upper section of the stomach, seen as the gastric bubble on X-rays
- The body is the main section and the pyloric antrum is the widest part of the pylorus. The pylorus forms the terminal section before it joins with the first part of the duodenum
- The pyloric sphincter (which regulates the flow of stomach contents into the duodenum) is a circular muscle around the lumen of the pylorus. Its position is marked by an external constriction of the alimentary tract at the junction of the duodenum and pylorus. This is also the point at which the constant vein of Mayo crosses it vertically. The different regions of the stomach have specialized functions which are described
- The omenta are two-ply regions of peritoneum which are created where viscera impress upon the peritoneal cavity:
  - The greater omentum is attached to the greater curve of the stomach and extends downwards like an apron before folding inwards and upwards towards the transverse colon
  - The lesser omentum is attached to the lesser curve of the stomach and extends to the liver
- The gastrosplenic omentum (ligament) runs from the stomach to the spleen.

The omenta contain the blood vessels and nerves supplying the stomach.

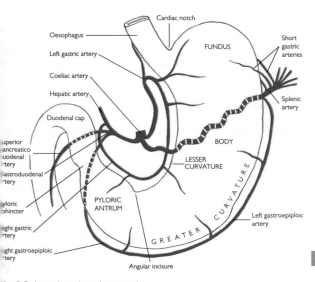

Fig. 8.5 Arterial supply to the stomach.
Reproduced from MacKinnon, Pamela and Morris, John, *Oxford Textbook of Functional Anatomy*, vol p138 (Oxford, 2005). With permission of OUP.

Anatomical relations

- Anteriorly: the anterior abdominal wall, the diaphragm, and the left lobe of liver
- Posteriorly: the stomach is divided by the lesser peritoneal sac from the pancreas, left kidney and adrenal gland, spleen, aorta, coeliac trunk, and the transverse mesocolon
- Superiorly: diaphragm.

Blood supply

See Fig. 8.5.

- Lesser curvature: left (from coeliac axis) and right (from hepatic artery) gastric arteries anastomose with each other and run in the lesser omentum
- Upper part of greater curvature: the short gastric arteries and the left gastroepiploic artery supply the upper part of the greater curvature of the stomach. Both arteries arise from the splenic artery and run in the gastrosplenic ligament
- Lower part of greater curvature: the right gastroepiploic artery (from the gastroduodenal branch of the hepatic artery) runs in the greater omentum. It sometimes anastomoses with the left gastroepiploic artery

Venous drainage

The course of the gastric veins largely mirrors that of the arteries and, ultimately, they drain into the portal veins.

- The left gastroepiploic and short gastric veins drain into the splenic vein
- The right gastroepiploic vein drains into the superior mesenteric vein
- The splenic and superior mesenteric veins merge, forming the portal vein. This receives drainage from the right and left gastric veins directly
- There is no gastroduodenal vein.

Lymphatic drainage

- Lesser curvature: left and right gastric nodes
- Upper left side of greater curvature: splenic and pancreatic nodes
- Lower greater curvature of the stomach: gastroepiploic and pyloric nodes.

  All lymph from the stomach ultimately drains into the coeliac nodes.

Innervation

- The coeliac plexus (T6–9) supplies the stomach with sympathetic fibres which run with blood vessels
- The vagus supplies parasympathetic fibres that innervate motor and secretory targets.
- The plexus has two trunks, anterior and posterior:
  - The anterior trunk (mainly from the left vagus) supplies the gastric branches, large hepatic branches, and branches to the pylorus
  - The posterior trunk (mainly from the right vagus) supplies the branches to the posterior stomach and a large coeliac branch, which forms the coeliac plexus.

The duodenum

- The duodenum is the first part of the small intestine, which also comprises the jejunum and ileum. The stomach passes chyme into the duodenum, which then continues the digestive process receiving bile from the gallbladder and pancreatic juices from the pancreas. It is protected against acidic contents from the stomach by alkaline secretions
- The duodenum is a C-shaped tube curving around the head of the pancreas. It is a retroperitoneal organ, with only its first part being covered in peritoneum: the rest is immobile and covered in a serous membrane
- It is commonly divided into four parts:
  - First part (5cm): starts at the gastroduodenal junction and is known radiologically as the duodenal cap. Posterior to it are the portal vein, common bile duct, gastroduodenal artery, and, moving further posteriorly, the inferior vena cava. Anterior to this section of the duodenum are the liver and gallbladder
  - Second part (7.5cm): this curves downwards, around the pancreatic head, anterior to the hilum of the right kidney and right ureter. It is crossed by the transverse colon anteriorly. The common bile duct and main pancreatic duct join, forming the hepatopancreatic ampulla which opens into the duodenum at the sphincter of Oddi. This structure opens into the posteromedial wall of this section of the duodenum at the duodenal papilla. The accessory pancreatic duct opens just superior to the sphincter of Oddi in the duodenum
  - Third part (10cm): the duodenum runs horizontally, with its superior margin running around the head of the pancreas. The root of the mesentery and the superior mesenteric vessels cross anteriorly to this part of the duodenum
  - Fourth part (2.5cm): runs superiorly and to the left. Termination of the duodenum is defined by a fibromuscular peritoneal fold from the right crus of the diaphragm called the suspensory ligament of Treitz. It attaches to the terminal part of the duodenum at the duodenal–jejunal junction. Contraction of this muscle (the ligament of Treitz) widens the flexural angle, aiding the movement of the contents of the duodenum. Running on the left side of this junction is the inferior mesenteric vein as it descends from behind the pancreas.

Blood supply

- Blood supply to the duodenum is regional and defined by its embryological development
- The foregut is supplied by the coeliac axis, and the midgut is supplied by the superior mesenteric axis. In the duodenum and pancreas, these two blood supplies anastomose

The inferior pancreatico-duodenal artery (from the superior mesenteric artery) anastomoses with the superior pancreatico-duodenal artery (from the coeliac axis) at the level of the duodenal papilla, where the common bile duct enters the duodenum. Thus, the duodenal papilla is the dividing line between the foregut and the midgut

Duodenal ulcers (🔁 *OHCM10* p.252) most commonly occur in the duodenal cap and, as a consequence of its rich blood supply, are devastating should they bleed.

enous drainage

The superior pancreatico-duodenal vein drains directly into the hepatic portal vein

The inferior pancreatico-duodenal vein drains into the superior mesenteric vein.

ymphatic drainage

uodenal lymph drains to the superior mesenteric and coeliac nodes.

nnervation

ll nerves reach the duodenum with blood vessels:

Sympathetic supply: coeliac and superior mesenteric plexus

Parasympathetic supply: vagus nerve.

The jejunum and ileum

- The remainder of the small intestine is divided into two parts: the jejunum (upper two-fifths) and the ileum (lower three-fifths). It is a long section of bowel 3–10m long and is responsible for terminal food digestion and nutrient absorption
- There is no obvious junction between the two halves of the small intestine: the bowel characteristics gradually change. The mucosa of the jejunum is thicker, with a smaller diameter, than the ileum
- The two parts lie in different areas of the abdomen: the jejunum in the umbilical region and the ileum in the hypogastrium and pelvis
- The mesentery carries blood vessels to the small intestine, nerve fibres (autonomic), and lymphatic vessels:
  - Starts at the duodenal–jejunal junction and attaches most of the small intestine to the posterior wall of the abdomen
  - Originates from the posterior abdominal wall at the same level and to the left of L2
  - It runs diagonally downwards from L2 towards the right sacroiliac joint along the left side of the second lumbar vertebrae
  - The root of the mesentery crosses the abdominal aorta, inferior vena cava, the third part of the duodenum, right psoas major muscle, right ureter, and right testicular/ovarian vessels.
- The ileum terminates at the ileocaecal junction.

Blood supply

See Fig. 8.6.

- The superior mesenteric artery sends many branches to the intestine, which anastomose and form arterial arcades
- Vasa recta are straight arteries that arise from the arcades and supply individual sections of the small intestine wall. These do not form anastomoses within the mesentery, but have extensive anastomoses among the blood vessels of the wall of the intestine
- The jejunum is more vascular than the ileum and the arterial arcades are more complex and shorter. This (along with the higher levels of fat found in the mesentery of the ileum) helps surgeons distinguish and localize parts of the small intestine when operating.

Venous drainage

Drained by the superior mesenteric vein, which forms the portal vein as it joins the splenic vein, posterior to the neck of the pancreas, anterior to the superior mesenteric artery.

Lymphatic drainage

The majority of the intestine drains into the mesenteric lymph nodes, which drain into the superior mesenteric lymph nodes. The terminal ileum drains into the ileocolic lymph nodes.

Innervation

Parasympathetic and sympathetic nerves form the myenteric and sub-mucosal plexuses in the intestinal walls.

- Parasympathetic: vagus
- Sympathetic: coeliac plexus via the sympathetic trunks and greater splanchnic nerves.

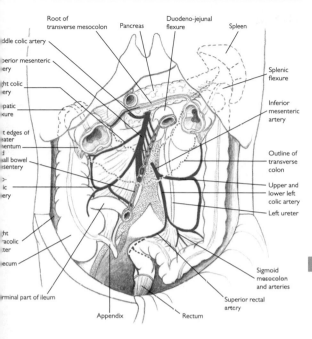

Fig. 8.6 Arterial supply to the small and large bowel derived from the superior and inferior mesenteric arteries.

Reproduced from MacKinnon, Pamela and Morris, John, *Oxford Textbook of Functional Anatomy*, vol 0147 (Oxford, 2005). With permission of OUP.

The large intestine

- The large intestine is the most distal part of the GI tract. It is usually considered to comprise four sections:
  - Caecum and appendix
  - Colon
  - Rectum
  - Anal canal
- The large bowel is functionally important for the absorption of water and storage of faeces prior to defecation
- Most of the large intestine can be distinguished from the small intestine by certain distinguishing features, including:
  - Taeniae coli (three flattened, thick, muscular bands which run across the entire wall of the caecum and colon)
  - Haustrae (pouches of mucosa between the taeniae coli)
  - Epiploic appendices (fat-filled pouches attached to the outer surface of the colon)
- These characteristics are useful landmarks during abdominal surgery.

The caecum

- The caecum (Fig. 8.7) lies in the right iliac fossa distal to the ileocaecal junction, over iliacus and psoas
- It projects downwards as a pouch at the start of the ascending colon, with which it is continuous
- It is covered in peritoneum but has no mesentery.

The appendix

- The appendix is devoid of taeniae coli, a feature which distinguishes it from the caecum and colon. It is a blind-ending tube on the posteromedial aspect of the caecum, although its position is highly variable, with 75% of cases lying behind the caecum
- The mesoappendix is a short, triangular-shaped mesentery from the mesentery of the terminal ileum, and contains the appendicular branch of the ileocolic artery and the ileocolic vein (a branch of the superior mesenteric vein). Lymph from the appendix (and caecum) drains to lymph nodes in the mesoappendix, and from there to ileocolic nodes. The superior mesenteric plexus supplies sympathetic and parasympathetic nerves to the appendix and caecum. The appendix is rich in lymphoid follicles
- Inflammation of the appendix (appendicitis, ⟶ OHCM10 p.608) is common, and usually caused by obstruction of the outlet of the appendix. The secretions cannot escape, causing swelling and stretching of the visceral peritoneum. Clinical features include acute, dull, generalized abdominal pain, which then localizes to McBurney's point (one-third of the way along a diagonal line from the right anterior superior iliac spine to the umbilicus). This localization and tenderness occurs when the parietal peritoneum around the appendix becomes inflamed.

The colon

The colon comprises four sections.

Ascending colon
Arises from the caecum
Lacks mesentery
Lies retroperitoneally on the posterior wall of the abdomen and is covered by peritoneum anteriorly and on its sides
Paracolic gutters exist on either side of the ascending colon
Travels upwards on the right side of the abdomen inferior to the liver, where it turns medially, forming the right colic flexure (hepatic flexure), and continues as the transverse colon.

Transverse colon
Completely covered in peritoneum
Mesentery starts at the inferior border of the pancreas, where it is continuous with the posterior wall parietal peritoneum
Runs from the right colic flexure to the left colic flexure at the spleen, at which point it becomes the descending colon. The diaphragm is attached to the left colic flexure by the phrenicocolic ligament.

Descending colon
A retroperitoneal organ, covered in peritoneum over its lateral and anterior surfaces
Descends to the sigmoid colon in the left lower quadrant of the abdomen
Paracolic gutters run along the medial and lateral aspects of the descending colon. The lateral gutter is separated from the spleen by the phrenicocolic ligament. Pus in this gutter (e.g. as a result of diverticular disease (⊕ OHCM10 p.628) causing bowel perforation) will drain into the pelvis.

Sigmoid colon
Connects the descending colon to the rectum
The pelvic brim marks the beginning of the sigmoid colon and the rectum starts where the taeniae coli terminate, at the level of the third sacral segment
Possesses a long mesentery (sigmoid mesocolon) and is therefore intraperitoneal. The attachment of the mesentery is V-shaped, running from the anterior aspect of the sacrum upwards to the bifurcation of the common iliac vessels, then turning laterally and crossing the external iliac vessels along the pelvic brim. Behind the apex of this mesentery lies the left ureter at the point at which it crosses the left common iliac artery—an important surgical landmark.

Blood supply
The blood supply to the large bowel is a result of its embryological origins
The midgut (distal duodenum, appendix, caecum, ascending colon, proximal part of the transverse colon) is supplied by the superior mesenteric artery

- The hindgut (distal part of the transverse colon, descending colon, sigmoid colon, rectum) is supplied by the inferior mesenteric artery
- In the rectum, there are anastomoses between branches of the inferior mesenteric and pudendal vessels, and an anastomosis also occurs between the superior and inferior mesenteric vessels via the marginal artery. These are important during bowel resection.

Venous drainage

- All blood drains into the portal venous system via the superior mesenteric vein and the inferior mesenteric vein (corresponding to arterial supply)
- The inferior mesenteric vein joins the splenic vein, which merges with the superior mesenteric vein to form the portal vein.

Lymphatic drainage

- The right side of the large bowel drains into the inferior mesenteric nodes, and from there to the superior mesenteric nodes and para-aortic nodes
- The left side of the large bowel drains directly into the superior mesenteric nodes.

Innervation

- Parasympathetic supply to the large bowel is partly via the vagus nerve which supplies the entire bowel up to the distal transverse colon. The distal large bowel receives parasympathetic fibres from sacral segments via pelvic splanchnic nerves
- Sympathetic nervous supply to the large bowel is from spinal cord segments T10–L2. This is via the lumbar sympathetic chain and the superior hypogastric plexus (presacral nerves)
- Autonomic innervation of the large bowel is important for the regulation of vascular tone and bowel motility. Autonomic efferents supply the bowel via myenteric and submucosal plexuses in the bowel wall
- Visceral sensory fibres run with the lesser splanchnic nerve and can give rise to referred pain of abdominal organs. This pain is poorly localized and corresponds with the skin supplied by the spinal cord segment where visceral pain afferents first synapse.

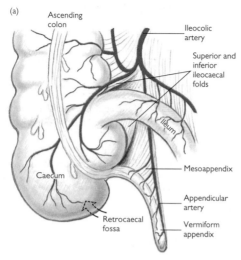

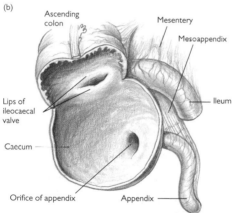

Fig. 8.7 (a) Ileo-caecal junction, caecum, and appendix with its mesentery and arterial supply. (b) Internal aspect of caecum, showing the ileo-caecal valve and appendicular orifice.

Reproduced from MacKinnon, Pamela and Morris, John, *Oxford Textbook of Functional Anatomy*, vol [?] p148 (Oxford, 2005). With permission of OUP.

The rectum and anal canal

- The rectum is the terminal, retroperitoneal section of the intestine which leads into the anal canal
- The rectum starts anterior to the sacrum (S3) and terminates as the dilated rectal ampulla, in front of the coccyx, having followed the curves of the sacrum and coccyx
- The rectal ampulla is extremely distensible and stores faeces prior to defecation
- The anterior and lateral sides of the upper third, and anterior part of the middle third of the rectum, are covered in peritoneum, whereas the inferior third has no such covering.

Anatomical relations

- The anatomical relations of the rectum are important and rectal examination is a common clinical procedure. In this way, local tumour invasion can be detected. Such growths usually originate from the rectum or prostate.
- Anterior:
  - Bladder
  - Pouch of Douglas (between the rectum and bladder) and posterior wall of the vagina and distal part of the uterus (in females)
  - Rectovesical pouch, seminal vesicles, and prostate (in males)
- Posterior:
  - Sacrum and coccyx
  - Middle sacral artery
  - Lower sacral nerves
- Lateral:
  - Levator ani
  - Coccygeus
- The anorectal junction is at the level of the pelvic floor, where the taeniae coli merge to form a continuous layer of muscle and the sphincters of the anal canal start. Puborectalis forms a muscular sling around the anorectal junction to create a 90° angle with the pelvic floor Puborectalis is continuous with the external anal sphincter
- The anal canal is the termination of the alimentary tract and forms the sphincter control for the excretion of waste products. The involuntary internal anal sphincter (smooth muscle) and voluntary external anal sphincter (skeletal muscle) are the two circular muscle layers which make up the wall of the anus. The intersphincteric groove is seen as an indentation on the wall of the anal canal. It marks the area between the internal and external sphincters
- Anal columns are longitudinal ridges formed in the upper anal canal and contain the terminal branches of the superior rectal arteries. Horizontal anal valves join the distal ends of the columns together and are formed by folding of the mucous membrane lining the walls of the canal. Above these valves, mucus is secreted by the submucosal anal glands. These glands can become infected, forming abscesses and fistulae

The pectineal (or dentate) line is the junction of the two embryological origins of the anal canal, where the epithelium changes from a columnar to a stratified squamous morphology. This occurs just above the anal valves and the intersphincteric groove. Between the intersphincteric groove and the dentate line is a region of transitional epithelium called the pecten. Embryologically, mucosa above the dentate line originates from the endoderm, while below, it originates from the ectoderm. This divide is important when considering the blood, lymphatic, and nervous supplies, and the histology of the epithelial lining.

Blood supply

The superior rectal artery (terminal branch of the inferior mesenteric artery) supplies the rectum and upper half of anal canal (above the dentate line)

The median sacral artery (from the internal iliac artery) supplies the rectum

The middle rectal artery (from the internal iliac artery) supplies the middle and inferior rectum

The inferior rectal artery (a branch from the internal pudendal branch of the internal iliac artery) supplies the lower half of the anal canal (below the dentate line)

Significant anastomoses exist between these blood vessels.

Venous drainage

The rectal venous plexus of rectal veins comprises an internal rectal plexus deep to the epithelial layer and an external rectal plexus external to the muscularis layer. There is free communication between these two plexuses, forming an anastomosis between the portal and systemic systems

The superior rectal vein drains the superior part of the internal rectal plexus (which drains the anal canal above the dentate line) and the superior part of the external rectal plexus. The superior rectal vein is the first part of the inferior mesenteric vein

The internal pudendal vein, via the inferior rectal veins, drains the inferior part of the external rectal plexus and the inferior part of the anal canal via the inferior part of the internal rectal plexus found below the dentate line

The internal iliac vein, via the middle rectal vein, drains the middle part of the external rectal plexus

The internal pudendal and internal iliac veins drain into the systemic circulation. The middle rectal veins anastomose with the superior and inferior rectal veins

In the anus, venous anastomotic cushions are found in the upper third of the anal canal. They are found at the 3, 7, and 11 o'clock positions and, when dilated (e.g. during portal hypertension), can become haemorrhoids. They help to maintain sphincter control.

Lymphatic drainage
- Lymphatic drainage from the rectum largely follows the blood vessels supplying the rectum and anal canal
- Lymph vessels from the upper rectum follow the superior rectal vein to drain into the abdominal lymph nodes (para-aortic) and, from the lower rectum, into the inguinal nodes
- Lymphatic drainage of the anal canal is separated by the dentate line. Above the line, lymph drains to the internal iliac nodes, whereas below it drains to the superficial inguinal nodes.

Innervation of the rectum
- Parasympathetic: pelvic splanchnic nerves (S2, S3, and S4)
- Sympathetic: plexuses of coeliac and hypogastric nerves.

Innervation of the anal canal
- Parasympathetic: pelvic splanchnic plexus (relaxes internal sphincter)
- Sympathetic: pelvic plexus (contracts internal sphincter)
- Somatic: pudendal nerve (S2) via the inferior rectal branch (external sphincter and sensation to the distal anal canal—below the dentate line).

The spleen

(⊖ OHCM10 p.373.)

- The spleen forms part of the reticuloendothelial system. It is roughly the shape of a fist and is ~12cm long and 7cm wide in the adult, fitting neatly under the left side of the diaphragm
- It is protected by the rib cage and lies inferior and posterior to the stomach
- The spleen is contained within a fibrous capsule which is continuous with a meshwork of trabeculae which carry nerves and blood vessels (trabecular arteries and veins) into the parenchyma of the organ. The fibrous capsule of the spleen is covered with visceral peritoneum, except at the hilum, where the blood vessels emerge
- The main substance of the spleen is referred to as splenic pulp, which is further divided into white pulp and red pulp:
  - White pulp is lymphoid tissue containing large numbers of immune system cells and is concentrated in periarticular lymphoid sheaths which surround smaller arterioles as well as splenic nodules. Different areas of lymphoid tissue in the spleen have specific immunological functions
  - Red pulp is so called because of the large amount of blood it contains. It consists of sinusoids (large diameter, thin-walled blood vessels) and splenic cords (composed of reticular cells, reticular fibres, and certain immune and blood cells, e.g. macrophages, platelets, and red blood cells). Certain blood cells, including red blood cells, can pass between the sinusoids and the splenic cords
- The exact functions of the spleen vary between species, but in humans it has several well-defined roles:
  - In the newborn, it produces lymphocytes
  - Splenic phagocytes act as a filter removing foreign material and worn out red blood cells from the circulation
  - In the foetus, it provides a source of red blood cells
  - It is a reservoir of blood.

Attachments

- Gastrosplenic ligament (omentum)—attaches spleen to the greater curvature of the stomach
- Splenorenal ligament—attaches spleen to the left kidney.

Relations

- Superiorly: diaphragm
- Posteriorly: spleen is divided from the 9th to 11th ribs by the diaphragm
- Anteriorly: stomach and tail of pancreas (at hilum of spleen)
- Medially: left kidney
- Laterally: splenic flexure of the descending colon and diaphragm.

Blood supply

The splenic artery (a branch of the coeliac trunk) divides into several branching arteries that supply the spleen at the hilum. The splenic artery runs in the splenorenal ligament behind the omental bursa.

Venous drainage

Several venous branches drain the spleen and merge to form the splenic vein. The splenic vein receives the inferior mesenteric vein behind the pancreas. Here, the splenic vein merges with the superior mesenteric vein to form the portal vein draining into the liver.

Lymphatic drainage

The spleen drains to nodes at the hilum, which drain into pancreaticosplenic lymph nodes on the posterior surface and superior border of the pancreas.

Innervation

The spleen is supplied with vasomotor fibres by the coeliac plexus. Generally, they follow the same route as branches of the splenic artery.

The liver

See Fig. 8.8.

- The liver is the largest internal organ, situated in the right upper quadrant of the abdomen. Superiorly, the liver apposes the diaphragm and, inferiorly, it reaches the costal margin
- The liver is divided into right and left lobes by the falciform ligament. Anterosuperiorly lies the umbilical fissure, within which is the round ligament (an embryological remnant of the umbilical vein). The gallbladder is attached anteroinferiorly
- Between the umbilical fissure and gallbladder is the quadrate lobe. The caudate lobe lies posterior to the quadrate lobe, separated by the portal vein, hepatic artery, and hepatic duct. These structures enter the liver at the porta hepatis. The inferior vena cava sits at the posterior surface of the liver.

Blood supply

- The liver receives two main sources of blood:
  - The hepatic artery (~20% of flow and 60% of oxygen supply)
  - The portal vein (~80% of flow and 40% of oxygen supply)
- Blood enters the sinusoids within the liver, and drains via tributaries into the hepatic vein, which joins the inferior vena cava
- Sinusoids are arranged between layers of hepatocytes. They are lined by endothelial cells and Kupffer cells, which have a phagocytic function. Between the layers there is the space of Disse, which is composed of basement membrane and stellate cells
- When blood flow is obstructed, porto-systemic anastomoses open. This results in the formation of varices at a number of sites, which include the oesophagus and rectum. This is clinically important as they can be the cause of major GI bleeds (◐ *OHCM10* p.256).

Lymphatic drainage

Lymph is formed in the perisinusoidal spaces of the liver and drains into hepatic ducts via smaller lymphatic vessels.

Biliary system

(◐ *OHCM10* pp.634–5.)

- Bile is formed in the liver and secreted via bile ductules into right and left hepatic ducts. The hepatic ducts join at the porta hepatis forming the common hepatic duct
- The gallbladder, which stores and concentrates bile, is connected to the common hepatic duct by the cystic duct, together forming the common bile duct
- The pancreatic duct and common bile duct join the second part of the duodenum at the ampulla of Vater.

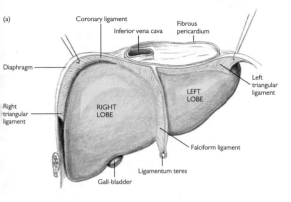

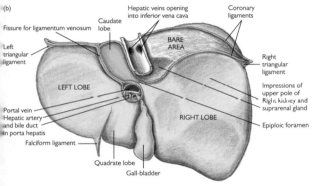

Fig. 8.8 (a) Anterior and (b) posterior views of the liver and its peritoneal reflections.

Reproduced from MacKinnon, Pamela and Morris, John, *Oxford Textbook of Functional Anatomy*, vol 2, p161 (Oxford, 2005). With permission of OUP.

The pancreas

- The pancreas is a retroperitoneal organ, located in the epigastrium. It can be anatomically divided into three parts: head, body, and tail (Fig. 8.9). The head is apposed to the duodenum and the tail contacts the spleen
- Microscopically, the pancreas is divided into exocrine and endocrine cells. The containing exocrine pancreas is made up of acinar glands, which produce digestive enzymes that drains an alkaline secretion into the pancreatic duct
- The endocrine pancreas is made up of islets of Langerhans. These contain three main cell types that produce, store, and secrete different enzymes: α cells (secrete glucagon); β cells (secrete insulin); δ cells (secrete somatostatin)
- The head of the pancreas receives arterial blood from the superior and inferior pancreatico-duodenal arteries, which are derived from the coeliac and superior mesenteric arteries. The remainder of the pancreas is supplied by the splenic artery, a branch of the coeliac artery. Venous blood drains into the portal vein behind the pancreatic neck. Lymphatic drainage follows blood vessels to lymph nodes around the coeliac axis
- The pancreas is innervated by sympathetic and parasympathetic fibres from the coeliac plexus. These regulate blood flow (sympathetic) and exocrine and endocrine secretions (parasympathetic)
- Pancreatitis is the inflammation of the pancreas. This can be acute or chronic and there are a number of causes (➲ *OHCM10* p.270 (chronic), p.673 (acute)).

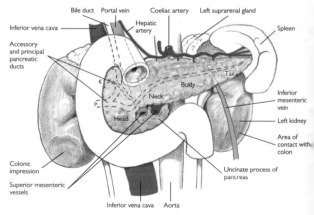

Fig. 8.9 Pancreas and its immediate relationships.

Reproduced from MacKinnon, Pamela and Morris, John, *Oxford Textbook of Functional Anatomy*, vol 2, p164 (Oxford, 2005). With permission of OUP.

Gastrointestinal motility

- GI motility churns the contents of the tract to promote digestion and absorption; propels the contents along the tract; and stages the progression of the contents
- Most motility derives from circular and longitudinal smooth muscle layers in the walls of the GI tract and sphincters that separate different segments. The first (upper oesophageal) and last (external anal) sphincters comprise striated muscle
- While striated muscle is under voluntary control, smooth muscle possesses an enteric nervous system (● p.274); it can function without extrinsic innervation. The enteric nervous system coordinates muscular activity using reflex arcs:
  - The mucosal layer possesses cells that detect chemicals (H^+, protein digestion products) or mechanical forces (tension, stretch)
  - The receptor cell bodies are in the submucosal (Meissner's) plexus
  - Interneurones transmit signals between the submucosal plexus and the myenteric (Auerbach's) plexus
  - Effector neurones in the myenteric plexus modulate motor activity in smooth muscle layers in response to sensory input
- Enteric neurones release a variety of neurotransmitters and neuromodulators, including ACh, substance P, vasoactive intestinal peptide (VIP), nitric oxide, cholecystokinin (CCK), serotonin, somatostatin. Neurones often release more than one transmitter. Enteric reflex arcs can be modulated by inputs from the autonomic nervous system:
  - Parasympathetic preganglionic fibres make cholinergic synapses with myenteric excitatory motor neurones (= postganglionic neurones, which release ACh, substance P) to increase motility
  - Sympathetic postganglionic fibres form noradrenergic synapses with myenteric inhibitory motor neurones (which release VIP, nitric oxide) to reduce motility
  - Autonomic afferent fibres relay information from mucosal receptors to the brain.

Swallowing

Swallowing is controlled by the swallowing centre in the medulla. Outputs pass through cranial nerves (V, IX, X). It has three phases:
- Oral phase: material moved to rear of mouth by tongue
- Pharyngeal phase:
  - Soft palate moves up and back to close nasal passages
  - Voluntary pharyngeal muscles contract—bolus propelled towards oesophagus
  - Epiglottis closes over larynx, upper oesophageal sphincter relaxes
  - Food enters oesophagus, sphincter constricts
- Oesophageal phase: peristaltic wave spreads along oesophagus, propelling bolus towards stomach. Secondary waves clear residual material. The lower oesophageal (cardiac) sphincter and proximal stomach relax and bolus enters the stomach. Sphincter constricts to prevent reflux (● OHCM10 p.254).

Peristalsis and receptive relaxation of sphincter and stomach can fail in achalasia (● OHCM10 p.250).

Gastrointestinal motor activity

Stomach

- Activity differs in fed and fasting states. In the fed state, motility patterns comprise:
  - Propulsion: food is gradually propelled by peristalsis towards the antrum of stomach. Peristalsis initiates in pacemaker cells in the middle of greater curvature. Slow waves of depolarization spread and may excite circular muscle if the threshold is reached
  - Mixing by smooth muscle in the antrum: contents are forced towards the pylorus, and small particles (<2mm) pass through the pyloric sphincter
  - Retropulsion: contraction of pylorus and antrum closes the sphincter, forces larger material back toward antrum
- Repeated propulsion, mixing, retropulsion cycles ensue. Contents become increasingly fluid over time: the presence of the semi-liquid mass of partially digested food (chyme) induces the sphincter to remain open for longer periods
- Emptying depends on intragastric pressure, on volume and osmolarity for fluids, and on particle size and calorific content for foodstuffs. Duodenal content feeds back to modulate gastric emptying
- In the fasting (or interdigestive) state, there are migrating motor complexes (MMCs). These spread from stomach into small intestine:
  - The MMC comprises long periods of inactivity, punctuated by periods when action potential frequency progressively increases until a sustained peak is achieved after which the frequency declines
  - Contractility increases in parallel with action potential frequency. The MMC sweeps stomach contents (including acid) into and through the small intestine
  - Motilin (28-amino acid peptide released from endocrine cells in duodenum) initiates MMC in the stomach
  - Feeding terminates MMC, possibly through neurotensin released from endocrine cells throughout the tract.

Small intestine

- Exhibits slow waves of depolarization, which initiate contraction of circular smooth muscle
- Waves are result of transient relief of enteric inhibition. Frequency decreases along small intestine
- In the fed state, isolated contraction of circular muscle causes segmentation or churning; peristalsis causes propulsion. Peristaltic waves propagate only short distances, generally no more than 10cm
- In the fasting state, MMC pulses propel contents
- The ileocaecal sphincter separates small and large intestines: distension of the ileum initiates relaxation, distension of colon inhibits
- Vomiting (⮕ OHCM10 pp.56, 250) is initiated by the vomiting centre in the medulla. There is reverse peristalsis of small intestine, relaxation of pyloric sphincter and stomach. Intestinal contents are swept into stomach

- Abdominal contraction then forces the contents into the oesophagus, and there is reflex relaxation of oesophageal sphincters. Vomiting is induced by diverse stimuli that include emotion, pain, rotation, chemical composition of food, distension, obstruction, alcohol.

Large intestine
- In the large intestine, longitudinal smooth muscle forms three bundles (taeniae coli) that define folds or haustra
- Slow wave depolarization induces contraction of circular smooth muscle to produce segmentation. Waves increase in frequency along the large intestine
- Segmental contraction of haustra produces pendular movements of contents. Concerted contraction of haustra propels contents over short distances
- Large peristaltic propulsions (mass movements, ~20cm) occur infrequently (up to three times a day). Distension of stomach (gastrocolic reflex) and standing (orthocolic reflex) initiate mass movements. The distal colon demonstrates non-propulsive segmentation: this retards flow.

Rectum
- Intermittently receives material and undergoes segmental contraction
- Distension of the rectum initiates rectosphincteric reflex: the internal anal sphincter transiently relaxes
- If the timing is not good, contraction of the external sphincter overrides the reflex. The urge passes for the moment. If appropriate, voluntary relaxation of the external anal sphincter leads to evacuation
- The rectum propels faeces into and through anal canal, assisted by voluntary contraction of the diaphragm and abdominal wall, which increases intra-abdominal pressure.

Salivary and pancreatic secretion

- There are a number of similarities between the salivary glands and the exocrine pancreas
- Both comprise a branching ductal arrangement into which epithelial cell secretions are released. Secretions are composed of water, electrolytes and some digestive enzymes
- Both secretions aid digestion. Saliva lubricates ingested food, forms a protective buffer, and initiates digestion of starch. Alkaline pancreatic juice neutralizes stomach acid and completes digestion of ingested foodstuffs
- Glands comprise secretory units (lobules) made of an acinus: up to 100 cells lining an intercalated duct. Intercalated ducts drain into intralobular ducts, then into interlobular ducts, and, finally, into the main salivary or pancreatic duct. Postganglionic autonomic fibres innervate the cells.

Salivary glands

- There are three types of salivary glands:
  - Parotid: produces watery (serous) secretion amounting to 25% of total
  - Submandibular: produces both serous and mucous secretions—70% of total
  - Sublingual: produces mucous secretion—5% of total
- Serous secretions are supplemented with the enzyme, α amylase; mucous secretions with mucin. Average daily production = 1.5L day^{-1}. Basal flow rate = 0.5mL min^{-1} rising to 5mL min^{-1} after stimulation
- Acinar cells secrete isotonic NaCl:
  - Basolateral Na$^+$-K$^+$-2Cl$^-$ co-transporter accumulates Cl$^-$ ions inside the cell
  - Cl$^-$ ions diffuse across the apical membrane through channels. Na$^+$ is pumped out at the basolateral membrane, K$^+$ diffuses out through basolateral channels
  - Na$^+$ ions diffuse between cells through tight junctions, along electrical gradient established by Cl$^-$ movement
  - H$_2$O follows by osmosis
- Duct cells modify the primary secretion. Na$^+$ and Cl$^-$ are reabsorbed and K$^+$ and HCO$_3^-$ secreted to a lesser degree. Permeability to water is low: water reabsorption is minimal and luminal fluid becomes hypotonic. At higher rates of secretion, the time available for modification is reduced: fluid is more similar in composition to plasma
- Different acinar and duct cells synthesize, store, and release proteins:
  - Enzymes: α-amylase digests starch to maltose; lingual lipase starts fat digestion
  - Mucins: glycoproteins
  - Kallikrein: cleaves proteins to yield vasodilator peptides (e.g. bradykinin)
  - Lysozymes, lactoferrin, lactoperoxidase, proline-rich proteins, IgA: antimicrobial
- Salivary secretion is regulated by the autonomic nervous system (see topics in ➲ Chapter 4, 'Synaptic transmission', pp.242–64):

- Parasympathetic outflow in cranial nerves V and VII is most important. ACh stimulates primary secretion, reduces secondary modification: produces large volumes of watery saliva. ACh binds M_3 receptors, induces IP_3 generation, Ca^{2+} mobilization in acinar and duct cells. In acinar cells, Ca^{2+} stimulates protein kinases which activate apical Cl^- channels and basolateral K^+ channels. Phosphorylation of cytoskeletal elements also induces export of protein-containing vesicles
- Sympathetic actions are less pronounced, although noradrenaline binding to α receptors also mobilizes Ca^{2+} through IP_3. Binding to β receptors raises cAMP, activates protein kinase A, stimulates amylase secretion from vesicles: produces viscous saliva
- Substance P also raises Ca^{2+}, initiates secretion.

Pancreas

About 90% of the pancreas is made up of exocrine cells; the remainder is made up of endocrine cells. The pancreas secretes 1.5L day^{-1} of alkaline, protein-rich fluid. The alkalinity helps neutralize stomach acid (€ OHCM10 p.636)

Acinar cells produce a primary, plasma-like secretion. The mechanism of primary secretion is similar to that in salivary glands. The primary secretion hydrates digestive proteins released from the acinar cells Proteins secreted can be precursors (zymogens, activated in small intestine) or active enzymes:
- Proteases—for protein digestion (trypsinogens, chymotrypsinogens, proproteases, procarboxypeptidases)
- Amylases—for carbohydrate digestion
- Lipases—for lipid digestion
- Nucleases—for RNA, DNA digestion

Constitutive secretion occurs but can be increased tenfold by ACh M_3 and CCK receptor activation, IP_3 generation, Ca^{2+} mobilization

Pancreatic duct cells perform secondary modification: there is secretion of isotonic $NaHCO_3$:
- HCO_3^- is generated in the cell from hydration of CO_2 and secreted by apical Cl^--HCO_3^- exchange. Na^+ and H_2O follow through the paracellular pathway
- Cl^- ions that enter the cell recycle across the apical membrane through the Cl^- channel CFTR (cystic fibrosis transmembrane conductance regulator). In cystic fibrosis, mutations in the *CFTR* gene compromise secondary secretion. Secretions are more viscous and clog the duct: the consequent deficiency in digestive enzymes can be treated with oral administration of enzymes (coated to prevent gastric digestion) with every meal

As for gastric acid secretion, pancreatic secretion regulated by three phases of digestion:
- Cephalic phase: mediated primarily by ACh
- Gastric phase: mediated by CCK, gastrin
- Intestinal phase: mediated by secretin

Goblet cells in ducts secrete mucus to facilitate lubrication, offer mechanical protection, bind pathogens.

Gastric secretion

- Secretions act in concert to facilitate digestion (and absorption) of ingested foodstuffs. Digestion is commenced by salivary secretions, continues in the stomach, and is later assisted by secretions from the liver and pancreas
- The environment of the stomach is acidic: pH can be as low as 1. This acidity assists protein digestion by denaturing proteins and activating pepsin enzymes. A mucous lining on the epithelial surface of the stomach and the presence of HCO_3^- ions prevent autodigestion of the stomach
- The gastric epithelial lining is characterized by gastric glands, which increase the surface area. Glands comprise pits which open into a neck which leads to a base. A variety of cell types make up the lining:
  - Parietal (oxyntic) cells in the base and neck secrete HCl and intrinsic factor (necessary for vitamin B_{12} absorption in the ileum)
  - Chief (peptic) cells in the base and neck cells secrete pepsinogen
  - Endocrine cells in the base secrete regulators such as gastrin and somatostatin via the bloodstream
  - Mucous neck cells secrete mucus; superficial epithelial cells in the pit and on the surface lining secrete mucus, along with HCO_3^- ions.

Parietal cells

- Parietal cells possess deep invaginations of apical membrane. In an unstimulated cell, there are large numbers of tubulovesicles in the subapical cytoplasm. Tubulovesicle membranes contain H^+-K^+-ATPase proteins
- Tubulovesicle fusion with the apical membrane leads to acid secretion:
  - Intracellular carbonic anhydrase catalyses hydration of CO_2 to yield H^+ and HCO_3^-
  - The ATPase pumps H^+ into the lumen in exchange for K^+. K^+ recycles into the cell through apical K^+ channels
  - HCO_3^- exits across the basolateral membrane into interstitial fluid, then blood, on Cl^--HCO_3^- exchange
  - Cl^- diffuses through apical channels, joins H^+ in the lumen
  - Net result: secretion of HCl, alkalinization of blood ('alkaline tide')
- Tubulovesicle insertion is initiated by:
  - Acetylcholine (neurocrine): from vagus nerve endings. Binds M_3 receptors, triggers IP_3 cascade to raise $[Ca^{2+}]$, activate kinases
  - Gastrin (endocrine): from endocrine G cells in response to gastrin-releasing peptide (GRP) from peptidergic vagus nerve endings or protein digestion products in lumen. Binds CCK_B receptors, triggers IP_3 cascade to raise $[Ca^{2+}]$, activate kinases
  - Histamine (paracrine): from enterochromaffin-like cells. Binds H_2 receptors, triggers cAMP generation, activates protein kinase A. ACh and gastrin bind M_3 and CCK_B receptors on enterochromaffin-like cells, induce histamine release. This 'common mediator' action explains effectiveness of H_2 antagonists (e.g. ranitidine) as inhibitors of gastric acid secretion.

There are three phases of gastric secretion (Fig. 8.10):
- Cephalic: thought, sight, smell, taste initiate vagal stimulation, ACh, GRP release. Responsible for 30% of secretion
- Gastric: distension initiates vagovagal reflex, protein digestion products promote gastrin release. Responsible for 60% of secretion
- Intestinal: protein digestion products stimulate duodenal G cells. Responsible for 10% of secretion

Omeprazole inhibits the gastric H^+-K^+-ATPase.

Endocrine D cells

Release somatostatin, which inhibits adenylyl cyclase and reduces gastric acid secretion. ACh inhibits, and low luminal pH stimulates, somatostatin release

Acid-induced release of secretin from duodenal S cells stimulates somatostatin release (and inhibits gastrin release)

Gastric *Helicobacter pylori* infection inhibits somatostatin release and is an important cause of ulcers.

Chief cells

Secrete inactive pepsin precursor, pepsinogen, which is activated by N-terminal truncation. Spontaneous activation occurs in acidic lumen. Low pH is also required for optimal activity of pepsin

Once activated, pepsins catalyse pepsinogen truncation. Pepsin digests one-fifth of ingested protein. Pepsin release involves fusion of secretory granules with apical membrane and is stimulated by:
- ACh, via M_3 receptors, Ca^{2+} signalling
- Gastrin and CCK, via CCK_B receptor, Ca^{2+} signalling
- Secretin, via adenylyl cyclase-coupled receptors, cAMP signalling.

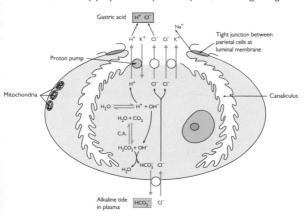

Fig. 8.10 Steps involved in the secretion of gastric acid by a parietal (oxyntic) cell.

Reproduced with permission from Pocock G, Richards CD (2006). *Human Physiology: The Basis of Medicine*, 3rd edn, p389. Oxford: Oxford University Press.

Mucous cells

- Secrete mucin, a large glycoprotein with high viscosity. This, together with ions, water, and phospholipids forms a mucus gel up to 200μm thick that acts as a barrier to H^+ ion diffusion from acidic bulk solution i the lumen
- Surface epithelial cells secrete HCO_3^- ions across the apical membrane, making the pH of the gel around 7 and neutralizing any H^+ ions that reach it
- Secreted acid bores through the mucus without lateral spread—stream of H^+ is termed a 'viscous finger'
- Secretion of mucous components is stimulated by ACh-induced Ca^{2+} signalling. Mucosa can be disturbed by aspirin, alcohol, and anti-inflammatory drugs.

Indigestion, peptic and duodenal ulcers

Dysfunction of one or more of the processes described can lead to dyspepsia (indigestion) and ulceration. Infection (*Helicobacter pylori* >80% of cases; ⊝ p.571), NSAIDs (⊝ p.904), and smoking are risk factors. Duodenal ulcers are more common than gastric ulcers.

NSAIDs and ulcers

Inhibition of prostaglandins that are thought to provide endogenous protection against acid-mediated damage of the gastric and duodenal mucosa through increased blood flow and mucus secretion is a major factor in induction of NSAID-mediated 'gastric toxicity' characterized by ulceration

Box 8.1 summarizes the treatments of ulcers.

Box 8.1 Treatments and drug therapies for ulcers

(⊝ *OHCM10* p.252; *OHPDT2* pp.32–37.)
- If *H. pylori* is suspected or confirmed as the primary cause of ulceration, it can be eradicated with antibiotics
- If NSAIDs are causing the problem, they should be withdrawn
- Alternatively, or sometimes as an adjunct, it is necessary to suppress gastric acid secretion. H_2 antagonists and proton pump inhibitors (⊝ see p.578 for mechanism) are the preferred choice of drug therapy.

Bile secretion

- Bile is secreted by the liver. It promotes lipid digestion, absorption, and elimination of endogenous (cholesterol) and exogenous (heavy metals) components. There is constitutive secretion, which is up-regulated in the fed state
- Bile components are dissolved in an alkaline solution, with similar composition to pancreatic juice. Bile comprises:
  - Bile salts
  - Water and electrolytes
  - Lecithin
  - Cholesterol
  - Pigments such as bilirubin
- Hepatocytes line bile canaliculi, which drain into bile ductules, then into a series of ducts that unite to form the common hepatic duct. Blood-to-lumen trafficking occurs across the cell, with compounds bound to proteins or contained within vesicles
- Hepatocytes synthesize primary bile acids cholic acid and chenodeoxycholic acid from cholesterol:
  - Acids are conjugated to glycine or choline. They ionize and exist as Na^+ or K^+ bile salts
  - Bile salts are actively secreted into the canalicular lumen by an ATPase in the hepatocyte apical membrane
  - Secondary bile acids result from bacterial deconjugation, dehydroxylation of primary bile salts in the intestine. Some bile acids are reabsorbed by ileum, bound to albumin, and returned to the liver in the blood. After dissociation from albumin, they are taken up into hepatocytes by transporters in the basolateral membrane. This is an example of enterohepatic circulation. After reconjugation, they are secreted again
  - Bile acids may recirculate up to three times before digestion of a meal is complete
- Conjugated bile acids are amphipathic: they possess water-soluble and lipid-soluble domains. Combined with phospholipid and cholesterol, they form mixed micelles in aqueous solution. They provide a vehicle for transport of lipid soluble substances in the aqueous environment of the small intestine
- The secretion of organic anions (e.g. thromboxanes), organic cations (e.g. choline, antibiotics), and the haemoglobin breakdown product, bilirubin, occurs by carrier-mediated transport in series, first across the basolateral and then across the apical membrane
- There is also secretion of inorganic ions (Na^+, K^+, Ca^{2+}, Cl^-, HCO_3^-) through channels and carriers in hepatocyte membranes, and by movement through the paracellular pathway. Secretion of bile salts and organic compounds establishes an osmotic gradient for water movement into the canaliculi. Most bile flow is dependent on bile acid secretion
- Cells lining bile ducts (cholangiocytes) secrete a watery $NaHCO_3$ solution using a mechanism similar to that in pancreatic duct cells. Secretion is stimulated by secretin, glucagon, VIP, which increase cAMP levels and activate CFTR. The solution secreted is water-rich and bile salt-poor. Somatostatin lowers cAMP, inhibits secretion

The gallbladder stores secreted bile, especially in the fasting state.
Gallbladder epithelial cells reabsorb an isotonic NaCl solution,
concentrating bile acids
CCK regulates release of gallbladder bile. It stimulates gallbladder
smooth muscle contraction and relaxes the sphincter of Oddi, which
allows bile to flow from the ducts into the duodenum.

Digestion and absorption of nutrients

- Carbohydrates, proteins, fats, and vitamins and minerals enter the body through the small intestine
- The small intestine surface area is enhanced by folds of mucosa and tiny finger-like projections (called villi). Villi are lined by epithelial cells that absorb nutrients, and possess ruffled apical ('brush border') membrane (⊃ p.60)
- Some substances can be absorbed directly from the lumen by the cell, without prior digestion (e.g. glucose). Alternatively, uptake can involve:
  - Hydrolysis in the lumen, absorption across the cell (e.g. proteins)
  - Hydrolysis on the brush border membrane, absorption across the cell (e.g. disaccharides)
  - Absorption into, and hydrolysis within, the cell (e.g. di- and tripeptides)
  - Luminal hydrolysis, uptake, and resynthesis within the cell (e.g. triglycerides).

Carbohydrates

Carbohydrates are provided predominantly in the form of starch.
- Salivary amylase initiates digestion
- Pancreatic amylase renders most starch in the form of disaccharides, trisaccharides, and α-limit dextrins
- Brush border disaccharidases (sucrase, lactase) release glucose, fructose, galactose
- Glucose and galactose are reabsorbed across the apical membrane by a Na^+-dependent carrier, SGLT-1; fructose is reabsorbed by the facilitated diffusion carrier GLUT5
- Monosaccharides exit the cell at the basolateral membrane on GLUT2
- Reabsorption of monosaccharides from interstitial fluid to blood completes the process. Absorption is complete by mid-jejunum.

Proteins

Digestion of proteins begins in the acidic environment of the stomach.
- Pepsin secreted and activated in the stomach initiates digestion
- Pancreatic peptidases (trypsin, chymotrypsin, elastase, carboxy-peptidases) responsible for the bulk of digestion of proteins to oligopeptides, amino acids
- Further digestion of oligopeptides to amino acids by brush border peptidases can occur
- Amino acids are absorbed by apical carriers, mostly Na^+-dependent. Different carrier proteins exhibit different specificities for different classes of amino acids (neutral, cationic, basic)
- Di- and tripeptides can be directly absorbed by a H^+-dependent apical carrier protein, PepT-1. Cytosolic peptidases then convert peptides to amino acids
- Amino acids exit the cell across basolateral membrane on Na^+-independent carriers
- Reabsorption of amino acids from interstitial fluid into the blood completes the process. Absorption is mostly complete by end of jejunum.

Lipids

Triglycerides are the most common dietary lipid. The absorption process is more complex, since fats are water insoluble.

- Muscular movements of the stomach emulsify fats (transformation into emulsion of oil droplets in water), aided by lingual lipases
- Most digestion occurs in small intestine. Pancreatic lipase digests triglycerides to monoglycerides and free fatty acids. Colipase, from the pancreas, coordinates binding of lipase to emulsion
- Monoglycerides—free fatty acids incorporated into bile micelles (spherical structures with the polar group facing outwards, the hydrophobic group facing towards the interior)—diffuse into an unstirred layer adjacent to the brush border
- Micelle components dissociate at the cell surface
- Digestion products enter cell by dissolving in lipid membrane—non-ionic diffusion
- Within the cell, the components are reassembled in smooth ER; they combine with apoproteins made in the RER to form microscopic particles of triglyceride called chylomicrons
- Chylomicrons are exported by the Golgi apparatus into interstitial fluid and pass into lymphatic capillaries. Lymph drains back into left subclavian vein via thoracic duct. In the fasting state, VLDLs, containing endogenous lipids, are secreted into lymphatic system.

Vitamins

Vitamins are organic substances that act as co-enzymes and/or regulators of metabolic processes. They can be water-soluble (e.g. thiamine (vitamin B_1), riboflavin (vitamin B_2), vitamin B_{12}, vitamin C, folic acid) or fat-soluble e.g. vitamins A, D, E, K). Water-soluble vitamins can be reabsorbed by:

- Passive diffusion
- Specific carrier proteins (vitamin B_1, vitamin C, folic acid)
- Binding to specific brush border receptors: vitamin B_{12} binds to intrinsic factor secreted by parietal cells of stomach. The complex formed binds an apical receptor which is then internalized. Vitamin B_{12} exits across the basolateral membrane bound to a second protein, transcobalamin.

Fat-soluble vitamins are presented for absorption dissolved in bile micelles. In most cases, they exit the epithelial cell, unmodified, in chylomicrons.

Minerals

Minerals are essential ions that need to be absorbed and include Ca^{2+}, Mg^{2+}, and Fe^{2+}.

Calcium

- Ca^{2+} is passively reabsorbed via the paracellular route throughout the intestine
- In addition, in the duodenum there is active transcellular absorption. Ca^{2+} enters across apical membrane through channels and is buffered in the cell by calbindin, exits at basolateral membrane on Ca^{2+}-ATPase or Na^+-Ca^{2+} exchange. Processes are stimulated by parathyroid hormone (PTH).

Magnesium

Absorption occurs through passive paracellular diffusion along the small intestine. Active absorption occurs in ileum.

Iron

- Absorption of non-haem iron occurs in the duodenum
- Fe^{2+} is the absorbed species:
  - Non-haem Fe^{3+} is reduced to Fe^{2+} by an apical membrane reductase
  - Fe^{2+} is taken up using the apical H^+-driven transporter DCT
  - Within the cell, Fe^{2+} binds mobilferrin
  - The complex is translocated to the basolateral membrane, where a carrier, IREG1, mediates efflux
  - An oxidase, hephaestin, oxidizes Fe^{2+} to Fe^{3+}
  - Fe^{3+} binds transferrin in plasma
- Haem iron is also absorbed in the duodenum. Haem is endocytosed at the apical membrane and split enzymatically. An intracellular reductase reduces Fe^{3+} to Fe^{2+}, which is then processed in the same way as non-haem Fe^{2+}
- The intestine is sole site for regulation of the body's iron levels. If iron levels are adequate, iron absorption can 'stall' within the intestinal cells. Cells turn over every 5 days, and any surplus iron stored in the cell is excreted (the 'iron curtain hypothesis').

Fluid and electrolyte movement

- Around 9L of fluid enters the GI tract each day—2L ingested, 7L secreted. Yet only 100mL is excreted in the stool each day
- The small intestine absorbs the majority of the fluid—up to 8.5L. Daily absorption by the colon is only 500mL—~10% of its maximum absorptive capacity
- Faeces are hypertonic. Water absorbed by the GI tract replaces that lost from urination, perspiration, respiration, and can be used for subsequent secretions
- The small intestine behaves as a leaky epithelium (◑ p.62) with paracellular transport of ions and water playing a significant role. NaCl transport is not regulated directly. The colon behaves as a tight epithelium, with transport through the cells dominating transepithelial transport. Colonic transport is regulated
- Water reabsorption occurs through osmosis and is the secondary consequence of solute (electrolyte or non-electrolyte) reabsorption. The movement of Na^+ is central to this process. It, along with co-transported osmolytes, establishes a hypertonic interstitial fluid that draws water, by osmosis, through the cell and tight junctions. The water then moves from the interstitial space to the blood. Water may move through tight junctions or transcellularly through membrane aquaporins.

Na^+

- There are a number of pathways for Na^+ reabsorption:
  - Passive diffusion through channels (ENaC)
  - In exchange for H^+ on Na^+-H^+ exchange (NHE)
  - Co-transported with glucose (SGLT), amino acids (e.g. system B)
  - Co-transport with Cl^- ions (Na^+-Cl^- co-transport, NCC, or by parallel operation of NHE and Cl^--HCO_3^- exchange (AE)
- Different processes play the dominant role in Na^+ absorption at different points along the GI tract:
  - Duodenum, jejunum: mediated by NHE, SGLT
  - Ileum: SGLT, NCC, or NHE/AE
  - Colon: NCC or NHE/AE
  - Rectum: ENaC
- The Na^+-K^+-ATPase in the basolateral membrane transports Na^+ ions into interstitial fluid. Other solutes exit the cell through channels and carriers. Aldosterone stimulates Na^+ absorption.

Cl⁻

- In addition to absorption by NCC and NHE/AE, Cl^- can be absorbed passively:
  - In the jejunum and ileum: through Cl^- channels in apical, basolateral membranes, and by paracellular route
  - In the colon: by AE alone
- Crypt cells at the base of villi can secrete Cl^-, through a mechanism that resembles that of salivary or pancreatic acinar secretion. There is a basal level of secretion, masked by reabsorption. Secretagogues (e.g. cholera toxin, VIP, histamine) stimulate apical Cl^- efflux mediated by CFTR. Cl^- secretion increases: Na^+, H_2O follow through paracellular route. Signalling is through kinase activation following cAMP generation or mobilization of Ca^{2+} by IP_3. Cholera activation is irreversible: it leads to massive Cl^--rich watery diarrhoea.

K⁺

- Both reabsorption and secretion of K^+ occur:
  - Passive absorption through paracellular route in jejunum, ileum
  - Active absorption by H^+-K^+-ATPase in colon
  - Passive secretion through paracellular route in colon
  - Active secretion through apical K^+ channels in colon
  - The relative importance of these pathways depends on K^+ balance:
    - Aldosterone stimulates K^+ secretion (by stimulating the Na^+-K^+-ATPase)
    - Hypokalaemia stimulates absorption.

Malabsorption

(⏎ *OHCM10* pp.266, 267.)

Definition

Insufficient absorption of nutrients for the body's needs, despite an adequate intake.

Mechanisms

- Loss of epithelium in the gut so that there is an inadequate surface area for absorption
- Lack of digestive enzymes to break the food down into absorbable molecules
- Lack of specific transport proteins to bind to some molecules and to facilitate their transcellular transport.

Causes

- Gluten-sensitive enteropathy (coeliac disease)—presentation of specific short peptides derived from gluten component of wheat in the blood of susceptible individuals triggers an autoimmune response which damages small bowel epithelium → villous atrophy and loss of absorptive surface area
- Post irradiation and chemotherapy—damage to epithelium leads to loss of absorptive surface area (e.g. after conditioning therapy before bone marrow transplant)
- Short bowel syndrome—loss of absorptive surface area due to resection of substantial amounts of small bowel or gross damage by inflammatory processes, most commonly Crohn's disease
- Pernicious anaemia—autoimmune atrophic gastritis destroys the cells in the stomach which produce intrinsic factor, so vitamin B_{12} is not absorbed in the terminal ileum
- Pancreatic insufficiency—if the exocrine glands of the pancreas are destroyed (e.g. in cystic fibrosis or chronic alcoholism), there is a deficiency of lipases, etc., to digest fats and, so, malabsorption of these and fat-soluble vitamins
- Enzyme deficiencies—e.g. disaccharidase deficiency producing lactose intolerance.

Complications of malabsorption

Deficiencies of whatever nutrients are not being absorbed, which may be specific or general, depending on the cause.

Intestinal obstruction

(➔ OHCM10 p.610.)

Definition

Obstruction of the intestine by any process, at any level, producing obstruction to passage of food or faeces at that point.

Mechanisms

- Obstruction within the lumen
- Dysfunction of the GI muscle wall
- Obstruction due to pressure from elements outside the GI tract.

Causes

- Obstruction within bowel lumen:
  - Polypoid tumour, e.g. adenomatous polyp/carcinoma in colorectum
  - Foreign body, e.g. hair ball mass in stomach
  - Fibrous diaphragms, e.g. with chronic use of NSAIDs or pancreatic enzyme supplements in cystic fibrosis
  - Intussusception, when the bowel herniates into itself, e.g. at the point of a small tumour or with gross lymphoid hyperplasia in the small bowel
- Dysfunction of the GI muscle wall:
  - Diffusely infiltrating tumour, e.g. diffusely infiltrating ('signet ring' type) adenocarcinoma in the stomach producing a rigid organ (well seen on barium meal—'leather bottle' stomach = linitus plastica)
  - Congenital abnormality of the bowel innervation, e.g. Hirschsprung's disease, where there is an absence of ganglion cells in the distal part of the colorectum, with loss of motility and neonatal intestinal obstruction
- Obstruction by elements outside the GI tract:
  - Intra-thoracic tumours obstructing the oesophagus, e.g. bronchial carcinoma
  - Intra-abdominal tumours obstructing the intestine, e.g. disseminated ovarian carcinoma.

Complications of intestinal obstruction

- Perforation due to increased pressure and ischaemic necrosis in the bowel proximal to the obstruction, e.g. caecal perforation in obstructing colorectal cancer—but only if the ileocaecal valve is competent (otherwise pressure is decompressed in the small bowel)
- Water/electrolyte imbalance due to gross pooling in the obstructed bowel
- Dysphagia if the obstruction is in the oesophagus
- Vomiting if the obstruction is in the stomach or lower, and obstruction is prolonged.

Diarrhoea

(⊕ *OHCM10* p.258.)

Definition

Rather vague. Usually means an increased frequency of more liquid stool.

Mechanisms

- Toxin—which causes increased secretion of fluid by gut epithelial cells into the gut lumen, e.g. the toxin of *Vibrio cholera*
- Damage to the gut epithelium—so that water is lost into the gut lumen and/or not resorbed:
  - Viral infection, e.g. enterovirus
  - Bacterial invasion, e.g. salmonella
  - Radiation damage, e.g. following conditioning for bone marrow transplantation
  - Toxic damage, e.g. during systemic chemotherapy
  - Immune damage, e.g. gluten-sensitive enteropathy
  - Idiopathic inflammatory damage, e.g. Crohn's disease
- Masking of the gut epithelium by organisms:
  - Protozoal infection, e.g. giardiasis
- Osmotic—passive movement into gut lumen:
  - Disaccharidase deficiency—undigested sugars left in gut lumen and thus greater osmolality
  - Laxative abuse—usually osmotically acting laxatives such as lactulose.

Complications

- Dehydration
- Anaemia—if blood is also being lost due to damaged epithelium
- Complications of specific cause, e.g. systemic infection.

  Box 8.2 summarizes the treatments for diarrhoea.

Box 8.2 Treatments and drug therapies for diarrhoea

(⊕ *OHPDT2* Ch. 1.)
- Most cases of diarrhoea are fairly short-lived and are dealt with simply in terms of maintenance of hydration and electrolytes
- 'Traveller's diarrhoea' is rapidly controlled by agents that reduce gut motility (primarily of the opiate class, e.g. diphenoxylate and loperamide—also known as Imodium®). Neither of these drugs cross the blood–brain barrier and are regarded to be relatively selective for the gut. Side effects are usually only experienced with prolonged use (constipation, drowsiness, and paralytic ileus). Bismuth subsalicylate is often used as a preventative measure against traveller's diarrhoea—its actions are thought to be mediated via the salicylate component.

Inflammatory diseases in the gut

Irritable bowel syndrome (IBS)

(➲ *OHCM10* p.266.)

- Thought to affect ~15% of the population to some degree
- Symptoms vary from mild and occasional abdominal discomfort, to frequent severe pain and irregular bowel motility → either constipation (IBS-C), or diarrhoea (IBS-D), or both (IBS-A)
- Stress, depression, and infections are thought to trigger the disease while specific constituents of the diet and some drugs (NSAIDs) can exacerbate the symptoms. Although a genetic predisposition to IBS has been suggested, there is little clear evidence to support this notion at present
- IBS is associated with enhanced inflammation and there is some evidence that it is mediated by chronic infections that are poorly detected by current best practice (e.g. *Blastocystis*, *Dientamoeba fragilis*, or simply intestinal flora overgrowth)

For treatments for IBS, see Box 8.3.

> Box 8.3 Treatments and drug therapies for IBS
>
> (➲*OHPDT2* p.227.)
> - In some cases, IBS can be managed by identification of the trigger (e.g. stress, avoidance of particular dietary constituents)
> - However, symptoms often have to be dealt with using medications—loperamide is often used in IBS-D, while laxatives (e.g. lactulose) can be helpful in IBS-C
> - Anticholinergic agents are occasionally used to inhibit gut motility (antispasmodics) for patients with cramps or diarrhoea
> - Some serotonin antagonists are also in clinical trials as antispasmodics
> - Serotonin agonists have been tried (with limited success) in IBS-C.

Inflammatory bowel disease

Not to be confused with IBS, inflammatory bowel disease (IBD) includes number of inflammatory conditions, of which Crohn's disease and ulcerative colitis are the most common.

Crohn's disease

(➲ *OHCM10* p.264.)

- Affects 25–50 people per 100,000 in industrialized countries
- Differential geographical and ethnic distribution, together with strong heritability suggests that there is a genetic component to the disease. However, there are also clear environmental factors involved, e.g. smoking is a recognized risk factor for the disease
- Widely regarded to be an autoimmune disease (i.e. driven by the damaging effects of the body's own inflammatory cells)
- Characterized by deep lesions in the ileum, the ileo-colonic junction (most common) and/or the colon

Symptoms include severe abdominal pain, diarrhoea, which might contain blood. Obstruction of the intestine can also occur (intestinal stenoses), particularly following surgery. The severe inflammation can also spill over into systemic effects, including, for example, depressed growth in children and adolescents.

Box 8.4 summarizes the treatments for IBD.

Box 8.4 Treatments and drug therapies for IBD

(● OHPDT2 p.48–50.)

There is no cure for Crohn's disease and interventions are restricted to managing the symptoms:

- Lifestyle changes can often help reduce the extent or frequency of episodes—smoking cessation, changing to frequent, regular small-portion meals and avoidance of known dietary triggers are all recognized to help control the symptoms
- Drug regimens are designed to help avoid inflammatory episodes (maintenance phase) and to moderate their effects when they occur (acute phase):
  - Aminosalicylates are the most commonly used anti-inflammatory agents in both the acute and maintenance phases, with glucocorticoids (e.g. hydrocortisone) used to help moderate severe attacks
  - Drugs to prevent diarrhoea (e.g. loperamide) can help in patients where this symptom is prevalent
- Antibiotics are sometimes prescribed to treat infections mediated by bacteria that take advantage of the depressed barrier function of the gut
- Surgery is necessary if inflammation is so severe as to cause a physical blockage of the bowel. In this case, the affected section of the intestine is removed; this is not seen as a cure in Crohn's disease, as repeat surgery is often required.

Ulcerative colitis

● OHCM10 p.262.)

- Ulcerative colitis is a severe form of colitis (inflammation of the colon) that affects ~0.1% of the population
- The cause is complex: no single gene defect has been identified, but a combination of gene defects are almost certainly involved in predisposing individuals to ulcerative colitis
- However, environmental factors are also closely associated with incidence of the disease; dietary constituents have received particular attention
- Ulcerative colitis has similar symptoms to Crohn's and IBS (abdominal pain or cramps, often associated with diarrhoea, which might contain blood). There might also be associated weight loss in the patient. Some patients complain of arthritic pain: on occasion this can pre-empt onset of the intestinal symptoms

- Diagnosis is best confirmed with endoscopy, although the most definitive diagnosis comes from microscopic analysis of biopsies: ulcerative colitis is characterized by shallow lesions and inflammation and crypt abscesses, while Crohn's disease features deep inflammation
- There is an increased risk of colon cancer in patients with ulcerative colitis.

For treatments for ulcerative colitis, see Box 8.5.

Box 8.5 Treatments and drug therapies for ulcerative colitis

(⊕ *OHPDT2* p.48)
- In some patients, changes in diet to include particular fibre and some fish oils can be beneficial
- From a drug therapy perspective, the aim is to first allow the mucosal lining to repair (instigate remission) prior to modifying the medication to facilitate full remission
- 5-amino salicylic acid is a mainstay in maintenance of remission of ulcerative colitis, with corticosteroids (e.g. prednisolone), immunosuppressive drugs (methotrexate), and, more recently biologicals (agents that directly bind to inflammatory mediators, such as tumour necrosis factor (TNF)-α are also commonly used
- Surgery is a successful means of treating ulcerative colitis—in many cases, it is considered a cure:
  - It involves removal of the affected section of the bowel, often with incorporation of an 'ileal pouch' in an effort to retain some bowel control and to avoid the necessity for colostomy bags.

Immune functions

Foods always contain a certain number of microbial organisms howeve carefully prepared and in the times of early humans, would have containe many more. These organisms include viruses and bacterial and eukaryote parasitic organisms. Thus, there has evolved an immune system in the gu which largely prevents the invasion or colonization of disadvantageous or ganisms. In the stomach, the acidic environment is the major non-immune mechanism of antimicrobial defence but, distal to this, an immune defence predominates.

Organization of the gut immune system

Although the organization of the mucosal immune cells in the gut does not at first, appear to be as precise as in lymph nodes, there is a very compart mentalized distribution of these cells (Fig. 8.11):

- Intraepithelial lymphocytes—are found in varying numbers throughout the small and large intestine. They are of cytotoxic/suppressor phenotype (CD8+)
- Lymphoid aggregates—are found in the mucosa of the small and large intestine and sometimes extend into the submucosa. They may be visible endoscopically as Peyer's patches or are sometimes mistaken for epithelial polyps. The epithelial cells overlying mucosal lymphoid aggregates contain membranous (M) cells as well as the usual absorptive cells. These M cells transcytose intact macromolecules from the lumen to be presented to the T lymphocytes
- B cells—in the gut tend to produce the dimeric IgA subclass of immunoglobulin rather than the more common IgG. This protein has a secretory chain which facilitates its secretion into the gut lumen where i will bind to specific antigens, and which also protects the protein against proteolysis.

The immune tissue in the gut is part of the mucosa-associated lymphoid tissue (MALT) system (◑ *OHCM10* p.362), which is found in other mucosa sites such as the bronchus and salivary glands.

Abnormalities of the gut immune system

- Neoplasia—lymphomas may develop in the lymphoid tissue of the gut, usually of B-cell origin and usually retaining some characteristics of MALT lymphoid cells, e.g. homing to MALT tissue in other sites in the body
- Immune deficiency:
  - IgA deficiency—specific deficiency of this class of immunoglobulin may be associated with increased frequency of GI infections
  - Generalized immunodeficiencies (e.g. post-chemoradiotherapy; ◑ *OHCM10* p.376)—will lead to an increased rate of GI infections.

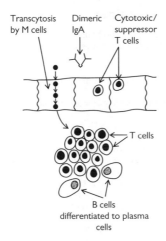

Fig. 8.11 Organization of the immune system in the gut.

The liver: overview

The liver is the metabolic workhouse of the body, playing a role in numerous biosynthetic and metabolic processes (see topics in ⊕ Chapter 2), including:

- Glucose homeostasis:
  - The liver stores glycogen for use for sustaining other tissues during times of fasting
  - In addition, the liver is the major site of gluconeogenesis during starvation
- Fat metabolism:
  - The synthesis of fatty acids takes place in the liver (for storage as triacylglycerides in adipose tissue), as does cholesterol synthesis
  - Ketogenesis is also performed by the liver during starvation
- Amino acid metabolism and urea synthesis:
  - Amino acids are metabolized in the liver and their carbon skeletons converted into glucose by the gluconeogenesis pathway
  - In order to maintain nitrogen balance, the body has to excrete excess nitrogen:
    - The liver makes urea—the non-toxic, water-soluble nitrogenous compound excreted by renal filtration
    - The ammonia groups incorporated into urea come from the breakdown of amino acids, especially glutamine.

More minor but still important roles of the liver include:
- Synthesis of plasma proteins, such as albumin and the blood-clotting factors
- Trace element homeostasis
- Detoxification
- Alcohol metabolism (⊕ p.603)
- Vitamin/co-factor metabolism
- Storage of iron.

Anything that affects these processes can have major detrimental effects on liver metabolism (⊕ *OHCM10* p.274).

Acute effects on liver metabolism

- Alcohol detoxification takes place primarily in the liver, converting ethanol → acetaldehyde → acetate:
  - Both reactions reduce NAD^+ → NADH, and even moderate alcohol ingestion produces more NADH than can be oxidized by the ETC
  - The high NADH:NAD^+ ratio is an allosteric inhibitor of a number of pathways (e.g. gluconeogenesis, fatty acid breakdown):
    - Consequences include hypoglycaemia and fatty acid accumulation in the liver
  - In addition, acetaldehyde is a reactive compound that covalently binds to biologically important functional groups of proteins in liver and blood
- Hepatitis is an inflammatory infection (⊃ *OHCM10* p.278) of the liver which interferes with plasma protein synthesis, causing clotting disorders.

Chronic effects on liver metabolism

(⊃ *OHCM10* p.274.)

- As the liver is the only organ capable of urea synthesis, serious liver damage will result in the rate of ammonia production exceeding the rate that it can be converted into urea for excretion:
  - Ammonia is toxic, and its build-up affects the functioning of the CNS
- In addition, there is a reduction in the amino acid metabolism capacity, and an accumulation of aromatic amino acids also affects CNS function
- There is also a reduction in capacity for gluconeogenesis, and if blood glucose levels drop, then hypoglycaemic coma and even death can occur in liver failure
- Finally, liver failure patients have muscle wasting due to a lack of insulin-like growth factor (IGF-1; made in the liver in response to growth hormone) compounding the problems.

Protein synthesis by the liver

Two of the most important proteins synthesized by the liver are albumin and the blood clotting factors.

Albumin

- Makes up ~50% of the protein found in plasma (concentration of ~40g l⁻¹):
  - Rest are globulins (α, β, γ) and fibrinogen
  - All except the γ-globulins are synthesized in the liver
- Albumin is a single molecular species of a 66kDa protein.

Two major roles for albumin

- The high concentration of albumin results in it making a large contribution to the osmolarity of the plasma, preventing oedema:
  - Oedema and ascites (accumulation of fluid in the peritoneal cavity causing swelling) are signs of chronic liver disease (➲ OHCM10 p.276), reducing the capacity for albumin synthesis
  - Protein malnutrition (kwashiorkor) can lead to depleted levels of albumin and subsequent oedema
  - Kidney disease (nephritis) can also deplete albumin through excessive excretion, with similar results
- Binding of small molecules—albumin is a major carrier of other molecules in the plasma, especially hydrophobic molecules and metal ions (trace elements):
  - Albumin has several sites capable of binding hydrophobic molecules:
    - The binding is non-specific and relatively weak, but this is still important quantitatively due to the high concentration of albumin in plasma
    - Important molecules carried by albumin include: free fatty acids; steroid hormones; bilirubin (breakdown product of haemoglobin); hydrophobic drug molecules
  - Trace elements such as Cu²⁺ and Zn²⁺ bind to the N-terminus region of the albumin protein
  - Another example of a plasma protein synthesized by the liver with a role in metal transport is transferrin (➲ p.604).

Blood clotting factors

- Blood clotting factors are proteases that act upon one another in a cascade when activated, ultimately converting fibrinogen to fibrin which is insoluble and forms the blood clot:
  - Prevents further blood loss, maintaining haemostasis (➲ p.488)
- Almost all of the blood clotting factors are synthesized by the liver:
  - Liver diseases (➲ OHCM10 pp.276–7, 282–8) can depress the clotting system enough to cause severe tendency to bleed
- Liver synthesis of four of the most important clotting factors (prothrombin, factors VII, IX, and X) is vitamin K dependent:
  - Lack of the fat-soluble vitamin K can also lead to a serious tendency to bleed
  - Normally, vitamin K is made by intestinal bacteria and absorbed with dietary fat

- Liver disease is one of the main causes of vitamin K deficiency:
  - Lack of bile secretion causes poor fat absorption and, thus, poor vitamin K uptake
  - Compounds the reduced synthetic capacity of diseased liver for producing clotting factors
- It is obviously important that blood clotting does not occur when not required:
  - Antithrombin III and α-macroglobulin are the major anticoagulation plasma proteins, with heparin co-factor II and α-1-antitrypsin:
    - Protease inhibitors prevent blood clotting factors acting inappropriately
    - The anticoagulation drug, heparin, strongly activates antithrombin
- Many of the proteins involved in blood clotting are modified, post-translationally, to have γ-carboxylglutamyl (Gla) residues
  - The proteins with Gla residues are effective at binding Ca^{2+} ions. Ca^{2+} is important for clotting factor interactions with other proteins and for enhancing their catalytic activity
  - The post-translational modification is vitamin K dependent:
    - Vitamin K is an essential co-factor for the carboxylase enzyme responsible, located on the luminal side of the RER
    - Coumarin anticoagulants (e.g. warfarin) inhibit the carboxylase reaction by interfering with the vitamin K co-factor.

Iron transport and storage

(→ *OHCM10* pp.324, 326.)

Iron is important in the body, being an essential constituent of haemoglobin, myoglobin, and cytochromes. It is ingested in the diet in two major forms—haem and non-haem iron—absorbed by separate pathways.

- Haem is absorbed across the enterocyte apical membrane by a specific haem transporter. Once in the cell, the iron is released by haem oxidase
- Non-haem (or free iron) is reduced from Fe^{3+} to Fe^{2+} by ferrireductase, present on the luminal surface of enterocytes:
  - The reduction of Fe^{3+} to Fe^{2+} is also favoured by ascorbic acid (vitamin C) in the diet (vitamin C is used in iron supplements to aid absorption)
  - Fe^{2+} is absorbed by the proton-coupled divalent metal transporter DMT1 (which can transport a variety of divalent cations).

Once inside the enterocyte, the iron can either be stored or transported across the basolateral membrane to enter the plasma.

- Storage is as a complex of Fe^{3+} with the protein ferritin
- Basolateral transport is by another transporter, regulatory protein 1 (REG1), and hephaestin. Hephaestin oxidizes Fe^{2+} back to Fe^{3+}, which is the form that binds the iron-transporting plasma protein, transferrin.

Absorption of iron by the intestine is tightly regulated, as there is no mechanism for excreting iron once it has been taken up into the body.

- Healthy individuals are in iron balance, with daily loss equalling uptake at about 1mg day^{-1}
- Most of iron (~60% or 2500mg) in the male body is in haemoglobin, with about 25% stored and the remainder in myoglobin and enzymes
- Females tend to have smaller stores (around 40% of that in males) and larger daily losses (averaging up to 2mg day^{-1}) due to blood loss during menstruation
- A poorly understood mechanism limits absorption of excess iron at the level of the enterocytes ('mucosal block').

Transferrin is a 76kDa β-globulin glycoprotein synthesized by the liver which is central to the control of iron metabolism by the body.

- 1 mole of transferrin binds 2 moles of Fe^{3+}
- To enter cells, transferrin binds to a cell surface transferrin receptor, which is then internalized by receptor-mediated endocytosis:
  - The low pH in the lysosome causes the release of the iron, and it enters the cytoplasm via DMT1
  - The transferrin receptor, with the transferrin still bound, is recycled, intact, to the plasma membrane. The transferrin is released and can bind more iron.

Synthesis of the intracellular storage protein, ferritin, and the transferrin receptor are reciprocally linked.

- Genes for both proteins have iron response elements in their promoter regions

When iron levels are high, ferritin mRNA synthesis is increased to promote iron storage in cells, and that of transferrin receptor is reduced When iron levels are low, transferrin mRNA levels rise, while ferritin mRNA is kept in an inactive form.

addition to iron deficiency, iron excess can also cause disease. Hereditary haemochromatosis is prevalent in Scotland, Ireland, and the USA (◉ OHCM10 p.288):
• Total body iron is >15g (compared with 2.5–3.5g normally)
• Tissues damaged include liver, pancreatic islets (→ diabetes), and heart:
  ○ Melanin and iron accumulate in skin, giving slate-grey colour
• Usually linked to a mutation in the gene HFE, which codes for a protein related to major histocompatibility complex (MHC) class I antigens
• It is thought that this affects mucosal block by enterocytes, → excessive absorption of iron by the intestine.

Detoxification of xenobiotics

The liver is the major site for metabolism of xenobiotics (foreign compounds, *xenos* = Greek for stranger), which can be both exogenous (e.g. drugs, carcinogens, pollutants) or endogenous (e.g. steroids, eicosanoids, fatty acids).

The main aim of the metabolism is to convert the mainly lipophilic xeno biotics into more water-soluble derivatives for excretion.

Xenobiotic metabolism can be divided into two phases:
- Phase 1 mainly involves hydroxylation, carried out by enzymes called mono-oxygenase or cytochrome P450:
  - Most tissues contain a number of different isoforms of cytochrome P450, especially the small intestine and liver. Present mainly in the membranes of smooth ER
  - Reaction catalysed:

$$RH + O_2 + NADPH + H^+ \rightarrow R-OH + H_2O + NADP$$

- Phase 2 involves the conjugation of the hydroxylated xenobiotics to various other compounds to further increase their hydrophilicity and enable excretion in bile or urine:
  - The most common conjugation is the addition of glucuronic acid. Examples of conjugated compounds include phenol, steroids, meprobamate (a tranquillizer)
  - Other conjugation reactions include sulphation, acetylation, methylation, and conjugation with glutathione.

Although xenobiotic hydroxylation and conjugation are referred to as de toxification, this is not always the outcome.
- Phase 1 reactions may make the xenobiotics more biologically active. This may be desirable if the xenobiotic is a prodrug which is being converted into its active form, or undesirable if it creates a carcinogen
- Phase 2 reactions can also theoretically increase activity, but this is rare.

Cytochrome P450 enzymes are often induced by the presence of their substrates.
- This can be the cause of drug interactions. For example, phenobarbital (for epilepsy) induces the cytochrome P450 that metabolizes warfarin (used to prevent blood clotting). Thus, the starting/discontinuing of phenobarbital will affect the dose of warfarin necessary to maintain the same level of effect
- Alcohol induces a cytochrome P450 isoform which metabolizes widely used solvents and components present in tobacco smoke. Alcohol consumption may therefore increase the risk of carcinogenicity.

Alcohol (ethanol) metabolism is catalysed by ethanol dehydrogenase.
- Ethanol + NAD^+ → acetaldehyde + NADH + H^+
- Subsequently, acetaldehyde is dehydrogenated to acetate (also NAD^+-dependent)
- The acetate can be conjugated to CoA to form acetyl-CoA. Acetyl-CoA is a substrate for fat synthesis in the liver, → obesity and/or a fatty liver in nutrient-replete alcoholics

* Large amounts of alcohol can deplete the liver of NAD⁺ (i.e. the NADH: NAD⁺ ratio increases):
  * This pushes the equilibrium mediated by lactate dehydrogenase towards lactate, reducing the amount of pyruvate available for gluconeogenesis
  * Similarly, malate dehydrogenase will be pushed in the direction of malate (not oxaloacetate), which is also a gluconeogenic substrate
  * If this coincides with a falling blood glucose level, hypoglycaemia can occur:
    ○ One of the most sensitive parts of the brain to hypoglycaemia controls body temperature, which will drop, resulting in hypothermia (⊜ OHCM10 p.848).

Haem catabolism takes place in the liver, the spleen, and bone marrow (Fig. 8.12).
* An average person metabolizes 6g of haemoglobin per day
* The globin protein is degraded and the amino acids recycled
* The iron from the haem group is released after it has been reduced with NADPH, and returned to the body iron pool for reuse
* The porphyrin ring is opened to give biliverdin (green)
* Biliverdin is reduced with NADPH to bilirubin (yellow).

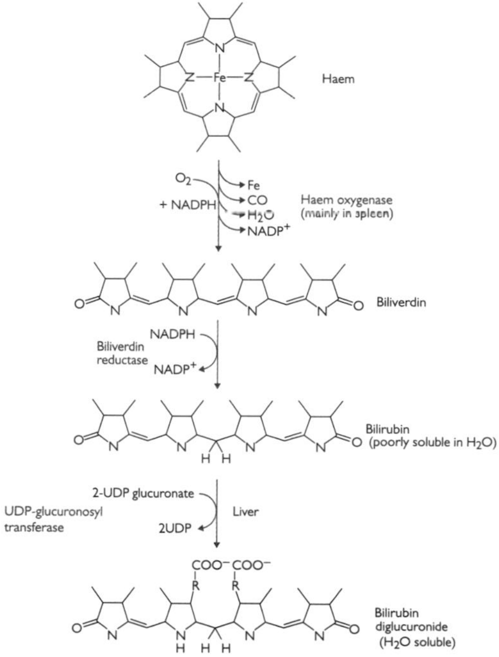

Fig. 8.12 Diagrammatic representation of the haem oxygenase reaction and the reduction of biliverdin.

Hepatic cirrhosis

(🔗 OHCM10 p.276.)

Definition

Formation of bands of fibrous tissue between portal tracts in the liver which surround regenerative nodules of liver cells.

Requirements

Continued damage to the liver while repair/regeneration is occurring—so is generally caused by chronic factors (e.g. decades of alcohol abuse, see Table 8.1) rather than a single acutely hepatotoxic episode (e.g. an overdose of paracetamol).

Table 8.1 Causes of hepatic cirrhosis

Generic group	Specific examples
Direct hepatotoxins	Alcohol, other drugs
Infective agents	Hepatitis B, C
Extra-hepatic obstruction	Gallstone obstruction of common bile duct, primary sclerosing cholangitis
Metabolic disorders	Wilson's disease (accumulation of copper), haemochromatosis (primary due to inherited disorder of iron metabolism, secondary due to iron overload from multiple blood transfusions or ingestion of excessive oral iron), A-1-antitrypsin deficiency
Autoimmune	Autoimmune hepatitis, primary biliary cirrhosis

Complications of hepatic cirrhosis

- Portal hypertension—due to obstruction of portal venous blood flow by the bands of fibrous tissue:
  - Oesophageal varices—may produce sudden and massive haematemesis
  - Rectal varices (🔗 OHCM10 p.257)—less recognized than oesophageal varices but can produce massive rectal bleeding. Distinguish from haemorrhoids (rectal varices are completely within the rectum)
- Hepatic failure (🔗 OHCM10 p.274)—when the majority of the parenchymal cells are dysfunctioning, often due to some precipitating event (decompensated hepatic failure) such as large bleed from oesophageal varices. A cirrhotic liver can provide adequate function for many years.

Hepatic failure

(→ OHCM10 p.274.)

Definition

Destruction or non-function of a sufficient amount of the liver cells so that it no longer performs its physiological functions.

Requirements

The liver has a large functional reserve, so about 90% of its mass has to be non-functioning before hepatic failure ensues.

Classification

- By time course:
  - Acute
  - Chronic
- By aetiology:
  - Toxin:
    - Alcohol
    - Drugs
  - Infection:
    - Viral hepatitis A, B, C, etc.
  - Metabolic disorder:
    - Wilson's disease
    - Haemochromatosis
    - α-1-antitrypsin deficiency
  - Biliary obstruction:
    - Gallstones in common bile duct
    - Primary sclerosing cholangitis
  - Autoimmune:
    - Autoimmune hepatitis
    - Primary biliary cirrhosis
  - Tumour:
    - Primary—hepatocellular carcinoma
    - Metastatic—commonly, colorectal cancer.

Complications of hepatic failure

- Failure of production of proteins:
  - Clotting factors → bleeding
  - Albumin → oedema due to decreased osmotic pressure
- Failure of detoxification:
  - Retention of bile salts → jaundice
  - Retention of other products → hepatic encephalopathy.

Endocrine organs

General principles

The endocrine system is one of the major control systems that use chemical messengers. Endocrinology is the study of the endocrine system.

Hormones are chemical messengers released from a cell to influence the activity of another/the same cell via a receptor. Hormones are normally present in very low concentrations (10^{-12}–10^{-7}M) in the blood/extracellular fluid.

- Endocrine glands are a well-defined collection of endocrine cells
- Diffuse endocrine systems: many hormone-producing cells are not aggregated in glands, but dispersed (e.g. in the gut)
- Neuroendocrine systems: some neurones release hormones into both the bloodstream and into the CNS.

Classical endocrine action is when a chemical messenger (the hormone), released by a cell, is transported, via the bloodstream, to its target cell. This is now known to be much too narrow a concept because hormones act via various routes (Fig. 9.1):

- Neuroendocrine: the hormone is released from a neurone into the bloodstream
- Paracrine: the hormone acts on local cells via the extracellular fluid
- Autocrine: the hormone acts on the cell producing the hormone.

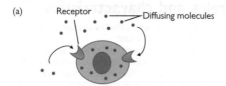

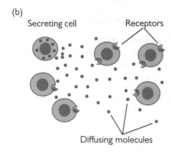

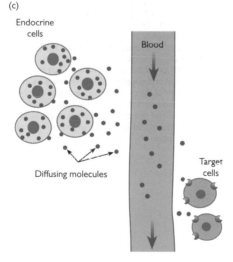

Fig. 9.1 A schematic comparison between autocrine, paracrine, and endocrine signalling. (a) Autocrine signalling: the signal acts on the cell which secreted it. (b) Paracrine secretion: the secreting cell releases a signalling molecule into the extracellular fluid, which binds to receptors on neighbouring cells. (c) Endocrine signalling: cells secrete hormone which enters the bloodstream, from where it can be distributed to other tissues to bind to cellular receptors. The target tissues may be at a considerable distance from the secreting cells.

Parts (b) and (c) Reproduced with permission from Pocock G and Richards CD (2006), *Human Physiology: The Basis of Medicine*, 3rd edn, p50. Oxford University Press.

Major roles and characteristics

The endocrine system promotes survival of the species by:
- Promoting survival of the individual:
  - Effects on development, growth, and differentiation
  - Help in preservation of a stable internal environment—homeostasis (but this is often disturbed in the short term for long-term gain)
  - Responds to an altered external environment—especially emergency 'stress responses'
- Control of the processes involved in reproduction.

There are different types of hormone

- Proteins/peptides/glycopeptides (hydrophilic) are translated on the RER and secreted by either the regulated pathway (e.g. insulin, prolactin) or the constitutive pathway (cytokines, growth factors):
  - The original translation product (the prohormone) is usually processed proteolytically to yield the active hormone(s). Some endocrine cells produce more than one active peptide hormone, in varying amounts.
  - Often stored in large amounts in intracellular granules, but some peptides (e.g. growth factors and cytokines) are not stored
- Steroids (hydrophobic, e.g. testosterone, oestrogen) are synthesized rapidly on demand (not stored) from cholesterol, via enzymes in the mitochondria and smooth ER
- Bioactive amines (hydrophilic, e.g. adrenaline, dopamine) are produced from tyrosine via intracellular enzymes. They are stored in large amounts in intracellular granules
- Thyroid hormones (hydrophilic, e.g. thyroxine (T_4)) are iodothyronines produced by iodination and coupling of tyrosyl residues in a protein (thyroglobulin), and are subsequently released by proteolysis:
  - Large amounts of iodinated thyroglobulin (the precursor for thyroid hormone synthesis), but not the free hormone, are stored in the thyroid. However, the blood contains a large reservoir of protein-bound thyroid hormone.

The pituitary gland

(➔ OHCM10 p.232.)

Pituitary structure

The pituitary gland is situated beneath the hypothalamus of the brain, in a depression (the pituitary fossa or sella turcica) of the skull. The human pituitary comprises anterior and posterior lobes.

The anterior pituitary consists of distinct endocrine cell types which produce and secrete the various hormones. The posterior pituitary is formed by the axons and terminals of magnocellular neurosecretory neurones originating in the hypothalamus. Pituicytes (a type of glial cell) surround and support the terminals. The posterior pituitary hormones are synthesized in the hypothalamus, packed into granules, transported down the axons, and released, by exocytosis, into the systemic veins.

The functions hormones and their functions are summarized in Fig. 9.2 and described in detail in the following sections.

Anterior pituitary

Thyroid-stimulating hormone (TSH)

- Actions: TSH acts in the thyroid. It stimulates tri-iodothyronine (T_3) and T_4 production; increases iodine uptake by the thyroid; and stimulates thyroid growth
- Control: TSH release is stimulated by thyrotropin-releasing hormone (TRH) from the hypothalamus and is inhibited by T_3 and T_4 negative feedback. Secretion of TRH is stimulated by cold and by stress, via the CNS.

Adrenocorticotropic hormone (ACTH)

Polypeptide hormone cleaved from the prohormone, proopiomelanocortin (POMC).

- Actions: stimulates production and therefore secretion of cortisol from the cortex of the adrenal gland. ACTH also produces some increase in adrenal sex steroids and stimulates growth of the adrenal cortex. Melanocyte-stimulating hormone is also cleaved from POMC and stimulates pigmentation of skin via actions on melanocytes
- Control: secretion of ACTH is increased by stress. Secretion is pulsatile with a diurnal rhythm (high at 7.00am, low at midnight). Release of ACTH is stimulated by corticotropin-releasing hormone (CRH) from the hypothalamus, inhibited by glucocorticoid negative feedback.

Gonadotrophins

Luteinizing hormone (LH), follicle-stimulating hormone (FSH)

- Actions:
  - Female—LH and FSH control growth and development of follicles; ovulation; synthesis of sex steroids by the ovary; growth and secretion of the sex steroid progesterone by the corpus luteum
  - Male—LH controls testosterone production by the Leydig cells; FSH stimulates the Sertoli cells and sperm production
- Control: hypothalamic—LH and FSH release is stimulated by hourly pulses of gonadotrophin-releasing hormone (GnRH) during reproductive life; inhibited by sex steroid oestrogen negative feedback.

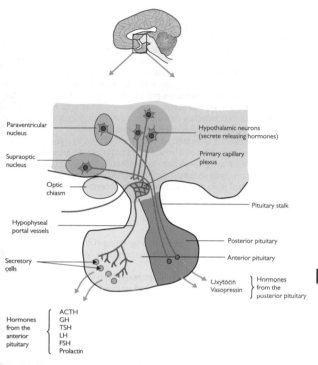

Fig. 9.2 The main features of the pituitary gland.

Reproduced with permission from Pocock G and Richards CD (2006), *Human Physiology: The Basis of Medicine*, 3rd edn, p193. Oxford University Press.

However, switch to 'positive feedback' triggers LH surge at ovulation. Cyclical variations in LH and FSH in menstrual cycle. In males, LH release is inhibited by negative feedback from testosterone; ovarian peptides (inhibin and follistatin) inhibit FSH.

Prolactin (PRL)

- Secreted by lactotroph cells
- Actions: principal role in preparation for lactation (→ p.680). PRL stimulates the development and growth of secretory alveoli in the breast and milk production. PRL also inhibits the reproductive system at the level of the gonads and pituitary (causes 'lactational amenorrhoea' in women after delivery of baby)
- Control: secretion of PRL is increased by suckling. PRL release is inhibited by dopamine from the hypothalamus. PRL synthesis is stimulated by circulating oestrogen.

Growth hormone (GH)
- Actions: stimulates long bone and soft tissue growth, both via stimulating the release of IGFs from the liver and by direct actions. It is essential for growth after 2 years postnatally, but only promotes growth if sufficient nutrition is available. Growth hormone also exerts complex actions on metabolism (amino acid, fatty acid, glucose). It has insulin-like effects to promote amino acid uptake by liver and muscle and, therefore, promotes protein synthesis. However, if growth hormone is chronically increased, it has anti-insulin effects. It is one of the hormones that switches metabolism away from glucose use and towards increased oxidation of fat (e.g. in starvation)
- Control: secretion is increased via the hypothalamus by hypoglycaemia, stress, and exercise. Hypothalamic factors that regulate growth hormone release are growth hormone-releasing hormone (GHRH), which stimulates, and somatostatin, which inhibits, its release. Systemic control is via negative feedback by growth hormone at the hypothalamus.

Posterior pituitary

ADH (or vasopressin)
- Actions: increases water reabsorption by acting in collecting ducts of kidney (V2 receptor) and vascular pressor effects constricting peripheral arterioles and veins (V1)
- Control: osmotic—sensitive to 1% increase from normal plasma osmotic pressure of 285mOsm. Sensed by hypothalamic osmoreceptors. Also sensitive to decreases in blood volume or blood pressure.

Oxytocin
- Actions: causes contraction of uterine myometrium in childbirth and contraction of breast myoepithelium to eject milk
- Control: stretch of cervix/vagina during parturition (the Ferguson reflex); suckling—stimulation of the nipple causes the milk ejection reflex.

Pathology—pituitary tumours

(➔ *OHCM10* p.234.)
- Hormonal effects: hormone-secreting and non-secreting tumours—effects depend on cell type
- Mechanical effects: effects on vision via pressing on optic chiasm.

Pituitary adenomas

(➲ *OHCM10* p.234)

The pituitary gland is composed almost entirely of cells which make hormones. Thus, a benign tumour of this gland—an adenoma—will often make the same hormone as its cell of origin, but the production of that hormone will not be under the control of the hypothalamus and will usually be produced in excess. Thus, very small tumours in the pituitary, only a few millimetres in diameter, can have extensive effects on the rest of the body. Pituitary adenomas arise predominantly in the anterior pituitary, which constitutes ~80% of the pituitary volume.

Before the development of immunohistochemistry, the cells in the pituitary were labelled according to their staining properties with various dyes, but this produced confusing names that were not obviously linked to their function (e.g. chromophobe adenoma). These terms may appear in some literature but it is better to refer to the cells by their products, (e.g. growth hormone). The effects of pituitary adenomas will be specific to the hormones which they produce:

• Adenomas of growth hormone-producing cells:
  • Pituitary gigantism, if occurring before puberty when the epiphyseal plates are still open and long bones can grow
  • Acromegaly, if after puberty, with disproportionate growth of the bones in the jaw, hands, and feet
• Adenomas of ACTH-producing cells:
  • Cushing's syndrome (➲ p.632)
• Adenomas of PRH-producing cells:
  • Galactorrhoea, amenorrhoea, loss of libido, infertility
• Adenomas of TSH-producing cells:
  • A rare (1%) cause of hyperthyroidism.

Adenomas of the other hormone-producing cells (such as FSH- and LH-producing cells) in the anterior pituitary can occur, but they are much less common than those listed.

Sometimes pituitary adenomas do not produce hormones but they expand within the confined space of the sella turcica and cause pressure atrophy of the remaining pituitary with resultant deficiencies of all the pituitary hormones. This leads to end endocrine organ deficiencies such as hypothyroidism, hypoadrenalism, etc.

Since the pituitary gland is confined in the pituitary fossa, with the only space available for expansion being superior to this, structures above it may be compressed by pituitary adenomas. The optic chiasma lies immediately above the pituitary fossa, so a pituitary adenoma may cause pressure atrophy on this with a resultant defect in the lateral fields of vision—a bitemporal hemianopia.

Pituitary adenomas treatment by surgical removal (transsphenoidal) and drug therapy

- Dopamine agonists, e.g. bromocriptine, cabergoline (→ *OHPDT2* p.359)
- Act at D_2 receptors to inhibit prolactin secretion from prolactinomas. Also used to inhibit growth hormone secretion from growth hormone-secreting tumours. Results in tumour shrinkage
- Somatostatin analogues, e.g. octreotide, lanreotide (→ *OHPDT2* p.532)
- Act mainly on somatostatin receptors 2 and 5 to inhibit growth hormone secretion from growth hormone-secreting tumours and TSH secretion from TSH-secreting pituitary tumours.

The thyroid

Lies anterior to the 2nd–4th rings of the trachea, and the lateral lobes extend upwards on either side of the trachea and larynx (Fig. 9.3).

- Epithelial cells of the thyroid (follicular cells) are arranged into follicles around a lumen filled with colloid (Fig 9.3). The cuboidal follicular cells synthesize thyroglobulin, which is released into the colloid
- C cells (parafollicular cells), which release the peptide hormone calcitonin (◑ p.648), are located in the base of the follicle epithelium (Fig. 9.3).

Control of thyroid hormone production and secretion

See Fig. 9.4.

- Production requires iodide. A sodium/iodide symporter on the basal membrane of follicular cells traps and pumps in iodide from the plasma
- A thyroperoxidase enzyme on the apical plasmalemma oxidizes the iodide to iodine, iodinates tyrosyl residues in the thyroglobulin, and couples tyrosyl residues to produce the thyroid hormones T_4 and T_3 still bound in the thyroglobulin and, hence, inactive
- TSH stimulates endocytosis of colloid and its digestion by lysosomes, to free T_4 and T_3
- Iodine deficiency can prevent formation of T_4 or T_3, whereas excess iodine inhibits thyroid activity
- The main thyroid product (T_4) is not the metabolically active hormone. Metabolism of T_4 to produce active T_3 occurs primarily in the liver by type I (5′)-deiodinase.

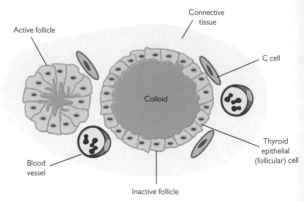

Fig. 9.3 The main features of the thyroid gland.

(a) Synthesis of iodinated thyroglobulin
(colloid)

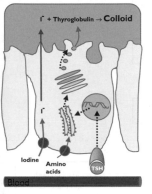

(b) Release of thyroid hormones

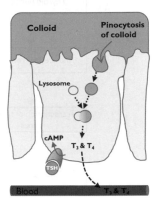

Fig. 9.4 The cellular processes involved in the synthesis and subsequent release of the thyroid hormones. Note that TSH stimulates both the synthesis of thyroglobulin (TG) and the secretion of T_3 and T_4.

Reproduced with permission from Pocock G and Richards CD (2006), *Human Physiology: The Basis of Medicine*, 3rd edn, p205 Oxford University Press.

Mechanism of action of thyroid hormones

See Fig. 9.5.
- Thyroid hormones are transported into cells and T_3 acts on nuclear receptors (TRs) which act on response elements (TREs) in gene promoters
- This interaction results in stimulation or inhibition of the production of many different mRNAs and, therefore, proteins. Sensitivity to T_3 is regulated via the number of TRs.

Effects of thyroid hormones

See Fig. 9.5.
- Thyroid hormones act on almost every tissue of the body but have little metabolic effect in the brain, spleen, or testis. They act to increase the basal metabolic rate, which increases oxygen use and heat production
- Stimulate production of Na^+/K^+ ATPase (which uses 20–45% of all ATP)
- Stimulate RNA polymerase I and II activity and, thereby, production of many proteins; stimulates protein degradation; when T_3 is excessive, degradation > production
- Potentiate β adrenoceptor effects, on glycogenolysis, gluconeogenesis; potentiate insulin effects, increasing glycogenesis and glucose usage
- Stimulate cholesterol breakdown and synthesis; increase number of LDL receptors on cell surface and enhances lipolysis
- Stimulate production of β adrenergic receptors and TRH receptors
- Affect cardiovascular system to increase cardiac output, rate, force, systolic blood pressure; but diastolic blood pressure falls because of vasodilatation
- Stimulate gut motility
- Stimulate bone turnover (breakdown > synthesis)
- Increase speed of muscle contraction.

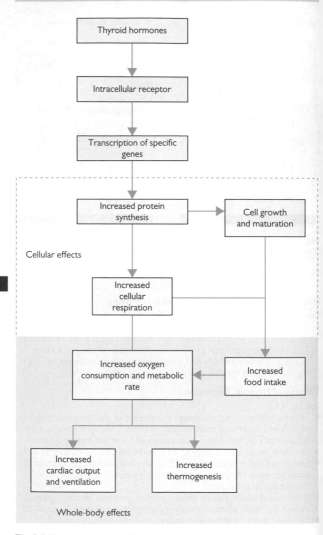

Fig. 9.5 The principal actions of the thyroid hormones.
Reproduced with permission from Pocock G and Richards CD (2006), *Human Physiology: The Basis of Medicine*, 3rd edn, p207 Oxford University Press.

Developmental effects
- Lungs: stimulate surfactant production and lung maturation (with glucocorticoids)
- CNS: essential for postnatal growth of CNS; stimulate production of myelin, neurotransmitters, axonal growth
- Bone: stimulate linear growth by effects on chondrocytes
- Stimulate normal development, maturation, and eruption of teeth, hair, epidermis.

Thyroid dysfunction

Goitre (swelling of the thyroid)

Causes

- Iodine deficiency
- Graves' disease
- A tumour—may be functioning (i.e. secreting T_4 and T_3) or non-functioning.

Hyperthyroidism (high T_3)

(❍ *OHCM10* p.218.)

Causes

- Graves' disease—autoantibody with stimulatory activity when it binds to the TSH receptor on thyroid cells:
  - Most common cause of hyperthyroidism
- Pituitary adenoma producing TSH
- Tumours of thyroid follicular cells can produce large amounts of T_4 and T_3
- Iatrogenic—over-administration of thyroxine.

Effects

- Increased basal metabolism rate—weight loss (despite increased appetite), increased resting heart rate and 'bounding' pulse, heat intolerance, increased sympathetic drive, eye protrusion
- Atrial fibrillation.

Hypothyroidism (low T_3)

(❍ *OHCM10* p.220.)

Causes

- Deficiency of iodine in diet—rare since introduction of iodinized table salt and greater distribution of seafood
- Hashimoto's thyroiditis—organ-specific autoimmune disease with destruction of thyroid epithelial cells
- Pituitary hypofunction with lack of production of TSH—non-functioning pituitary adenoma, Sheehan's syndrome
- Iatrogenic—surgical removal of thyroid, damage of thyroid by radioactive iodine, antithyroid drugs
- Thyroid hormone resistance—a number of genetic defects in the thyroid hormone receptor reduce hormone binding. The individual is hypothyroid but will have normal levels of plasma hormone.

Effects

- In the neonate—cretinism leads to gross deficits in CNS myelination and stunting of postnatal growth
- In the adult: myxoedema—decreased basal metabolic rate (tiredness, lethargy, weight gain), slow mentation, hypothermia, constipation.

Thyroid dysfunction

See Box 9.1.

**Box 9.1 Treatment and drug therapies
for thyroid dysfunction**

(⮒ *OHPDT2* pp.478, 482.)

- Thyroid carcinoma—thyroidectomy, radioiodine therapy (^{131}I)
- Hyperthyroidism—antithyroid drugs carbimazole or propylthiouracil to restore euthyroid state and radioiodine treatment to ablate thyroid or thyroidectomy
- Hypothyroidism—thyroid hormone replacement with T_4.

The adrenal gland

The adrenal gland comprises an inner medulla that secretes the catecholamines, noradrenaline and adrenaline, and an outer cortex that secretes steroid hormones. The adrenal glands are located just medial to the upper pole of each kidney.

Structure

The adrenal medulla is made up of chromaffin cells packed with granules which store large amounts of adrenaline and noradrenaline. The adrenal cortex is made up of sheets of cells surrounded by capillaries and arranged in three zones: the outer zona glomerulosa, which makes aldosterone; middle zona fasciculata, which makes cortisol; and inner zona reticularis, which makes small amounts of androgens (Fig. 9.6).

Adrenal medulla

- Actions: preparation for emergency physical activity. The adrenal medulla contributes 10% of the total sympathetic nervous system response to stress and so, thus, is not vital ($\bigodot$ p.652)
- Receptors: adrenaline and noradrenaline act at adrenergic receptors ($\bigodot$ p.251)
- Stimuli: any stressful stimuli which activate the sympathetic nervous system (e.g. low blood pressure, haemorrhage, pain, low blood glucose, exercise, surgery, asphyxia)
- Pathology: tumours of the adrenal medulla (phaeochromocytoma) constantly secrete catecholamines causing hypertension, tremor, anxiety, forceful heartbeat.

Adrenal cortex

($\bigodot$ OHCM10 p.224.)

Cortisol

- Actions: provides protection of the body in prolonged stress—primarily to preserve glucose for the brain. Exerts widespread actions on many tissues ($\bigodot$ p.654)
- Receptors: glucocorticoid receptors (GRs) are present in almost all cells. GRs are located in the cytoplasm of cells and migrate to the nucleus to regulate gene transcription when cortisol binds
- Control of output
- Abnormal function:
  - Cortisol insufficiency—Addison's disease ($\bigodot$ OHCM10 p.226)
  - Cortisol excess—Cushing's syndrome ($\bigodot$ OHCM10 p.224).

Aldosterone (regulation of body sodium and fluid volume)

- Receptors: mineralocorticoid receptors (MRs) are present in the nuclei of only a few cell types—kidney collecting tubule epithelia, salivary and sweat glands
- Actions: in the kidney, aldosterone regulates ion transport in the kidney collecting tubules in order to stimulate reabsorption of Na^+ in exchange for secretion of K^+, H^+, NH_4^+. There is a 2h lag in the response to aldosterone as MR effects are via stimulating transcription of the Na^+/K^+-ATPase protein. In salivary and sweat glands, aldosterone regulates ion transport to retain sodium

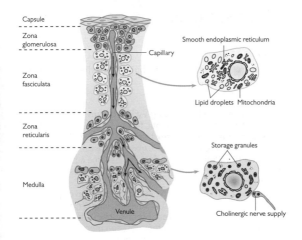

Fig. 9.6 Diagram of the adrenal gland, showing the organisation of the cortex (composed of three layers: zona glomerulosa (secreting aldosterone), zona fasciculata (cortisol), zona reticularis (androgens) and the medulla (adrenaline, noradernaline). The cellular structure of a steroid secreting cell from the zona fasciculata and a catecholamine-secreting chromaffin cell from the medulla are shown in more detail.

Reproduced with permission from Pocock G and Richards CD (2006), *Human Physiology: The Basis of Medicine*, 3rd edn, p211 Oxford University Press.

• Control of output: the renin–angiotensin system (➔ p.650)
• Abnormal function:
  • Hypoaldosteronism (in adrenal failure) results in sodium loss, low blood volume, and low blood pressure
  • Hyperaldosteronism (Conn's syndrome; ➔ *OHCM10* p.228) results in excess sodium retention, water retention, and increased blood pressure. Spironolactone (an aldosterone antagonist) is used as an antihypertensive.

Adrenal androgens (DHEA)
DHEA (dehydroepiandrosterone) is produced and released from the adrenal cortex zona reticularis. DHEA is a weak androgen which is a very minor component of adrenal secretions.

Adrenal insufficiency
ACTH used in diagnosis of adrenal insufficiency (Addison's disease) (*OHPDT2* pp.496–7).

Cushing's syndrome

(→ *OHCM10* p.224.)

Definition

Excess glucocorticoid hormones in the body, from whatever source.

Causes

See Fig. 9.7.
- Pituitary adenoma producing excess ACTH because it no longer responds to the normal homeostatic feedback loop. The excess ACTH then stimulates the adrenal cortex to produce excess glucocorticoid hormones. This is also known as Cushing's disease
- Adenoma of the adrenal cortex which has become autonomous from the pituitary–adrenal feedback loop and produces excess glucocorticoids
- Excess ACTH administered by the medical profession or produced endogenously by a tumour (such as small cell lung cancer)
- Excess glucocorticoid steroids (e.g. prednisolone) administered by the medical profession. This was the most common mechanism for a few decades, since steroids were powerful drugs that could control severe allergic and inflammatory processes such as asthma or rheumatoid arthritis. Fortunately, more specific drugs are now available for some conditions, (e.g. asthma) and in others, 'steroid-sparing' immunosuppressants, (e.g. azathioprine) are used to reduce the risk or magnitude of Cushing's syndrome.

Complications of Cushing's syndrome

- Diabetes mellitus
- Proximal muscle weakness
- Decreased immunity to infections
- Osteoporosis
- Truncal obesity
- Hirsutism
- Depression, psychosis.

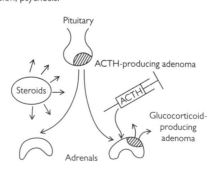

Fig. 9.7 Mechanisms of Cushing's syndrome. Arrows indicate either an excess of ACTH or glucocorticoids.

For treatments for Cushing's syndrome, see Box 9.2.

Box 9.2 Treatments and drug therapies for Cushing's syndrome

- In the case of adrenal or pituitary adenomas, surgical removal of the adenoma is the preferred option. The impending loss of endogenous steroids following surgery is overcome by replacement therapy (hydrocortisone or prednisolone). In the event that surgery is not possible, cortisol synthesis inhibitors can be prescribed, but they have weak efficacy
- In the event that over-prescription of steroids is responsible for the effect, the treatment simply involves slowly reducing doses until they can eventually be removed altogether
- As well as treating the cause, a range of therapies might be required to treat the various consequences (diabetes, osteoporosis, mental disorders) that are associated with the syndrome.

Hyperaldosteronism

(→ OHCM10 p.228.)

Definition

Excess production of aldosterone hormone by the adrenal cortex.

Causes

- Primary:
  - Adrenal cortical adenoma with autonomous production of aldosterone
  - Primary adrenal cortical hyperplasia
- Secondary—normal response to activation of the renin–angiotensin system:
  - Congestive heart failure
  - Decreased renal perfusion (e.g. renal artery stenosis), so the juxtaglomerular apparatus senses a lack of perfusion and produces more renin
  - Hypoalbuminaemia—and, thus, reduced osmotic pressure within the vascular system, so reduced plasma volume and reduced renal perfusion
  - Pregnancy—oestrogen induces increased plasma renin substrate.

Consequences

- Sodium and water retention with potassium excretion → hypertension and hypokalaemia
- In primary hyperaldosteronism, renin levels will be low (useful in the diagnostic process).

For treatments for hyperaldosteronism, see Box 9.3.

> **Box 9.3 Treatments and drug therapies for hyperaldosteronism**
>
> (→ OHPDT2 Ch. 2.)
> - Spironolactone is an aldosterone receptor inhibitor, acting primarily on receptors in the collecting duct, with the result of increased sodium excretion in the urine and increased potassium retention.

The endocrine pancreas

Structure—the islets of Langerhans

- The islets are diffusely distributed throughout the pancreas
- Islets comprise insulin cells (β cells; Fig. 9.8), glucagon cells (α cells), somatostatin cells (δ cells), and pancreatic polypeptide cells
- The endocrine pancreas regulates the availability of metabolic substrates.

Plasma glucose concentrations

The 'normal' morning fasting blood glucose concentration is 4–5mM and rises transiently up to ~8mM after a meal.

- Hypoglycaemia (<3mM) causes dizziness, confusion, hunger, convulsions, coma; sympathetic activation
- Hyperglycaemia >8mM glucose causes osmotic effects (if >10mM glucose is lost in urine with water; excess urine production causes thirst in individuals with diabetes), abnormal glycation of proteins, increased urinary tract infections (→ p.640).

Insulin

Synthesis

Insulin is a protein hormone from the prohormone, proinsulin, cleaved to insulin and C peptide in secretory granules. Insulin comprises two peptide chains linked by two S–S bonds; stored with zinc in granules.

Mechanism of insulin secretion

See Fig. 9.8.

- Glucose enters the β islet cell (diffusion is facilitated by GLUT2)
- Metabolism of the glucose in the β cell produces ATP
- ATP closes ATP-dependent potassium channels present in the β islet cell membrane, which in turn depolarizes the β islet cell

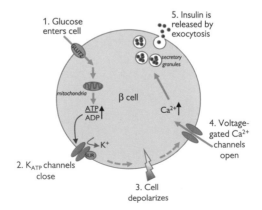

Fig. 9.8 Mechanism by which raised plasma glucose stimulates the release of insulin from β cells in the pancreatic islets of Langerhans.

- Depolarization causes Ca^{2+} entry; this in turn stimulates exocytosis of insulin
- Sulphonylurea drugs inhibit the ATP-dependent K^+ channels and promote insulin release.

Mechanism of insulin action
- The insulin receptor is a tyrosine kinase-linked receptor
- Phosphorylation of insulin receptor substrate (IRS) proteins occurs which activates intracellular signalling cascades.

Insulin promotes anabolism and controls use of metabolic substrates
- Acts to lower elevated plasma glucose
- Controls transport of glucose and amino acids into many types of cell
- Directs the use of glucose and amino acids and augments their oxidation to ATP
- Increases protein synthesis and inhibits proteolysis
- Supports growth and proliferation of many cell types.

Specific actions
(➲ p.182.)

Pathology
- Diabetes mellitus type 1 (formerly insulin-dependent diabetes (IDDM))—no insulin secretion as result of autoimmune destruction of β islet cells. Treatment with insulin (➲ *OHCM10* p.209)
- Diabetes mellitus type 2 (maturity onset, obesity-associated, formerly non-insulin dependent (NIDDM)) (➲ *OHCM10* p.206):
  - Early stages: peripheral resistance to insulin as down-regulation of insulin receptors—leads to increased plasma insulin; disordered insulin secretion:
    - Treatment: regulation of diet (no sugar-rich foods), weight loss, administration of metformin to regulate metabolism, stimulation of insulin release by sulphonylurea drugs which bind to and inhibit the K^+-ATP channel subunit
  - Later stages: pancreatic amyloid formation and islet destruction occurs:
    - Insulin treatment may be necessary
- Insulinoma—unregulated hypersecretion of insulin by tumour of β islet cells.

Glucagon
Peptide hormone that protects against a lack of metabolic substrates.
- Synthesis: produced as prohormone, proglucagon, then cleaved to glucagons in secretory granules
- Produced: by α cells of the pancreatic islets
- Release: increased in response to low plasma glucose. The mechanism of release is poorly understood. α cells do have ATP-sensitive K^+ channels; fall in insulin inhibition
- Actions: virtually all in the liver (via G_s, cAMP); protects against hypoglycaemia:
  - When low plasma glucose
  - When high amounts of protein ingested or when high demand for glucose (e.g. exercise).

Effects in the liver depend on the concentration of glucagon in the plasma
- Low amount of glucagon—stimulates glycogenolysis
- Medium amount of glucagon—stimulates gluconeogenesis
- High secretion of glucagon—stimulates lipolysis, fatty acid oxidation, ketogenesis.

NB: synergism with other hormones involved in metabolic control—catecholamines, glucocorticoids, and growth hormone all stimulate liver conversion of glycogen to glucose (Fig. 9.9).

Somatostatin
Peptide hormone made in δ cells of pancreas.
- Actions: inhibits the secretion of both insulin and glucagons (paracrine action); role in the physiology of the islets is as yet uncertain.

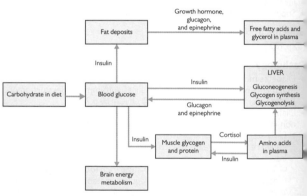

Fig. 9.9 An overview of the hormonal regulation of plasma glucose concentration.

Diabetes mellitus

(➜ *OHCM10* pp.206–13.)

Diabetes mellitus is, as yet, an incurable condition that is diagnosed as a chronic increase in blood glucose levels—hyperglycaemia. The condition is broadly divided into two types, depending on the underlying cause:

* Type 1 diabetes (➜ *OHCM10* p.206) is caused by an inability to synthesize sufficient insulin—the hormone responsible for stimulating uptake of glucose into cells. Insulin is usually synthesized in β cells of the islets of Langerhans in the pancreas, but they are destroyed by an autoimmune response, usually early in the life of susceptible individuals (onset typically occurs before children reach 10 years of age, but can occur much later). The trigger for this response is believed to be an environmental stimulus in individuals with a genetic predisposition to the condition. Although the child of a parent with type 1 diabetes is at increased risk of developing the disease, the risk is relatively small (<2% if the mother is affected, <6% if the father is affected) unless both parents have the disease, in which case genetic counselling should be sought

* Type 2 diabetes (➜ *OHCM10* p.206) includes a wide range of disorders that develop over many years, often later in life, ultimately → hyperglycaemia. In cases where a specific cause can be identified, reduced secretion of insulin or a reduction in the effectiveness of insulin to facilitate uptake of glucose into cells (so-called insulin resistance) is implicated. However, up to 98% of cases are 'idiopathic', meaning that no specific cause has been identified.

There is a far clearer genetic link to type 2 than to type 1 diabetes, with people of Asian or African Caribbean ethnic origin, and those with a family history of diabetes or gestational diabetes at increased risk of developing the disease. Furthermore, there are a number of rare inherited diseases including MODY (dominantly inherited type 2 diabetes), mitochondrial diabetes, and insulin-resistant diabetes (due to genetic defects in the insulin receptor (type A) or autoimmune destruction of the receptor (type B)).

An increasingly prominent risk factor for diabetes is obesity, probably due to dysregulation of insulin receptor signalling. The rise in prevalence of obesity in Western countries is thought to be responsible for the emergence of childhood type 2 diabetes.

Gestational diabetes is a specific term relating to pregnant women who are diagnosed with hyperglycaemia during routine plasma glucose tests at ~28 weeks of gestation. Blood glucose levels typically return to normal within 6 weeks of birth, but the condition is indicative of a predisposition to type 2 diabetes later in life for both mother and child. Babies of mothers with gestational diabetes are often born overweight because of the growth stimulating effects of increased foetal insulin secretion in response to high glucose levels derived from the mother.

Complications of diabetes

The major complications associated with persistent hyperglycaemia relate to the cardiovascular system (➜ *OHCM10* pp.210–13). Broadly, the effect of diabetes on the cardiovascular system can be divided into microvascular complications in the eyes, kidneys, and nerves, and complications pertaining to the major arteries, including coronary artery disease, stroke, and peripheral vascular disease. Diabetes-induced hypertension and advanced glycated end-products together with oxidative stress and inflammation are crucial in vascular disease. The precise mechanism by which persistent hyperglycaemia induces hypertension has yet to be fully elucidated, but increased generation of reactive oxygen species and the associated dysfunction of protective endothelial effects are believed to be important.

Diabetic ketoacidosis (DKA)

DKA is a potentially dangerous complication of diabetes (usually type 1 diabetes, but occasionally in advanced type 2, when insulin production is low). The condition is driven by a lack of glucose entering cells via insulin-modulated GLUT transporters; in the absence of carbohydrates to metabolize, cells turn to fatty acids as a primary energy source. The by-products of high levels of fatty acid metabolism are ketones, which can accumulate to reduce the pH of the blood on account of their acidity overwhelming buffering capacity. While normally associated with hyperglycaemia, euglycaemic ketoacidosis paradoxically can occur in people with persistent hypoglycaemia on account of insufficient carbohydrate and/or insufficient insulin—the underpinning mechanism of fatty acid metabolism → ketone production is nevertheless responsible for the ketoacidotic outcome.

Symptoms:
• 'Pear drops' breath odour
• Vomiting
• Dehydration
• Heavy breathing
• Tachycardia
• Confusion
• Coma (if untreated).

Diagnosis and treatment of DKA

DKA is diagnosed by measurement of ketones in the blood or urine. It is treated by replacing body fluids lost through increased urinary output and by injecting insulin. This is a dangerous condition that can cause death if not treated rapidly.

Diagnosis

Diabetes is notoriously difficult to diagnose because early symptoms are varied (● *OHCM10* pp.206–7):
• Patients generally present having suffered weight loss, but other symptoms include tiredness, dry mouth, ketoacidosis
• More advanced cases may present with foot or leg ulcers and sepsis.

Routine screening of patients' blood and urine has improved diagnosis—the following criteria apply for diagnosis of diabetes:
• Symptoms as previously described and
• Random venous plasma glucose ≥11.1 mmol L^{-1} or
• Fasting plasma glucose ≥7.0 mmol L^{-1} or
• Fasting whole blood glucose ≥6.1 mmol L^{-1} or

- Two-hour plasma glucose ≥11.1 mmol after glucose load in glucose tolerance test (see following topic)
- For adult individuals suspected of type 2 diabetes only, glycated haemoglobin in venous blood is now the gold standard diagnostic test in the UK (see 'Haemoglobin A1c (HbA1c)' later in this topic).

Oral glucose tolerance test (OGTT)

OGTT is a confirmatory test used in diagnosis of diabetes, but is more commonly used to detect impaired glucose tolerance, which can be a fore runner for diabetes (i.e. 'pre-diabetes'—see later in this topic). The OGTT consists of the following:

- Individuals fast for 12h, after which a baseline blood sample is taken
- Individuals are then given an oral load of 75g of glucose; a second sample is taken 2h after the oral glucose load
- Generally accepted limits for diagnosis of diabetic patients are given in Table 9.1 and patients whose glucose levels are elevated but do not exceed the threshold for full-blown diabetes are said to have impaired glucose tolerance.

Ultimately, diabetes is an incurable condition; treatments aim to improve the lifestyle of patients and to reduce the risk of serious diabetes-related conditions (Box 9.4).

Haemoglobin A1c (HbA1c)

HbA1c is an advanced glycated end-product that increases under conditions of hyperglycaemia. HbA1c is increasingly being used in the diagnosis of type 2 diabetes, unless contraindicated: the threshold for diagnosis is 48mmol mol^{-1} (6.5%).

HbA1c is also the primary method used to monitor glucose control in patients with type 2 diabetes and the level of this marker is associated with disease severity, prognosis, and outcome.

Table 9.1 Limits for diagnosis of impaired glucose tolerance and diabetes mellitus

	Glucose concentration (mmol L^{-1})		
	Venous plasma	Venous whole blood	Capillary whole blood
Impaired glucose tolerance			
Fasting sample	<7.0	<6.1	<6.1
2h sample after glucose load	≥7.8<11.1	≥6.7<10.0	≥7.8<11.1
Diabetes mellitus			
Fasting sample	≥7.0	≥6.1	≥6.1
2h sample after glucose load	≥11.1	≥10.0	≥11.1

1. A 'casual' plasma sample (i.e. taken without regard for time of last meal) of ≥11.1mmol L^{-1} can also be indicative of diabetes but usually needs to be confirmed by a glucose tolerance test or by repeating on another day.

2. In the absence of symptoms, a positive glucose tolerance test might have to be supported by elevated glucose levels at a second time point (e.g. ≥11.1mmol L^{-1} at 1h).

Box 9.4 Treatment and drug therapies for diabetes

(➲ OHCM10 pp.208–9.)

- Impaired glucose tolerance and mild gestational diabetes are often treated through changes in diet and lifestyle, in an effort to avoid exacerbating the problem through high carbohydrate intake and weight gain and to slow the progression of insulin resistance
- The diet and exercise approach is recognized to be more successful the earlier the diagnosis is made, but in most cases this approach has to be supplemented by drug therapies as the disease progresses
- Metformin is often used to slow carbohydrate absorption and reduce insulin resistance, while sulphonylureas increase insulin secretion
- The glitazones are peroxisome proliferator-activated receptor G (PPAR-G) activators that reduce insulin resistance and can be used alongside either metformin or a sulphonylurea
- Incretin mimetics reproduce the effects of endogenous incretins— peptides (e.g. glucagon-like peptide (GLP-1)) that reduce blood glucose by stimulating insulin release and reducing glucagon release. These drugs also reduce the rate of stomach emptying, increasing satiety and slowing glucose absorption into the bloodstream. These require injection
- Dipeptidyl peptidase-4 (DPP-4) inhibitors reduce the effect of the glycoprotein enzyme DPP-4, which normally metabolises proline-rich peptides, including incretins (GLP-1—see earlier in list). Inhibition of DPP-4 therefore increase the persistence of GLP-1, inducing insulin release and inhibiting glucagon release
- Sodium-glucose transporter 2 inhibitors (SGLT-2i; gliflozins) inhibit SGLT-2—a cotransporter central to glucose reabsorption in the proximal tubule of the kidney. Inhibition of SGLT-2 increases urine excretion, reducing blood glucose. These drugs also appear to reduce blood pressure and weight via unexplained mechanisms. Side effects include increased risk of genital infections probably because of the increase in urinary glucose
- Ultimately, however, patients can progress to requiring insulin injections to control their blood glucose. The peptide nature of insulin prevents the use of an oral preparation, requiring patients to inject subcutaneously on a daily basis
- Insulin therapy is the only treatment available to patients with type 1 diabetes. The inherent danger of this form of treatment is that, should a patient fail to eat after injecting, blood glucose can fall fatally low (hypoglycaemia). Insulin pumps are an alternative to multiple daily injections and are increasing in popularity.

Pre-diabetes

Pre-diabetes is a term commonly used to describe individuals who have indications of hyperglycaemia (e.g. impaired glucose tolerance, as indicated by an OGTT, or any other diagnostic marker that is just below the recognized threshold for diagnosis). HbA1C of 42–47 mmol mol^{-1} can signal 'pre-diabetes' and is likely to be monitored on an annual basis. The HbA1c range recognized to indicate pre-diabetes can differ in different countries.

Gastrointestinal hormones

- Functions: GI hormones, with the enteric and autonomic nervous systems, integrate and coordinate the mechanisms which move, digest, and absorb the various meals that are ingested. They control GI tract exocrine and endocrine secretion, motility, growth, and blood flow
- Routes: hormones released from gut endocrine cells act via endocrine, paracrine, neurocrine routes, and possibly also via the gut lumen. Peptides are also released from nerves of the enteric nervous system
- Gut endocrine cells are part of the GI tract epithelium, variably positioned in crypts or villi; the hormone is secreted basally; most cells have sensory microvilli on an apex open to the gut lumen
- Most gut hormones are peptides produced from larger precursors, so different molecular forms of the hormones are found.

Gastrin

- Distribution: G cells of gastric antrum crypts; some duodenal cells
- Synthesis: produced as prohormone preprogastrin, cleaved to progastrin and, in turn, to gastrin
- Stimuli:
  - Luminal protein digestion products (especially tryptophan, phenylalanine; also calcium, beer, wine, coffee)
  - Vagus, via acetylcholine (M_3 receptors) and gastrin-releasing peptide
  - Distension of stomach
  - Hypercalcaemia (as in hyperparathyroidism)
- Inhibitors:
  - Inhibited by stomach pH <2.5; alkali short term has little effect; long term causes hyperplasia
  - Inhibited by somatostatin (local negative feedback)
- Actions:
  - Stimulates gastric acid secretion (direct and indirect, via histamine H_2 receptor); stimulates parietal cell growth
  - Stimulates pepsin secretion (also water and electrolyte secretion in liver, pancreas, intestine)
  - Stimulates antral motility, mucosal blood flow
  - Trophic to body of stomach
- Pathology: gastrinoma (usually in pancreas) causes repeated peptic ulceration due to high acid and pepsin secretion.

Histamine

- Distribution: enterochromaffin-like cells of stomach wall
- Synthesis: from histidine by histidine decarboxylase
- Stimuli: vagal stimulation, gastrin
- Actions: local paracrine action stimulates gastric acid (HCl) secretion by parietal cells via H_2 receptors (hence the use of H_2 receptor antagonists for treatment of peptic ulcerations).

Secretin

Distribution: S cells, from duodenum to distal ileum; in neck region of intestinal glands

Stimuli: acid in proximal duodenal causing lumen pH <4.5; inhibited by somatostatin

Actions:

- Stimulates pancreatic secretion of HCO_3^- and water—this washes pancreatic enzymes into the gut
- Stimulates liver secretion of HCO_3^- and water into bile:
  - Potentiates CCK; calcitonin, and PTH secretion.

Cholecystokinin

Distribution: I cells in the duodenum and jejunum

Stimuli: protein and fat digestion products in duodenum

Actions:

- Stimulates secretion of pancreatic enzymes and potentiates action of secretin
- Stimulates contraction of gallbladder:
  - Inhibits gastric emptying; increases small bowel transit
  - High levels potentiate secretion of calcitonin
- In CNS, CCK is linked to satiety (the feeling that one has had enough to eat).

Hormones influencing calcium, phosphate, and bone

Ca^{2+} has very important extracellular effects on excitable tissues and blood clotting; it is also an essential component of bone.

- Hypocalcaemia (→ OHCM10 p.678) (i.e. total Ca^{2+} <1.2–1.5mM)—very dangerous:
  - Increases neuronal membrane excitability by increasing sodium permeability causing tetany (involuntary nerve-induced spasm of skeletal muscles)
  - Heart QT is increased ('prolonged QT') and heart failure may occur
- Hypercalcaemia (→ OHCM10 p.676)—only dangerous in the long term
  - Decreases neurone excitability
  - Calcium salts are rather insoluble, so urinary stones and tissue calcification occurs
  - Attempts to secrete excess calcium via urine causes polyuria, which leads to dehydration and exacerbates the problem.

A fall in plasma Ca^{2+} is therefore more dangerous, short term, than a rise; endocrine control mechanisms reflect this.

Amounts of calcium in the body

(→ OHCM10 p.676.)
- The main controlled parameter is plasma-ionized Ca^{2+}—1.2mM. Plasma contains 2.5mM total (2.4–2.6mM); ~50% is ionized
- Total body calcium: normal adult contains 25 moles (1kg) of calcium
- Bone calcium: 1% (250mmol) is rapidly exchangeable; 99% is hydroxyapatite bound to collagen—slowly exchangeable. Interstitial fluid calcium is freely exchangeable with plasma calcium.

Phosphorus

- In the skeleton, 85% as hydroxyapatite; 15% in soft tissues
- Plasma: phosphate 0.2mM; varies widely with age, sex, and diet.

Hormone control of calcium and phosphate

PTH

(→ OHCM10 pp.222, 676.)

Principal control; essential for life (Fig. 9.10).
- Effects:
  - PTH, a peptide hormone, is secreted from the parathyroid glands in response to falling plasma Ca^{2+}
  - PTH restores low plasma Ca^{2+} to normal and causes increased phosphate loss
  - Kidney—PTH inhibits sodium–phosphate co-transport and phosphate reabsorption in the kidney; phosphate loss causes an increase in Ca^{2+} mobilization:
    - Increases Ca^{2+} reabsorption in the distal kidney tubule and collecting duct (independent of Na^+)
    - Increases activity of 1α-hydroxylase and production of the active vitamin D metabolite $1,25(OH)_2D_3$ (calcitriol)

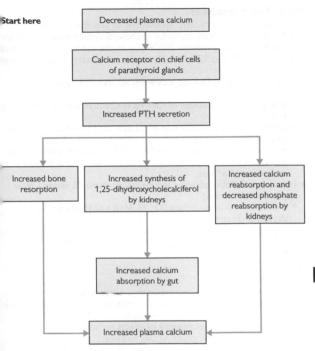

Start here

Decreased plasma calcium

↓

Calcium receptor on chief cells
of parathyroid glands

↓

Increased PTH secretion

Increased bone
resorption

Increased synthesis of
1,25-dihydroxycholecalciferol
by kidneys

Increased calcium
reabsorption and
decreased phosphate
reabsorption by
kidneys

Increased calcium
absorption by gut

Increased plasma calcium

Fig. 9.10 A flow diagram summarizing the principal actions of PTH.

- Bone: normal intermittent secretion of PTH is necessary for healthy bone growth and remodelling
- Rapid effects: stimulates Ca^{2+} flux from bone across osteoblasts which lay down new bone and line the bone surface; PTH reorganizes osteoblasts so that they become separated to allow: calcium efflux from matrix; and access of osteoclasts (which break down bone). In the long term, osteoclasts are activated via osteoblasts (protein synthesis needed) which break down bone; excess PTH limits growth of osteoblasts and bone matrix synthesis, causes destruction of bone

Pathology—Ca^{2+} can be disturbed, especially when PTH is insufficient:
- Deficiency: hypoparathyroidism (⊕ *OHCM10* p.222)—low plasma Ca^{2+}, tetany; pseudohypoparathyroidism—resistance to PTH due to receptor defect (PTH levels are normal or increased)
- Excess hyperparathyroidism (tumours) (⊕ *OHCM10* p.222)—raised plasma Ca^{2+}, bone destruction, urinary stones, sluggish CNS.

Vitamin D and its metabolites

Increases whole-body calcium. Synthesized in skin from cholesterol. UV light converts a cholesterol derivative to cholecalciferol = vitamin D_3. In the liver, 25-hydroxylase produces $25(OH)D_3$. In the kidney, 1α-hydroxylase converts $25(OH)D_3$ to $1,25(OH)_2D_3$. Also some dietary intake of D_3 (egg yolks, fish oils).

- Effects of $1,25(OH)_2D_3$—via nuclear receptors which regulate transcription:
  - Intestine: increases uptake of Ca^{2+} via synthesis of the calcium binding protein, calbindin
  - Kidney: facilitates conservation of calcium and phosphate for growth and repair
  - Bone: necessary for the action of PTH; inhibits synthesis of collagen by osteoblasts; action on osteoclasts to increase bone breakdown and increase Ca^{2+} loss
- Pathology—in both deficiency and excess of vitamin D, serum Ca^{2+} is approximately normal:
  - Deficiency: children—rickets (⊃ OHCM10 p.684); renal rickets; vitamin D-resistant rickets; adults—osteomalacia. Increasingly implicated in a range of chronic diseases (e.g. MS, diabetes)
  - Excess—vitamin D poisoning which leads to symptoms of hypercalcaemia.

Calcitonin

(⊃ OHCM10 p.676.)

Principal function is to prevent hypercalcaemia and excessive bone breakdown. Peptide hormone produced by C cells of thyroid and secreted in response to serum Ca^{2+} above normal range.

- Effects—prevents hypercalcaemia by effects on bone. Causes acute reduction in plasma phosphate, uptake to bone:
  - Bone: decreases activity of osteoclasts; inhibits mineral/matrix reabsorption; greatest effect when bone resorption is rapid (e.g. in the young)
  - Gut:
    - Meal related—helps to control rise in plasma Ca^{2+} due to absorption, also inhibits absorption
    - Pregnancy/lactation related—protection against demands of foetus/infant
  - Kidney and intestine: pharmacological doses affect ion fluxes
- Pathology:
  - Deficiency: compensation by changes in PTH
  - Excess: uncontrolled secretion from 'medullary' carcinoma of thyroid (⊃ OHCM10 p.600).

Other sites of hormone production

Adipose tissue: leptin

- Production: produced in fat cells throughout the body
- Stimuli: constitutive release parallels the accumulation of fat; correlates with body mass index; reflects the availability of metabolic substrates
- Actions:
  - Acts on receptors in basal hypothalamus; inhibits neuropeptide Y action in hypothalamus and influences many other neuro-transmitters involved in feeding behaviour
  - Decreases appetite (signals satiety)
  - Increases energy expenditure, sympathetic activity
  - Decreases insulin secretion
  - Mutations in leptin (ob) protein or receptor are a rare cause of human gross obesity; leptin resistance may be involved in human obesity
  - Effects on reproductive system, signalling an adequate reserve of metabolic fuels for reproduction
  - Probably evolved to protect against starvation ('thrifty genotype').

Heart: atrial natriuretic peptide

- Production: produced by atrial myocytes
- Stimulus: atrial dilatation (i.e. increased venous return, right heart failure; ➔ OHCM10 p.137)
- Actions:
  - Stimulates loss of sodium (and water) in urine, probably by effect on glomeruli
  - Inhibits renin–angiotensin–aldosterone system
  - Reduces blood pressure (reduces venous return and depresses cardiac output).

Kidneys: erythropoietin

- Production: glycoprotein produced by epithelial cells of glomeruli
- Stimuli: reduced oxygen saturation of the blood; androgens, β-adrenergic stimulation
- Actions: stimulates production of erythrocytes in the bone marrow.

Renin–angiotensin system

Renin is an acid protease produced by juxtaglomerular cells in the afferent arterioles of glomeruli, in response to sodium depletion, hypotension, dehydration, poor renal artery blood flow, sympathetic stimulation. This cleaves plasma angiotensinogen to angiotensin I, which is then cleaved by angiotensin-converting enzyme (in the lungs) to the active angiotensin II which stimulates aldosterone secretion, thirst, and vasoconstriction.

The stress response

Stress

Any change/event that either disrupts, or threatens to disrupt, homeostasis to an unusual degree.

Stressor

Any severe disturbance.

Acute stressors: extreme heat/cold, blood volume depletion by heavy bleeding, dehydration, hypoglycaemia, pain, surgical procedures, toxins from a bacterial infection, severe exercise, sleep deprivation

Chronic stressors: may be obvious or more subtle, e.g. chronic infection, chronic pain, housing problems, marital problems, financial worries, difficulties at work, commuting.

The stress response

The stress response is, at least in the short term, counter-homeostatic. It raises blood pressure, blood sugar, ventilation rate, etc.

The purpose of these changes is to prepare the body to meet an emergency situation. It has evolved in order to allow the individual to survive the emergency, and return to normal homeostasis when the stress is no longer present

It involves a short-term alarm reaction; and a more long-lasting resistance reaction.

The hypothalamus controls and coordinates the stress response via its actions on the autonomic nervous system and the endocrine systems. It receives inputs from:

The brainstem, e.g. from nucleus tractus solitarius; non-specific from raphe, locus coeruleus

Higher centres fornix from hippocampus; amygdala; orbitofrontal and septal cortex; dorsomedial and midline thalamic nuclei convey potentially stressful information from the external world, the internal organs, and the 'psyche'.

The acute alarm reaction

The acute 'alarm reaction' (fight-or-flight) involves hypothalamic control b
activation of:
- CNS outputs
- The sympathetic nervous system and adrenal medulla.

The responses are immediate and mobilize the body's resources for imme
diate physical activity, and cause arousal of the cerebral cortex.

CNS outputs

- Increased respiratory rate/depth; increased cardiac output via
  cardiovascular centre and autonomic nervous system
- Secretion of CRH to activate the pituitary adrenal axis
- Secretion of ADH to conserve body water
- Arouses the cerebral cortex by stimulation of the locus coeruleus and
  widespread central release of noradrenaline
- Blunts pain by release of endorphins and enkephalins.

Sympathetic activity

(➲ pp.251–3.)
  Increased sympathetic activity causes changes which:
- Increase the circulation
- Increase availability of energy substrates (promote catabolism)
- Decrease non-essential activities.

These changes include:
- Increased heart rate, force of heart contraction—increases supply of
  O_2 and substrates to where they are most needed
- Vasodilatation of muscle vascular beds—preparation for intense
  physical exertion
- Vasodilation of the adrenal vasculature
- Vasoconstriction of visceral and skin arteries and veins—redistributes
  blood away from organs that do not have an immediate role;
  contraction of veins reduces pooling of blood:
  - NB: no constriction of heart, lung, cerebral blood flow
- Dilation of respiratory tract by relaxing smooth muscle—permits
  greater O_2 and CO_2 exchange
- Increased sweat production—'cold sweat', but then improves heat loss
  if physical exertion
- Decreased secretion from the GI tract; decreased GI motility, closes
  sphincters
- Increased glycogenolysis—from adrenaline (β-adrenoreceptors);
  reduction of insulin and increase of glucagon secretion
- Decreased insulin causes reduced use of glucose by muscle and fat,
  conserving glucose for the brain (muscle runs on glycogen, free fatty acids
- Pupil dilation, eyelid retraction, accommodation for distant vision
- Increased activity of preganglionic sympathetic nerves stimulates
  the adrenal medulla to release adrenaline, which supplements and
  prolongs all the above-listed reactions; it also causes increased liver
  glycogenolysis—mobilizes glucose to avoid risk of brain hypoglycaemia
- Stimulates hormone-sensitive lipase—mobilizes free fatty acids for use
  as energy substrates by many organs.

The resistance response

If the stress is more sustained, more prolonged effects of the acute response and more chronic responses occur to produce the resistance reaction. These involve the slower results of the sympathetic nerve stimulation and actions of various hormones that are more prolonged than those of the catecholamines.

Sympathetic nerves stimulate juxtaglomerular cells of the kidney. This results in the production of angiotensin II which both causes vascular constriction and also stimulates mineralocorticoid (aldosterone) release which

- Increases Na$^+$ reabsorption which causes water retention, maintains a high blood pressure, and counteracts fluid loss
- Increases elimination (exchange) of H$^+$ ions (accumulate as a result of the increased catabolism).

Glucocorticoids (e.g. cortisol)

- Metabolic substrate metabolism
- Cortisol stimulates metabolism of:
  - Carbohydrates—stimulates glucose production; opposes insulin actions
  - Lipids—stimulates lipolysis and ketogenesis (results in redistribution of fat to trunk if fatty acids are in excess)
  - Proteins—stimulates gluconeogenesis
- Cardiovascular effects:
  - Maintains the circulation via increased myocardial contraction; increases vascular tone
  - Maintains plasma volume by preventing increased vascular permeability
- Skeletal muscle: maintains ability to give sustained contractile responses
- Ion control: promotes Na$^+$ retention and K$^+$ excretion (actions at mineralocorticoid receptors when cortisol is high; ⊃ p.630).
- In the CNS: varied effects on mood and behaviour
- Haemopoiesis: increases red blood cell production, so enhances oxygen-carrying capacity of blood
- Inflammatory response immune system—immunosuppressive actions:
  - Inhibits leucocyte translocation from blood to sites of tissue damage or infection
  - Stimulates lymphocyte destruction
  - Glucocorticoid selective drugs are used therapeutically to treat inflammatory diseases such as asthma and eczema (e.g. prednisone, betamethasone) but have some mineralocorticoid effects
- Reproduction and lactation: inhibits, in part by inhibition of LH and PRL release from the anterior pituitary gland (pregnancy is a non-essential metabolic drain on resources).

Hypothalamic TRH

Stimulates the release of TSH and, therefore, thyroid hormones. Thyroid hormones increase the metabolic rate and the catabolism of glucose, fat and proteins (⊃ p.624).

- In starvation (which is a particular form of stress), glucocorticoids reduce the conversion of T$_4$ to T$_3$ in the liver, blunting the catabolic response.

Hypothalamic GHRH

GHRH stimulates the release of growth hormone which, when secreted in a prolonged manner (normally it is secreted in short pulses every 4–6h), causes:

Increased liver breakdown of fats to fatty acids and glycerol
Increased liver conversion of glycogen to glucose.

By these means, the resistance reaction allows the body to continue fighting a stressor long after the effects of the acute alarm reaction. It produces the energy and circulatory changes required for the performance of strenuous tasks, fighting infection, avoiding fatal haemorrhage. If there is greatly increased metabolism, blood glucose returns nearly to normal during the resistance reaction as input = output; blood pH is controlled by the kidney. However, blood pressure remains raised because of retention of water.

When the stress is prolonged

Within the hypothalamus, CRF, glucocorticoids, opioid peptides reduce GnRH secretion—avoid risk of pregnancy and a further drain on metabolic resources
In the hippocampus, glucocorticoids act on GRs to modify emotional reactions—induce mild euphoria, diminishing the psychic effects of the stress.

Chronic stress at different ages

In utero: the stress of undernourishment → low birthweight is associated with a significant increase in risk of hypertension, diabetes mellitus, and lower life expectancy
Childhood: chronic stress in childhood is associated with retarded growth
Old age: the morning peak of cortisol occurs earlier in the aged; the cortisol response to the stress of an operation is also prolonged.

Chapter 10

Reproduction and development

Anatomy of the pelvis and perineum

The pelvis

The pelvis comprises bony and soft tissues. The bony pelvis forms a ring that protects the pelvic contents (rectum; ureters; bladder; urethra; male reproductive system—ductus deferens, seminal vesicles, prostate; female reproductive system—uterus, uterine tubes, ovary, vagina) and articulates with the femurs. See Table 10.1 for male/female differences.

- The pelvis is formed by the two pelvic (innominate or hip) bones, the sacrum, and the coccyx. The pelvic bones meet anteriorly at the pubic symphysis (a secondary cartilaginous joint, usually immobile) and posteriorly they articulate with the sacrum at the sacroiliac joints (a synovial joint but one which allows only minimal movement and transmits weight of upper body to the hip bones)
- The pelvis is divided into the greater pelvis which lies above the pelvic brim (pelvic inlet) and the lesser pelvis which lies between the pelvic inlet and outlet
- The sacrum consists of the fused five sacral vertebrae and contains foramina for the passage of the sacral spinal nerves
- The pelvic bones are formed by the fusion of the ilium, the ischium, and the pubic bone shortly after puberty. All three bones contribute to the formation of the acetabulum
- The superior ramus of the pubic bone forms the superior border of the obturator foramen.

The pelvic floor

The pelvic floor is a group of muscles around the terminal part of the rectum and the prostate and urethra in the male and vagina and the urethra in the female. Damage to the pelvic floor results in prolapse of the pelvic contents and incontinence.

The perineum

The perineum overlies the pelvic outlet. It can be divided into an anterior urogenital triangle and a posterior anal triangle. The perineum is the area between the skin and levator ani.

The anal triangle
- Is similar in both sexes and contains the two ischiorectal fossae separated by the anal canal, the anococcygeal ligament, and the perineal body.

The female urogenital triangle
- Includes the perineal membrane, vagina, clitoris, perineal body (knot of tissue between the vagina and anal canal to which muscles are attached), vestibule, and labia.

The male urogenital triangle
- Includes the erectile tissue of corpora cavernosa and corpus spongiosum that form the penis, and scrotum.

The urogenital region in both sexes is supplied by the internal pudendal artery (branch of internal iliac artery) and branches, and innervated by the pudendal nerve.

Table 10.1 Male–female differences in the pelvis

	Male	Female
Acetabulum	Large	Small
Build	Robust	Thin
Inferior pelvic aperture	Relatively small	Relatively large
Obturator foramen	Round	Oval
Pubic arch	Narrow	Wide
Superior pelvic aperture	Heart-shaped	Oval or rounded

Male genital system

The male reproductive system functions to produce a continuous supply of functional spermatozoa from puberty to old age.

- Spermatozoa are produced in the testes, which are located in the scrotum so that the temperature is correct for spermatogenesis
- Spermatozoa are modified, concentrated, and stored in the epididymis and transported towards the urethra by the vas deferens (which passes into the abdominal cavity through the inguinal canal)
- Each vas descends into the pelvis where its terminal dilated ampulla joins the duct of the seminal vesicle in the interval between the rectum and prostate
- The combined ducts form an ejaculatory duct, which pierces the prostate and empties into the urethra
- The prostate surrounds the first part of the urethra as it leaves the bladder and its secretions drain into the urethra via a number of small ducts. Together, the secretions of the seminal vesicle and the prostate form the main part of the ejaculate
- The prostatic urethra leaves the pelvis by piercing the perineal membrane that separates the pelvis from the perineum
- In the perineum, the urethra dilates to form the bulb of the urethra which is surrounded by cavernous erectile tissue (corpus spongiosum) and by striated muscle (bulbospongiosus), contraction of which causes ejaculation of semen
- Ducts of two small bulbo-urethral mucous secreting glands also enter the urethral bulb
- On either side of the perineum, a second mass of erectile tissue (corpus cavernosum), covered with smooth muscle, joins with the corpus spongiosum and urethra to form the shaft of the penis.

Blood supply and innervation

See Fig. 10.1.

- The testes and vas deferens receive a dense plexus of autonomic fibres which reach them by running along their arteries. The prostate and seminal vesicles receive autonomic fibres from the pelvic plexuses
- The testes are supplied by testicular arteries which arise from the abdominal aorta. The vas deferens, prostate, and seminal vesicles are supplied by branches of the internal iliac artery. The prostate also receives blood from the internal pudendal artery
- Arteries of the penis derive from the internal pudendal artery.

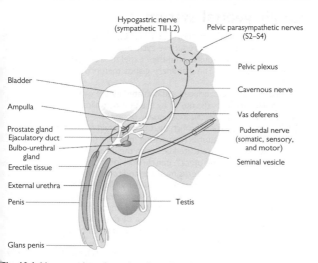

Fig. 10.1 Nerve supply to the penis and accessory sex organs.

Reproduced with permission from Pocock G, Richards CD (2006). *Human Physiology: The Basis of Medicine*, 3rd edn, p452. Oxford: Oxford University Press.

Female genital system

The female reproductive system (Figs 10.2, 10.3) functions to produce (usually) one ovum in each reproductive cycle and also to activate the spermatozoa and to house, nourish, and then expel the resultant offspring at the end of the pregnancy. The mammary glands then provide milk to nourish the infant.

- Fertilizable ova are produced and ovulated by the ovaries, then transported towards the uterus by the uterine (fallopian) tubes (oviducts), in which fertilization usually occurs
- Each uterine tube runs through the upper margin of the broad ligament to join the uterus (comprising muscular fundus and body and fibromuscular cervix), situated between the bladder and the rectum. Benign tumours of the uterine muscle wall form 'fibroids'
- During pregnancy, the uterus must expand to accommodate the developing foetus so that, at term, it nearly fills the abdominal cavity
- The neck of the uterus is guarded by the cervix which opens into the anterior wall of the vagina. The cervix and, through it the uterus, are held in place by the fibromuscular tissue of the pelvic floor
- The vagina pierces the perineal membrane to open onto the vestibule immediately behind the urethra
- Within the vestibule, smaller masses of cavernous tissue form the bulb of the vestibule and clitoris and greater vestibular glands secrete mucus for lubrication
- The vestibule is bounded on either side by labia (major and minor)
- Because, during childbirth, the foetus must pass through the birth canal within the bony pelvis, the cavity of the female pelvis differs from that of the male in ways that facilitate the passage of the foetus
- The mammary glands (breasts), situated on the anterior chest wall, develop during pregnancy so that after childbirth their secretory alveoli produce milk (→ pp.680, 681)
- This milk is expelled through lactiferous ducts to collect in the lactiferous sinuses, which open onto the nipple and from where milk is extruded during suckling by the milk-ejection reflex.

Blood supply and innervation

The uterus is supplied by the uterine artery (a branch of the internal iliac artery), which runs in the broad ligament and anastomoses with branches of the ovarian artery. The uterine artery also supplies the upper part of the vagina. The ovary is supplied by the ovarian artery (a branch of the abdominal aorta). Veins run in the broad ligament to drain into the internal iliac vein.

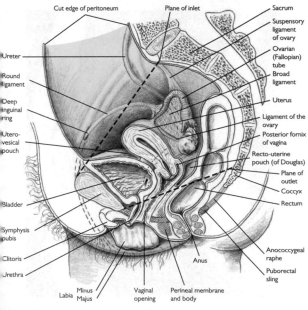

Fig. 10.2 Sagittal section of female pelvis. Note: the rectum is not normally distended until just before defecation.

Reproduced from MacKinnon, Pamela and Morris, John, *Oxford Textbook of Functional Anatomy*, vol 2, p193 (Oxford, 2005). With permission of OUP.

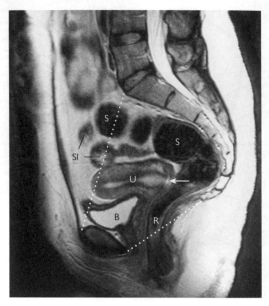

Fig. 10.3 Magnetic resonance imaging of the female pelvis: sagittal section Normal uterus (U), cervix (arrow), vagina. The planes of the inlet and outlet are marked. B, bladder; S, sigmoid colon; R, rectum; SI, small intestine.

Reproduced from MacKinnon, Pamela and Morris, John, *Oxford Textbook of Functional Anatomy*, vol 2, p198 (Oxford, 2005). With permission of OUP.

Sexual differentiation

(See also ➔ p.676.)

- In the fourth week of foetal life, primordial germ cells can be distinguished close to the root of the yolk sac
- The primordial germ cells migrate up the posterior abdominal wall, multiplying en route until they reach the developing gonadal ridge, medial to each mesonephros.

Chromosomal sex

XX = female; XO = female; XY = male; XXY = male.
Generally, a Y chromosome nearly always leads to a male phenotype.

Male differentiation

- Embryos with XY sex chromosomes express the gene SRY (sex-determining region Y) on the Y chromosome, which causes the gonad to develop as a testis
- Mesoderm of the gonadal primordium differentiates into Sertoli cells, enclosed by basement membrane to form primitive seminiferous tubules
- In the interstitial tissue between the tubules, Leydig cells develop and begin to secrete testosterone by the 7th week
- Testosterone diffuses locally to stimulate growth and differentiation of the male reproductive tract from the mesonephric/Wolffian system.
- The Sertoli cells also secrete anti-Müllerian hormone, which suppresses development of the paramesonephric Müllerian system, which forms the female reproductive tract
- Testosterone is converted to dihydrotestosterone in target tissues to stimulate the differentiation and growth of male external genitalia via actions at androgen receptors encoded by the X chromosome
- Testosterone converted to oestrogen in the brain influences differentiation of regions associated with male sexual behaviour.

Female differentiation

- In the absence of a Y chromosome, the mesonephros ducts begin to disappear and an invagination of the coelomic epithelium forms along each gonadal ridge to form the paramesonephric/Müllerian duct
- The Müllerian duct differentiates into the vagina, ovary, uterus, and uterine tubes.

Production of male gametes

The primordial germ cells that give rise to the gametes in both males and females originate within the primary ectoderm of the embryo during week 2. They then detach from the ectoderm and migrate to the yolk sac. Between 4 and 6 weeks, the primordial germ cells migrate from the yolk sac to the posterior body wall via the gut and mesentery (◕ p.666).

Dual function of the testes

The adult testes:
- Produce sperm—the male gametes
- Secrete androgens—the steroid hormones which regulate full male development. The principal androgen is testosterone.

Structure of the testes

- The testes consist of large numbers of tightly packed seminiferous tubules, each surrounded by a basement membrane and a sheath of peritubular myoid cells separated by an interstitial space containing blood vessels, lymphatics, and the endocrine Leydig cells
- Leydig cells produce and secrete androgens synthesized from cholesterol
- Within each seminiferous tubule are cells derived from primordial germ cells—spermatogonial stem cells, spermatocytes, spermatids, spermatozoa, and specialized epithelial Sertoli cells
- The Sertoli cells control and coordinate the development of germ cells. Between the Sertoli cells are numerous tight junctions, which divide the tubule into a basal and an adluminal compartment. The tight junctions form a *blood–testis barrier* which prevents the uptake of compounds that could disrupt spermatogenesis. Spermatogenesis, which begins at puberty, produces numerous gene products which are foreign to the immune system. Sertoli cells could protect against autoimmune reactions.

Spermatogenesis produces many spermatozoa from one stem cell

- The production of a spermatozoon capable of fertilization takes ~64 days
- The diploid stem germ cells (the spermatogonia) lie on the basement membrane of the tubule. These divide, first by mitosis, and then meiosis, to become spermatocytes, spermatids, and spermatozoa. Mature spermatozoa are designed to deliver the haploid complement of DNA to an oocyte surrounded by the cumulus oophorus and zona pellucida within the female oviduct. The essential features of spermatogenesis are:
  - Production of a rearranged haploid set of chromosomes by meiosis
  - Condensation of DNA to protect it from damage
  - Production of a flagellum to propel the spermatozoon into the female tract
  - Production of the acrosome—an enzymatic knife—to allow the spermatozoon to penetrate the layers surrounding the oocyte
  - Development of the capacity to fertilize an ovum and development of recognition molecules to bind to the oocyte

- Each spermatogonium undergoes five mitoses to yield a primary spermatocyte committed to meiosis. The first division of meiosis produces two secondary spermatocytes which then undergo a second division of meiosis to produce four round haploid spermatids. These mature into spermatozoa by spermiogenesis
- When spermatozoa leave the testis they are incapable of directional movement or fertilization. They take up to a week to pass through the epididymis, in which further maturation takes place so that they are capable of traversing the female tract and fertilizing an oocyte
- Each testis produces 30,000–70,000 spermatozoa—large numbers are required for fertile sexual intercourse
- Infertility can be caused by abnormalities in spermatozoa but, if there is only a partial problem, active sperm can be separated from inactive and used in artificial insemination.

Control of testis function by gonadotrophins (LH, FSH) and testosterone

During childhood, the testes are inactive. At puberty, LH and FSH are secreted by the anterior pituitary gland in response to hypothalamic GnRH (⟳ p.618). These are the main hormones governing testis function.

- LH receptors are only found on Leydig cells and FSH receptors only on Sertoli cells
- LH acts on the Leydig cells to stimulate androgen production
- Testosterone stimulates the development of secondary sexual characteristics: deepening voice, increased muscle mass, body hair, behaviour, etc. Androgens also pass into the seminiferous tubules to affect spermatogenesis by local paracrine mechanisms
- Excess secretion of androgens is prevented by a negative feedback loop acting at both the hypothalamus and the pituitary to inhibit LH release
- FSH acts on Sertoli cells to stimulate their cell division and is important in spermatogenesis and spermiogenesis. In humans, mutations of the FSH receptor are associated with reduced fertility, but not infertility
- Testosterone and a peptide from Sertoli cells—inhibin—exert negative feedback on FSH.

Functions of the epididymis, seminal vesicles, and prostate

- The epididymis serves as a reservoir for sperm with their passage through it taking 1–21 days. The spermatozoa and testicular secretions are then transported along the vas deferens and into the ejaculatory ducts
- The seminal vesicles (60%) and prostate (20%) contribute to the seminal fluid
- The seminal fluid provides nutrients and is alkaline, helping to neutralize the acidic fluid of the vagina and, thus, to increase the motility and fertility of sperm.

Erection, emission, and ejaculation

Sexual intercourse involves penile erection, penetration of the vagina by the penis, emission of fluid from the seminal vesicles, vas deferens, and prostate gland, and ejaculation.

- Erection is a result of sacral parasympathetic reflexes (via pelvic nerve–pelvic plexus–cavernous nerve). Via increased NO synthesis, parasympathetic stimulation triggers dilation of the internal pudendal artery resulting in increased blood flow and erection
- Emission is stimulated by sympathetic activation (T11–L2 hypogastric nerve–pelvic plexus–cavernous nerve)
- Ejaculation is stimulated by somatic fibres running in the pudendal nerves (motor and sensory). Activation of the pudendal nerves stimulates contractions of the bulbocavernosus and ischiocavernosus muscles, resulting in expulsion of semen from the penis.

Production of female gametes

Dual function of the ovary

The adult ovary:
- Produces ova (Fig. 10.4), released at ovulation, throughout the reproductive years of women
- Produces oestrogens (predominantly 17β-oestradiol) and progesterone—steroids important in the regulation of female reproductive function.

Maturation of primordial follicles to pre-ovulatory follicles: the follicular phase

See Fig. 10.4.
- The first half of the menstrual cycle is the follicular phase (days 1–14) and is the period during which a follicle undergoes growth and development which culminates in ovulation—rupture of the follicle and release of the oocyte from the ovary
- Normally, only one 'dominant follicle' matures to ovulation in each cycle. The remainder become atretic and die
- Follicle development is regulated by pituitary LH and FSH and also by oestrogens produced by the follicle in response to LH and FSH
- The follicle is made of three components: the theca interna and externa (several layers of cells), the granulosa cells within the follicle, and the oocyte
- LH causes thecal cells to grow, divide, and secrete androgens. FSH stimulates the division of granulosa cells and activates them to convert this androgen to oestrogens by stimulating production of the aromatase enzyme. Both LH and FSH together elicit full folliculogenesis
- During the pre-ovulatory stage, under the influence of high-circulating LH concentrations, the meiotic division of the oocyte is completed, progesterone secretion commences, and the follicle ruptures to release the oocyte—ovulation (day 14)
- Ovulation is suppressed in starvation, severe exercise, and emotional stress
- Oestrogens produced during the follicular phase stimulate:
  - Contractile activity in the fallopian tubes
  - Uterine endometrium proliferation, to prepare for gamete transport and implantation respectively
- Also, oestrogens stimulate the production of thin, watery cervical mucus that is easily penetrated by sperm.

The luteal phase: formation of corpus luteum; secretion of progesterone and oestrogen; luteolysis

See Fig. 10.5.
- The second half of the menstrual cycle (days 14–28), following ovulation, is the luteal phase. The post-ovulatory follicle becomes a corpus luteum under the influence of LH
- The corpus luteum secretes progesterone which prepares the uterus to receive and nourish an early embryo in the event of fertilization and maintains the endometrium in a condition suitable for implantation and placentation. Progesterone also reduces myometrium excitability. Oestrogens are also produced

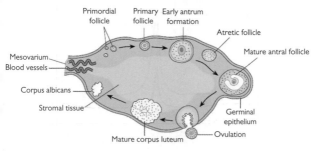

Fig. 10.4 Internal structure of ovary showing the stages of follicular development, ovulation, the formation of the corpus luteum, and its subsequent regression.

Reproduced with permission from Pocock G, Richards CD (2006). *Human Physiology: The Basis of Medicine*, 3rd edn, p439. Oxford: Oxford University Press.

- In the absence of fertilization, the corpus luteum degenerates around days 24–28 and steroid production stops. This is the process of luteolysis and marks the end of one menstrual cycle
- As progesterone concentrations fall, the endometrium built up during the cycle is shed, together with blood from spiral arteries (days 1–4). This process is menstruation and marks the start of another menstrual cycle.

The hypothalamic–pituitary axis; endocrine control of menstrual cycle; gonadal hormone feedback

See Figs 10.5 and 10.6.

Cyclical variations in the concentrations of steroids, LH, and FSH act to-gether to ensure the regular release of mature ova and to prepare the body for fertilization and pregnancy.

- Oestrogens and progesterone secreted by the ovary regulate the secretion of LH and FSH by both positive and negative feedback
- Very high concentrations of oestrogens stimulate the anterior pituitary to initiate an LH surge which is crucial for the pre-ovulatory phase and ovulation. For the remainder of the cycle, negative feedback prevails and LH and FSH output is relatively low.

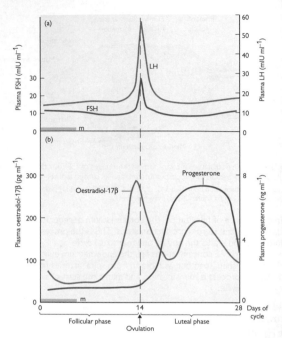

Fig. 10.5 Changes in: (a) pattern of secretion of gonadotrophins (FSH and LH); (b) plasma levels of 17_β-oestradiol and progesterone.

Reproduced with permission from Pocock G, Richards CD (2006). *Human Physiology: The Basis of Medicine*, 3rd edn, p441. Oxford: Oxford University Press.

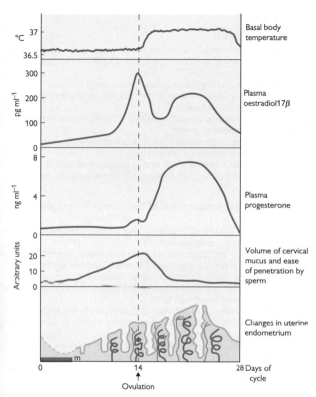

Fig. 10.6 Cyclical changes in body temperature, cervical secretions, and uterine endometrium in relation to the circulating levels of oestradiol 17β and progesterone.

Reproduced with permission from Pocock G, Richards CD (2006). *Human Physiology: The Basis of Medicine*, 3rd edn, p443. Oxford: Oxford University Press.

Pregnancy

Coitus and orgasm

Coitus is the act of sexual intercourse which results in the ejaculation of spermatozoa (male gametes) into the vagina. The sexual response can be divided into four phases ('EPOR'):

1. Excitement—increasing sexual arousal causes penile erection, thickening of the labia, secretion of cervical mucus (all activated by parasympathetic stimulation).
2. Plateau phase—sexual arousal intensified, distension of penis and testes, mucus secretion greater.
3. Orgasmic phase—ejaculation in the male, contractions of uterus and anal and urethral sphincters in the female.
4. Resolution phase—loss of pelvic vasocongestion. Followed by a short refractory period in the male during which further arousal is not possible.

Fertilization and preparation of the uterus for pregnancy
(➔ p.688.)

- A sperm is only capable of fertilizing an egg if it first undergoes the acrosome reaction (capacitation)
- The first stage of fertilization occurs when an activated sperm fuses with an oocyte in the fallopian tube. The newly fertilized egg (the zygote) then completes its second meiotic division and undergoes the cortical reaction to create a fertilization membrane which prevents further sperm from fusing with it
- Sperm retain fertility up to 48h after ejaculation, while ova are viable 12–24h after ovulation. Therefore, a short time is available during which coitus must take place for pregnancy to occur
- The zygote signals its presence by secreting human chorionic gonadotrophin (hCG) which prolongs the secretory life of the corpus luteum. This ensures that progesterone and the specialized endometrial layers of the uterus are maintained until the pregnancy can be supported by placental progesterone (around 6–8 weeks of gestation)
- Once the zygote begins to divide, it is termed an embryo. It continues to divide as it is transported along the fallopian tube towards the uterus
- Oestrogen levels, high at ovulation, promote the embryo transport to the uterus and induce endometrial proliferation
- Progesterone induces secretory changes in the endometrium that enable the blastocyst/embryo to be successfully implanted.

The process of implantation; status of the foetus as an allograft

- At implantation, the trophoblast tissue of the fertilized egg invades the endometrial tissue of the uterus by the growth of chorionic villi containing the foetal capillaries. As a result of the invasion, the maternal spiral arteries of the uterus are eroded and spill their blood into the spaces between adjacent chorionic villi. In this way, a dialysis pattern of blood flow is set up within the placenta such that foetal capillaries dip into maternal blood spaces

- The foetus is an allograft—a mother would not accept a skin graft from a child, yet allows the foetus to remain in the uterus for 9 months. The uterus is not an immunologically privileged site and, during pregnancy, there is no overall immunosuppression of the mother. The foetus expresses unique histocompatibility proteins and it has been proposed that there is a local barrier around the trophoblast, local hormonal down-regulation of lymph nodes, and the presence of suppressor cells.

Structure and function of the placenta; placental villi

During foetal life, the placenta (Fig. 10.7) performs the functions normally carried out by the lungs, kidneys, and GI tract in the adult.

- Oxygen diffuses passively from maternal to foetal blood, and carbon dioxide in the opposite direction. Glucose and amino acids move across the placenta from maternal to foetal blood by carrier-mediated transport, while free fatty acids diffuse passively across the lipid-rich placental barrier
- Foetal waste products such as urea and bilirubin diffuse from foetal to maternal plasma down their concentration gradients
- The surface area available for exchange is immense due to the extensive branching of the chorionic villi.

Production and roles of placental hormones during pregnancy

The placenta is also an important endocrine organ, secreting a variety of steroid and peptide hormones:

- hCG—prevents regression of the corpus luteum to ensure continued progesterone secretion for up to 10 weeks of gestation (⊃ p.676)
- Human placental lactogen—secreted from week 10 of gestation. It contributes to proliferative changes in the breast in preparation for lactation and exerts important metabolic effects in the mother. It

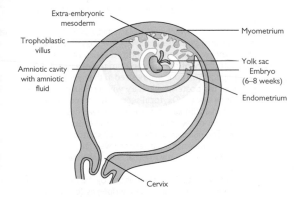

Fig. 10.7 The placenta in relation to adjacent structures during early pregnancy.
Reproduced with permission from Pocock G, Richards CD (2006). *Human Physiology: The Basis of Medicine*, 3rd edn, p456. Oxford: Oxford University Press.

stimulates an increase in the maternal plasma levels of glucose, amino acids, and free fatty acids, ensuring that placental availability of essential metabolites for the foetus is optimal
- Relaxin—promotes the relaxation of the pelvic ligaments to allow the foetus to pass through the pelvis
- Progesterone—the placenta takes over from the corpus luteum as the major source of progesterone at week 10 to maintain the endometrium and reduce myometrial excitability (via down-regulation of oxytocin and oxytocin receptors) as well as to stimulate mammary development in readiness for lactation
- Oestrogens are secreted by the placenta (made from precursors of both foetal and maternal origin). They prepare the body for labour and lactation.

Parturition

The onset of labour is marked by regular, painful uterine contractions and progressive dilation of the cervix. Labour consists of three stages:
- First stage: the time from onset to full cervical dilation. Uterine contractions become progressively stronger to propel the foetus down the birth canal. At the same time, the amniotic sac ruptures ('waters break')
- Second stage: the time from full cervical dilation until birth. It usually lasts 40–60min
- Third stage: the time from birth to the delivery of the placenta. Ergometrine, a smooth muscle stimulant, is often administered to enhance contractions and promote placental delivery (afterbirth).

Mechanism of parturition; hormonal control—role of oxytocin and prostaglandins

- The nature of the trigger for parturition is poorly understood but it is believed that the foetus plays a role in determining the time of its birth and involves mechanisms in both the maternal and foetal nervous and endocrine systems
- Foetal cortisol initiates a switch in the placenta away from progesterone synthesis to oestrogen synthesis in the last few days of pregnancy. Oestrogens, together with oxytocin and prostaglandin $F_{2\alpha}$ ($PGF_{2\alpha}$), increase the contractility of the myometrium to bring about delivery
- Oxytocin is a peptide hormone synthesized in the hypothalamus and released from the posterior pituitary which stimulates uterine contractions during labour. Oxytocin secretion is stimulated by uterine distension and vaginal stimulation as the foetal head descends which, in turn, stimulates more oxytocin release (positive feedback—the Ferguson reflex). The number of myometrium oxytocin receptors increase in late pregnancy as progesterone levels fall
- Gap junctions form between myometrial cells, ensuring a rapid spread of contractions
- $PGF_{2\alpha}$ (stimulated by oxytocin) is synthesized by the myometrium and stimulates muscle contraction and cervical ripening (cervix softens and dilates). Labour can be induced at term using vaginal pessaries containing $PGF_{2\alpha}$.

Premature and delayed parturition

- Normal 'term' in humans is 40 weeks after the last menstrual period. Both pre-term and post-term labour are hazardous. Pre-term because the foetus is not yet prepared for extra-uterine life and post-term because continued foetal growth and placental insufficiency pose problems for both delivery and foetal nutrition
- Towards the end of pregnancy, a number of changes must occur in the foetus in preparation for postnatal life. Most of these are induced by the secretion of glucocorticoids which increase markedly towards term. Such changes include:
  - Production of surfactant in the lungs to allow lung expansion when air is first breathed
  - Changes to gut and liver enzymes to allow the foetus to metabolize its postnatal milk diet
- Post-term, it is usual to induce labour by 42 weeks of gestation (induction).

Induction of parturition

- Physical separation of the amniotic sac from the cervix by an obstetrician or midwife (a sweep)
- Application of prostaglandin cream/gel on the cervix to encourage dilation
- Oxytocin can be administered to help induce myometrial contractions
- Rupture of the membranes (also known as breaking of the waters), → an increase in prostaglandin release, which accelerates development of full contractions.

Mammary glands and lactation

Lactation is the synthesis and secretion of milk by the mammary gland.

Structure of the mammary gland

See Fig. 10.8.
- In the embryo, the mammary glands develop from modified skin glands
- At birth, the breasts are the same in the male and female, with very few acini. Further development does not occur in males
- At puberty, in females oestrogens stimulate lactiferous glands to branch and lobules containing abundant acini develop. Fat deposition and connective tissue growth increase the size of the breast
- Cyclic breast changes occur with the menstrual cycle due to fluctuations in oestrogens and progesterone (premenstrually, breasts become swollen and tender).

Breast development in pregnancy and lactation, and its hormonal control

- Mammary glands develop and the alveoli mature in response to placental hormones (placental lactogen, oestrogen, and progesterone) during pregnancy
- Massive development of the tubero-alveolar structure of the epithelium and ducts occurs to give 12–20 galactopoietic units, emptying into a common sinus at the nipple.

Lactogenesis—prolactin and other hormones

Prolactin stimulates milk production. Maternal anterior pituitary prolactin (● p.619) release is controlled negatively by dopamine, and dopamine agonists can be used to inhibit lactation.
- The maintenance of lactation depends on maternal prolactin release elicited by the suckling stimulus
- The first milk (colostrum) formed in the mother's breast contains an abundance of antibodies to protect the foetus against any local infections. The antibodies are of the IgA type, which are able to readily pass across the gut epithelium of the baby.

Milk-ejection reflex—oxytocin

- The ejection of milk is stimulated by oxytocin (● p.620) release from the posterior pituitary, elicited by suckling
- Oxytocin acts to contract the myoepithelial cells surrounding the alveoli within the breast which, together with the negative pressure of suckling at the nipple, ensures milk ejection.

Lactation, raised prolactin, and fertility

- The daily suckling stimulus reduces the pulsatile release of GnRH. This is probably the mechanism by which lactation can act to suppress fertility
- Hypersecretion of prolactin by pituitary tumours (prolactinomas; ● OHCM10 p.236) may cause galactorrhoea and infertility.

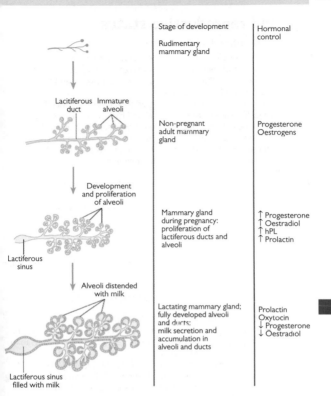

Fig. 10.8 Structure of the mammary gland at different developmental stages.
Reproduced with permission from Pocock G, Richards CD (2006). *Human Physiology: The Basis of Medicine*, 3rd edn, p471. Oxford: Oxford University Press.

Age and reproductive status

Females: childhood and puberty

- In females, the fertile years are defined by menarche (the commencement of menstrual cycles) and the menopause (the cessation of menstrual cycles). Menarche occurs around 12 years and the menopause around 50 years of age
- During childhood, LH and FSH rise gradually. Prior to menarche, around age 10, pulsatile release of LH and FSH is established and spurts of secretion occur during sleep. This results in increased oestrogen secretion from the ovary (gonadarche), which triggers breast development and changes in body composition (adult females have twice the body fat and reduced skeletal muscle mass compared with males)
- Increased androgen secretion by the adrenal cortex (zona reticularis) stimulates pubic hair growth (adrenarche)
- The trigger for menstruation is not understood but is thought to include:
  - Increased ovarian sensitivity to LH and FSH
  - Increased pituitary sensitivity of positive feedback effects of oestrogens
  - Attainment of either a critical body mass (around 47kg) or ratio of fat to lean mass
- Menstrual cycles are disrupted in girls who lose a large amount of weight through anorexia, excessive exercise, or starvation.

Females: the menopause

Menopause is the progressive decline in the female reproductive system.
- Numbers of oocytes in the ovaries are depleted by atresia and ovarian responsiveness to LH and FSH decreases
- Cycles become anovular and irregular before ceasing altogether
- Oestrogen and progesterone concentrations fall. FSH and LH circulating concentrations are high because of the loss of negative feedback inhibition by oestrogen but no LH surge is seen
- The loss of ovarian steroids results in vaginal dryness, uterine muscle fibrosis, loss of breast tissues, depression, night sweats, hot flushes, increased risk of myocardial infarction, increased bone resorption, and resulting bone weakness
- The changes can be treated with hormone replacement therapy (HRT).

Males: childhood and puberty

Between 10 and 16 years of age, boys show a growth spurt and develop full reproductive capacity.
- Infancy to start of puberty: LH and FSH secretion and testosterone secretion is low
- Start of puberty: increase in secretion of pituitary LH resulting in maturation of the Leydig cells and initiation of spermatogenesis
- Testosterone plasma concentrations rise and trigger the secondary sexual characteristics: enlargement of the testes and penis, growth of pubic hair, appearance of facial hair, deepening of the voice due to thickening of the vocal cords and enlargement of the larynx.

Is there a male menopause?

A small decline in male reproductive function occurs with age: sperm production declines between age 50 and 80; plasma testosterone in men >70 years decreases, and plasma concentrations of LH and FSH increase (though much less dramatically than in females).

Precocious puberty and delayed puberty

- Precocious puberty can arise from tumours of the posterior hypothalamus; activating mutations of the LH receptor; thecal, granulosa, or Leydig cell tumours; gonadotrophin-secreting tumours
- Delayed puberty can arise from Kallmann syndrome—a deficit in formation and migration of GnRH neurones in the developing brain which results in GnRH, LH, and FSH deficiency; tumours of the anterior hypothalamus.

Pharmacology of sex hormones

Oral contraceptives

(❂ *OHPDT2* p.504.)

The combined pill

- Contains oestrogen and progesterone and is taken for 21 consecutive days out of 28
- Oestrogen inhibits FSH release from the pituitary and, therefore, inhibits follicle development
- Progesterone inhibits LH release from the pituitary and, therefore, inhibits ovulation
- Both oestrogen and progesterone alter the endometrium such that the receptivity to blastocyst implantation is reduced. Cervical mucus thickens such that sperm penetrability is decreased
- Withdrawal of the progesterone precipitates a withdrawal bleed that simulates a period.

The mini-pill

Contains only progesterone and acts by thickening cervical mucus and decreasing receptivity of the endometrium to blastocyst implantation.

The 'morning-after' pill

Contains a high concentration of progesterone and inhibits implantation by accelerating the transport of the embryo in the fallopian tube so that it reaches the uterus before the secretory phase has occurred.

Side effects

Side effects of oral contraceptives include minor symptoms of early pregnancy—nausea, breast tenderness, fluid retention, hypertension, and, in rare cases, thromboses.

Hormone replacement therapy

Prevents symptoms of the menopause (❂ p.682) by oestrogen replacement with or without progesterone.

Treatment of female infertility: *in vitro* fertilization (IVF)

- The patient is given a GnRH agonist to suppress spontaneous gonadal activity and is then treated with gonadotrophins to produce many large mature follicles
- Just before ovulation, several ova are obtained by laparoscopy. The ova are cultured for several hours and sperm (washed free of seminal fluid) added. The injection of a single sperm into the oocyte can also be performed (intracytoplasmic sperm injection) where the man has very few sperm. After 2–3 days, several four- to six-cell conceptuses are transferred to the uterus via the cervix
- Problems with this approach include the high incidence of multiple pregnancies and frequency of early miscarriage.

Male contraception

Not currently available. Long-acting GnRH agonists combined with testosterone have been studied as a possible male contraceptive but are not always effective in blocking spermatogenesis. Progesterone implants combined with monthly testosterone injections are currently in clinical trials.

Male infertility

Male hypogonadism (→ OHCM10 p.231) may be due to hypothalamic–pituitary disorders (e.g. LH and FSH deficiency), gonadal abnormalities (e.g. Klinefelter syndrome), or androgen insensitivity (e.g. testicular feminization). Erectile dysfunction (impotence → OHCM10 p.230) is the inability to achieve or maintain an erection that is adequate for sexual intercourse. Causes may be psychogenic or due to disturbances in the nerve or vascular supply to the penis due to, for example, diabetes, hypertension, smoking, ageing.

Box 10.1 summarizes the treatment for erectile dysfunction.

Box 10.1 Treatment of erectile dysfunction: sildenafil (Viagra®)

(→ OHCM10 p.231; OHPDT2 p.236.)
- A type V phosphodiesterase inhibitor that potentiates the effect of NO by inhibiting the breakdown of cyclic guanosine monophosphate. This causes enhanced dilation of the internal pudendal artery and its branches, resulting in increased blood flow and erection
- Side effects include headache and visual disturbances.

Human embryology: introduction

Embryology covers the period of prenatal development, from fertilizatio (day 0) up to birth 38 weeks later. This continuous process is divided int three phases:
- Pre-embryonic development (weeks 1 and 2):
  - Period of initial cleavage of the zygote, implantation into the uterine wall, and formation of the bilaminar embryo
- Embryonic development (weeks 3–8):
  - When all the major body systems are laid down and established
  - This phase is a time of rapid and dramatic developmental change
- Foetal development (weeks 9–birth):
  - When tissues and organs formed during the embryonic phase grow, differentiate, and mature so they are ready to function in postnatal life.

Nomenclature of embryonic axes

These are equivalent to the following adult anatomical terms:

Adult	Anterior–posterior	Superior–inferior	Left–right
=	=	=	=
Embryonic	Ventral–dorsal	Anterior–posterior	Left–right
		=	
		or rostral–caudal	
		or cranio–caudal	

Cell–cell interactions

Embryological development depends on inductive interactions between dif ferent cell/tissue layers, e.g. epithelial–mesenchymal interactions.

An inductive process is an interaction between non-equivalent cell popu lations whereby the responding cells undergo a change in fate.

Such interactions require cell–cell communication, e.g. by direct contac via gap junctions, cell–cell contact, or the production of diffusible factor by one cell type, which impinge on a responding population. Examples o inductive interactions:
- Mesoderm induction
- Neural induction
- Limb bud initiation
- Nephron formation.

Congenital malformations

- Defects in body structure
- Incidence: ~3% live births (estimate influenced by definition, rate of detection, geographic variation). This value represents a fraction of the whole since ~75% structurally abnormal foetuses will be aborted spontaneously. Each part and organ of an embryo has a critical or sensitive period during which its development may be disrupted.

Causes

Environmental

• Environmental factors cause 7–10% congenital abnormalities.
  Teratogens are environmental agents that cause, or raise the incidence
  of, a developmental anomaly following maternal exposure. Include:
  • Ionizing radiation
  • Drugs, e.g. thalidomide
  • Viruses, e.g. rubella
  • Maternal effects, e.g. metabolic disorders
  • Trauma.

Genetic

• Estimated to cause at least one-third of all birth defects. Major genetic
  errors cause failure in late embryonic or early foetal stages and include:
  • Numerical defects, e.g. trisomy, monosomy
  • Structural defects: translocations, deletions, inversions, duplications
  • Single gene defects: account for 7–8% congenital abnormalities.

Multifactorial

• Many common congenital abnormalities have distributions that suggest
  multifactorial inheritance, i.e. multiple genes may be involved and/or
  there is a gene/environmental interaction (e.g. cleft/lip palates).

Fertilization and pre-implantation development (weeks 1 and 2)

Fertilization between the female oocyte and the male sperm creates the new individual—the zygote. The zygote begins a series of mitotic cell divisions, called cleavage, producing daughter cells, called blastomeres, which remain totipotent up to the eight-cell stage. Clinical implications:
• IVF
• Pre-implantation prenatal diagnosis.

The embryo moves down the oviduct, reaching the uterus by day 4. It now consists of 16–32 tightly aligned blastomeres, called the morula. Compaction occurs, tightly aligning blastomeres and segregating them into two groups:
• Inner cell mass (ICM) (or embryoblast)—give rise to the embryo
• Outer cell mass (OCM) (or trophoblast)—contribute to the placenta.

Fluid is accumulated, which collects inside the blastocoele cavity. The ICM becomes displaced to one side—the embryonic pole. The embryo is now called the blastocyst (Fig. 10.9a).

Implantation

Begins about day 6, when the embryo attaches to the uterus endometrium. Endometrial cells adjacent to the implanting blastocyst undergo the decidual reaction and start to secrete growth factors and metabolites to support the embryo. Trophoblastic cells invade the uterine wall and differentiate into two populations:
• An inner layer of mononuclear cells—the cytotrophoblast
• A highly invasive outer layer, without distinct cell boundaries—the syncytiotrophoblast.

The trophoblastic cells expand, surrounding the blastocyst so it becomes embedded in the endometrium. During week 2, these trophoblastic cells begin to collaborate with uterine tissue to form the placenta. Implantation is complete by day 13. Embryos that implant outside the uterus cause ectopic pregnancies (95% in uterine tubes).

Formation of the bilaminar germ disc (day 14)

See Fig. 10.9b.
During implantation, the icm differentiates into two layers:
• The epiblast
• The hypoblast—adjacent to the blastocoele.

These comprise the bilaminar germ disc.

A cavity forms in the epiblast (the amniotic cavity), splitting it into two layers. The epiblast layer abutting the cytotrophoblast will form the amnion which eventually envelops the embryo.

Two cell populations (extra-embryonic endoderm and extra-embryonic mesoderm) migrate from the hypoblast to line the blastocoele cavity, transforming it first into the primary yolk sac and then the definitive/secondary yolk sac. This does not contain yolk, but will be an important site of blood formation.

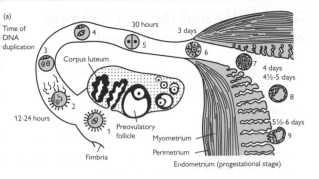

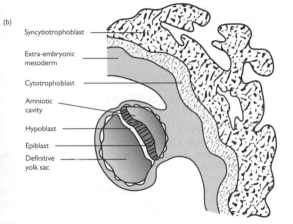

Fig. 10.9 (a) Events of the first 6 days of development of a human embryo. 1: oocyte immediately after ovulation; 2: fertilization 12–24h later results in the zygote; 3: zygote contains male and female pronuclei; 4: first mitotic division; 5: two cell stage; 6: 3-day morula made up of up to 16 blastomeres; 7: morula stage (16–32 blastomeres) reaches the uterine lining; 8: early blastocyst; 9: implantation occurs at around day 6. (b) The site of implantation at the end of the second week.

Extra-embryonic mesoderm splits, forming the chorionic cavity. The bilaminar embryo, and its associated amniotic and yolk sac cavities, is suspended in it, connected to the outer wall by the connecting stalk.

Gastrulation and the establishment of germ layers (weeks 3 and 4)

Gastrulation

- Converts the bilaminar embryo to the three-layered embryo
- Involves a complex series of cell movements
- Begins with formation of the primitive streak.

Primitive streak

See Fig. 10.10.

- Forms on day 14 when epiblast cells pile up at the future posterior edge of the blastodisc
- Clearly defined by day 15 or 16; composed of a narrow groove with the primitive node (or organizer) at the anterior end
- The node has neural inducing and organizing properties
- Provides the first morphological sign of the main embryonic body axes
- Epiblast cells move through the streak, in a process called invagination:
  - First wave of invagination displaces the hypoblast creating embryonic endoderm
  - Second wave spreads between the endoderm and epiblast, forming embryonic mesoderm
  - Epiblast cells that do not go through the streak form embryonic ectoderm
  - These three germ layers give rise to all the tissues and organs in the embryo.

Germ layer derivatives

Ectoderm: skin and central nervous system

Neural-inducing signals from the node and notochord cause ectodermal cells to thicken, forming the neural plate which rolls up and closes (neurulation) creating the neural tube, separating from non-neural ectoderm (Fig. 10.11).

Neural crest is induced at the lateral edges of the neural plate; it forms diverse mesodermal and neural derivatives, including melanocytes, dorsal root ganglia, and enteric nervous system.

Non-neural ectoderm forms skin and associated structures, e.g. nails and tooth enamel.

Mesoderm: all skeletal and connective tissue, blood, and muscle

Subdivided into:

- Axial mesoderm (notochord):
  - Forms the basis of the midline axial skeleton
  - Source of neural-inducing and patterning signals
- Paraxial mesoderm:
  - Condenses in segments on either side of the notochord
  - In the head, these form somitomeres, which contribute to head mesenchyme
  - In the trunk, they form 37 pairs of somites
  - Somites reveal the underlying segmental organization of the embryo.

Somites differentiate into (Fig. 10.12):

- The sclerotome which migrates medially forming:
  - The vertebral body (surrounding notochord)
  - The vertebral arch (surrounding neural tube)

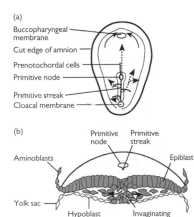

Fig. 10.10 (a) Diagrammatic view of the dorsal side of a 16-day embryo germ disc showing the movement of surface epiblast cells (solid black lines) through the primitive streak and node and the subsequent migration of cells between the hypoblast and epiblast (dashed lines). (b) Diagrammatic cross-section at 15 days through the cranial region of the primitive streak showing invagination of epiblast cells. The first cells to move inwards displace the hypoblast to create the definitive endoderm. Once definitive endoderm is established, inwardly moving epiblast forms mesoderm.

- The dermomyotome, which splits into:
  - Dermatome: contributes to dermis (fat and connective tissue of the neck and back)
  - Myotome: differentiates into muscles (dorsal epimere and ventral hypomere)
- Intermediate mesoderm:
  - Differentiates into urogenital structures
- Lateral plate mesoderm:
  - Divides to line the intra-embryonic coelom (primitive body cavity) forming:
    - Somatic (parietal) mesoderm—adjacent to ectoderm, forms dermis and limbs
    - Splanchnic (visceral) mesoderm—adjacent to endoderm; gives rise to gut wall and vascular system.

  Mesoderm is absent at two sites:
- Buccopharyngeal membrane (anterior)
- Cloacal membrane (posterior).

These mark the future ends of the gut tube.

Endoderm

- Folds extensively to form the gut tube and associated structures (→ p.708).

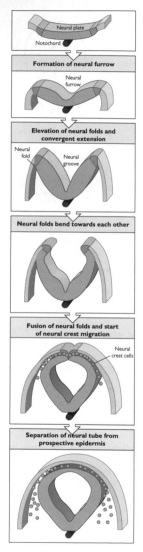

Fig. 10.11 Neurulation: neural tube formation results from formation and fusion of neural folds. The tube then detaches from the ectoderm.

Reproduced with permission from Wolpert, L. et al, Principles of Development 6e, 2019 Oxford University Press.

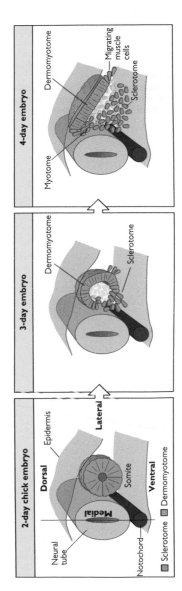

Fig. 10.12 Diagram of a transverse section through an embryo showing somite differentiation. The ventral medial quadrant of the somite gives rise to the sclerotome cells, which migrate medially forming the vertebral body. The rest of the somite, the dermomyotome, forms the dermatome and myotome, which gives rise to the dermis and all the trunk muscles. It also gives rise to muscle cells that migrate into the limb bud.

Reproduced with permission from Wolpert, L. et al. Principles of Development 6e, 2019, Oxford University Press.

Primitive streak regression

- During gastrulation, the embryonic disc grows and elongates along the anteroposterior axis
- The primitive streak gradually shortens (regresses) towards the posterior end of the embryo, disappearing by the end of week 4
- Gastrulation is now complete
- The extended period of gastrulation creates an anteroposterior gradient of development—thus anterior structures begin to differentiate, while gastrulation continues posteriorly.

Embryonic folding and formation of body cavities (weeks 4 and 5)

- Folding turns the flat, three-layered embryonic disc into a 3D structure
- It is driven by differential growth, since the embryo grows more than the yolk sac
- Incorporates the primitive body cavity called the intra-embryonic coelom
- Takes place in three directions (Figs 10.13, 10.14).

Rostral (anterior) folding

Results in ventral positioning of:
- Buccopharyngeal membrane/mouth
- Cardiogenic regions/heart
- Septum transversum/diaphragm.

Caudal (posterior) folding

Results in:
- Ventral positioning of cloacal membrane/anus and urogenital openings
- Displaces the connecting stalk so that it merges with the neck of the yolk sac (both contribute to umbilical cord)
- Formation of the allantois (outpocket of hindgut), which forms part of the bladder.

Lateral folding

Results in fusion of all three germ layers at the ventral midline.
- Endoderm forms the gut tube

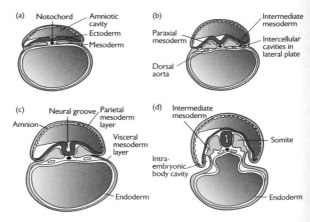

Fig. 10.13 Transverse sections showing development of the mesodermal germ layer at days 17 (a), 19 (b), 20 (c), and 21 (d). The thin mesodermal sheet gives rise to paraxial mesoderm (future somites), intermediate mesoderm (future excretory units), and lateral plate, which is split into parietal and visceral mesoderm layers lining the intra-embryonic cavity.

- Somatic (parietal) and splanchnic (visceral) lateral plate mesoderm fuse, lining and enclosing the intra-embryonic coelom. Somatic mesoderm coats the inside of the body wall; splanchnic mesoderm coats the endodermal gut tube and associated structures
- Ectoderm fuses, covering the outside of the body with skin.

Subdivision of the body cavity (5th week)

The intra-embryonic coelom is partially divided into:
- Thoracic (pericardial) and
- Abdominal (peritoneal) portions

by a mesodermal bar called the septum transversum, which later forms part of the diaphragm. Two spaces at the dorsolateral edges of the septum transversum—the pericardial-peritoneal canals—are later sealed by pleuroperitoneal membranes.

The diaphragm is made up of four elements:
- Septum transversum (central tendon)
- Pleuroperitoneal membranes (posterior diaphragm)
- Paraxial mesoderm of the body wall (T7–12)
- Dorsal mesentery surrounding the oesophagus (the crura).

Incidence of diaphragmatic hernias: 1/2000.

The primitive pericardial cavity splits into:
- Pleural cavities
- Pericardial cavity

by pleurocardial folds that originate along the lateral body wall and grow medially towards each other, between the lungs and heart, fusing at the end of week 5.

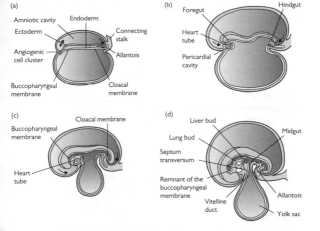

Fig. 10.14 Sagittal midline sections of embryos at various stages of development demonstrating cephalocaudal folding and its effect on position of the endoderm-lined cavity. Presomite embryo (a), seven-somite embryo (b), 14-somite embryo (c), 1-month embryo (d). Note the position of the angiogenic cell clusters in relation to the buccopharyngeal membrane.

Limb development

Limb specification and identity

- Paired limb buds appear at the end of week 4
- Upper limb buds appear and develop slightly ahead of hind limbs
- Limb identity (i.e. upper versus lower) is specified before they grow out

Limb bud initiation and bud outgrowth

- The early limb bud is composed of a core of somatic lateral plate mesoderm covered in a jacket of ectoderm
- Reciprocal epithelial–mesenchymal interactions between these cell layers govern limb initiation and outgrowth
- Signals from the mesoderm cause the overlying ectoderm to thicken forming the apical ectodermal ridge (AER), which edges the distal tip of the bud
- The AER maintains a region of dividing cells at the tip of the limb called the progress zone
- Signalling between the AER and the progress zone is required for bud outgrowth. Disrupting this signalling halts outgrowth, causing limb truncations.

Limb patterning

Limbs are asymmetrical and patterned in three dimensions (Fig. 10.15):
- Proximal–distal (shoulder to fingertips): limbs grow progressively; digits form last. The progress zone model proposes cells measure time spent in this zone—those staying a short time form proximal structures, those staying longer, distal structures
- Anteroposterior (thumb to little finger): patterned by long-range signals emanating from a group of mesenchymal cells at the posterior limb margin called the zone of polarizing activity. Cells closest to it form posterior structures; those further away form anterior structures
- Dorsal–ventral (back to palm of hand; extensor–flexor): patterning across this axis is controlled by non-ridge ectoderm.

Differentiation of limb structures

- Skeletal elements: condense from lateral plate mesoderm
- Skin and associated structures: form from ectoderm
- Blood vessels: differentiate from angioblasts in mesoderm
- Muscles: arise from somitic mesoderm which invades the limb as dorsal and ventral muscle masses giving rise to dorsal extensor and ventral flexor muscles
- Nerves: ventral branches of spinal nerves C5–T2 upper, L4–S3 lower, innervate limb muscles. They come together at the limb bud base forming the plexus (brachial, upper limb; lumbosacral, lower limb) before separating to enter the limb. Sensory nerves follow motor nerves (Figs 10.16, 10.17).

Limb shaping

Sculpting of the axilla (armpit) and separation of the digits depends on programmed cell death. During weeks 6–8, limbs rotate from their initial position (roughly at right angles to the trunk) to assume their adult position.

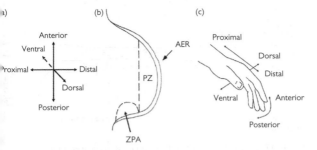

Fig. 10.15 Signalling regions in the early limb bud. Proximal–distal outgrowth is regulated by epithelial–mesenchymal interactions between the apical ectodermal ridge (AER) and progress zone (PZ). Anteroposterior patterning is controlled by cells in the zone of polarizing activity at the posterior edge of the bud. Dorsal–ventral patterning is directed by non-ridge ectoderm. (a) Limb axes, (b) early limb bud, (c) adult hand.

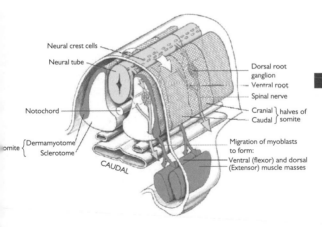

Fig. 10.16 Origin of tissues in the limb. The myoblasts migrate from the dermomyotome into the limb, where they form ventral (flexor) and dorsal (extensor) muscle masses. Motor innervation from the ventral root region of the neural tube migrates through the cranial half of each sclerotome, where it joins sensory (dorsal root ganglion) nerve cells and their processes and Schwann cells, both derived from neural crest cells. Other neural crest cells forming melanocytes migrate dorsal to the somites. Surface ectoderm forms skin, hair, nails, and sweat and sebaceous glands.

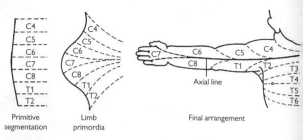

Fig. 10.17 Development of segmental sensory innervation (dermatomes) of skin of upper limb.

Cardiovascular system development

- Cardiovascular system (CVS) anomalies = most common congenital defects—average at least 1 in 200
- This reflects the complex demands of having to combine development with efficient CVS function
- CVS develops early because by week 3 the embryo cannot get sufficient oxygen and nutrients or remove waste products by diffusion alone.

Formation of early blood vessels and simple heart tube

- CVS is derived from mesoderm
- In the yolk sac mesoderm, blood islands form, creating a network of vessels
- These link with left and right dorsal aorta that condense in the embryo
- Paired endocardial tubes develop in the horseshoe-shaped cardiogenic region, anterior to the gastrulating embryo, from splanchnic mesoderm
- Embryonic folding draws these tubes together, which fuse in the midline forming the primitive heart tube by day 22
- Contractile activity begins just before fusion
- Heart tube has an inner endocardial layer coated in myocardium, enclosed in epicardium.

From inflow to outflow the heart tube consists of:
- Sinus venosus
- Primitive atria
- Ventricle
- Bulbus cordis.

Three pairs of veins drain blood to the heart via the sinus venosus—the common cardinal, the vitelline, and umbilical veins. Blood leaves the heart through the first aortic arch, which hooks over the foregut, joining the paired dorsal aorta. Vitelline arteries branch off to supply the yolk sac, and umbilical arteries reach the placenta.

- Primitive heart tube loops to form the basic definitive shape by 4½ weeks (Fig. 10.18)
- Looping = first morphological sign of left/right asymmetry
- Looping affects blood flow through the heart tube.

Circulatory system development

Asymmetrical changes in the blood outflow tract (aortic arches) and inflow tract (venous system) accompany changes in heart shape, ultimately producing systemic and pulmonary circulatory systems (Fig. 10.19).

Development of the aortic arches
- Initially, blood leaves the heart through one pair of aortic arches (first)
- As the embryo grows, a series of additional arches form, called 2, 3, 4, and 6
- These are never all present at once, but instead form progressively in a rostral/caudal sequence
- Important derivatives are:
  - Carotid arteries (third arch)
  - Aortic arch (left fourth arch)
  - Pulmonary artery (sixth arch), connected to aorta via the ductus arteriosus in the foetus.

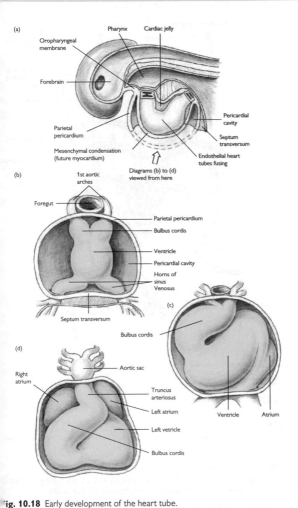

Fig. 10.18 Early development of the heart tube.

Reproduced from MacKinnon, Pamela and Morris, John, *Oxford Textbook of Functional Anatomy*, vol , p68 (Oxford, 2005). With permission of OUP.

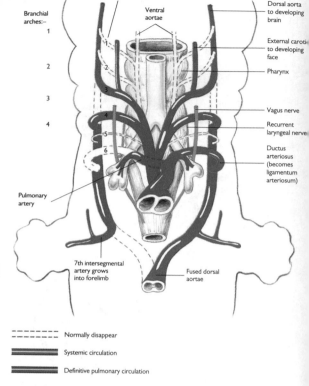

Fig. 10.19 Development of the great arteries.
Reproduced from MacKinnon, Pamela and Morris, John, *Oxford Textbook of Functional Anatomy*, vol 2, p70 (Oxford, 2005). With permission of OUP.

Development of the venous system
- Initially, bilaterally symmetrical but asymmetrical changes during weeks 5 and 6 bias blood flow to enter just the right side of the heart
- Rostrally, returning blood flows through a new vessel (left brachiocephalic vein) linking the left and right anterior cardinal veins, which drains into the right side of the heart, via the superior vena cava
- Caudally, a capillary network develops in the growing liver from vitelline and umbilical veins. To bypass this, the ductus venosus forms, shunting blood from left umbilical vein to enter the right side of the heart, via the inferior vena cava. The right umbilical vein regresses
- Pulmonary veins form as outgrowths of the left atrial wall.

Division of the heart

See Fig. 10.20.

Haemodynamic influences

Blood entering the asymmetrical heart tube creates differential pressures that act on the internal walls of the heart. These haemodynamic forces fundamentally affect heart shape during septation, and distortions in them e.g. caused by defective inflow vessels) may cause septation defects.

Division of the atrioventricular canal and outflow tract (truncus arteriosus) (early 5th week)

- Divided by endocardial cushions, derived from either endocardium (AV canal) or neural crest (truncus arteriosus)
- Fusion between inferior and superior endocardial cushions splits the single AV canal into right and left. The cushions also contribute to the mitral (bicuspid) valve on the left-hand side, and the tricuspid valve on the right-hand side
- A spiral septum forms in the truncus arteriosus, which separates the future aorta and pulmonary artery
- Aortic and pulmonary semilunar valves form at the distal tips of the ventricles from three endocardial swellings
- Abnormal blood flow through the AV canal or truncus arteriosus may cause septation defects (e.g. unequal partition → Fallot's tetralogy).

Atrial septation (late 5–7th weeks)

- Common atrium is divided by the septum primum growing down from the roof
- It perforates superiorly as a result of apoptosis just before reaching the fused endocardial cushions at the AV canal
- Septum secundum grows to the right of the septum primum overlapping the perforations
- Together, these create an interatrial valve—the foramen ovale
- At birth, this is sealed by blood pressure changes pressing the septa together
- Incidence of atrial septation defects: 6/10,000 live births.

Ventricular septation (late 5–7th weeks)

- The muscular interventricular septum forms between the enlarging ventricles, leaving a small interventricular gap above
- This closes by fusion between the membranous interventricular septum with the base of the spiral septum in the outflow tract
- Failure is common, → an interventricular septal defect: 12/1000.

Changes at birth

At birth, the source of oxygen switches from placenta to lungs, due to which:
- Foramen ovale closes—becomes fossa ovalis
- Umbilical arteries close—become medial umbilical ligaments
- Ductus arteriosus closes—becomes ligamentum arteriosum
- Umbilical vein closes—becomes ligamentum teres
- Ductus venosus closes—becomes ligamentum venosum.

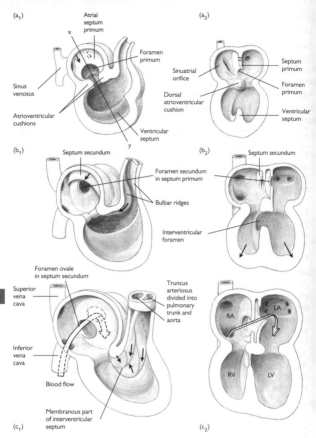

Fig. 10.20 Septation of the heart: the developing heart bisected sagittally and viewed from the right (a_1, b_1, c_1) and bisected coronally and viewed from the front (a_2, b_2, c_2). The diagrams show the division (septation) of the right and left atria and ventricles and the formation of the aorta and pulmonary artery from the bulbus cordis.

Reproduced from MacKinnon, Pamela and Morris, John, *Oxford Textbook of Functional Anatomy*, vo 2, p69 (Oxford, 2005). With permission of OUP.

Development of the gut and associated structures

- Embryonic folding creates the endodermal gut tube and incorporates the intra-embryonic coelom
- The gut is initially connected to the body wall by ventral and dorsal mesenteries formed from splanchnic mesoderm
- This ventral connection (except at stomach and liver levels) breaks down, leaving the majority of the gut tube suspended by the dorsal mesentery.

Connective tissue, muscles, and blood vessels surrounding the gut tube are derived from splanchnic lateral plate mesoderm. The gut is initially blind-ended, sealed by the buccopharyngeal membrane rostrally and cloacal membrane caudally. These membranes mark the borders between ectodermally derived portions of the digestive tract, stomodeum (first part of oral cavity), and proctodeum (last part of anal canal). The gut is divided into:

- Pharyngeal gut—gives rise to the pharynx and related structures
- Foregut, midgut, and hindgut.

Foregut development

- Blood supply = coeliac axis
- Gives rise to oesophagus, trachea, and lung buds (◆ p.712), stomach, and cranial duodenum
- Associated structures: liver, gallbladder, and pancreas—which all bud off the duodenum
- The stomach is suspended by the dorsal mesogastrium (part of the dorsal mesentery) and connected to the ventral body wall by the ventral mesentery (derived from the septum transversum), that also encloses the developing liver. As it grows, it rotates along its longitudinal axis pressing the duodenum against the dorsal body wall, creating a space = lesser sac. The rest of the peritoneal cavity = greater sac
- The endodermal liver bud sprouts into the ventral mesentery at the foregut–midgut junction; remains linked to foregut by the bile duct from which buds off the future gallbladder and cystic duct. The liver bud forms hepatic cords that become parenchyma. Bile formation begins during the 12th week
- The pancreas develops from the dorsal and ventral pancreatic buds, which fuse together. Both buds branch extensively. The tips of the endodermal branches form the acini of the exocrine pancreas. The endocrine pancreas consists of ~1 million islets of Langerhans, scattered among the acini. The pancreatic and common bile ducts enter the duodenum together at the duodenal papilla
- Spleen: considered a gut-associated structure but originates from mesodermal cells, which condense in the dorsal mesogastrium.

Midgut development

See Fig. 10.21.
* Supplied by the superior mesenteric artery
* Forms the caudal duodenum, jejunum, ileum, caecum, appendix, ascending colon and two-thirds of the transverse colon
* Lengthens rapidly and, by early 6th week, herniates, also rotating 90° anticlockwise, into the umbilical cord as a hairpin loop, which at its apex remains connected to the yolk sac via the vitelline duct. During 10th week, it retracts into the peritoneal cavity and rotates anticlockwise 180° (i.e. total 270° rotation). Failure to retract fully, or rotate incorrectly, are common defects
* The vitelline duct normally regresses, but persists as a Meckel's diverticulum in ~2% of live births.

Hindgut development

* Supplied by inferior mesenteric artery and middle rectal artery
* Forms the distal third of transverse colon, descending colon, rectum, and upper part of the anal canal
* Ends at the cloaca, sealed by the cloacal membrane
* Cloaca is divided by the urorectal septum, so when the cloacal membrane breaks down there are two exterior openings—the urogenital sinus and anus. Abnormalities in separation or cloacal size can cause the anus to exit abnormally via the vagina or urethra (fistulas) or be absent (atresia).

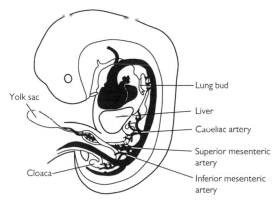

Fig. 10.21 Embryo during the 6th week of development, showing blood supply to the segments of the gut and formation and rotation of the primary and intestinal loop. The superior mesenteric artery forms the axis of this rotation and supplies the midgut. The coeliac and inferior mesenteric arteries supply the foregut and hindgut, respectively.

Gut fixation and recanalization

See Fig. 10.22.

- After gut retraction and rotation, some parts of the gut fuse with the dorsal body wall (fixation)—abnormalities in rotation and fixation may cause gut strangulation
- During the 6th week, proliferation of the epithelial endoderm transiently blocks the gut lumen. Recanalization, forming the definitive gut tube, is completed by week 9—failure leads to stenosis (narrowing) or duplications.

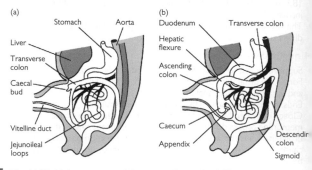

(a)

Stomach   Aorta

Liver

Transverse colon

Caecal bud

Vitelline duct

Jejunoileal loops

(b)

Duodenum   Transverse colon

Hepatic flexure

Ascending colon

Caecum

Appendix

Descending colon

Sigmoid

Fig. 10.22 (a) Anterior view of the intestinal loops after 270° anti-clockwise rotation. Note the coiling of the small intestinal loops and the position of the caecal bud in the right upper quadrant of the abdomen. (b) Similar view as in (a), with the intestinal loops in their final position. Displacement of the caecum and appendix caudally places them in the right lower quadrant of the abdomen.

Respiratory system development

- Respiratory system buds off the ventral wall of the foregut during week 4 at the level of the oesophagus
- Epithelia of the alveoli, bronchi, trachea, and larynx are endodermal
- Muscles, cartilage, and connective tissue that surround this layer are derived from splanchnic mesoderm
- Respiratory bud elongates forming the trachea, and a lung bud appears at the caudal end. It is separated from the oesophagus by the tracheo-oesophageal septum, only joining the gut tube at the larynx. Defective separation leads to fistulas and oesophageal atresia.

The lung bud initially splits into the right and left primary bronchi which divide further, showing left–right asymmetry.

- Right forms three secondary bronchi and three lung lobes
- Left bud forms two secondary bronchi and two lung lobes
- Lung endoderm goes on developing by branching morphogenesis and budding forming the entire bronchial tree, which expands into the pleural cavities.

Respiratory system differentiation

- Lung endoderm becomes coated with mesenchymal tissue derived from splanchnic mesoderm. In the upper respiratory tract, this differentiates into cartilaginous rings around the trachea and bronchi
- Lung endoderm differentiates into the respiratory epithelium, eventually forming terminal sacs called alveoli
- This epithelium matures late (95% of ~300 million alveoli develop after birth)
- Gaseous exchange between blood and air only becomes possible in the 7th month, so lung maturation limits survival of premature babies
- Insufficiency causes respiratory distress syndrome.

Head and neck development

The vertebrate cranial region has:
- A neurocranium associated with the brain and major sense organs (nose/eye/ear)
- A viscerocranium, formed from the pharyngeal (branchial) arches, associated with the oral region and pharynx.

Embryonic components of the head and neck

See Fig. 10.23.
- Neural tube and ectodermal placodes form the brain and special sense organs
- The neural crest contributes to face, palate, tongue, pharynx, larynx, external and middle ear, and also forms the intrinsic eye muscles
- Cranial paraxial mesoderm forms the muscles of the head, including extrinsic eye muscles
- Occipital somites form the occipital part of the skull and tongue muscles.

Structure and derivatives of the pharyngeal arches

See Fig. 10.24.
- Five bilateral pairs of pharyngeal arches appear during weeks 4–5
- Composed of bars of mesenchyme sandwiched between surface ectoderm and pharyngeal endoderm
- Each arch has its own artery, cranial nerve, muscles, and skeletal element
- Separated by pharyngeal clefts on the outer surface and by pharyngeal pouches on inner surface (outpockets of the pharyngeal foregut)
- Pharyngeal arch-derived structures include face, palate, tongue, thyroid gland, pharynx, larynx, and the external and middle parts of the ear.

Facial development

The face and jaw originate from primordia that surround the primitive mouth (Fig. 10.25).
- Neural crest from the cranial neural folds migrates ventrally and rostrally forming paired mandibular and maxillary processes and nasal swellings
- Mandibular processes fuse medially, forming the lower jaw
- Medial and lateral nasal swellings grow and partially surround the paired nasal placodes, which deepen into two nasal cavities. These remain separated by the nasal septum but are continuous with the oral cavity
- Medial nasal swellings fuse together creating the intermaxillary segment, which forms the median part of the nose, philtrum, and primary palate
- Maxillary processes fuse with the lateral nasal swellings (forming the side of the nose) and the nasolacrimal duct forms along this line of fusion
- Maxillary and mandibular processes partially fuse forming cheeks
- Inside the oral cavity, the maxillary segments project downwards, either side of the tongue, as palatal shelves which elevate and fuse together in the midline, and also with the primary palate, creating the definitive palate
- If any of these fusions fail, congenital facial clefting occurs—most commonly cleft lip and palate.

Skull development

- Base of the skull (chondrocranium) formed by endochondral ossification
- Skull vault (neurocranium) formed by direct dermal ossification
- Skull bones do not fuse together until early childhood, allowing the cranium to deform during birth then expand during childhood as the brain enlarges.

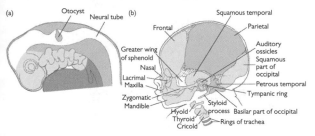

Fig. 10.23 Contributions of neural crest, paraxial, and lateral plate mesoderm to (a) the connective and skeletal tissues of the developing head in a 7mm embryo, and (b) the skull and anterior neck skeleton at term.

Reproduced from MacKinnon, Pamela and Morris, John, *Oxford Textbook of Functional Anatomy*, vol 3, p37 (Oxford, 2005). With permission of OUP.

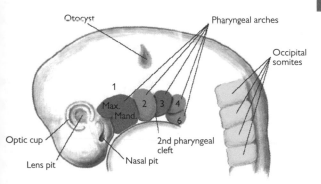

Fig. 10.24 Views of 7mm embryo, to show the early development of nose, eye, ear pharyngeal arches, and somites.

Reproduced from MacKinnon, Pamela and Morris, John, *Oxford Textbook of Functional Anatomy*, vol 3, p35 (Oxford, 2005). With permission of OUP.

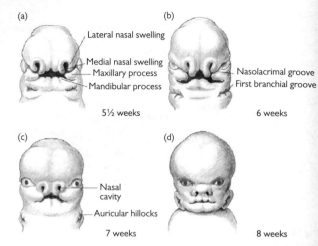

Fig. 10.25 Development of face; frontal views at (a) 5½ weeks, (b) 6 weeks, (c) 7 weeks, and (d) 8 weeks of foetal life.

Kidney development

Kidney development is intertwined with the genital system (➔ p.720), as both develop from intermediate mesoderm. Three overlapping kidney systems form in rostral–caudal sequence (Fig. 10.26).

Pronephros
- Forms and regresses during week 4
- Rudimentary and non-functional.

Mesonephros
- Forms in week 4; extending from T1 to L3
- Has S-shaped tubules draining into mesonephric ducts, which in the male participates in genital system formation, but regresses in females.

Metanephros
- Definitive kidney appears in week 5
- Forms from the ureteric bud that branches off the mesonephric duct and metanephric blastema.

Collecting system
- Ureteric bud forms collecting ducts, which project into the metanephric blastema
- Reciprocal epithelial–mesenchymal interactions between these tissues cause the bud to dilate and divide forming the renal pelvis and major calyces
- Further branching eventually forms minor calyces and approximately 1–3 million collecting ducts.

Nephrons
- = functional units of the kidney
- Develop from metanephric blastema
- Reciprocal inductive interactions induce the blastema tissue to form renal vesicles which lengthen into S-shaped tubules, each supplied by a capillary that differentiates into a glomerulus
- The tubule and glomeruli form the nephron
- The proximal end of the tubule forms Bowman's capsule; the distal end links to the collecting duct
- Tubule elongation creates the proximal and distal convoluted tubule and loop of Henle
- Nephrons are formed until birth, when there are ~1 million in each kidney.

Ascent of kidneys

The kidneys ascend from the pelvis to the upper lumbar region. Occasionally, they partially fuse together during their ascent, forming a horseshoe kidney.

Bladder formation
- Endodermally derived from dilation of the allantois—a hindgut diverticulum
- Distal ends of the ureters and mesonephric ducts (if male) drain into the bladder at the trigone, entering at an oblique angle to prevent urine reflux back to the kidneys.

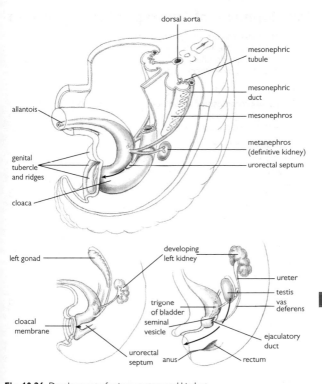

Fig. 10.26 Development of urinary system and hindgut.
Reproduced from MacKinnon, Pamela and Morris, John, *Oxford Textbook of Functional Anatomy*, vol 2, p170 (Oxford, 2005). With permission of OUP.

Development of the reproductive system

- Intimately linked to urinary system development (→ p.718)
- Originates from intermediate mesoderm
- Reproductive development is initiated by differential gene expression and propagated by endocrine signals.

Sex determination

- Normal sex chromosome karyotypes: XX = female; XY = male
- Sex-determining region on Y chromosome = SRY
- SRY = transcription factor that initiates a gene cascade.

Gonad development

See Fig. 10.27.

- Gonads sexually indifferent until week 7
- Appear in week 6 as a pair of longitudinal tubes called genital ridges formed from mesonephric mesenchyme covered in epithelium, medial to the mesonephric (or Wolffian) and paramesonephric (or Müllerian) ducts
- They are populated by primordial germ cells, which migrate from the endodermal wall of the yolk sac
- Before and during primordial cell arrival, genital ridge epithelium proliferates, penetrating the underlying mesenchyme forming primitive sex cords, and maintains a connection with the surface epithelium.

Testis differentiation

- SRY transcripts are present in XY genital ridges
- Sex cords proliferate forming horseshoe-shaped testis cords connected to the rete testis at their apex
- The cords are composed of primitive germ cells and Sertoli cells, which start to secrete anti-Müllerian hormone (or Müllerian inhibitory substance)
- Between the cords, mesenchymal Leydig cells begin to synthesize testosterone from week 9
- Cords remain solid until puberty, when they differentiate into seminiferous tubules
- A dense, fibrous layer—the tunica albuginea—separates the cords from the surface epithelium.

Ovary differentiation

- No SRY transcripts
- Primitive sex cords degenerate
- Instead, surface epithelium proliferates forming cortical cords, which surround the primordial germ cells, now oogonia
- Cortical cells differentiate as follicular cells and oogonia first proliferate (~2 million present at birth), then, by week 20, arrest at meiotic prophase.

Genital duct system

Indifferent stage (week 7)

- Mesonephric ducts draining into bladder
- Paramesonephric ducts, with funnel-shaped openings rostrally, fused together caudally—they do not enter the bladder.

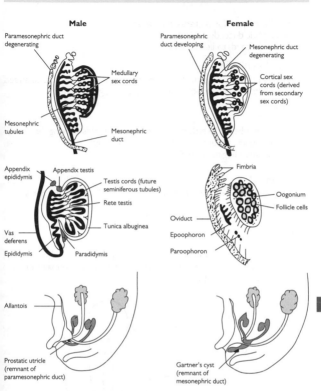

Fig. 10.27 Male and female gonadal development. The male and female genital systems are virtually identical through the 7th week. In the male, SRY protein produced by the pre-Sertoli cells causes the medullary sex cords to develop into presumptive seminiferous tubules and rete testis tubules and causes the cortical sex cords to regress. Anti-Müllerian hormone produced by the Sertoli cells then causes the paramesonephric ducts to regress and Leydig cells also develop, which in turn produce testosterone, the hormone that stimulates development of the male genital duct system, including the vas deferens and the presumptive efferent ductules.

Male
- anti-Müllerian hormone causes paramesonephric ducts degeneration
- Testosterone mediates differentiation of mesonephric ducts (future vas deferens) and tubules (some connect to duct = vas efferens)
- Associated glands are seminal vesicles, prostate, and bulbourethral
- Testes descend, entering the scrotal sacs via the inguinal ring.

Female
- No anti-Müllerian hormone, so paramesonephric ducts persist
- Cranially become fallopian tubes

- Caudally joined—form uterus and upper vagina
- Mesonephric ducts degenerate
- Ovaries descend, to lie in the pelvis.

External genitalia

Depends on conversion of testosterone to dihydrotestosterone, catalysed by 5α-reductase, synthesized by tissue around urogenital sinus.

Indifferent

See Fig. 10.28.
- Has genital tubercle, which divides into urethral folds flanking the urogenital sinus, all surrounded by genital swellings
- Differentiation begins week 9.

Male

See Fig. 10.29.
- Genital tubercle extends rapidly to form the phallus, later the penis
- Urethral folds close the urogenital sinus
- Genital swellings become scrotal swellings.

Female

See Fig. 10.30.
- Genital tubercle elongates slightly, forming clitoris
- Urethral folds become labia minora and urogenital sinus remains open
- Genital swellings form the major labia.

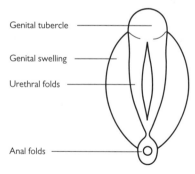

Genital tubercle

Genital swelling

Urethral folds

Anal folds

Fig. 10.28 Indifferent stages of the external genitalia, ~6 weeks.

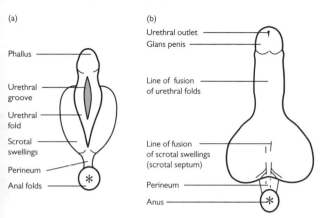

Fig. 10.29 (a) Development of external genitalia in the male at 10 weeks. (b) newborn.

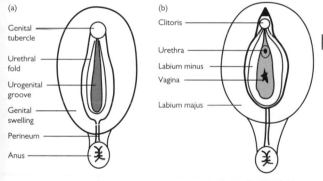

Fig. 10.30 Development of the external genitalia in the female at 5 months (a) and in the newborn (b).

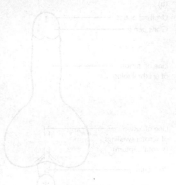

Neuroanatomy and neurophysiology

The skull and cervical vertebrae

The bones of the skull form the cranium and facial skeleton (Figs 11.1–11.8)

The cranium contains the brain and immediate relations and is divided into:

- The upper vault—comprising four flat bones:
  - The frontal bone anteriorly
  - The occipital bone posteriorly
  - Two lateral parietal bones
- The lower base, characterized by stepped fossae:
  - Anterior (containing the frontal lobes of the brain) formed from:
    - Orbital plates of frontal bone
    - Cribriform plate of the ethmoid bone
    - Lesser wing of the sphenoid bone
  - Middle (containing the temporal lobes of the brain) formed from:
    - Greater wing and body of the sphenoid bone
    - (Vertical) squamous and (horizontal) petrous parts of temporal bone
  - Posterior (containing cerebellum, pons. medulla oblongata) formed from:
    - Petrous part of temporal bone
    - Squamous part of occipital bone
    - Body of the sphenoid bone.

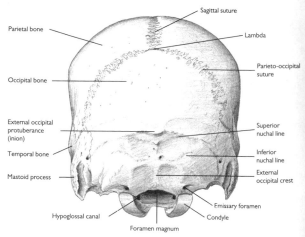

Fig. 11.1 Posterior aspect of skull.

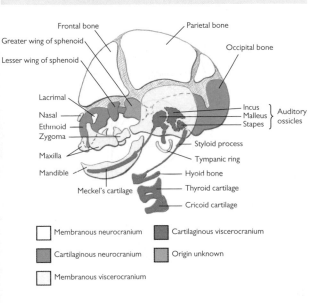

Fig. 11.2 Skeletal components of skull and anterior part of the neck. Cervical vertebrae (not shown) are formed from cervical sclerotomes.

Reproduced from MacKinnon, Pamela and Morris, John, *Oxford Textbook of Functional Anatomy*, vol 3, p44 (Oxford, 2005). With permission of OUP.

There are a number of important features in the base of the skull (Figs 11.3, 11.4, 11.6):
- Anterior fossa:
  - In the cribriform plate of the ethmoid bones:
    - Foramina convey the olfactory nerve[cranial nerve I]
- Middle fossa:
  - In the greater wing of the sphenoid bone:
    - The foramen ovale conveys the mandibular division of trigeminal nerve[VII] and the lesser petrosal nerve
    - The foramen rotundum conveys the maxillary division of trigeminal nerve[VIII]
    - The foramen spinosum conveys middle meningeal artery and vein
  - At the medial edge of the petrous part of the temporal bone:
    - The upper part of the foramen lacerum conveys the internal carotid artery (from the carotid canal)
  - Between the body and lesser wing of the sphenoid bone:
    - The optic canal conveys the optic nerve[II] and ophthalmic artery
  - Between the greater and lesser wings of the sphenoid:
    - The superior orbital fissure conveys the oculomotor nerve[III], trochlear nerve[IV], ophthalmic branch of trigeminal nerve[VI], and abducent nerve[VI]

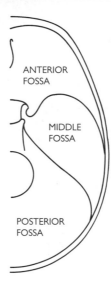

Fig. 11.3 Cranial fossae.

Reproduced from MacKinnon, Pamela and Morris, John, *Oxford Textbook of Functional Anatomy*, vol 3, p48 (Oxford, 2005). With permission of OUP.

- In the body of the sphenoid:
  - The sella turcica is a depression in which sits the pituitary gland
  - The junction of the frontal, parietal, and temporal bones—the pterion—is the thinnest and weakest point of the lateral skull (Fig. 11.5)
- Posterior fossa:
  - In the occipital bone:
    - The foramen magnum conveys the medulla oblongata, spinal part of accessory nerve[XI], upper cervical nerves, vertebral arteries
    - Anterior to the foramen magnum the brainstem lies on the clivus (fused basiocciput and basisphenoid)
    - The hypoglossal canal conveys the hypoglossal nerve[XII]
  - In the petrous temporal bone:
    - The internal acoustic meatus conveys facial nerve[VII], vestibulocochlear nerves[VIII], labyrinthine artery
  - Between the petrous temporal bone and the occipital bone:
    - The jugular foramen conveys the glossopharyngeal nerve[IX], vagus nerve[X], accessory nerve[XI], and the sigmoid sinus.

The pyramidal pterygopalatine fossa is defined by the sphenoid, palatine, and maxilla bones: it contains the maxillary division of the trigeminal nerve, maxillary artery, and accompanying veins and lymphatics.

The bones forming the facial skeleton (Fig. 11.5) are suspended below the anterior cranium and comprise:

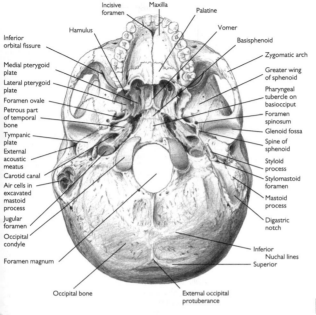

Fig. 11.4 Basal aspect of skull. The hypoglossal canal is covered by the occipital condyles.

Reproduced from MacKinnon, Pamela and Morris, John, *Oxford Textbook of Functional Anatomy*, vol 3, p47 (Oxford, 2005). With permission of OUP.

- Two nasal bones, joined to form the ridge of the nose
- Two maxillary bones, which form the floor of orbit, lateral wall of the nose, floor of the nasal cavity, and carry upper teeth
- Two lacrimal bones, which form medial wall of orbit
- One ethmoid bone, which forms the roof of the nose:
  - The cribriform plate of the ethmoid, along with the vomer and septal cartilage, form the nasal septum
- Two zygomatic bones, which form lateral wall of the orbit, cheek bone
- Two palatine bones with:
  - A perpendicular plate, which contributes to lateral wall of orbit
  - A horizontal plate, which, together with the palatine processes of the maxillary bones, forms the hard palate.

There are four paired paranasal air sinuses, contained within the frontal, maxillary, ethmoid, and sphenoid bones.

The mandible carries the lower teeth. It articulates with the cranium at the temporomandibular joint, which is:

- Between the head of the mandible and the mandibular fossa of the temporal bone

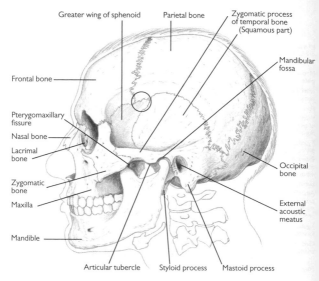

Fig. 11.5 Lateral aspect of skull, pterion circled.

Reproduced from MacKinnon, Pamela and Morris, John, *Oxford Textbook of Functional Anatomy*, vol 3, p47 (Oxford, 2005). With permission of OUP.

- A synovial joint containing a fibrocartilaginous disc
- Encapsulated, with reinforcement by temporomandibular, sphenomandibular, and stylomandibular ligaments.

Seven cervical vertebrae form the skeleton of the neck. C1 (the atlas), C2 (the axis), and C7 are atypical (Fig. 11.9). The long spine of C7 is the ver- tebra prominens—the superior-most process which can be palpated.

The lateral masses of the C1 vertebra articulates with condyles on the occipital bone at the atlanto-occipital joint, which is a loosely encapsulated synovial joint that permits flexion and extension (nodding movements).

Lateral masses of the atlas articulate with superior facets of the axis at atlanto-axial joints to permit rotation. In addition, the odontoid process or dens makes a midline articulation with an anterior facet.

The transverse part of the cruciate ligament (Fig. 11.10) of the atlas holds the dens in place, prevents the dens impinging on the spinal cord. Alar liga- ments from the dens to the margin of the foramen magnum prevent exces- sive rotation. Longitudinal ligaments attach to the anterior and posterior aspects of the vertebral bodies.

Joints between the C2 and T1 vertebrae possess anterior and posterior longitudinal ligaments. Supraspinous and infraspinous ligaments are replaced by the nuchal ligament between occipital bone and C7; these joints permit flexion and extension.

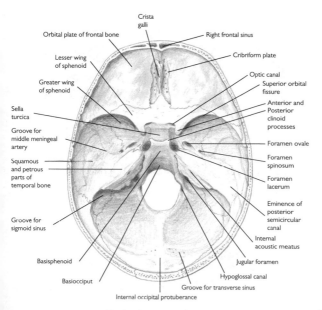

Fig. 11.6 Interior of skull. The foramen rotundum is hidden by the anterior clinoid process.

Reproduced from MacKinnon, Pamela and Morris, John, *Oxford Textbook of Functional Anatomy*, vol 3, p48 (Oxford, 2005). With permission of OUP.

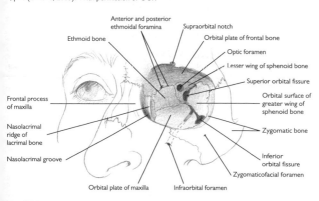

Fig. 11.7 Bones of orbit.

Reproduced from MacKinnon, Pamela and Morris, John, *Oxford Textbook of Functional Anatomy*, vol 3, p120 (Oxford, 2005). With permission of OUP.

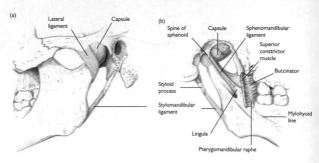

Fig. 11.8 Temporomandibular joint and its ligaments: (a) lateral aspect; (b) medial aspect.

Reproduced from MacKinnon, Pamela and Morris, John, *Oxford Textbook of Functional Anatomy*, vol 3, p82 (Oxford, 2005). With permission of OUP.

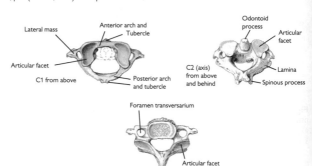

Fig. 11.9 Atlas (C1), axis (C2), and C7 vertebrae.

Reproduced from MacKinnon, Pamela and Morris, John, *Oxford Textbook of Functional Anatomy*, vol 3, p56 (Oxford, 2005). With permission of OUP.

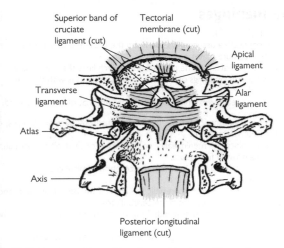

Fig. 11.10 Ligaments of the occipito-atlanto-axial region: midline section.

Reproduced from MacKinnon, Pamela and Morris, John, *Oxford Textbook of Functional Anatomy*, vol 3, p58 (Oxford, 2005). With permission of OUP.

The meninges

See Figs 11.11–11.13.

Three connective tissue layers surround the brain and spinal cord. The inelastic dura mater comprises:

- Outer endosteal layer (the periosteum), which lines the bones of the cranium and at openings of the skull and is continuous with that on outer surface
- Inner meningeal layer, which is continuous with that of the spinal cord.

Meningitis is inflammation of the meninges caused by bacteria, viruses, o fungi (→ OHCM10 p.822).

At some points, the meningeal layer doubles-back on itself to form dura folds. These provide four septa:

- The midline falx cerebri, between the cerebral hemispheres, which meets
- The horizontal tentorium cerebelli, the roof of the posterior fossa, and separates the cerebrum from the cerebellum
- The falx cerebelli which descends from the tentorium and separates the cerebellar hemispheres
- The diaphragm sellae which provides the roof of the sella turcica.

The middle layer—the arachnoid mater—follows the folds of the menin geal layer of dura mater, to which it is loosely attached. Between the two i the subdural space; the innermost layer—the pia mater—closely envelope

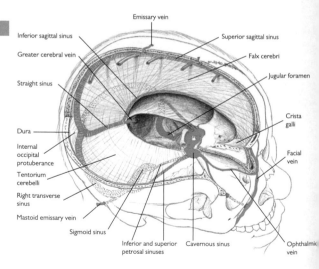

Fig. 11.11 Dural venous sinuses.

Reproduced from MacKinnon, Pamela and Morris, John, *Oxford Textbook of Functional Anatomy*, vol 3, p144 (Oxford, 2005). With permission of OUP.

the brain and spinal cord. It is separated from the arachnoid mater by the subarachnoid space, filled with CSF, which cushions the brain (Fig. 11.12).

- Bleeding can occur into the potential spaces created by the dura folds
- Extradural haemorrhage can occur following head trauma, and results from laceration of middle meningeal artery and vein (⊕ OHCM10 p.482)
- Subdural bleeds can occur insidiously following minor trauma and result from damage to bridging veins between cortex and venous sinuses (⊕OHCM10 p.482)
- Subarachnoid haemorrhages are spontaneous, most commonly due to rupture of an aneurism (⊕ OHCM10 p.478).

CSF is secreted by the choroid plexus (a vascularized epithelial structure) into each of the ventricles of the brain.

- CSF escapes from the fourth ventricle of the brain into the subarachnoid space
- It exchanges freely with the extracellular fluid surrounding neurones across the pia mater covering the surface of the brain and across the epithelial lining (ependyma) of the ventricles. Failure for this to happen is a cause of hydrocephalus (⊕ OHCM10 pp.467, 478, 486)
- In the choroid plexus, the epithelial cell barrier dictates the composition of CSF and insulates it from the blood. Elsewhere, the tight capillary endothelium prevents free exchange between the blood and the brain extracellular fluid. In these ways, the blood–brain barrier is established.

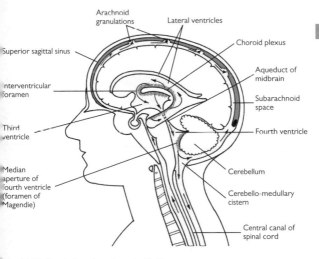

Fig. 11.12 Circulation of cerebrospinal fluid.
Reproduced from MacKinnon, Pamela and Morris, John, *Oxford Textbook of Functional Anatomy*, vol 3, p144 (Oxford, 2005). With permission of OUP.

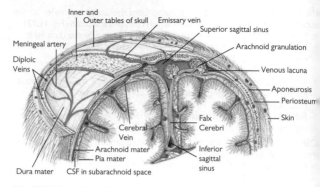

Fig. 11.13 Coronal section through cranium to show scalp, meninges, and arachnoid granulations.

Reproduced from MacKinnon, Pamela and Morris, John, *Oxford Textbook of Functional Anatomy*, vol 3, p145 (Oxford, 2005). With permission of OUP.

Blood and lymphatic vessels of the head and neck

See Figs 11.14–11.19.

Arterial supply

Blood is supplied to the brain by:
- The vertebral arteries (from subclavian artery):
  - They unite in the midline on the clivus as the basilar artery, and together these supply the brainstem and cerebellum
  - The basilar artery divides into left and right posterior cerebral arteries; these provide the posterior communicating arteries, which anastomose with the internal carotid artery
- The internal carotid artery (from the common carotid artery; Figs 11.20, 11.21):
  - Passes into the middle fossa in the carotid canal, through the foramen lacerum, then in the medial wall of the cavernous sinus
  - Gives rise to the ophthalmic artery, the central artery of the retina, and ciliary arteries
  - Terminates as the anterior and middle cerebral arteries, which supply the medial and lateral cerebral hemispheres, respectively
  - The anterior communicating artery unites the anterior cerebral arteries, completing the anastomosis between carotid and vertebral systems (the circle of Willis, which equalizes blood pressure)

The second division of the common carotid artery is the external carotid artery (Fig. 11.22). This artery has many branches which supply the structures of the head and neck other than the brain: superior thyroid; ascending pharyngeal; superficial temporal; lingual; facial; occipital; posterior auricular and superficial temporal.

It has two terminal branches which arise within the parotid gland:
- The (larger) maxillary artery, which gives off three groups of branches that supply the temporal fossa, infratemporal fossa, cranial dura, nasal cavity, oral cavity, and pharynx
- The superficial temporal artery.

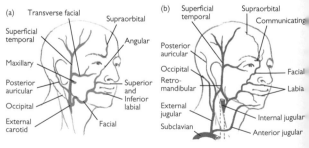

Fig. 11.14 (a) Arteries, (b) veins of the face.

Reproduced from MacKinnon, Pamela and Morris, John, *Oxford Textbook of Functional Anatomy*, vol 3, p169 (Oxford, 2005). With permission of OUP.

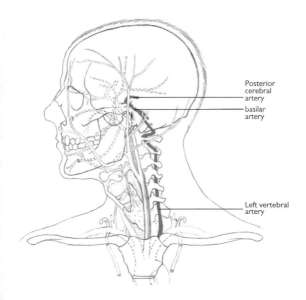

Fig. 11.15 Vertebral artery and its main branches.

Reproduced from MacKinnon, Pamela and Morris, John, *Oxford Textbook of Functional Anatomy*, vol 3, p155 (Oxford, 2005). With permission of OUP.

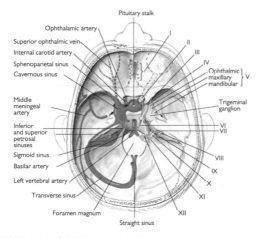

Fig. 11.16 Interior of skull base: vessels and nerves.

Reproduced from MacKinnon, Pamela, and Morris, John, *Oxford Textbook of Functional Anatomy* Vol 3, p48 (Oxford, 2005). With permission from OUP.

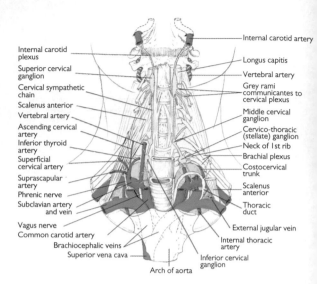

Fig. 11.17 Deep dissection of neck, showing prevertebral muscles, brachial plexus, root of neck, and sympathetic chain.

Reproduced from MacKinnon, Pamela, and Morris, John, *Oxford Textbook of Functional Anatomy* Vol 3, p190 (Oxford, 2005). With permission from OUP.

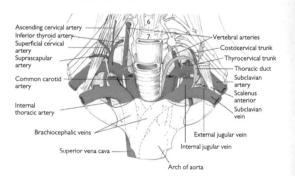

Fig. 11.18 Vessels in the root of neck.

Reproduced from MacKinnon, Pamela, and Morris, John, *Oxford Textbook of Functional Anatomy* Vol 3, p154 (Oxford, 2005). With permission from OUP.

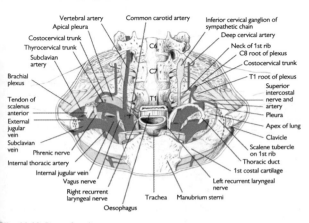

Fig. 11.19 Root of neck.

Reproduced from MacKinnon, Pamela, and Morris, John, *Oxford Textbook of Functional Anatomy* Vol 3, p191 (Oxford, 2005). With permission from OUP.

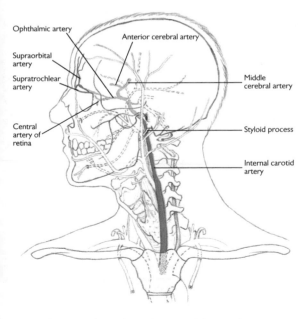

Fig. 11.20 Common and internal carotid arteries and their main branches.

Reproduced from MacKinnon, Pamela and Morris, John, *Oxford Textbook of Functional Anatomy*, vol 3, p156 (Oxford, 2005). With permission of OUP.

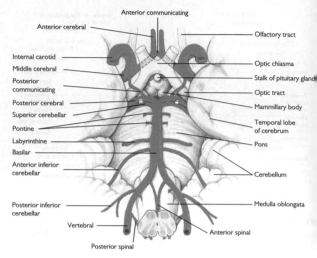

Fig. 11.21 Intracranial branches of the vertebral artery; circle of Willis.

Reproduced from MacKinnon, Pamela and Morris, John, *Oxford Textbook of Functional Anatomy*, vol 3, p155 (Oxford, 2005). With permission of OUP.

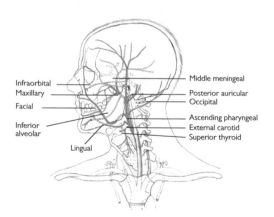

Fig. 11.22 External carotid artery and its branches.

Reproduced from MacKinnon, Pamela and Morris, John, *Oxford Textbook of Functional Anatomy*, vol 3, p158 (Oxford, 2005). With permission of OUP.

Between the two layers of the dura mater, intracranial venous sinuses lie in grooves on the overlying bone. These drain tributary veins from the brain, eye, and skull interior, diploic veins from the marrow of the cranial bones, and CSF from the subarachnoid space. The superior sagittal sinus runs in the attached margin of the falx cerebri, with lacunae along its length containing arachnoid granulations to reabsorb CSF. The sinus drains into:

* The transverse sinus which runs along the attached margin of the tentorium cerebelli, then turns inferiorly to become the sigmoid sinus
* The inferior sagittal sinus which runs in the free margin of the falx cerebri
* The straight sinus which forms from the unification of the inferior sagittal sinus and the great cerebral vein
* The intercommunicating cavernous sinuses are positioned either side of the sphenoid and pituitary gland and drain into the superior and inferior petrosal sinuses, which drain into the transverse and sigmoid sinuses, respectively
* The sigmoid sinus, which drains into the internal jugular vein.

There are also extracranial veins:
* The supratrochlear vein and supraorbital vein drain the forehead
* The facial vein, superficial temporal vein, and postero-auricular veins drain the face and scalp
* The superior temporal vein unites with the maxillary vein to form the retromandibular vein, which bifurcates to unite with:
  * The facial vein (anteriorly) as it enters the internal jugular vein
  * The posterior auricular vein to establish the external jugular vein.

In the neck:
* The external jugular vein receives the anterior jugular vein which has drained the superficial chin and neck, then itself drains into the subclavian vein
* The internal jugular vein receives tributaries from the neck, then unites with the subclavian vein to establish the brachiocephalic vein.

Lymphatic drainage
See Fig. 11.23.

Superficial vessels accompanying superficial veins drain into a collar of nodes around the neck, the including submental, submandibular, parotid, mastoid, and occipital groups. The nodes in turn drain to deep cervical nodes, which drain deeper structures. These nodes drain through jugular trunks into the venous circulation at the junction of the internal jugular and subclavian veins.

The lingual and palatine tonsils, together with the retropharyngeal lymphatic tissue, form a ring of lymphatic tissue in the mucosa and submucosa of the nose, pharynx, and mouth.

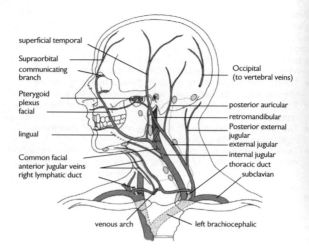

Fig. 11.23 Venous drainage of head and neck (intracranial venous sinuses not shown); position of major groups of lymph nodes.

Reproduced from MacKinnon, Pamela and Morris, John, *Oxford Textbook of Functional Anatomy*, vol 3, p160 (Oxford, 2005). With permission of OUP.

Muscles of the head and neck

Nerve supply is indicated by [superscript].

Muscles of the neck

See Figs 11.17 and 11.24–11.26.
- Trapezius[XI] elevates, retracts, and laterally rotates scapula
- Platysma[VII] depresses skin of lower face and mouth, depresses mandible
- Sternomastoid[XI] flexes and rotates neck
- Scalene muscles[C3–8] laterally flex, rotate neck
- Suprahyoid muscles elevate the hyoid: digastric[VII]; stylohyoid[VII]; mylohyoid[VII]; geniohyoid[C1]
- Infrahyoid ('strap') muscles:
  - Depress the hyoid: sternohyoid[C1,3]; thyrohyoid[C1]; omohyoid[C1,2,3]
  - Depress the larynx: sternothyroid[C1,2,3].

The muscles of the neck define two triangles:
- The posterior triangle:
  - Boundaries:
    - Anterior—sternomastoid
    - Posterior—trapezius
    - Base—clavicle
  - Contents:
    - Roots of brachial plexus
    - Subclavian artery and branches
    - Spinal part of accessory nerve
    - Branches of cervical plexus
    - Subclavian vein
    - Lymph nodes
- The anterior triangle:
  - Boundaries:
    - Superior—mandible
    - Lateral—sternomastoid
    - Medial—midline of the neck
  - Contents:
    - Pharynx, larynx, oesophagus, trachea
    - Thyroid, parathyroid, submandibular, parotid glands
    - Suprahyoid and infrahyoid muscles
    - Glossopharyngeal, vagus, hypoglossal nerves, and sympathetic chain
    - Strap muscles and their nerve supply (ansa cervicalis[C1,2,3])
    - Carotid arteries and branches
    - Internal and external jugular veins.

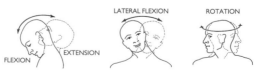

Fig. 11.24 Anatomical relations used to describe movements of the head and neck.
Reproduced from MacKinnon, Pamela and Morris, John, *Oxford Textbook of Functional Anatomy*, vol 3, p4 (Oxford, 2005). With permission of OUP.

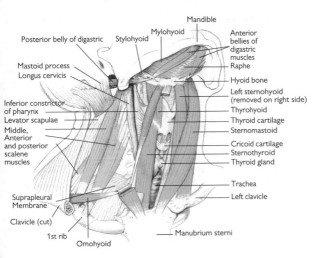

Fig. 11.25 Muscles of the neck.
Reproduced from MacKinnon, Pamela and Morris, John, *Oxford Textbook of Functional Anatomy*, vol 3, p60 (Oxford, 2005). With permission of OUP.

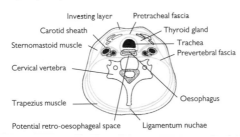

Fig. 11.26 Fascial planes of the neck.
Reproduced from MacKinnon, Pamela and Morris, John, *Oxford Textbook of Functional Anatomy*, vol 3, p61 (Oxford, 2005). With permission of OUP.

Within the anterior triangle, four smaller triangles—the superior carotid, inferior carotid, suprahyoid (submandibular), and submaxillary (submental)—are created by the digastric and omohyoid muscles.

Muscles of facial expression[VII, with exceptions noted]

See Fig. 11.27.

Sphincters and dilators, between bone and overlying skin:

- Of the orbit:
  - Sphincter—orbicularis oculi:
    - Palpebral portion in the eyelids
    - Orbital portion around the orbital margin
  - Dilator: occipitofrontalis, levator palpebrae scapularis (skeletal[III] and smooth muscle[sympathetic nervous system])

SUPERFICIAL MUSCLES

DEEPER MUSCLES

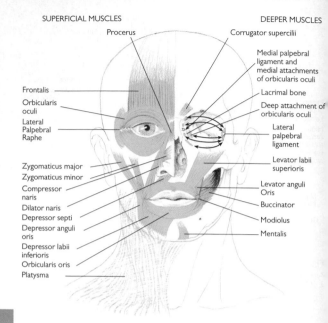

Procerus

Corrugator supercilii

Medial palpebral ligament and medial attachments of orbicularis oculi

Lacrimal bone

Deep attachment of orbicularis oculi

Lateral palpebral ligament

Levator labii superioris

Levator anguli Oris

Buccinator

Modiolus

Mentalis

Frontalis

Orbicularis oculi

Lateral Palpebral Raphe

Zygomaticus major

Zygomaticus minor

Compressor naris

Dilator naris

Depressor septi

Depressor anguli oris

Depressor labii inferioris

Orbicularis oris

Platysma

Fig. 11.27 Muscles of facial expression.

Reproduced from MacKinnon, Pamela and Morris, John, *Oxford Textbook of Functional Anatomy*, vol 3, p67 (Oxford, 2005). With permission of OUP.

- Of the nose:
  - Compressor and dilator nares
- Of the mouth:
  - Sphincter: orbicularis oris
  - Dilator:
    - ○ Levator labii superioris, levator anguli oris, zygomaticus major and minor
    - ○ Depressor labii inferioris, depressor anguli oris
  - Buccinator defines the size of the cavity between cheek and teeth.

Platysma in the lateral neck pulls the mouth downwards.

Muscles of the eye

See Figs 11.28 and 11.29.

- Extrinsic muscles, made of striated (skeletal) muscle fibres. Actions on cornea:
  - Superior rectus[III]: up, medial
  - Inferior rectus[III]: down, medial

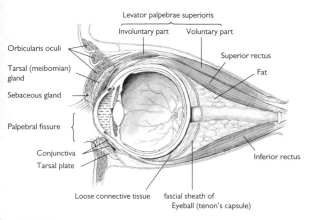

Fig. 11.28 Sagittal section through orbit.

Reproduced from MacKinnon, Pamela and Morris, John, *Oxford Textbook of Functional Anatomy*, vol 3, p121 (Oxford, 2005). With permission of OUP.

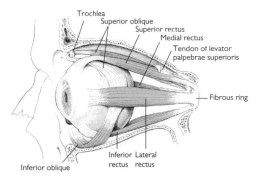

Fig. 11.29 Extrinsic muscles of eyeball.

Reproduced from MacKinnon, Pamela and Morris, John, *Oxford Textbook of Functional Anatomy*, vol 3, p123 (Oxford, 2005). With permission of OUP.

- Medial rectus[III]: medial rotation
- Lateral rectus[VI]: lateral rotation
- Superior oblique[IV]: down, lateral
- Inferior oblique[III]: up, lateral
- Intrinsic muscles, made of smooth muscle. Actions:
  - Sphincter pupillae of iris[III]: constriction of pupil
  - Dilator pupillae of iris[sympathetic nervous system]: dilation of pupil
  - Ciliary muscle[III]: fattens lens.

Muscles of mastication

See Fig. 11.30.

Acting on the mandible:
- Elevate the mandible, close the mouth:
  - Masseter[VIII]
  - Temporalis[VIII] (also retracts)
  - Medial pterygoid[VIII]
- Depress the mandible, open the mouth:
  - Lateral pterygoid[VIII].

In addition, digastric, stylohyoid, and mylohyoid depress and geniohyoid elevates the mandible.

Muscles of the tongue

See Fig. 11.30.

Extrinsic muscles:
- Genioglossus[XII] provides the bulk, draws forward, retracts tip
- Styloglossus[XII] elevates, retracts
- Hyoglossus[XII] depresses
- Palatoglossus[pharyngeal plexus] elevates.

Intrinsic muscles:
- Interlaced longitudinal, transverse, vertical fibres[XII] alter shape of tongue.

Muscles of the soft palate

- Tensor palatini[VIII] tenses soft palate
- Levator palatini[X, XI via pharyngeal plexus] elevates soft palate
- Palatopharyngeus[pharyngeal plexus] depresses soft palate
- Palatoglossus depresses soft palate
- Uvular muscle[pharyngeal plexus] elevates uvula.

Muscles of the pharynx

There are three circular overlapping muscles—superior, middle, and inferior constrictor muscles[pharyngeal plexus]—which contract sequentially during swallowing to propel the bolus of food downwards.

Inner, longitudinal muscles—palatopharyngeus[pharyngeal plexus], salpingopharyngeus[pharyngeal plexus], and stylopharyngeus[IX]—shorten the pharynx and elevate the larynx to close the laryngeal inlet against the base of the tongue during swallowing.

Muscles of the larynx

Intrinsic muscles[X] modify the shape of the airway through the larynx. They include:
- Cricothyroid
- Thyroarytenoid
- Cricoarytenoid
- Interarytenoid.

These muscles have roles in regulating airway diameter during swallowing, coughing, and vocalization.

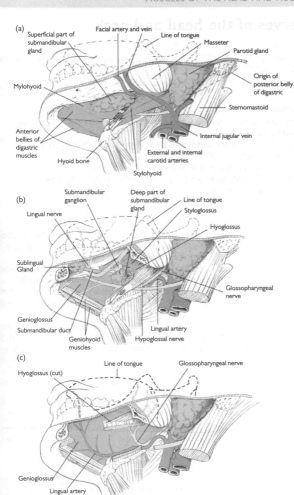

Fig. 11.30 Progressively deeper dissections (a), (b), and (c) of the side of the floor of the mouth; viewed from below and to the left.

Reproduced from MacKinnon, Pamela and Morris, John, *Oxford Textbook of Functional Anatomy*, vol 3, p84 (Oxford, 2005). With permission of OUP.

Nerves of the head and neck

The cranial nerves

See Figs 11.31 and 11.32 (for examination of cranial nerves, see ⦿ *OHCM10* p.70).

There are 12 cranial nerves—see Table 11.1. See Figs 11.16, 11.17 and 11.19.

Functions

- Sensory functions are performed by I (smell); II (vision); VIII (balance and hearing)
- Motor functions are performed by IV (eye); VI (eye); XI (pharynx, larynx, shoulder, neck); XII (tongue)
- Mixed sensory, motor, and autonomic (parasympathetic: III—motor; VII and IX—secretomotor (⦿ p.219) functions are performed by the remaining five nerves.

The cervical plexus

See Fig. 11.33.

Formed from anterior rami of C1–5 and located behind the carotid sheath. C1 emerges above the atlas, while C2–4 pass through the intervertebral foramina above the corresponding cervical vertebra.

Segmental branches supply the prevertebral muscles. In addition, the ansa cervicalis supplies the strap muscles though upper (C1) and lower limbs.

Sensory fibres carried in C2–4 are arranged as the lesser occipital nerve[nerve root of C2], the great auricular nerve[C2,3], the transverse cutaneous nerve[C2,3], and the supraclavicular nerve[C3,4].

The phrenic nerve formed from C3–5 passes to the diaphragm to provide motor supply and sensory supply to the overlying pleura and peritoneum.

Posterior primary rami of cervical nerves segmentally supply the extensors of the neck.

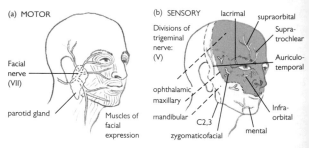

Fig. 11.31 (a) Motor and (b) sensory nerves of the face.

Reproduced from MacKinnon, Pamela and Morris, John, *Oxford Textbook of Functional Anatomy*, vol 3, p69 (Oxford, 2005). With permission of OUP.

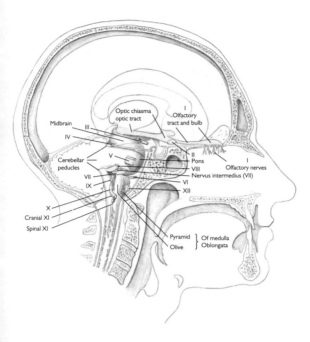

Fig. 11.32 Origin of the cranial nerves from the brainstem.
Reproduced from MacKinnon, Pamela and Morris, John, *Oxford Textbook of Functional Anatomy*, vol 3, p167 (Oxford, 2005). With permission of OUP.

Table 11.1 Cranial nerves

Origin	Cranial nerve	Component fibres	Structures innervated	Functions
Forebrain	I Olfactory	Sensory	Olfactory epithelium	Olfaction
	II Optic	Sensory	Retina	Vision
Midbrain	III Oculomotor	Motor	Superior, medial and inferior rectus, inferior oblique and levator palpebrae eye muscles	Movement of the eyeball
		Parasympathetic	Pupillary constrictor and ciliary muscle of the eye	Pupil constriction and accommodation
	IV Trochlear	Motor	Superior oblique eye muscle	Movement of the eyeball
Pons	V Trigeminal	Sensory	Face, scalp, cranial dura mater, nasal and oral cavities, cornea	Sensation
		Motor	Mastication muscles, tensor tympani muscle	Opening and closing the mouth, mastication, tension on tympanic membrane
	VI Abducens	Motor	Lateral rectus eye muscle	Movement of the eyeball
	VII Facial	Sensory	Anterior 2/3 tongue	Taste
		Motor	Muscles of facial expression, stapedius muscle	Movement of the face, tension on middle ear bones
		Parasympathetic	Salivary and lacrimal glands	Salivation and lacrimation
	VIII Vestibulocochlear	Sensory	Vestibular apparatus of ear	Position and movement of head
			Cochlea	Hearing

Medulla	IX Glossopharyngeal	Sensory	Pharynx, posterior 1/3 of tongue	Sensation and taste
			Eustachian tube, middle ear	Sensation
			Carotid body and sinus	Chemo- and baroreception
		Motor	Stylopharyngeus muscle	Swallowing
		Parasympathetic	Parotid salivary glands	Salivation
	X Vagus	Sensory	Pharynx, larynx, oesophagus, external ear	Sensation
			Aortic bodies, aortic arch	Chemo- and baroreception
			Thoracic and abdominal viscera	Visceral sensation
		Motor	Soft palate, pharynx, larynx, and upper oesophagus	Speech and swallowing
		Parasympathetic	Thoracic and abdominal viscera	Control of GI, cardiovascular, and respiratory systems
	XI Accessory	Motor	Trapezius and sternomastoid muscles	Head and shoulder movement
	XII Hypoglossal	Motor	Intrinsic and extrinsic muscles of the tongue	Movement of the tongue

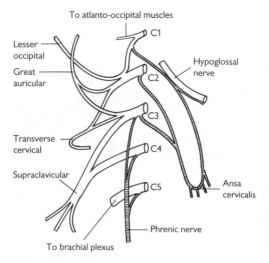

Fig. 11.33 Cervical plexus. The ansa cervicalis has been deflected medially; it normally lies anterior to the plexus.

Reproduced from MacKinnon, Pamela and Morris, John, *Oxford Textbook of Functional Anatomy*, vol 3, p189 (Oxford, 2005). With permission of OUP.

Structure of the eye and ear

The eye

See Figs 11.34 and 11.35.

A sphere, ~2.5cm in diameter, with the cornea bulging forwards. The anterior chamber is situated between the cornea and the iris. The iris defines the pupil, which leads to the posterior chamber between the muscular iris and lens.

The anterior and posterior chambers contain aqueous humour. The vitreous chamber behind the lens contains vitreous humour.

In coronal section, three layers can be discerned:

* The sclera, which anteriorly gives rise to the transparent cornea covered by stratified epithelium
* The pigmented choroid, lining the posterior eyeball, which becomes the iris and, in the posterior chamber, establishes a ciliary body from which ciliary processes secrete aqueous humour
* The retina (the innermost layer), comprises a pigmented epithelium under which lie receptor cells the rods and cones. Neuronal ganglion cells from the rods and cones converge on the optic disc to unite as the optic nerve.

The intrinsic muscles of the eye sphincter and dilator pupillae control pupil diameter. The lens is suspended from the ciliary body by the circular suspensory ligament; the tension of the ligament, and hence the curvature of the lens, is determined by the ciliary muscle.

The macula lies lateral to the optic disc; it is the site of sharpest vision (the fovea) and contains only cones.

Retinal arteries and veins, derived from the central artery of the retina and associated veins, pass with the optic nerve, and run on the vitreous aspect of the retina.

Two folds of skin constitute the eyelids, separated by the palpebral fissure. The inner surface of the eyelids is covered by a mucosal layer (the conjunctiva) which is continuous with the surface of the eyeball to form

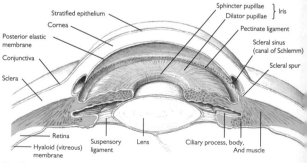

Fig. 11.34 Coronal section of the front eye.

Reproduced from MacKinnon, Pamela and Morris, John, *Oxford Textbook of Functional Anatomy*, vol 3, p119 (Oxford, 2005). With permission of OUP.

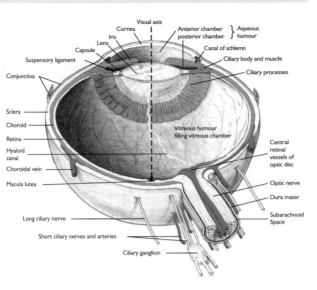

Fig. 11.35 Internal features of the eye: horizontal section.

Reproduced from MacKinnon, Pamela and Morris, John, *Oxford Textbook of Functional Anatomy*, vol 3, p118 (Oxford, 2005). With permission of OUP.

the conjunctival sac. The fibrous orbital septum acts as a framework for the eyelid and is thickened at the margins to form tarsal plates and medial and lateral palpebral ligaments.

Lacrimal gland secretions (tears) enter the conjunctival sac at the lateral upper eyelid. Tears drain into the lacrimal puncta, through canals to the lacrimal sac, and then, via the nasolacrimal duct, to the nose (Fig. 11.36).

The ear

The ear is divided into three structures—the outer ear, the middle ear, and the inner ear.

The outer ear

Comprises the auricle—a fold of skin reinforced by cartilage from which the external auditory meatus, made of cartilage and then bone, extends to the tympanic membrane.

The middle ear

See Figs 11.37 and 11.38.

Lies within the petrous temporal bone, and comprises the vertical tympanic cavity which is fluted at its upper end as the epitympanic recess.

The oval and round windows (fenestra ovale and rotundum) provide connections with the inner ear. Three articulated bones (the ossicles) are present:

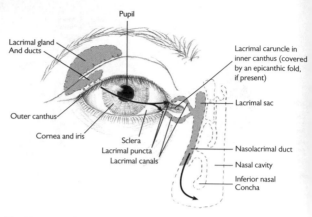

Fig. 11.36 Eye and lacrimal apparatus.
Reproduced from MacKinnon, Pamela and Morris, John, *Oxford Textbook of Functional Anatomy*, vol 3, p122 (Oxford, 2005). With permission of OUP.

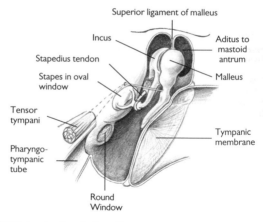

Fig. 11.37 Anterior view of right middle ear cavity showing ossicles.
Reproduced from MacKinnon, Pamela and Morris, John, *Oxford Textbook of Functional Anatomy*, vol 3, p134 (Oxford, 2005). With permission of OUP.

- The malleus (hammer), connected to the tympanic membrane articulates with
- The incus (anvil) which in turn articulates with
- The stapes (stirrup) which is attached to the oval window.

The chain of synovial joints transmits vibration from the tympanic membrane to the oval window. The tensor tympani muscle dampens vibrations of the tympanic membrane; stapedius limits vibration of the stapes.

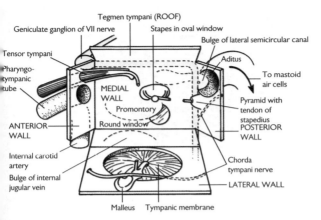

Fig. 11.38 Walls of the middle ear in the form of an opened-out box. The incus is not shown.

Reproduced from MacKinnon, Pamela and Morris, John, *Oxford Textbook of Functional Anatomy*, vol 3, p134 (Oxford, 2005). With permission of OUP.

The tympanic cavity is continuous with the nasal cavity through the bony and then cartilaginous auditory tube, which is lined by mucosa. This connection equalizes the pressure in the middle ear with the atmospheric pressure.

The inner ear

See Fig. 11.39.

Located within the petrous temporal bone, which comprises a bony labyrinth lined with endosteum and filled with perilymph (continuous with CSF through the perilymphatic duct—the aqueduct of the cochlea).

Within the bony labyrinth lies a membranous labyrinth, filled with endolymph, which resembles intracellular fluid. The bony labyrinth comprises:

* The cochlea, containing the organ of hearing
* The vestibule and semi-circular canals, for perception of orientation.

The cochlea makes 2.5 turns around the central modiolus in which the cochlear nerve travels. The vestibule is continuous with the cochlea and with the three semi-circular canals which lie perpendicular to each other. The aqueduct of the vestibule reaches the posterior cranial fossa at the internal auditory meatus.

The membranous labyrinth forms a series of ducts and sacs. The spiral cochlear duct (scala media) is wedge-shaped. It defines two channels—the vestibular and tympanic canals—which meet at the tip of the cochlea.

* The vestibular membrane separates the duct from the vestibular canal
* The basilar membrane separates the duct from the tympanic membrane.

Vibrations of the oval window initiate vibrations of the perilymph in the canals and then of the endolymph in the duct. The organ of Corti lies on the basilar membrane and detects these vibrations.

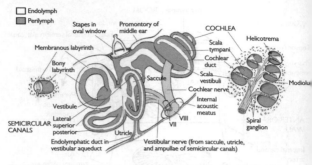

Fig. 11.39 Inner ear with sectional views of a semicircular canal (left) and the cochlea (right).

Reproduced from MacKinnon, Pamela and Morris, John, *Oxford Textbook of Functional Anatomy*, vol 3, p136 (Oxford, 2005). With permission of OUP.

The three semi-circular canals contain semi-circular ducts; these are enlarged to form ampullae, where they join with a sac-like structure, an otolith organ (the utricle). The utricle is continuous with the other otolith organ the saccule, which in turn is continuous with the cochlear duct.

Endolymph within this network is reabsorbed into the bloodstream from the endolymphatic duct within the aqueduct of the vestibule.

The structures of the central nervous system

See Figs 11.40–11.45.

The structures of the CNS can be grouped as:

- The cerebrum
- The diencephalon
- The mesencephalon
- The rhombencephalon
- The spinal cord.

The cerebrum

- The cerebrum comprises two lateral cerebral hemispheres, defining a horizontal fissure but connected by white matter, the corpus callosum. Each hemisphere extends from frontal to occipital bones. It lies above the anterior and middle cranial fossa and then above the tentorium cerebelli. The falx cerebri descends into the horizontal fissure
- The frontal lobe is the largest component and is separated from the parietal lobe by the central sulcus. The lateral sulcus separates the temporal lobe from the parietal and frontal lobes. The parietal lobe is separated from the most caudal cerebrum, the occipital lobe, by the parieto-occipital sulcus
- Grey matter (cell bodies and myelinated axons) on the surface of the cerebrum constitutes the cortex. A central mass of white matter (largely myelinated axons) lies within and contains a number of clusters of grey matter (basal ganglia or nuclei). These nuclei are:

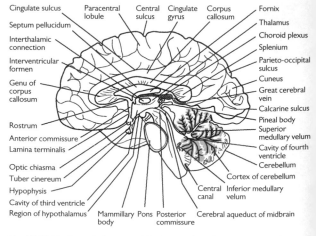

Labels (left side, top to bottom): Cingulate sulcus, Septum pellucidum, Interthalamic connection, Interventricular formen, Genu of corpus callosum, Rostrum, Anterior commissure, Lamina terminalis, Optic chiasma, Tuber cinereum, Hypophysis, Cavity of third ventricle, Region of hypothalamus

Labels (top centre): Paracentral lobule, Central sulcus, Cingulate gyrus, Corpus callosum

Labels (right side, top to bottom): Fornix, Thalamus, Choroid plexus, Splenium, Parieto-occipital sulcus, Cuneus, Great cerebral vein, Calcarine sulcus, Pineal body, Superior medullary velum, Cavity of fourth ventricle, Cerebellum, Cortex of cerebellum, Inferior medullary velum

Labels (bottom): Mammillary body, Pons, Posterior commissure, Central canal, Cerebral aqueduct of midbrain

Fig. 11.40 Median sagittal section of the brain to show the third ventricle, the cerebral aqueduct, and the fourth ventricle.

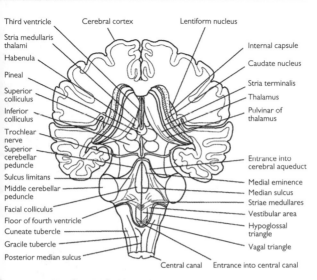

Fig. 11.41 Posterior view of the brainstem showing the two superior and the two inferior colliculi of the tectum of the midbrain.

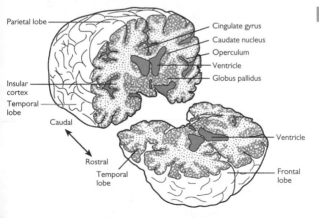

Fig. 11.42 Some structures of the cerebral hemispheres cannot be seen from the surface of the brain. For example, the basal ganglia (caudate nucleus and globus pallidus) and insular cortex can be seen only after the brain has been sectioned. Large cavities in the brain called ventricles are filled with CSF.

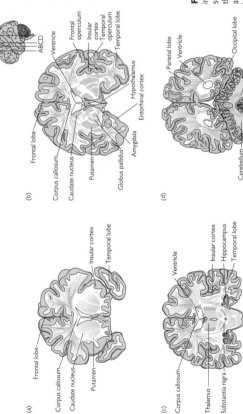

Fig. 11.43 Several brain regions are shown in these sections of the human brain. The sections are from rostral (a) to caudal (d) and the approximate location of these sections are shown on the lateral surface view of the brain shown above.

Reproduced with permission from Kandel ER et al. (2000). *Principles of Neural Science*, 4th edn. ©The McGraw-Hill Companies, Inc.

- The corpus striatum, situated laterally to the thalamus and composed of the caudate and lentiform nuclei
- The amygdaloid nucleus, situated in the temporal lobe
- The claustrum, lateral to the lentiform nucleus
- Within the white matter, a fan of nerve fibres (the corona radiata) runs between the cortex and the brainstem. The lateral ventricles located within each hemisphere are continuous with the third ventricle within the thalamus via the interventricular foramina, which in turn communicates with the fourth ventricle anterior to the cerebellum via the cerebral aqueduct.

The diencephalon

- The largely inaccessible diencephalon comprises the ovoid dorsal thalamus and ventral hypothalamus. The thalamus is formed of grey matter and is expanded at its posterior end as the pulvinar. The subthalamus contains cranial parts of the substantia nigra and red nucleus. The epithalamus contains the habenular nuclei and the pineal gland
- The hypothalamus is located between the optic chiasma and the caudal border of the mammillary bodies. It consists of a number of interposed clusters of cells (the hypothalamic nuclei) which include the supraoptic and paraventricular nuclei.

The mesencephalon

The midbrain, or mesencephalon, connects the forebrain to the hindbrain and contains the cerebral aqueduct.

- Posterior to the aqueduct lies the tectum, the surface of which demonstrates four raised structures—the superior and inferior colliculi
- The cerebral peduncles, comprising the anterior crus cerebri and the posterior tegmentum, lie anterior to the aqueduct
- The anterior and posterior components are separated by the substantia nigra
- At the level of the superior colliculus, the tegmentum contains the red nucleus.

The rhombencephalon

- The hindbrain, or rhombencephalon, comprises:
  - The medulla oblongata
  - The pons
  - The cerebellum
- The medulla oblongata (or myelencephalon) connects the superior pons to the inferior spinal cord. The anterior surface has a median fissure, on either side of which is a pyramid. Posterior to the pyramids are the bulges of the olivary nuclei (the olives). The cerebellum is connected to the medulla by the inferior cerebellar peduncles, which lie posterior to the olives
- The posterior aspect of the medulla shows medial gracile tubercles of the gracile nucleus and the laterally placed cuneate tubercles of the cuneate nucleus
- The pons (or metencephalon) lies inferior to the midbrain, superior to the medulla, and on the anterior surface of the cerebellum. It contains transverse fibres connecting the two hemispheres of the cerebellum

- The cerebellum is situated posterior to the medulla and pons, within the posterior cranial fossa. Two hemispheres are united in the midline by the vermis. Superior and middle cerebral peduncles provide connections to the midbrain. The cerebellum displays a highly ridged cortex of grey matter within which lies white matter. Cerebellar nuclei of grey matter are situated within the white matter, comprising the dentate, emboliform, globose, and fastigial nuclei.

The spinal cord

See Figs 11.44 and 11.45.
- The spinal cord arises from the medulla and runs from the foramen magnum to the lower border of the first lumbar vertebra. It is cylindrical and lies within the vertebral canal defined by the vertebral column. An outer coat of white matter (in anterior, lateral, and posterior columns) envelopes an inner core of grey matter. The grey matter defines a cross shape, with anterior and posterior horns
- Cervical (C7–8) and lumbosacral (L4–5) enlargements occur at the points at which the cervical plexus and lumbosacral plexus arise. At its termination, the spinal cord tapers into the conus medullaris. There is a deep anterior median fissure and a shallower posterior median sulcus along the longitudinal length of the cord. Thirty-one pairs of spinal nerves arise along the cord as anterior (motor) and posterior (sensory) roots. Posterior roots display a posterior root ganglion

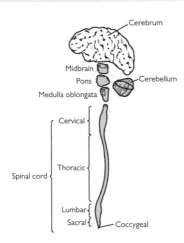

Fig. 11.44 Main divisions of the central nervous system.

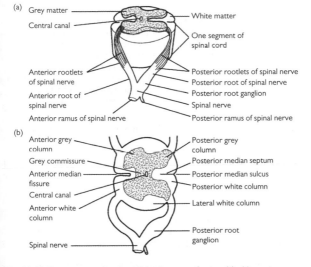

Fig. 11.45 Transverse section through lumbar part of spine: (a) oblique view; (b) face view, showing anterior and posterior roots of a spinal nerve.

Neurones

There are ~10^{20} neurones in the human nervous system, of which 10^{11} are in the brain. The basic morphology of a neurone consists of a cell soma, from which arise a dendritic tree and an axon. Different types of neurone can be distinguished by morphology, especially in the number and configuration of the neurone's processes.

Neuronal morphology

Dendritic tree

- Comprises branch-like processes (dendrites) containing cytoskeletal elements radiating from the cell soma
- Accounts for up to 90% of the neurone's surface area and is the main region for receiving synapses from other neurones
- Dendrites may be spiny (e.g. pyramidal cells) or non-spiny (e.g. most interneurones)
- Spines are the primary region for receiving excitatory input; each spine generally contains one asymmetrical synapse (under electron microscopy synapses are asymmetrical and excitatory, or symmetrical and inhibitory)
- Dendrites cannot propagate action potentials, but Ca^{2+} signalling may be involved in dendritic processing of incoming information.

Cell soma

- Contains most of the neurone's organelles (nucleus, Golgi apparatus, RER, and mitochondria, plus neurofilaments and microtubules)
- Macromolecules required in the rest of the neurone are synthesized in the cell soma and transported by axoplasmic transport
- Receives very few synapses compared with dendrites
- Contains the axon hillock (where the axon arises from the soma) which is the point where action potentials are initiated for propagation along the axon (i.e. where integration of synaptic signals occurs).

Axon

- Carries the output of the neurone to other neurones or to effector organs (e.g. muscle and glands)
- Contains smooth ER and a prominent cytoskeleton
- Is of variable length and is generally unmyelinated when short, as in local circuit neurones (e.g. inhibitory interneurones), and myelinated in longer neurones
- Myelination increases the speed of action potential propagation
- All axons are sheathed by Schwann cells (PNS) or oligodendrocytes (CNS), whether myelinated or not
- Several branches arise at end of axon, the telodendria, each with one or more synaptic boutons containing synaptic vesicles for storing neurotransmitter
- The number and form of the branches of the telodendria depends on the type of neurone
- In local circuit neurones there can be many short axonal collaterals, while in neurones projecting to subcortical centres, such as motor neurones that extend axons to the ventral horn of the spinal cord, there is a single long axon to the distant site with a small number of recurrent collaterals.

Most nervous system cells are multipolar (cell body gives rise to several processes—an axon and multiple dendrites). Cell bodies of bipolar neurones (e.g. retinal bipolar cells and dorsal root ganglion cells) give rise to two processes (one true axon and one which eventually branches forming dendrites).

Bipolar cells form no synapses on their cell body.

Unipolar cells have a single process from the soma, and are occasionally found in the ganglia of the autonomic system, but generally are only found in invertebrates.

Examples of neuronal cell types

Pyramidal cells

- The main type of excitatory neurone in the brain
- All cortical output is carried in pyramidal cell axons
- Roughly pyramidal cell soma, the axis of which lies perpendicular to the cortical surface
- Dendrites are either short and arise from the base of the cell soma, or long and arise from the apex
- Form of the dendritic tree and axonal projections depends on the laminar location of the cell soma:
  - The pyramidal cells of cortical layer V and deep in layer III have extensive dendritic trees, and axons that project to distant cortical and (layer V only) subcortical sites (e.g. basal ganglia, brain stem, and spinal cord)
  - Pyramidal cells of cortical layers II and III have a small soma and dendritic tree, and their axon gives rise to many recurrent collaterals that extend into neighbouring areas of the cortex, thus providing a major intrinsic excitatory input to other cortical areas.

Spiny stellate cells

- Glutaminergic neurones, which are another main source of intrinsic excitatory input
- Found mostly in cortical layer IV of primary sensory areas
- High density of dendritic spines
- Influence extends only locally through small dendritic and axonal trees
- Suggested role in local cortical circuits between the neurones in layer IV that receive thalamic input, and neurones in layers III, V, and VI which carry the cortical output.

Inhibitory interneurones

- Majority of other interneurones are GABAergic and inhibitory
- Regulate pyramidal cell function (GABA antagonists such as picrotoxin are potent convulsants)
- Many subtypes distinguishable by their morphology
- Most common are:
  - Basket cells, axonal endings of which form a 'basket' around the somas of pyramidal cells with which they synapse. They synapse with many pyramidal cells as their axons extend horizontally for up to 2mm across cortex; several basket cells may contribute to a single pyramidal cell's 'basket'
  - Chandelier cells, which have characteristic axonal endings that consist of a string of synaptic boutons spread along the distal segment of the axon forming axo-axonal synapses with pyramidal cells and can inhibit pyramidal cell firing

- Double bouquet cells, which have strictly vertically oriented dendritic and axonal trees which thus extend little into neighbouring areas of cortex. Their axons form tight bundles that project across cortical layers II–V and their axonal terminals contain a wide range of neurotransmitters
- Chandelier and double bouquet cells vary morphologically depending on their connectivity and cortical location.

Purkinje cells

- Found only in the cerebellar cortex
- Characteristic dendritic tree—highly branched and in only two dimensions like the veins on a pressed leaf
- The dendritic trees of all Purkinje cells are aligned in parallel across the whole of cerebellar cortex.

Neuroglia

Glia are a relatively poorly understood part of the CNS. From the Greek word for glue, since it was originally thought that their role was simply to glue the brain together, 'glia' is an umbrella term for all the non-neuronal cell types in the CNS. In fact, the different types of glial cell have widely differing morphologies and functional roles which extend far beyond simply providing neuronal scaffolding. There are two major classes of glia: macroglia and microglia.

Macroglia

- Derived embryologically from precursor cells that line the neural tube constituting the inner surface of the brain
- Comprise astrocytes, oligodendrocytes, and ependymal cells.

Astrocytes

- Irregular star-shaped cells, often with relatively long processes
- Make up between 20% and 50% of the CNS by volume
- Subdivided into fibrous astrocytes, which are found among bundles of myelinated fibres in CNS white matter, and protoplasmic astrocytes, which contain less fibrous material and are found around cell bodies, dendrites, and synapses in the grey matter
- Embryonically, astrocytes develop from radial glial cells, which provide the framework for the migration of neurones and their subsequent organization, and thus play a critical role in defining the cytoarchitecture of the CNS. Once the CNS has matured, radial glial cells retract their processes and serve as progenitors of astrocytes. Some specialized astrocytes retain their radial morphology in the adult cerebellum (Bergmann glial cells) and the retina (Müller cells)
- Astrocytes have multiple roles in the CNS:
  - Isolation of the brain parenchyma: the glia limitans is formed by long processes projecting to the pia mater and the ependyma, and astrocytes processes ensheath capillaries and the nodes of Ranvier. However, astrocytes do not themselves form the blood–brain barrier, but induce and maintain the tight junctions between the endothelial cells that create the barrier. Astrocytes are a major source of extracellular matrix proteins and adhesion molecules in the CNS that help to maintain connections between nerve cells
  - CNS homeostasis: astrocytes are connected to each other by gap junctions, forming a syncytium that allows ions and small molecules to diffuse across the brain parenchyma. Astrocyte processes around synapses are thought to be important for removing transmitters from the synaptic cleft. They contain transport proteins for the reabsorption of many neurotransmitters, such as glutamate. Glutamate is converted into glutamine by astrocytes and then released into the extracellular space. Glutamine is taken up by neurones as a precursor for both glutamate and GABA
  - Astrocyte processes come into close apposition with the axonal membrane at the nodes of Ranvier. Astrocyte membranes seem to act as perfect K^+ electrodes, and their resting potential is determined purely by their high permeability to K^+. They are thus able to buffer excess K^+ released by neurones when their activity is high. There is

a greater concentration of K⁺ channels at the end-feet of astrocyte processes, which contact blood vessels and the pial membrane, than at any other point on their surface membrane, so they can balance out high K⁺ uptake in one part of the cell, in an area of high neuronal activity, by extruding it through their end-feet. High K⁺ concentrations are also distributed further across the parenchyma via the syncytial gap junctions between neighbouring astrocytes. Furthermore, a raised K⁺ concentration in the extracellular space between the capillaries and the astrocyte end-feet causes local vasodilation, thus providing a mechanism for autoregulation of the blood flow to maintain appropriate oxygen and nutrient delivery

- Response to infection and injury: astrocytes have been shown to interact with T lymphocytes, whose activity they can stimulate or suppress. Astrocytes therefore qualify as inducible, facultative antigen-presenting cells. In addition, they help microglia remove neuronal debris and seal off damaged brain tissue after injury

- Production of growth factors: astrocytes produce a great many growth factors, which act singly or in combination to regulate selectively the morphology, proliferation, differentiation, or survival, or all four, of distinct neuronal subpopulations. Most of the growth factors also act in a specific manner on the development and functions of astrocytes and oligodendrocytes. The production of growth factors and cytokines by astrocytes and their responsiveness to these factors are a major mechanism underlying the developmental function and regenerative capacity of the CNS. The growth factors are also important for angiogenesis, again especially in development and repair of the CNS. This role is poorly understood.

Oligodendrocytes

- Form sheaths around nerve axons, like Schwann cells in the PNS. However, unlike Schwann cells, which sheathe single axons, the pressure for space in the CNS means that oligodendrocytes sheathe several axons. Around larger diameter fibres the sheath is myelinated to increase the speed of axonal transmission with nodes of Ranvier at regular intervals of 1mm in most nerve fibres. Around smaller diameter fibres oligodendrocytes do not produce myelin. How the CNS determines which fibres will be myelinated is not well understood. The signal between nerve axons and myelin-producing glia is thought to occur early in development. Schwann cells that do not produce myelin in their normal environment can if 'transplanted', so the neurones appear to provide the switch signal

- Mammalian CNS does not regenerate well. CNS neurones *can* regenerate in an environment provided by Schwann cells, however. During the regeneration of the PNS, Schwann cells act as a conduit for regenerating axons to grow along, and also attract neurones to grow towards them from a distance. Thus, it appears to be an inhibitory influence from oligodendrocytes that prevents regeneration, although the role and mechanism is not understood.

Ependymal cells

- Line the inner surface of the brain in the ventricles
- No known physiological role.

Microglia

- Develop from bone marrow-derived monocytes that enter the brain parenchyma during early stages of brain development
- Numerous processes extending symmetrically from a small rod-shaped cell body
- Primary function is as immune response mediators in the CNS:
  - Most are derived from monocytes early during brain development— they retain the ability to divide and have the immunophenotypic properties of monocytes and macrophages
  - Respond rapidly to immune activation or injury in the CNS, and play a role in the immune-mediated response itself and also in scavenging debris from dying cells
  - Reactive microglia divide more rapidly than resting microglia, and differ both in their morphology and in the increased expression of monocyte-macrophage molecules
- Also thought to secrete cytokines and growth factors that are important in fibre-tract development, gliogenesis, and angiogenesis.

Central synaptic transmission

CNS neurones often form up to 10,000 synapses, rather than the single synapse that a motor neurone forms at the NMJ. However, activation of any single synapse is not sufficient to trigger a neurone to fire. Instead, excitatory and inhibitory postsynaptic potentials (EPSPs and IPSPs) from each of the many synapses over the neurone's dendrites are summed together over time and space (known as temporal and spatial summation; ➔ pp.246–8). The point where the cell body meets the axon is called the axon hillock, and it is here that action potentials are triggered, if the sum total of the depolarization over the neurone's dendritic tree is great enough.

There are many more neurotransmitters in the CNS than in the PNS and they can be:

- Ionotropic channel (directly gated):
  - Glutamate (NMDA and quisqualate A AMPA (α-amino-3-hydroxy-5-methyl-4-isoxazole propionic acid))
  - Nicotinic ACh
  - $GABA_A$
  - Glycine
- Metabotropic channel (indirectly gated via a G protein):
  - Glutamate (quisqualate B AMPA)
  - $GABA_B$
  - Adrenaline and noradrenaline
  - Muscarinic acetylcholine
  - Serotonin (5-HT)
  - Dopamine
  - Various neuropeptides such as substance P and opioids such as enkephalin (➔ p.801).

Excitatory synapses

- Glutamate is the most widespread excitatory neurotransmitter in the CNS. Glutamate receptors have been typed as either AMPA or NMDA receptors, named after the agonists that were first used to distinguish them
- There are three subtypes of AMPA receptor:
  - Kainate receptors are directly gated Na^+- and K^+-permeable cation channels
  - Quisqualate activates both ionotropic and metabotropic glutamate receptors, thus:
    - The quisqualate A receptor is a directly gated Na^+ and K^+ channel
    - The quisqualate B receptor is G-protein coupled and opens a Na^+ and K^+ channel via a phosphoinositide-linked second-messenger system (IP_3)
- The NMDA receptor is a cation channel that is permeable to Ca^{2+} ions in addition to Na^+ and K^+. The main effect of the NMDA receptor appears to be Ca^{2+} mediated, however, since they make little contribution to the EPSP when the neuronal membrane is at resting potential, when extracellular Mg^{2+} blocks the NMDA receptor channel. As the membrane becomes depolarized, the Mg^{2+} is removed and the channel can open, allowing Na^+, K^+, and Ca^{2+} ions through. Thus, in effect, the NMDA receptor is both ligand and voltage gated
- NMDA receptors have been implicated in the excitotoxicity that occurs following stroke or ischaemia, and in persistent seizures in status epilepticus. Excessive influx of Ca^{2+} ions through NMDA receptors, caused by the continuously elevated glutamate levels that appear to occur in these conditions, allows intracellular Ca^{2+} to reach catatonic levels. Cell damage and death may result from the subsequent activation of Ca^{2+}-dependent proteases and production of toxic free radicals
- NMDA receptors play a critical role in the induction of long-term potentiation (LTP), at least in the dentate gyrus and area CA1 of the hippocampus where LTP has been most studied, and also appear to be involved in the formation of the eye-blink conditioned reflex.

Inhibitory synapses

- GABA and glycine are the primary inhibitory neurotransmitters of the CNS:
  - GABA is predominant in the brain and its main function is in local circuit interneurones, although there are also long-axoned GABAergic projection neurones
  - The functional morphology of GABAergic interneurones varies widely, with their effects extending over as small an area as a single cortical column, or to regions many columns wide
- GABA receptors can be divided into ionotropic and metabotropic subtypes:
  - The ionotropic $GABA_A$ receptor is the most common GABA receptor and opens a Cl^- anion channel. The anion selectivity is produced by positively charged amino acids positioned near the ends of the ion channel
  - The metabotropic $GABA_B$ receptor is G-protein coupled and can block Ca^{2+} channels or activate K^+ channels
- $GABA_A$ receptors also bind benzodiazepines, such as diazepam and chlordiazepoxide, and barbiturates, such as phenobarbital and secobarbital. The effect of GABA is allosterically modulated by benzodiazepines, increasing the frequency of channel opening, and thus also GABA-induced Cl^- current:
  - Barbiturates act by increasing the length of time that a Cl^- channel remains open. This dampening, inhibitory effect on generalized CNS activity underlies the use of benzodiazepines and barbiturates as anticonvulsants and anxiolytics
- Picrotoxin and bicuculline inhibit GABA receptor function and produce widespread and sustained seizure activity due to a generalized dampening of inhibitory synapses throughout the CNS. Penicillin inhibits GABA receptors in a similar way and, at a high enough concentration, is also a potent convulsant
- Glycine is the main inhibitory neurotransmitter in the spinal cord. It activates a Cl^- anion channel that is functionally very similar to the $GABA_A$ receptor. It is blocked by strychnine
- Glutamate and GABA metabolism is similar, since the molecules are chemically closely related and both synthesized from the same precursor, α-ketoglutarate, which is a product of the Kreb's cycle. α-Ketoglutarate is transaminated to glutamate by GABA α-oxoglutarate transaminase (GABA-T). The conversion of glutamate to GABA is catalysed by glutamic acid decarboxylase (GAD). GABA-T is mitochondrial, while GAD is cytosolic, and it is not clear how transport is arranged in and out of the mitochondria for vesicular storage. Once glutamate or GABA have been released into the synaptic cleft, they are deactivated by reuptake into the surrounding glia or the presynaptic neurone. There is no enzymatic deactivation in the synaptic cleft. Glutamate receptors are more densely expressed in astrocytes than in neurones. Glia convert glutamate and GABA to glutamine, which is recycled to the presynaptic neurone
- Up- or down-regulation of transmitter release from glutaminergic and GABAergic neurones is mediated by metabotropic autoreceptors.

Motor and sensory cortex

Motor cortex

- 'Conscious movement' is a collective term that includes those movements that we actually think about before we carry them out, as well as those with which we are so familiar that we don't apparently think about, but that still originate from the brain (e.g. movements associated with walking—→ p.796). Instructions for these movements originate in the motor cortex, an area of the cerebral cortex, located just anterior to the somatosensory cortex (Fig. 11.46)
- Movement of a particular part of the body is controlled from a clearly defined area of the motor cortex—the more highly used is a particular muscle, the bigger is the area devoted to that muscle in the motor cortex
- The relative size of areas devoted to different regions of the body are often represented as a 'map' or motor homunculus (Fig. 11.47), distorting those areas of the body that have large areas of the cortex devoted to them so that they are considerably larger than those that are poorly represented in the motor cortex
- The motor homunculus for humans shows that most of the cortex is devoted to the hands, face, and tongue, indicative of the importance of these features to us. Different species have vastly different homunculi, which reflect those motor skills that are most important to each species.

Somatosensory cortex

- Primary somatosensory cortex (area S1) is located on the postcentral gyrus. The representation of the body in area S1 is somatotopic, with the head represented laterally, and the feet medially. Area S1 is divisible into four distinct cytoarchitectonic areas—Brodmann's areas 1, 2, 3a, and 3b. Although all four areas receive thalamic input, most thalamic fibres terminate in areas 3a and 3b. Areas 3a and 3b have been implicated by lesion studies in discrimination of texture, size, and shape. Cells in areas 3a and 3b then project to area 1, which is associated with texture discrimination, and area 2, which is associated with size and shape
- Each area has its own somatotopic map, in which the representations of fast and slowly adapting receptors are kept separate. Within these fast and slowly adapting regions, different receptor types are represented in separate cortical columns. Thermal and nociceptive sensitivity is usually not affected by lesions of area S1. Beyond area S1 is area S2, the secondary somatosensory cortex, which receives projections from all four areas of S1, and the posterior parietal cortex. Area S2 is involved with higher order aspects of touch, such as stereognosis, while in the posterior parietal cortex, somatosensory information is integrated with information from other sensory modalities such as vision and audition.

Descending motor pathways

Signals originating in the motor cortex are transmitted by either the pyramidal (corticospinal) system or the extrapyramidal (extracorticospinal)

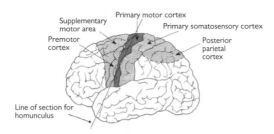

Fig. 11.46 Location of motor cortex and associated areas of the brain.

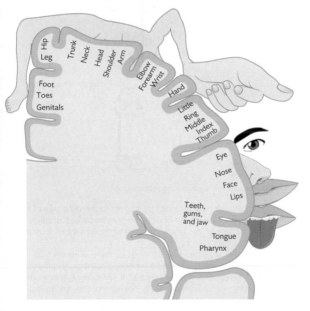

Fig. 11.47 Motor homunculus (right hemisphere).

Reproduced under the Creative Commons Attribution 4.0 International License from Betts, J. G., Young, K. A., Wise, J. A. et al, Anatomy and Physiology, Openstax, 2013 (https://openstax.org/details/books/anatomy-and-physiology)

system to α and γ motor neurones in the relevant segment of the spinal cord for a specific muscle.

- The pyramidal system (Fig. 11.48) employs a single neurone with an axon that passes all the way from the cerebral cortex to the relevant segment of the spinal cord without any intervening synapses. The

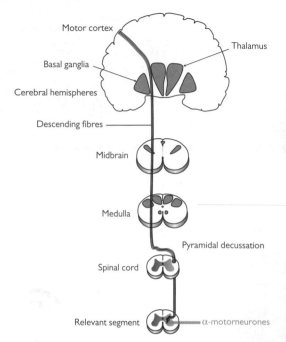

Fig. 11.48 The pyramidal system for motor control.

nerves from the left motor cortex cross over to the right side of the spinal cord at the base of the brain, in the pyramidal decussation. Thus, the left hemisphere of the brain controls the movement of muscles on the right side and vice versa

• The extrapyramidal system (Fig. 11.49) is far more complex than the pyramidal system, largely because it includes a number of different pathways and involves a greater number of synapses. Although the extrapyramidal system is anatomically distinct from the pyramidal system, the two interact because the extrapyramidal system assimilates information from a wide range of sensory inputs, whereupon it modifies the motor signals in the pyramidal tract. For this reason, the extrapyramidal tract is seen to be central to producing smooth, controlled movements and maintaining posture. The extrapyramidal system includes nerves in the basal ganglia of the brain (⟶ p.794), the reticular formation in the brainstem (which determines the level of consciousness—⟶ p.828), and the brainstem nuclei.

Somatosensory pathways

Lateral inhibition through inhibitory interneurones operates at all levels of the somatosensory pathways from the dorsal column nuclei upwards. Thus, secondary and tertiary afferent neurones have antagonistic centre-surround receptive fields. At each synaptic relay on the pathway, several presynaptic neurones converge on a single postsynaptic neurone, so that at each level, the size of the neurone's receptive fields grows. Receptive field sizes depend on the area of the body in question, due to the density of somatic receptors. Thus, without lateral inhibition, two-point discrimination (the minimum distance between two points on the skin that can be discriminated apart) would be very poor. However, lateral inhibition at each synaptic level keeps the central, excitatory zone of the neurone's receptive field small, so that two-point discrimination is not affected by neuronal convergence. Distal inhibition of afferent fibres by regions of cerebral cortex occurs in the thalamus, the dorsal column, and trigeminal nuclei and may contribute to selective attention.

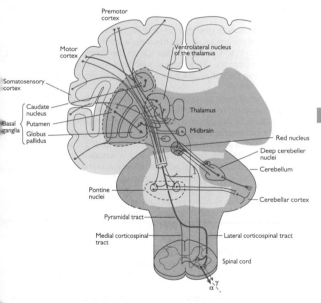

Fig. 11.49 Main structures of the extrapyramidal system.
Reproduced with permission from Kandel E, Schwartz JH, Jessell TM (2000). *Principles of Neural Science*, 4th edn. © The McGraw-Hill Companies Inc.

Thalamus and hypothalamus

The thalamus

- The thalamus (Fig. 11.50) is a bilateral structure located in the diencephalon, a brain structure found between the midbrain (the most rostral part of the brainstem) and the cerebral hemispheres
- Almost all of the sensory and motor information that reaches the cortex is processed by the thalamus. As such, it is composed of several sensory nuclei that receive input from distinct sensory stimuli such as vision, hearing, and somatic sensation. The main exception is olfaction, that has a direct pathway to the sensory cortex
- Projections from the thalamus to the sensory cortex initiate the processing of sensory information, while projections to the association cortex (a region associated with movement, perception, and motivation) initiate a behavioural response to sensory input
- Other thalamic nuclei relay information concerning motor activity to the motor cortex. For example, extrapyramidal motor information from the cerebellum and basal ganglia is relayed, via the thalamus, to the primary motor cortex
- A large fibre bundle (the internal capsule) carries thalamic projections to and from the cortex. As a consequence, the function of the thalamic nuclei is also modulated by feedback from the cortex through recurrent projections from the same cortical regions to which they project
- Nuclei within the thalamus function either as relay nuclei or diffuse projection nuclei:
  - Relay nuclei generally process either sensory information resulting from a specific type of stimulus (e.g. vision, hearing, somatic sensation) or input from a particular part of the motor system. These relay nuclei send axonal projections to a region of the cerebral cortex that is also anatomically restricted and defined by the source of input it receives. For example, visual information from the retina is processed by the lateral geniculate nucleus in the thalamus and this projects to a spatially restricted region of the cortex, termed the visual cortex
  - Diffuse projection nuclei are perceived to be involved in mechanisms that regulate the state of arousal of the brain. Their connections are more widespread than the relay nuclei and they include projections to other thalamic nuclei.

The functional anatomy of the thalamus

- Nuclei of the thalamus are anatomically divided into six groups, separated by a Y-shaped collection of fibres termed the internal medullary lamina
- Lateral, medial, and anterior nuclei are named by their location relative to this lamina, the remaining groups being the intra-laminar, reticular, and midline nuclei
- The lateral nuclei are subdivided into dorsal and ventral groups and each subdivision can be defined by its restricted connections with a specific region of the motor or sensory cortex. These relay nuclei therefore perform processing of specific sensory or motor input.

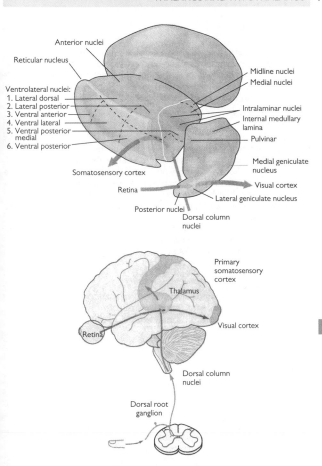

Fig. 11.50 The major subdivisions of the thalamus. The thalamus is the critical relay for the flow of sensory information to the neocortex. Somatosensory information from the dorsal root ganglia reaches the ventral posterior lateral nucleus, which relays it to the primary somatosensory cortex. Visual information from the retina reaches the lateral geniculate nucleus, which conveys it to the primary visual cortex in the occipital lobe. Each of the sensory systems, except olfaction, has a similar processing step within a distinct region of the thalamus.

Motor relay nuclei
See Fig. 11.51.

Motor signals are principally relayed by the ventral lateral and ventral anterior nuclei. The ventral lateral nucleus receives information from the cerebellum and sends axons to both the motor and premotor cortices, while the ventral anterior nucleus is innervated by the globus pallidus and projects primarily to the premotor cortex.

Sensory relay nuclei
See Fig. 11.52.

* Somatic sensation is processed in the ventral posterior nucleus. The lateral division of the ventral posterior nucleus receives input via the spinothalamic tract and dorsal column, whereas the medial division receives input from the sensory nuclei of the trigeminal nerve. Thus, the former is involved in processing of somatic sensation from the body, while the latter is involved in facial sensory information processing. Both nuclei project to appropriate regions in the somatosensory cortex (parietal lobe)
* The medial geniculate nucleus processes auditory information, receiving its major input from the inferior colliculus, and sends axons to the auditory cortex (temporal lobe)
* Visual processing is performed by the lateral geniculate nucleus. It receives a direct input from the retina via the optic nerve and projects to the visual cortex
* The lateral posterior and pulvinar nuclei play primary roles in the integration of sensory information by virtue of reciprocal connections between the parietal lobe and the temporal, occipital, and parietal lobes, respectively.

Limbic relay nuclei
See Fig. 11.52.

* The anterior group receives inputs from the hypothalamus and sends projections to the cingulate gyrus. This pathway constitutes part of the limbic system contributing to awareness and emotional aspects of sensory processing (➲ p.830)
* The lateral dorsal nucleus is involved in emotional expression and receives signals from, and sends them to, the cingulate gyrus

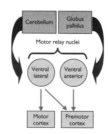

Fig. 11.51 Principal thalamic motor relay nuclei and their connections.

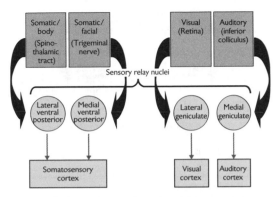

Fig. 11.52 Principal thalamic sensory relay nuclei and their connections.

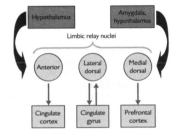

Fig. 11.53 Principal thalamic limbic relay nuclei and their connections.

- The medial dorsal nucleus is also involved in limbic processing, receiving input from the amygdala, hypothalamus, and olfactory system and sending axons to the prefrontal cortex.

Diffuse projection nuclei

See Fig. 11.54.

- The thalamic midline and intra-laminar nuclei are diffuse projection nuclei receiving input from the reticular formation, hypothalamus, globus pallidus, and several cortical areas
- The reticular nucleus receives information from the cortex, other thalamic nuclei, and the brainstem. This information is largely distributed to other thalamic nuclei, where it is used in the modulation of thalamic activity.

Cellular physiology of the thalamus: relationship with the sleep–wake cycle

- In the waking state, thalamic neurones that project to the cortex (TC neurones) are tonically depolarized by cholinergic, aminergic (noradrenergic, serotonergic, and histaminergic), and peptidergic (orexins) inputs from the brainstem and hypothalamus. This allows TC neurones to respond faithfully to incoming sensory and motor signals and, thus, ensures accurate transfer of this information to higher cortical areas

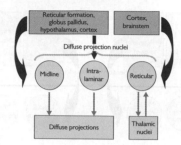

Fig. 11.54 Thalamic diffuse projection nuclei and their connections.

- During non-rapid eye movement (NREM) sleep (slow-wave sleep; ⊃ p.825, Fig. 11.65), these brainstem and hypothalamic inputs diminish and TC cells first hyperpolarize, then switch to a pattern of slow membrane potential oscillations driven by intrinsic voltage-dependent ionic conductances in their neuronal membrane
- Under these conditions, processing of externally sourced sensory and motor information is depressed. Thus, cortical processing is regulated by intrinsic excitation within the cortex and, possibly to some extent, waves of activity known as 'spindle' waves (because of their relatively high frequency) generated in the thalamus as a result of disinhibition of inhibitory GABA neurones in the thalamic reticular nucleus
- During REM sleep, it is thought that an internal representation of the external world becomes the input to the thalamo-cortical circuitry. This change in thalamic and cortical activity is thought to be important in the iteration of information and essential for processes such as learning and memory.

The hypothalamus

- The hypothalamus (Fig. 11.55), is also located in the diencephalon. Observed in a frontal section, three divisions—lateral, medial, and periventricular (immediately bordering the third ventricle)—can be identified. In medial section it can be divided into anterior, middle, and posterior areas along the rostral–caudal axis:
  - The medial forebrain bundle acts as a pathway from the hypothalamus to the neocortex and other parts of the brain. However, it also carries fibres that are not of hypothalamic origin, such as aminergic fibres from brainstem nuclei
  - The lateral region sends both long-distance connections to the cortex and spinal cord and shorter axons to ascending and descending pathways
  - Defined nuclei in the medial region of the hypothalamus include:
    ○ The preoptic and suprachiasmatic (SCN) nuclei in the anterior region (sleep and circadian regulation)
    ○ The dorsomedial, ventromedial, and paraventricular (PVN) nuclei in the middle region
    ○ The posterior nucleus and mammillary bodies in the posterior region

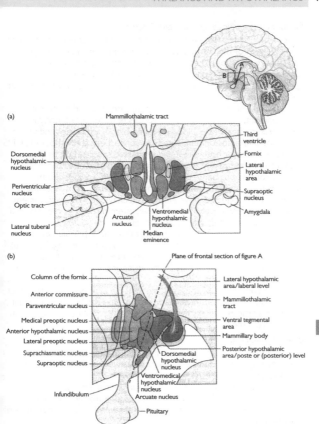

Fig. 11.55 The structure of the hypothalamus. (a) Frontal view of the hypothalamus (section along the plane shown in part b). (b) A medial view shows most of the main nuclei. The hypothalamus is often divided analytically into three areas in a rostrocaudal direction: the preoptic area, the tuberal level, and the posterior level.

Reproduced with permission from Kandel E, Schwartz JH, Jessell TM (2000). *Principles of Neural Science*, 4th edn. © The McGraw-Hill Companies Inc.

• Most nuclei have bidirectional fibre systems, sending and receiving signals via the medial forebrain bundle, mamillotegmental tract, and dorsal longitudinal fasciculus. There are two exceptions—supraoptic and paraventricular neurones project, via the hypothalamohypophyseal tract, to the posterior pituitary (⊙ p.792), while the SCN receives a unidirectional input from the retina (⊙ pp.793, 806).

Afferent hypothalamic connections and function

- The hypothalamus has a major role in maintaining homeostasis through its regulation of the autonomic nervous system, the endocrine system, and visceral function. These functions are carried out through both conventional, direct, synaptic contacts and indirect neurohormonal pathways that involve an intimate regulatory influence over the function of the pituitary
- By receiving information from, and acting directly on, the internal environment, the hypothalamus regulates body functions including temperature, heart rate, blood pressure, blood osmolarity, and water and food intake. In addition, extensive connections throughout the CNS provide routes for both direct and processed sensory information to modulate hypothalamic activity and influence behavioural processes
- Direct connections between the limbic system (⊖ p.830) and the hypothalamus regulate autonomic and visceral responses associated with motivation and adaptive emotional behaviour. The limbic system also exerts control over the endocrine system through its regulation of the secretion of hypothalamic hormones
- The influence of the cortex on the hypothalamic expression of emotional behaviour is through a pathway involving the cingulate gyrus and hippocampus, while a reciprocal influence of the hypothalamus exists via the mammillary bodies, the anterior nucleus of the thalamus, and the cingulate gyrus
- Another component of the limbic system, the amygdala, also has direct projections to the hypothalamus and this region has been implicated in learning, particularly tasks that might involve associating stimuli with an emotional response.

Efferent hypothalamic connections and function

- The hypothalamus exerts a major influence over the pituitary gland thereby regulating metabolism, feeding, water homeostasis, and the circadian timing of physiology and behaviour
- Connected to the hypothalamus by the infundibulum, the anterior pituitary is controlled by substances secreted primarily from small, peptide-releasing neurones found in the basal part of the middle region of the hypothalamus
- Parvocellular (small) neurones located primarily in the PVN, as well as the periventricular region and the preoptic nucleus, release peptides into the local plexus of blood vessels that drain into the vasculature of the anterior pituitary. These hormones then act to promote or inhibit the production of anterior pituitary hormones. For example, CRH promotes the release of ACTH from the anterior pituitary and this induces cortisol release by the adrenal cortex—an important mediator of the stress response. Other hypothalamic hormones that act on the pituitary in this way include TRH, → thyrotropin production and regulation of growth and metabolism, GnRH, stimulating FSH and LH release and, thus, controlling gametogenesis; GHRH; somatostatin; and dopamine
- In addition to this indirect pathway, magnocellular (large) neurones located in the PVN and supraoptic nucleus project to the posterior pituitary and directly release hormones into the circulation. In this

way, water homeostasis is maintained by releasing the peptide, ADH (vasopressin), while the release of another peptide, oxytocin, controls uterine contraction and milk ejection
• The temporal organization of hormonal release and behavioural patterns is dictated by the SCN. This nucleus generates an endogenous circadian pattern that entrains many physiological processes to the light–dark cycle. Unidirectional input to the SCN from the visual system synchronizes the circadian clock to day length. The SCN itself has minimal projections, mainly to other parts of the hypothalamus. Thus, temporal synchronization of biological processes involves intra-hypothalamic interactions and output from other hypothalamic nuclei.

Other functions of the hypothalamus
• Direct connections from the PVN to sympathetic preganglionic neurones in the intermediolateral cell columns of the thoracic and lumbar spinal cord as well as projections to parasympathetic nuclei in the brainstem indicate direct neuronal control of the autonomic motor nervous system. This control is regulated by the sensitivity of the hypothalamus to stimuli in the blood such as glucose and insulin
• Visceral sensory information from the major organs is directed to the paraventricular and lateral hypothalamic nuclei via brainstem sensory nuclei, including the nucleus of the solitary tract
• Hypothalamic involvement in sleep appears to involve at least three nuclei. Histamine-releasing neurones of the tuberomammillary nuclei that project to the cortex are inhibited during sleep by ventrolateral preoptic neurones, thereby reducing histamine release, a neurotransmitter implicated in arousal. Other hypothalamic nuclei are involved in wider serotonergic and noradrenergic pathways that regulate arousal. In addition, the circadian timekeeping associated with the SCN is also implicated in the timing and modulation of sleep patterns

Sexual dimorphisms in the hypothalamus of animals implicate a role in sexual behaviour as well as gametogenesis. Although well established in animals, evidence for similar dimorphisms in the human hypothalamus is less secure. Nevertheless, structural differences have been identified between the male and female hypothalamus, particularly in the preoptic nucleus.

Basal ganglia and cerebellum

- The basal ganglia at the base of the brain receive nervous input from both the sensory and motor cortices. Nervous outflow from the basal ganglia is exclusively to the areas involved in higher processing of movement, namely the prefrontal, premotor, and supplementary motor cortices; there is no output to the spinal cord. The primary role of basal ganglia is thought to be related to the planning and control of complex motor behaviour by selectively activating some movements and suppressing others. These functions are highlighted in patients with Parkinson's disease, where the cells of the basal ganglia degenerate and motor control is compromised (➔ p.841; *OHCM10* p.495)
- The cerebellum acts in concert with the motor cortex and basal ganglia to coordinate movement. In simple terms, the cerebellum can be seen to control fine motor tasks by constantly comparing 'intention' with 'performance'. It receives input from a vast array of sensory nerves that allow it to compute the actual position of limbs and compare this information to the intended position. The net result is an error calculation, which is translated into an output signal via the Purkinje fibres to make fine adjustments to correct the error
- The role of the cerebellum in modulating motor output from the pyramidal and extrapyramidal systems, rather than instigating movement itself, is highlighted in patients with damage to the cerebellum. These individuals are not paralysed and still receive sensory information as normal, but they are unable to coordinate movements in complex or even quite simple tasks requiring motor control and timing. Lesions to the cerebellum give rise to a characteristic set of clinical signs (➔ *OHCM10* p.499). It is now recognized that the cerebellum is also important in some aspects of speech recognition and learned responses, but not memory.

Locomotion

- Walking may seem an automatic and straightforward task, but from a neuromuscular standpoint, it is highly complex and involves motor control of the majority of our skeletal muscles
- The key to the neuronal control of locomotion is so-called pattern generation, where a set of commands are relayed to all the relevant muscles so that they contract and relax in the correct order. On completion of one cycle of the task, the same set of commands is repeated so that a pattern of movement is instigated. The neural circuit must also include suitable facilities to modify the pattern to take into account any changes in the environment and to stop the pattern of movement.

Pattern generators

- Neural control of locomotion stems from the spinal cord—it does not necessarily require instructions from the brain but does require sensory information from the limbs that are involved. Locomotion can be best described as a series of reflexes in response to incoming sensory information (e.g. from the Golgi tendon organs and the muscle spindles) about the position of the limb in motion
- Much of the experimental work relating to pattern generation has been conducted in primitive vertebrates, where neural circuits can be more easily identified and characterized. These studies clearly show that although basic pattern generation originates in the spinal cord in response to sensory information, higher centres in the brain, including the cerebellum, basal ganglia, sensory and motor cortices, are necessary to accommodate variation in locomotion and adaptability in response to the environment (e.g. response to a trip) (Fig. 11.56):
  - The cerebellum sends modulating signals to motor neurones, via the descending pathways, to fine-tune the movements of the crude, ungainly gait that originates from pattern generators in the spinal cord. It also modulates activity to adjust the walking pattern to accommodate changes in the terrain (e.g. going up or down stairs)
  - The basal ganglia are required to initiate movement and to help compute visual and sensory information from the parietal cortex, and generate appropriate spatially directed movement towards a given goal
  - The parietal cortex collects visual and somatosensory information and collates it into a 3D representation of position with respect to environment.

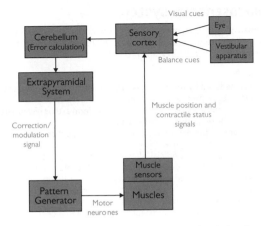

Fig. 11.56 Pattern generation for locomotion originates in the spinal cord; modulation is controlled by the cerebellum.

Somatosensory system

The somatosensory system comprises touch, nociception (pain), tempera-ture, and proprioception (→ pp.266, 804). Each relies on different types of receptor, pathway (Fig. 11.57), and processing.

Receptors

- Touch is mediated by mechanoreceptors. In non-glabrous skin, there are four types of mechanoreceptor:
  - Pacinian corpuscles are found deep in the dermis, thus having large receptive fields, and are rapidly adapting
  - Meissner's corpuscles are also rapidly adapting, but are superficial and thus have small receptive fields
  - Merkel's discs are superficial, slowly adapting receptors
  - Ruffini receptors are deep, slowly adapting receptors. No conscious sensation is associated with direct stimulation of Ruffini receptors and a role has been postulated for them in proprioception
- In glabrous skin, there are three types of follicle receptor (hair-guard, hair-tylotrich and hair-down receptors) and the same deep receptors as in non-glabrous skin
- Temperature is sensed by two types of thermoreceptor, one for heat (responding in the range of 30–45°C) and one for cold (responding in the range of 1–20°C). Cold receptors are also activated by extreme heat (at >45°C), resulting in the sensation of 'paradoxical cold'. Thermal sensitivity is punctate: hot and cold thermoreceptors are grouped in non-overlapping heat- or cold-sensitive zones. Heat nociceptors also respond to extreme heat
- Pain is mediated by nociceptors, which can be mechanical, thermal, or polymodal (mechanical nociceptors that can be activated or sensitized by inflammatory mediators and other indicators of tissue damage such as bradykinin, K^+ ions, 5-HT, prostaglandin, and histamine).

Transduction

Transduction by mechanoreceptors relies on a simple mechanical deform-ation of the nerve ending, which opens mechanically gated cation channels, ion flow through which triggers an action potential that is transmitted along the nerve axon. Rapidly adapting mechanoreceptors have a large capsule around the nerve ending, formed by multiple foldings of a Schwann cell. These act to absorb a constant pressure, so that rapidly adapting mech-anoreceptors are only activated by changes in pressure. This can be dem-onstrated by stripping the capsule off the nerve ending, which no longer adapts to constant pressure. Slowly adapting mechanoreceptors have a much smaller capsule around the nerve ending, so they respond with a more constant rate of firing to the absolute level of pressure. Nociceptors are all free nerve endings.

Fibre types

Fibre types are different for different modalities. All mechanoreception is carried by fast myelinated Aβ (type II) fibres. Mechanical nociception is carried by thinner, slower, myelinated Aδ (type III) fibres. Polymodal nociception is carried by thin, unmyelinated C (type IV) fibres, which are the

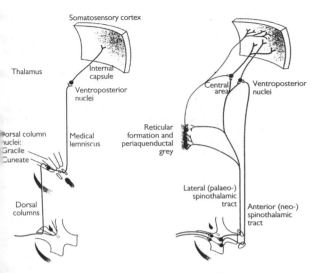

Fig. 11.57 The main ascending somatosensory pathways: (left) the lemniscal system; (right) the neo-spinothalamic and paleo-spinothalamic divisions of the anterolateral system.

Reproduced with permission from Carpenter R (2003). *Neurophysiology*, 4th edn, Edward Arnold.

lowest of all. Thus sharp, acute pain, mediated by mechanical nociceptors, is felt faster than dull inflammatory pain, which is mediated by polymodal nociceptors. The sensation of cold from thermal receptors is carried by Aδ fibres, while that of heat is carried by slower C fibres.

Central pathways: sensory

All sensory fibres enter the spinal cord through the dorsal roots, with their cell bodies lying just before the dorsal root in the dorsal root ganglia. Axons from mechanoreceptors and proprioceptors ascend the spinal cord in the dorsal columns. These carry fibres from the ipsilateral side of the body and continue to the dorsal column nuclei at the level of the medulla

The dorsal columns are divided into the gracile and cuneate fascicles, which contain fibres representing the lower limb and trunk, and the upper limb and neck, respectively. The fascicles are divided by the dorsal intermediate septum. Similarly, the dorsal column nuclei are divided into the gracile and cuneate nuclei. At this level, the sensory axons synapse onto internal arcuate fibres, which decussate and continue to ascend on the contralateral side in the medial lemniscus. Fibres from the sensory trigeminal nucleus, representing touch sensation from the face and head, ascend from this level in the neighbouring trigeminal lemniscus. The medial lemniscus fibres synapse in the ventral posterior lateral nucleus of the thalamus, and trigeminal lemniscus fibres synapse in the ventral posterior medial nucleus, whence all tertiary fibres proceed to the primary somatosensory cortex on the postcentral gyrus. Through all levels of the pathways, fibres remain somatotopically arranged

- Nociception and temperature sensation are carried by the anterolateral system. Axons entering the spinal cord through the dorsal horns may ascend or descend one or two spinal segments in the tract of Lissauer, before synapsing onto secondary fibres and interneurones in the dorsal horn laminae. Aδ fibres synapse in laminae I (marginal zone) and V (part of the nucleus proprius), while C fibres synapse in lamina II (substantia gelatinosa). Referred pain is thought to be a result of cutaneous and visceral nociceptive afferents converging on the same secondary fibres through interneurones in the dorsal horn laminae

- The secondary fibres decussate at the same spinal segment as they arise, and ascend in one of three tracts that form the anterolateral system on the contralateral side:
  - The spinothalamic tract is the largest tract, arising from dorsal horn laminae I and V–VII (the nucleus proprius and Clarke's nucleus), and ascends to the central lateral and the ventral posterior lateral nuclei of the thalamus
  - The spinoreticular tract arises from laminae VII (Clarke's nucleus) and VIII. Some fibres terminate in the reticular formation of the pons and medulla, while others continue to the central lateral nucleus of the thalamus. Fibres that terminate in the thalamus, from both spinothalamic and spinoreticular tracts, synapse onto fibres that project to primary somatosensory cortex
  - Fibres in the spinomesencephalic tract arise from laminae I and V, and project to the periaqueductal grey and the mesencephalic reticular formation. These projections provide inputs to the limbic system via the hypothalamus

Nociception and pain

Pain is a sensory function that deserves special attention because of the dramatic impact it can have on wellbeing, both in response to injury (acute) and chronic illness. Nociception is the term given to the perception of a noxious stimulus, but it is not necessarily painful: pain is much more subjective and is influenced by aspects other than the strength of the stimulus (e.g. emotion).

Sensing pain

- The sensory endings that trigger pain sensation fall into the same categories as those described previously, but they have a higher threshold for activity, so are only stimulated by stimuli of sufficient (noxious) intensity to potentially cause damage

- Stimulation of the receptors triggers action potentials along afferent fibres to the dorsal horn of the spinal cord. The nerve type and speed of conduction is dependent on the type of receptor involved (Table 11.2)

Table 11.2 Nerve fibres transmitting pain sensations

Fibre	Sensory ending(s)	Conduction speed	Pain sensation
Aδ	Mechanoreceptor	Fast (myelinated)	Sharp, localized
	Nociceptor		
C	Nociceptor	Slow; non-myelinated	Dull (ache)
	Thermoreceptor		
	Mechanoreceptor		

- C fibres release neuropeptides (calcitonin gene-related peptide; CGRP) and substance P at both the peripheral and central nerve endings; the important role played by these peptides in pain has attracted attention as possible new targets for drugs to reduce pain
- Impulses from pain sensors are transmitted up the spinal cord via transmission neurones (ascending pathway) to the thalamus and on to the sensory cortex, resulting in pain sensation.

Modulation of the pain pathway
- As well as being subject to local interneurone modulation, nociception pathways can also be modulated by distal inhibition from the periaqueductal grey matter in the midbrain
- Noradrenergic and serotonergic efferent fibres from the raphe nucleus, driven by neurones from the periaqueductal grey matter, descend through the dorsolateral funiculus to inhibit nociceptive afferent fibres via enkephalinergic interneurones in laminae I and II of the dorsal horn.
- In another mechanism of pain modulation, called dorsal horn gating, Aβ afferent fibres inhibit secondary nociceptive afferents arising in lamina V via inhibitory interneurones.

Neuropathic pain
- Diseases that cause damage to neurones in the sensory pathway (e.g. stroke, multiple sclerosis, diabetes, shingles) are often associated with chronic and debilitating pain, most likely mediated via hyperactivity of damaged nerves

  Box 11.1 summarizes current available therapies for management of pain.

Box 11.1 Treatments and drug therapies for pain

(➲ *OHPDT2* Ch. 15; *OHCM10* p.574)
- The most comprehensive means of preventing pain is through anaesthesia. General anaesthetics render the patient unconscious and oblivious to pain, whereas local anaesthetics inhibit nerve conduction and prevent the sensation of pain. Anaesthesia is covered in detail in ➲ Chapter 4 (pp.238–40)
- Analgesia is the term used to describe suppression of pain. Drugs that are currently administered to suppress pain fall into two broad categories: opiates (e.g. morphine), which act at several points in the pain pathway to prevent nociception, and anti-inflammatory drugs (e.g. NSAIDs).

Opioids
- Opioids (endogenous or synthetic agents with morphine-like characteristics) have been used for >200 years as analgesics. Morphine is the best-known opioid, gaining particular notoriety during the First World War in the treatment of injured soldiers, when both its usefulness in pain relief and the problems associated with withdrawal became apparent. The following are common opioids:
  - Diamorphine (heroin)
  - Codeine
  - Methadone
  - Pethidine
  - Endorphin
  - Enkephalins
- Opioids act on G-protein-coupled opioid receptors, of which there are three sub-types (μ, δ, κ) Activation of μ-receptors induces analgesia at both central and peripheral levels, it also is responsible for the side effects, including euphoria, respiratory depression, constipation, and physical dependence. For this reason, some of the opioids used for pain relief are relatively κ-selective. Nausea is a common side effect that is common to all opioids. Opioids are usually administered orally, by injection or in drip lines
- Opiates is a term that refers specifically to synthetic agents with morphine-like properties and excludes endorphins and enkephalins, for example, which are endogenous.

Non-steroidal anti-inflammatory drugs
- The inflammatory process is central to the propagation of chronic pain and NSAIDs (➲ p.904) are the mainstay of pain relief in this context, many of which can be obtained over the counter. The following are examples of common NSAIDs that are self-administered for pain relief (headaches, muscle strain), and are also frequently prescribed for chronic pain in conditions such as arthritis, neuropathies, and after surgical operations:
  - Aspirin (acetylsalicylic acid)
  - Ibuprofen
  - Paracetamol (acetaminophen)
  - Naproxen
  - Diclofenac

- Unfortunately, specific COX-2 inhibitors, which were hoped to be useful anti-inflammatory drugs in arthritic conditions, while avoiding the gastric toxicity associated with traditional NSAIDs, were deemed to increase cardiovascular risk and were subsequently withdrawn.

Analgesics designed for specific painful conditions
- Ergotamine—for migraine
- Carbamazepine—for trigeminal neuralgia.

Future developments

Despite the power of opiates and NSAIDs in treating many manifestations of pain, the issue remains a serious impediment to quality of life in a wide range of diseases, not least neuropathic and neurogenic pain and that associated with chronic illnesses including cancer, multiple sclerosis, and rheumatoid arthritis.

The recreational drug, cannabis, which is illegal in most jurisdictions, has long been suspected to have analgesic qualities through activation of endocannabinoid G-protein-coupled receptors (particularly CB_1, but also potentially CB_2 and $TRPV_1$ receptors). The downstream mechanisms that are activated are numerous and complex, but it is understood that the following pathways and mediators are affected by CB and $TRPV_1$ receptor activation:

- Inhibition of neuronal activity in the mid-brain (particularly the periaqueductal grey area)
- Inhibition of inflammatory cytokines and chemokines in the periphery.

As a result, a number of synthetic cannabinoids have been developed for use in a range of applications, including chronic pain. The main cannabinoids under investigation are:
- Δ-9-tetrahydrocannabinol
- Cannabidiol
- Nabilone (Cesamet®)
- Ajulemic acid (CT3, IP-751).

The reflex arc

Reflex movements are involuntary; they do not require a command from the brain to happen. Voluntary movements are entirely dependent on a signal from the motor cortex to be implemented. The reflex arc describes an integrated response to a stimulus that involves both somatosensory and motor neurones.

- Reflex responses are movements that are made very rapidly (~20ms; 0.02s) in response to a stimulus that is sensed by specific sensory receptors. These tend to be defensive reflexes that must be made quickly in order to avoid injury (e.g. hand withdrawal when placed on a hot surface)
- The speed of the response is facilitated by limiting the number of neurones (and hence synapses) involved. The fastest responses are mediated by monosynaptic reflexes (Fig. 11.58), whereby there is only one synapse, located in the spinal cord, between the sensory neurone and the motor neurone (e.g. the knee-jerk reflex)
- Polysynaptic reflex arcs include interneurones in the spinal cord (Fig. 11.59), which connect the sensory neurone to the motor neurone. These reflexes are marginally slower than monosynaptic reflexes but are still considerably faster than conscious movements that require input from the brain. The role of the interneurone is often to modulate the signal to the motor neurone; an excitatory signal from a sensory neurone can stimulate an inhibitory interneurone, resulting in inhibition of the motor neurone (e.g. Golgi tendon organs)
- Far more complex reflexes are constantly called upon to react to assimilated information from sensory organs such as the eyes and balance centre of the ear, just to stay upright.

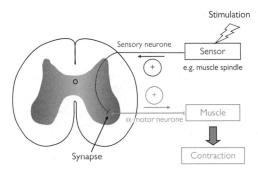

Fig. 11.58 Monosynaptic reflex arc.

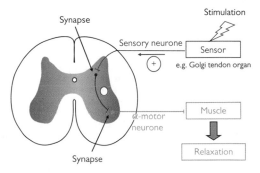

Fig. 11.59 Polysynaptic reflex. An interposed inhibitory interneurone results in reduced stimulation of the α-motor neurone → muscle relaxation.

Visual system

- The retina is composed of five layers (starting nearest the lens):
  - The ganglion cell layer contains the cell bodies of the ganglion cells
  - The inner plexiform layer contains the dendrites of amacrine, bipolar, and ganglion cells, and synapses between them
  - The inner nuclear layer contains bipolar cell axons and the cell bodies of amacrine, bipolar, and horizontal cells
  - The outer plexiform layer contains the dendrites of horizontal cells and synapses between rods, cones, horizontal cells, and bipolar cells
  - The outer nuclear layer contains the rods and cones
- Cells in the retina are unmyelinated, in order to reduce visual distortion as light passes to the photoreceptors at the back of the retina. However, at the centre of the retina, in the fovea, maximal acuity is achieved by displacing to the side all but the photoreceptors themselves. Behind the retina, adjacent to the outer nuclear layer, is the pigment epithelium.

Photoreceptive cells

- The photoreceptive cells of the retina are the rods and cones. These cells contain photosensitive pigment made up of a light-absorbing molecule called retinal, bound to a membrane protein called opsin. Each cell type, rods, and the three types of cone, contains a different visual pigment, made up of retinal and a different opsin molecule, each of which has a different peak light absorption energy
- Colour vision relies on the three types of cone which are maximally excited by light with short (blue: 437nm), middle (green: 533nm), and long (red: 564nm) wavelengths. In contrast, the rod system is achromatic, since there is only one visual pigment, rhodopsin, with a peak absorbency at 498nm
- Rods and cones have an inner and an outer segment. The inner segment is where the nucleus, mitochondria, and other cellular apparatus are found. The outer segment comprises a comb-shaped series of folds of the cell's membrane, which, in rods, often seal off to form free-floating discs within the rod. These folds provide a large surface area over which to collect photons with the photosensitive transmembrane proteins. The outer segments of rods are larger than those of cones, making the rods much more sensitive than cones in dim light.

Transduction

See Fig. 11.60.
- Transduction occurs via a cascade of events triggered by the absorption of a photon by a pigment molecule, retinal. In rods, the retinal usually exists as the 11-*cis* isomer. Excitation of rhodopsin by a photon causes a series of conformational changes in the retinal, ending ultimately in the all-*trans* state. Rhodopsin is unable to bind all-*trans* retinal, so the retinal becomes detached. A semi-stable intermediate configuration called metarhodopsin II activates transducin, a G-protein, which in turn activates phosphodiesterase to convert cyclic GMP (cGMP) to $5'$GMP. The lowered cGMP concentration decreases Na^+ influx through a cGMP-gated channel. Thus, in the dark, cGMP concentrations

Photoreceptor

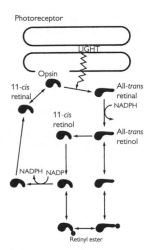

Fig. 11.60 Pigment epithelium. The cyclical sequence by which light leads to isomerization of retinal and its dissociation from opsin, followed by the relatively slow processes that lead to the final regeneration of rhodopsin.

Reproduced with permission from Carpenter R (2003). *Neurophysiology*, 4th edn, Edward Arnold.

are raised, and so there is a constant Na⁺ current, known as the dark current. The resting membrane potential is therefore relatively depolarized, at around 40mV, and light causes a hyperpolarization to around 70mV. A similar cascade operates in cones
- Detached retina is recycled in the pigment epithelium by converting the all-*trans* retinal to all-*trans* retinol (vitamin A). All-*trans* retinol is the precursor for 11-*cis* retinal and is not synthesized in the body. Dietary vitamin A deficiency can therefore cause night-blindness or, if severe, even total blindness
- The dynamic range of photosensitivity of cones is maintained through light adaptation. The transduction cascade also inactivates a cGMP-gated Ca²⁺ channel: Ca²⁺ inhibits guanylate cyclase (GTP → cGMP), thus a negative feedback loop operates to maintain a fairly constant cGMP concentration through a range of absolute levels of brightness, and so keep the cell maximally sensitive to changes in light level.

Retinal pathways

- Cones synapse onto bipolar cells and horizontal cells. Bipolar cells synapse onto ganglion cells, which are the output cells of the retina. Horizontal cells and amacrine cells are interneurones that affect the information passed between photoreceptors and bipolar cells, and between bipolar cells and ganglion cells, respectively. Horizontal cells are inhibitory and synapse back onto the same cones that supply them. However, neighbouring horizontal cells are also all connected to each other through electrical synapses, forming one continuous sheet along which cone inputs are conducted electrotonically. Each cone is therefore inhibited by the activity of a number of neighbouring cones, producing the centre-surround antagonism that characterizes the receptive fields of cells in the very early stages of the visual system

- Cones release glutamate, which has one of two effects, depending on the type of bipolar cell:
  - In off-centre bipolar cells, glutamate opens Na⁺ channels, which causes depolarization
  - In on-centre bipolar cells, glutamate closes Na⁺ channels and opens K⁺ channels, causing hyperpolarization
- Bipolar cells signal using electrotonic conduction, so there is a continuous, graded membrane potential rather than an all-or-nothing action potential. The ganglion cells onto which bipolar cells synapse are the first cells in the visual system to fire action potentials
- In photopic vision (under well-lit conditions), rods synapse via gap junctions onto nearby cones. However, they are sufficiently sensitive that normal levels of brightness bleach them of visual pigment and their input is negligible
- For scotopic vision (in low light), there is a process called dark adaptation, during which the rod pathways change. The gap junctions onto nearby cones close and, instead, rods synapse onto rod bipolar cells, which in turn synapse onto AII amacrine cells. AII amacrine cells synapse both onto off-centre ganglion cells and on-centre bipolar cells, which synapse onto on-centre ganglion cells
- As discussed, colour vision is mediated by the cone system. The principle of univariance means that it is the ratio of activities between the three cone types that carries colour information. A single cone type (or rod) cannot alone convey colour information
- Colour blindness is a result of a defect in or absence of one or more cone pigments. Defects of a visual pigment are classified as protanomaly, deuteranomaly, or tritanomaly, whereas an absence of a pigment is classified as protanopia, deuteranopia, or tritanopia (with prot-, deuter-, and trit- denoting the long-, middle-, and short-wavelength pigments). An absence of two cone pigments is classified as atypical monochromatopsia, while an absence of all cone pigments is classified as typical monochromatopsia. All these are congenital defects of colour vision. However acquired, retinal disease can also damage colour vision. Tetrachromatopsia is sometimes found in women, who have an extra visual pigment

Visual system: central visual pathways

See Figs 11.61 and 11.62.

- The output of the retina is via the ganglion cells, which have circular, centre-surround antagonistic receptive fields. They can be either on-centre, off-surround, meaning that they are excited by light falling on the receptive field centre and inhibited by light falling on their receptive field surround, or off-centre, on-surround

- Ganglion cell axons converge at the optic disc and leave the retina as the optic nerve. The ganglion cells are closer to the lens than the rods and cones, so where these axons pass through the retina at the optic disc, there are no photoreceptors. The resulting small scotoma is known as the blind-spot

- The optic nerves from each eye meet at the optic chiasm. Fibres arising from the nasal retinae from each eye cross, so that when the fibres continue as the optic tract, each tract carries information about left and right hemifields, rather than from left and right retinae

- The optic tract continues to the lateral geniculate nucleus (LGN), when the ganglion cell fibres synapse. The LGN is divided into six layers, with layer 1 being the most ventral and layer 6 the most dorsal

- Layers 1, 4, and 6 contain fibres arising from the contralateral retina, while layers 2, 3, and 5 contain ipsilateral fibres. The six layers of the LGN are also split into magnocellular ('M'—layers 1 and 2) and parvocellular ('P'—layers 3–6) streams. M cells have larger receptive fields and have good temporal resolution, but have poor spatial resolution and are monochromatic. P cells have smaller receptive fields, carry colour information, and have good spatial resolution but poor temporal. M cells are thought to be the substrate for processing of motion in the visual cortices, while P cells are thought to contribute to fine feature processing and colour vision

- The visual pathway continues from the LGN as the optic radiation, which passes back to the primary visual cortex at the occipital pole. Some fibres initially travel forward around the front of the temporal horn of the lateral ventricle. The loop that is formed is known as Meyer's loop. The LGN also projects to the pretectum, which controls pupillary reflexes via the IIIrd (oculomotor) cranial nerve, and to the superior colliculus, which controls saccadic eye movements via the IIIrd, IVth (trochlear), and VIth (abducens) cranial nerves

- The primary visual cortex, also known as striate cortex or area V1, is located at the occipital pole continuing along the calcarine sulcus on the medial surface of each cerebral hemisphere. The visual field is represented contralaterally in strict retinotopic order, with the central visual field represented most posteriorly, and the upper half of the visual field on the inferior banks of the calcarine sulcus and the lower half on the superior banks.

- As in all neocortex, area V1 is divided into six layers. Inputs from the LGN arrive in layer 4, which is further subdivided into layers 4A, 4B, 4C, and 4D. Fibres from the magnocellular layers of the LGN arrive in layer 4C, and fibres from the parvocellular layers of the LGN arrive in layer 4C

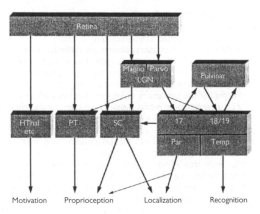

Fig. 11.61 The functional destinations of fibres in the optic nerve. LGN (lateral geniculate nucleus); HThal (hypothalamus); PT (prectum and other visual proprioceptive areas); SC (superior colliculus); Par, Temp (parietal and temporal cortex).

Reproduced with permission from Carpenter R (2003). *Neurophysiology*, 4th edn. Edward Arnold.

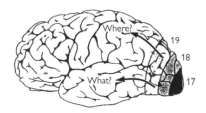

Fig. 11.62 Cerebral cortex, showing the general location of visual areas 17, 18, and 19.

Reproduced with permission from Carpenter R (2003). *Neurophysiology*, 4th edn. Edward Arnold.

- Receptive fields in area V1 have 'on' and 'off' bands running side-by-side, rather than the centre-surround structures of LGN cells. These cells are therefore selective for the orientation of visual features. Visual cortex is compartmentalized into columns of cells, through all six layers, that have similar receptive field structures. Neighbouring columns in area V1 will have selectivities for slightly different orientations and, thus, as one travels over a length of cortex there is a trend for a smooth rotation of orientation selectivity. Orientation columns are achromatic
- Primary visual cortex is the first stage in the visual pathway at which cells receive inputs from both eyes. Some cells have a preference for one eye over the other, and so area V1 is also compartmentalized by ocular dominance. Neighbouring columns will usually have a similar ocular dominance, and thus there is often a complete representation

of all orientations within each region of a particular ocular dominance. A complete set of orientation columns for both left and right ocular dominance has been termed a hypercolumn

- Interspersed between orientation columns are columns in which all the cells are selective for particular wavelengths. These columns of colour-sensitive cells are known as blobs, after their appearance when visual cortex is stained for the enzyme cytochrome oxidase
- Another aspect of the processing of information from left and right eyes in area V1 is a sensitivity to retinal disparity—a slight mismatching of the relative positions of a visual feature on each retina. This difference in retinal position of the feature is a strong cue to the feature's depth
- This early processing of visual information about feature orientation, colour, and depth is further refined in higher visual areas. Visual information from area V1 passes to the first extrastriate area, known as area V2, and thence to a multitude of other visual areas. These visual areas are broadly divisible into dorsal and ventral streams, deriving from the earlier magnocellular and parvocellular pathways:
  - The dorsal stream includes areas V5 and MST, where cells have strong selectivities for direction of motion and motion-in-depth, and has been labelled the 'where' pathway
  - The ventral stream includes areas V4 and IT, where cells have strong selectivities for colour and depth contours, and has been labelled the 'what' pathway.

Auditory system

- A sound is a pressure wave transmitted through a medium, such as the air, consisting of alternate compression and rarefaction of the medium caused by the vibration of an object, such as vocal cords or the strings on a violin. The auditory system gathers information about the frequency composition and intensity of a sound and the direction of the source of the sound
- The auditory system is composed of the ear itself, in which the sound is transduced into neuronal impulses, and the central pathways in which the transduced signal is analysed
- The ear comprises three compartments:
  - Outer ear—the pinna and external auditory meatus change the intensity of frequencies in the range of 2–7kHz ('colouration'). This is important for zenith and front/back localization. It is separated from the middle ear by the tympanic membrane (ear drum)
  - Middle ear—the ossicles (malleus, incus, and stapes—in order from outer to inner ear) are necessary for mechanical impedance matching between the tympanic membrane and the oval window, because the outer ear is filled with air, while the inner ear is filled with fluid. The tensor tympani and stapedius muscles alter the efficiency of impedance matching, and can thus protect the inner ear from extremely loud sounds. They are also active during speech. The middle ear is separated from the inner ear by the oval window, and joined to the nasopharynx by the eustachian tube
  - Inner ear—the inner ear is divided into two portions: bony and membranous labyrinths. The sensory organs of the auditory system are in the auditory division of the membranous labyrinth (the cochlea). The cochlea consists of three scalae, in a spiral around the modiolus:
    - The scala vestibuli begins at the oval window and joins the scala tympani at the helicotrema, at the head of the spiral, which runs back to another membrane between the middle and inner ears, just beside the oval window (the round window)
    - In between these runs the scala media, which is completely enclosed by Reissner's membrane towards the scala vestibuli and by the basilar membrane towards the scala tympani. The scala media is filled with endolymph, which has a similar chemical constitution to intracellular fluid
    - The scalae vestibuli and tympani are filled with perilymph, which has a similar chemical constitution to CSF
- Transduction of sound waves into neuronal impulses occurs in the organ of Corti, which is found on the basilar membrane in the inner ear (Fig. 11.63). When the oval window is moved by the ossicles in the middle ear, a travelling wave is set off along the basilar membrane. The basilar membrane is thin and stiff at its base, at the oval window, and becomes increasingly wide and less stiff towards its apex, at the helicotrema. Thus, high-frequency travelling waves have a maximum amplitude near the oval window, while low-frequency travelling waves have a maximum amplitude near the helicotrema

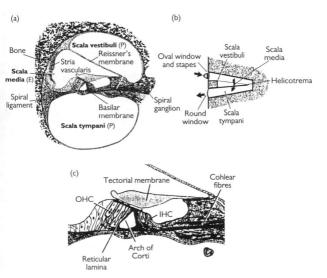

Fig. 11.63 (a) Section through the cochlea, showing the organ of Corti—shown in (c) in more detail; (b) a representation of the scala vestibuli, scala media, and scala tympani, and the path of sound through them. OHC, IHC (outer and inner hair cells); E (endolymph); P (perilymph).

Reproduced with permission from Carpenter R (2003). *Neurophysiology*, 4th edn, Edward Arnold.

- The organ of Corti contains rows of hair cells, which have stereocilia arranged either in linear rows (inner hairs cells) or in V or W formations (outer hair cells). There are three rows of outer hair cells and one row of inner hair cells along the length of the basilar membrane
- The stereocilia rest against the tectorial membrane, which lies above the organ of Corti. The tectorial membrane is stiff, so the stereocilia are bent sideways by a shearing force when the basilar membrane vibrates. The stereocilia on each hair cell are arranged in graded height order, with the tallest being thicker and known as the kinocilium. The tips of the stereocilia are linked, such that when the basilar membrane moves downwards, the stereocilia bend in one direction and the tip-links slacken, and when the basilar membrane moves upwards, the stereocilia bend the other way and the tip-links tighten
- Tightening of the tip-links opens stretch-activated non-selective cation channels, which allow K^+ influx from the endolymph in the scala media. There is no chemical gradient, but there is an electrical gradient, resulting in an oscillatory membrane potential. At the basal end of the hair cells the oscillatory membrane potential causes oscillatory transmitter release

- Innervation of the inner ear is by the VIIIth cranial (vestibulocochlear) nerve. It carries both afferent and efferent fibres:
  - ~90% of afferent fibres are from the inner hair cells, which synapse one-to-one. The rest innervate the outer hair cells, which are pooled ~20 outer hair cells per afferent fibre
  - Efferent fibres are from the superior olive and innervate mostly the outer hair cells. They are thought to contribute to the sharpening of the frequency tuning along the basilar membrane
  - Outer hair cells contribute to sharpening frequency tuning by changing the length of their stereocilia. Destruction of the outer hair cells (e.g. by drugs such as furosemide and gentamicin) can lead to hearing deficits
  - Electrical resonance does not contribute to the sharpness of frequency tuning in mammals
- Afferent fibres in the auditory nerve have a characteristic frequency, determined by the hair cells they innervate. Other nearby frequencies will also stimulate the nerve fibre but require a greater sound intensity, resulting in a V-shaped frequency tuning curve. The afferent fibres carry the frequency of a sound in two ways, the combination of which is known as Duplex theory. With frequencies of up to ~10kHz, phase-locking is used. The firing of the nerve fibre coincides with the rising of the basilar membrane causing a depolarization of the hair cell. The firing of the fibre is thus at the same frequency as that of the sound whose signal it carries. This is important because the fibre may not fire on every cycle, especially at the higher frequencies. The position of the hair cell along the basilar membrane is the other way in which frequency is encoded in the afferent fibres (known as tonotopicity). Intensity is encoded simply in the number of action potentials. An increased amplitude of the sound wave causes an increased depolarization in the hair cells, and thus an increased amount of transmitter is released
- Sound localization, although contributed to by colouration in the outer ear, is primarily mediated by two binaural processes:
  - The sound-shadow of the head causes intensity differences between the two ears. Intensity differences are measured by cells in the lateral superior olive that receive excitatory inputs from the contralateral cochlea and inhibitory inputs from the ipsilateral cochlea. Thus, the cells are activated if the sound source is on the contralateral side. Intensity differences are used for frequencies in the range of ~1.5–20kHz
  - Phase differences occur because the distance to the sound source is different from each ear when the head is turned towards or away from the source. This cue is used for lower frequencies, in the range of ~40Hz–3kHz, and is measured by coincidence detectors in the medial superior olive (⊕ *OHCM10* p.464)
- Primary auditory cortex (area A1) is located on the superior temporal gyrus. Area A1 contains several tonotopic maps, in a typical columnar organization. The columns are arranged in patterns of summation (binaural) and suppression (one ear dominant). Pitch is represented bilaterally, so that unilateral lesions have little effect on pitch discrimination, while sound source location is represented contralaterally. Higher auditory areas concerned with speech are:

- Wernicke's area, which is located in the temporal lobe (usually left) and is involved in speech perception
- Broca's area, which is located in the frontal lobe (usually left) and is involved in speech expression
- Lesions of these sites result in receptive and expressive dysphasias, respectively (→ *OHCM10* p.86). There is much feedback to earlier stages of the auditory pathways (to the medial geniculate nucleus; inferior collicular nucleus; dorsal and ventral cochlear nucleus, but not to the cochlea), most likely having a role in attention, giving rise to the 'cocktail party effect' of being able to single out a sole voice amid a noisy crowd.

Vestibular system

- The vestibular system measures linear and angular acceleration of the head. The sensory organs of the vestibular system are found in the inner ear, which is divided into two portions: bony and membranous labyrinths. There are two principal structures in the vestibular part of the membranous labyrinth:
  - The otolith organs (utricle and saccule)
  - The semicircular ducts
- Both the otolith organs and the semicircular ducts are filled with endolymph, which has a similar chemical constitution to intracellular fluid
- Each semicircular duct arises and joins back to the utricle in a circular loop. Each duct is oriented at 90° to the other two. At one end of each duct there is a dilation of the duct (the ampulla) containing the hair cells. These are similar to those of the auditory system, each having a row of stereocilia arranged in order of height at their apex, of which the tallest is thicker and known as the kinocilium. The apices of the hair cells are embedded in the cupula, a gelatinous membrane stretching from the ampullary crest to the roof of the duct, thus completely occluding the lumen of the duct
- Angular acceleration of the head causes the cupula to move, due to the inertia of the endolymph in the ducts and, therefore, the stereocilia on the hair's cells are pushed to one side. The tips of the stereocilia on each hair cell are linked, so when the cupula moves and the stereocilia bend, the tip-links slacken or tighten, depending on which way the head is rotated. Tightening of the tip-links opens stretch-activated non-selective cation channels, which allow K^+ influx from the endolymph down an electrical gradient (there is no chemical gradient)
- The utricle has a zone called the macula, which is covered with hair cells. The otolithic membrane—a layer of a gelatinous substance in which are embedded crystals of calcium carbonate (the otoliths)—lies over the stereocilia of these hair cells. The macula lies in the horizontal plane such that when the head is upright, the otoliths rest on it. When the head tilts, the otoliths deform the gelatinous membrane and therefore bend the stereocilia on the hair's cells. The utricle is thus sensitive to horizontal, linear forces. The axes of the stereocilia on the hair cells form an approximately radial pattern over the macula, in order to sense linear forces in any horizontal direction. The saccule has a similar structure, but is vertically orientated. Thus it is sensitive to any vertically directed linear force.

Pathways

The vestibular organs are innervated by the VIIIth cranial (vestibulocochlear) nerve. These fibres terminate in the vestibular nuclei. The vestibular nuclear complex comprises four nuclei:

- The lateral vestibular nucleus is functionally related to the control of posture, with vestibular inputs primarily in the ventral portion of the nucleus and cerebellar and spinal inputs going to the dorsal portion

- The medial and superior vestibular nuclei control the vestibulo-ocular reflexes. These reflexes are important for maintaining a stable eye position when the head is moved
- The inferior vestibular nucleus (along with input from the cerebellum) mediates the vestibulospinal and vestibuloreticular pathways.

Disorders of vestibular system or its central connections results in vertigo, an illusion of rotatory movement (→ *OHCM10* p.462).

Gustatory system

- Taste is mediated by chemical receptor cells. The chemoreceptors are epithelial cells clustered in sensory organs called taste buds. Each taste bud contains ~50–150 taste receptors. Taste buds are found in the papillae (numerous projections) of the tongue. There are three types of papillae:
  - Fungiform papillae contain between one and five taste buds and are found on the anterior two-thirds of the tongue
  - Foliate papillae contain thousands of taste buds and are found on the posterior edge of the tongue
  - Circumvallate papillae also contain thousands of taste buds and are found on the posterior third of the tongue
- Receptors have microvilli on their apical membrane that extend through an opening in the surface of the tongue, known as a taste pore, to reach the oral cavity
- There are four classes of taste receptor: bitter, sweet, sour, and salty. There is also mixed evidence for a fifth class for monosodium glutamate. Different receptor classes are distributed in broadly distinct regions around the tongue. There is a maximum density of each type in different locations, but each type can be found all over
- Taste receptors contain voltage-gated Na^+, K^+, and Ca^{2+} channels, similar to those in neurones, which are capable of generating action potentials. However, receptors appear to signal taste through graded subthreshold membrane potentials. The mechanism of transduction differs between the four submodalities:
  - Bitter stimuli cause release of Ca^{2+} from intracellular stores, triggered by IP_3 or cAMP second-messenger pathways. The raised Ca^{2+} concentration leads to transmitter release
  - There are two putative mechanisms for transduction of sweet tastes:
    - In the first, the 'sweet' molecule directly activates a Na^+-specific channel that depolarizes the cell. This hypothesis is supported by evidence that the depolarization has a reversal potential near the Na^+ equilibrium potential, and that the depolarization is blocked by amiloride
    - In the second, adenylate cyclase is activated via a G-protein, causing an increase in intracellular cAMP concentration. cAMP closes a voltage-dependent K^+ channel that is open at the normal resting potential, causing depolarization. It is possible that both mechanisms act in 'sweet' receptors, or that there are two types of sweet receptor, one with each mechanism
  - Sourness is the taste associated with acids, which are thought to pass directly into the receptor cells through the cell membrane. The acids block voltage-dependent K^+ channels in the apical membrane, but do not diffuse across the cell in sufficient quantities to block voltage-gated Na^+ and Ca^{2+} channels in the basolateral membrane. This alteration in the balance of the resting ion flux results in depolarization of the cell
  - Saltiness is thought to be mediated through a direct binding of the active chemical species to a voltage-independent, amiloride-sensitive, cation channel. Opening of this channel allows Na^+ influx and thus depolarization.

Central pathways

- Receptor cells form chemical synapses with primary afferent fibres. Primary afferent fibres innervate several receptor cells within a number of taste buds in several papillae. Fibres thus have highly distributed innervations with complex and overlapping receptive fields
- A branch of the VIIth (facial) cranial nerve (the chorda tympani) innervates the anterior two-thirds of the tongue, while the lingual branch of the IXth (glossopharyngeal) nerve innervates the posterior third. Cell bodies of the chorda tympani are in the geniculate ganglion, and those of the glossopharyngeal nerve in the petrosal ganglion
- There are also some taste buds on the palate, innervated by the greater superficial petrosal branch of the VIIth cranial nerve, and on the epiglottis and oesophagus, innervated by the superior laryngeal branch of the Xth (vagal) cranial nerve, whose cell bodies lie in the nodose ganglion
- Gustatory afferents from the VIIth, IXth, and Xth cranial nerves enter the solitary tract in the medulla, synapse in the gustatory nucleus, in the rostral and lateral part of the solitary nuclear complex, and project, via the central tegmental tract, to the ventral posterior medial nucleus of thalamus, and thence to the area of primary somatosensory cortex that represents the tongue on the postcentral gyrus and to the insular cortex deep in the Sylvian sulcus
- Taste is represented in a specialized area of primary somatosensory cortex, next to the area that represents touch sensation on the tongue. Spatial segregation of receptor subtypes in the tongue is preserved in the gustatory nucleus, thalamus, and cortex.

Olfactory system

- Smell is mediated by chemoreceptors that have a very high sensitivity to odorant molecules. They can detect molecules at concentrations as low as a few parts per trillion
- Receptors are located in a specialized region of olfactory epithelium, deep in the nasal cavity, which consists of receptor cells, supporting cells, and basal cells
- Olfactory receptors are bipolar cells (Fig. 11.64). One end is short and extends to the mucosal surface where it expands into an olfactory knob. The other end is longer and forms an unmyelinated axon that projects through the cribriform plate to the ipsilateral olfactory bulb, on the undersurface of the frontal lobe. These axons join together in bundles of 10–100 surrounded by Schwann cells, to pass through the cribriform plate as olfactory nerves
- Olfactory receptors have a lifetime of ~60 days and so, unusually for neurones, are constantly replaced. However, the postsynaptic cells in the olfactory bulb do not regenerate, and must continually make fresh connections with each new generation of receptors.

Transduction

- The olfactory knob has several cilia, which form a dense mat at the mucosal surface. Odorants in the mucus are bound by an olfactory binding protein to enable cilia to interact with odorant molecules in the layer of mucus that covers this surface
- Binding of odorants by membrane proteins causes an increase in intracellular cAMP concentration via a G-protein effect on adenylate cyclase. Depolarization arises through a cAMP-gated Na^+ channel, in a process homologous to phototransduction in the retina. There is also evidence for depolarization mediated through an IP_3 (inositol triphosphate) second-messenger system, without a change in the cAMP concentration
- There is a large family of olfactory receptors which are thought to act together, rather like the three cone types in colour vision, to give a very wide range of smells that can be discriminated.

Central pathways

- The olfactory bulb is divisible into four layers:
  - In glomeruli, receptor axons synapse onto mitral cells, tufted cells, and periglomerular cells
  - The external plexiform layer contains tufted cell bodies and the dendritic trees of granule cells
  - The mitral body layer contains mitral cell bodies
  - The granule layer contains the axons of mitral cells and granule cells, and granule cell bodies
- Mitral and tufted cells form the output from the olfactory bulbs, and their axons form the olfactory tract. Periglomerular cells and granule cells are inhibitory interneurones. Granule cells also receive efferent fibres from both olfactory nuclei

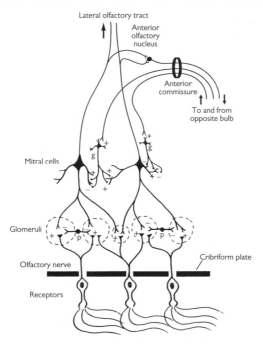

Lateral olfactory tract

Anterior
olfactory
nucleus

Anterior
commissure

To and from
opposite bulb

Mitral cells

g

Glomeruli

p

Cribriform plate

Olfactory nerve

Receptors

Fig. 11.64 Simplified representation of the cell types of the olfactory bulb and their connections—p (periglomerular cells); g (granule cells).

Reproduced with permission from Carpenter R (2003). *Neurophysiology*, 4th edn. Edward Arnold.

- Axons in the olfactory tract project to the anterior olfactory nucleus, the olfactory tubercle, the pyriform cortex, the cortical nucleus of the amygdala, and the entorhinal cortex:
  - The anterior olfactory nucleus projects to the contralateral olfactory bulb via the anterior commissure, and thence back to the contralateral olfactory bulb
  - The olfactory tubercle and pyriform cortex project to other olfactory cortical regions and to the medial dorsal nucleus of the thalamus, and thence to orbitofrontal cortex, which is associated with conscious perception of smell
  - The amygdala and entorhinal cortex are part of the limbic system, and are involved in the affective components of smell perception.
- There is no topographic central representation of olfaction as there is with other sensory modalities. However, glomeruli seem to have preferential sensitivities to certain odours over others, so there is clearly a distributed representation of odour.

Sleep

See Fig. 11.65.

The passive electroencephalogram (EEG) reveals rhythms of activity at four different frequencies:

- Alpha rhythm (8–13Hz) characterizes the awake but resting EEG
- Faster beta waves are associated with mental activity (>13Hz)
- Higher frequency gamma rhythm (35–45Hz) may be a signature of the waking state
- Slower rhythms, theta (4–7Hz) and delta (<3.5Hz), are more common during reduced arousal in adults.

Two types of sleep can be defined using electrophysiological features detected with EEG, electro-oculography (EOG), and electromyography (EMG)—REM (paradoxical) and non-REM (NREM, slow-wave sleep), see Fig. 11.65.

NREM sleep

- NREM sleep occupies ~75% of sleep time and is predominantly associated with delta activity, while activity more like that seen during the waking state characterizes REM sleep. This is despite the individual being difficult to arouse at this point—hence the alternative term of paradoxical sleep
- In general, NREM sleep is characterized by synchronized activity in the brain, while activity during waking and REM sleep is desynchronized
- Four stages of NREM sleep have been defined related to the depth of sleep and the increasing dominance of low-frequency synchronized wave activity in the EEG:
  - Stage 1 shows a slowing of the frequency of wave activity
  - In stage 2, the EEG displays distinctive bursts of high-frequency activity (spindles)—thought to be of thalamic origin (◒ pp.789–90)
  - In stages 3 and 4 (slow-wave sleep), low-frequency, synchronized wave activity dominates the EEG.

REM sleep

REM sleep is characterized by 'wake-like' EEG activity (low-amplitude, high-frequency gamma waves, 30–80Hz), clusters of eye movements on the EOG, and low muscle tone (atonia) on the EMG.

Processes

- Neuronal networks in the pons, midbrain, thalamus, hypothalamus, and basal forebrain (all of which regulate the rostral cerebral hemispheres) control wakefulness, NREM sleep, and REM sleep
- The areas of the brain implicated in arousal lie predominantly, but not exclusively, in the reticulum of the brainstem and midbrain (mesopontine junction—a region of the rostral brainstem containing the laterodorsal tegmental nucleus, pedunculopontine nucleus, dorsal raphe nucleus, and the locus coeruleus). An increase in activity in this ascending reticular activating system (ARAS) is suggested to underlie arousal
- The neurotransmitter systems that are prominently involved include the noradrenergic (locus coeruleus), serotonergic (5-HT, dorsal raphe)

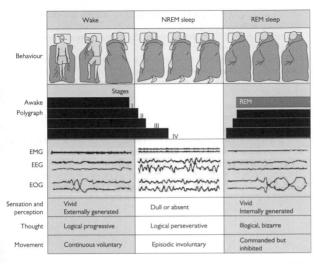

	Wake	NREM sleep	REM sleep

Fig. 11.65 Behavioural states in humans. States of waking, NREM sleep and REM sleep have behavioural, polygraphic, and psychological manifestations. In the row labelled behaviour, changes in position (detectable by time-lapse photography or video) can occur during waking and in concert with phase changes of the sleep cycle. Two different mechanisms account for sleep immobility. The first is disfacilitation (during stages I–IV of NREM sleep). The second is inhibition (during REM sleep). During dreams, we imagine that we move, but we do not. Sample tracings of three variables used to distinguish the state are shown: an EMG, an EEG, and an EOG. The EMG tracings are highest during waking, intermediate during NREM sleep, and lowest during REM sleep. The EEG and EOG are both activated during waking and inactivated during NREM sleep. Each sample shown is ~20s long. The three bottom rows describe other subjective and objective state variables.

Reprinted by permission from Macmillan Publishers Ltd: *Nature* 437: 7063, 1254–1256, Copyright (2005).

histaminergic (hypothalamus), cholinergic (pons/midbrain), and orexin/hypocretin (hypothalamus)
- In general, the release of these neurotransmitters is under the control of both circadian (light–dark cycle—SCN in the hypothalamus) and homeostatic (fatigue) influences
- The major projection of the ARAS is the thalamus—a critical relay for sensory and intra-cerebral pathways (⊙ p.786). In the absence of activity in the ARAS, the thalamus and the cortex tend towards slow-wave activity and unconsciousness
- The transfer into NREM sleep is associated with a reduction in activity of these networks
- The length of the NREM/REM cycle is also determined by signals from cholinergic and aminergic neurones. During NREM sleep, aminergic signalling dominates. However, the transition to REM sleep involves reduced activity of aminergic systems and an increase in cholinergic

activity. During REM sleep, aminergic signalling is silent and cholinergic excitatory activity is dominant
- The termination of REM sleep is driven by increased noradrenergic and serotonergic system activity
- NREM and REM activity alternate in each of four or five ~90min cycles each night. NREM is deeper and longer early in the night while, in later cycles, REM sleep occupies progressively longer periods of the cycle (p.824)
- NREM sleep appears associated with conservation of energy and repair mechanisms, thereby suggesting a restorative function, although it has also been suggested to function in the iteration of information
- The brain activity associated with REM sleep may be important in brain development and plasticity. A widely supported idea is that memory consolidation is a major function of REM sleep
- The disruption of metabolic homeostasis that results from severe sleep deprivation can lead to death, emphasizing the crucial role of sleep.

Consciousness

- In healthy individuals, three main states of consciousness are recognized—wakefulness, NREM sleep, and REM sleep
- Awareness might also be regarded as an aspect of consciousness, as we are aware, whether awake or dreaming. Nevertheless, analysis is difficult because someone other than the 'aware' individual interprets awareness on the basis of objective parameters such as behavioural or neuronal activity
- The conscious state can be usefully assessed using objective criteria such as those in the Glasgow Coma Scale (GCS; ⊃ p.829)—though it may be limited if the patient is prevented from making the usual responses to consciousness (e.g. by paralysis). The GCS includes numerically graded assessment of three parameters—eye opening, motor function, and verbal responsivity
- Altered consciousness states, as a result of some pathology, range from coma to vegetative state to brainstem death:
  - Coma is a state of continuous unconsciousness in the absence of a sleep–wake cycle. The degree to which an individual responds varies significantly, as does the level of cerebral metabolism. Coma may represent a transitional state between full recovery, re-establishment of the sleep–wake cycle with impaired awareness, or brainstem death
  - Brainstem death is associated with irreversible loss of all brainstem functions
  - The vegetative state illustrates distinctions between the underlying brain systems for wakefulness and awareness. In this state, several behaviours suggest wakefulness but any sense of purpose or attempts to communicate are absent. The vegetative state is often the result of severe brain injury (trauma/hypoxia/ischaemia) yet, while autopsies reveal damage to any or all of the cortical mantle, cerebral white matter, and thalamus, the brainstem networks associated with wakefulness are unaffected.

The Glasgow Coma Scale

(⊙ OHCM10 p.788.)

This gives a reliable, objective way of recording the conscious state of a person. It can be used by medical and nursing staff for initial and continuing assessment. It has value in predicting ultimate outcome. Three types of response are assessed.

Best motor response—this has six grades:

6 *Carrying out request ('obeying command')*: the patient does simple things you ask (beware of accepting a grasp reflex in this category).

5 *Localizing response to pain*: put pressure on the patient's fingernail bed with a pencil then try supraorbital and sternal pressure—purposeful movements towards changing painful stimuli is a 'localizing' response.

4 *Withdraws to pain*: pulls limb away from painful stimulus.

3 *Flexor response to pain*: pressure on the nail bed causes abnormal flexion of limbs—decorticate posture.

2 *Extensor posturing to pain*: the stimulus causes limb extension (adduction, internal rotation of shoulder, pronation of forearm)—decerebrate posture.

1 *No response to pain*.

Note that it is the best response of any limb which should be recorded.

Best verbal response—this has five grades:

5 *Oriented*: the patient knows who they are, where they are, and why, and the year, season, and month.

4 *Confused conversation*: the patient responds to questions in a conversational manner but there is some disorientation and confusion.

3 *Inappropriate speech*: random or exclamatory articulated speech, but no conversational exchange.

2 *Incomprehensible speech*: moaning but no words.

1 *None*.

Record the level of best speech.

Eye opening—this has four grades:

4 *Spontaneous eye opening*.

3 *Eye opening in response to speech*: any speech, or shout, not necessarily request to open eyes.

2 *Eye opening in response to pain*: pain to limbs as above.

1 *No eye opening*.

An overall score is made by summing the score in the three areas assessed. For example, no response to pain + no verbalization + no eye opening = 3. Severe injury, GCS ≤8; moderate injury, GCS 9–12; minor injury, GCS 13–15.

NB: an abbreviated coma scale, AVPU, is sometimes used in the initial assessment ('primary survey') of the critically ill:

- A = alert
- V = responds to vocal stimuli
- P = responds to pain
- U = unresponsive.

Some centres score GCS out of 14, not 15, omitting 'withdrawal to pain'.

NB: the GCS scoring is different in young children.

The limbic system

- The limbic system functions as an integrator of information from the external world and from within the body. Its proposed role is in the generation of emotional experience
- Anatomically, the limbic system consists of the limbic lobe, a ring of cortex around the brainstem comprising the parahippocampal gyrus, the cingulate gyrus, and the subcallosal gyrus (Fig. 11.66). In addition, the deeper lying hippocampal formation (including the hippocampus, dentate gyrus, and subiculum), the amygdala, parts of the hypothalamus, the septal area, the nucleus accumbens, and parts of the orbitofrontal cortex also contribute to the limbic system
- Pathways between the association cortex, entorhinal cortex, the hippocampus, and amygdala provide a link between the neocortex and the limbic system. Thus, connections between the limbic system and higher cortical areas enable the integration of emotional processing with cognitive functions such as attention, memory, and reasoning. In addition, it is well established that the hippocampus is involved in processes of memory storage, while the amygdala has also been implicated in learning, particularly tasks that might involve the association of stimulus and emotional response
- Both the amygdala and the hippocampus have direct, reciprocal connections with the hypothalamus. Pathways between the cortical cingulate gyrus, hippocampus, and hypothalamus extend the influence of the cortex over the hypothalamic expression of emotional behaviour
- The amygdala has two major projections:
  - The stria terminalis that innervates the hypothalamus and nucleus accumbens
  - The ventral amygdalofugal pathway that innervates the hypothalamus, dorsal medial nucleus of the thalamus, and the cingulate gyrus
- The connections between the limbic system and the hypothalamus regulate autonomic and visceral responses associated with motivational drives (food, water, sex) and emotional expression; they exert control over the endocrine system through its regulation of the secretion of hypothalamic hormones; and, through extensive connections with the sympathetic and parasympathetic nuclei of the brainstem and spinal cord, they exert influence over the autonomic motor system.

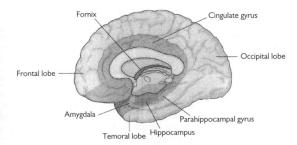

Fig. 11.66 The limbic system consists of the limbic lobe and deep-lying structures. This medial view of the brain shows the prefrontal limbic cortex and the limbic lobe. The limbic lobe consists of primitive cortical tissue that encircles the upper brain stem as well as underlying cortical structures (hippocampus and amygdala).

Memory

Much of our understanding of memory systems and their neurological basis originates from observations of patient HM, who, after undergoing bilateral medial temporal lobe resection (including the hippocampal formation) in an attempt to cure intractable epilepsy, displayed an inability to learn or retain memories of events after the transection (anterograde amnesia) along with impaired recollection of memories in the few years preceding transection (retrograde amnesia). Nevertheless, longer-term memories could be retrieved and were intact. Interestingly, HM could learn some motor tasks and was able to retain some immediate and short-term information indicating a distinction between short- and long-term memory systems.

Memory has been categorized into two types—declarative and non-declarative.

Declarative memory

The formation and retrieval of explicit memories of facts and events specific to an individual and involving the conscious recollection of past experiences.

- Episodic memory is the recollection of specific events occurring at a particular time and place. As seen in patient HM, the medial temporal lobe, including the hippocampus, perirhinal cortex, and parahippocampal cortex, is involved in both the formation and retrieval of this form of memory. Also involved is the prefrontal cortex. Damage in these regions can impair the ability to:
  - Retrieve the time and place at which an event occurred
  - Distinguish temporally, two or more events
  - Recollect where or when a new task was learned
- Semantic memory involves the recollection of information not associated with time or place but relating more to meaning and function (e.g. of objects, words). It is also dependent on the medial temporal lobes, particularly the anterior and lateral regions of the left hemisphere—a localization that, despite its close relationship to episodic memory, clearly distinguishes semantic memory.

Non-declarative memory

- Procedural memory relates to the acquisition of skills or habits. An intact corticostriatal system, motor cortex, and cerebellum all appear important mediators of different types of procedural memory. Damage to the striatum, such as that seen in Huntington's disease, impairs motor task learning without affecting declarative memory
- The perceptual representation system is important in the recognition of objects by their structure or form. This is distinct from the meaning or function that can be attributed to semantic memory. This form of memory also includes priming, or the ability to recognize a partial object as a result of prior exposure to the whole object. The posterior cortical regions that are activated during this form of memory are also distinct from those involved in semantic memory. Thus, visual object form involves the extrastriate occipital cortex, while global object structure involves areas at the interface between temporal and occipital cortices

- Working memory, a form of memory used for the short-term retention of information important for problem-solving or reasoning, can be severely disrupted by damage to the orbital frontal and medial frontal cortex, suggesting involvement of these regions in memory processes. Two subsystems appear to operate in the transfer of working memory into long-term memory:
  - A phonological loop that allows rehearsal of speech-based information
  - A visuo-spatial map, located in the visual association cortex, inferior parietal lobule, and prefrontal cortex of the right hemisphere, which retains the visual and spatial information.

Associative and non-associative learning in non-declarative memory

- Associative learning describes the linking of two events, whereas non-associative learning involves the influence of a single event on the probability that it elicits a response. Associative learning is typified by the classical conditioning experiments of Pavlov. Repeated pairing of a neutral (unconditioned) stimulus (bell ring) preceding one that evokes a response (conditioned stimulus, food) eventually leads to salivation in response to the bell ring—a conditioned response
- The motor component of procedural associative memory, typified by the example of an eye-blink in response to a conditioned stimulus, is dependent on the cerebellum, even though the reflex eye-blink to an unconditioned stimulus is independent of cerebellar involvement. The hippocampus and amygdala, on the other hand, are both involved in storing the learning experience but not the acquisition of the motor response
- Associative learning involving an unpleasant conditioned stimulus invokes mechanisms associated with fear and links emotion with memory. The amygdala plays a major role in emotional associative memory, along with the hippocampus and medial prefrontal cortex (➲ p.836)
- Non-associative learning, often expressed through reflex pathways, can involve a reduction in response (habituation), enhancement of a response (sensitization), or the removal of habituation by another, more powerful, stimulus (dishabituation).

Molecular and cellular mechanisms of learning and memory

Current ideas of the cellular basis of learning and memory revolve around mechanisms that lead to the selective strengthening or weakening of synaptic connections between neurones. Experimentally-induced cellular mechanisms such as long-term potentiation (LTP) and long-term depression (LTD), involving the strengthening or weakening of synapses, respectively, have been proposed as cellular correlates for learning and memory. The changes in synaptic strength in LTP and LTD are mediated by alterations in transmitter release and/or the insertion or removal of postsynaptic receptors. Both types of molecular event have been observed at synapses in pathways involved in the acquisition of particular tasks.

Ageing

With normal ageing (senescence), most individuals experience some form of cognitive impairment that, while falling short of dementia, can be problematic for quality of life. Mental capabilities most likely to show age-related impairment include aspects of memory, executive functions, and reasoning. It is significant that when one of these mental functions declines, so do the others. The impairment of these mental functions can be at least partially accounted for by a reduced speed of information processing that can be first detected in the third decade. In addition, sleep, motor, and sensory systems and brain circulation and metabolism may be affected.

Interconnectivity between temporal, prefrontal, and parietal cortical areas and the pathways between the entorhinal cortex, dentate gyrus (hippocampal formation), and the principal neurones of the hippocampus (all forming the hippocampal formation) play prominent roles in cognitive function (⊃ p.830).

As people grow older, there is a gradual decrease in brain volume. Traditionally, neuronal death spread across the cortex was proposed to underlie the reduced brain volume and, thus, explain cognitive impairment with ageing. This hypothesis was apparently supported by the large cortical cell loss seen in age-related disorders such as dementia and Alzheimer's disease (⊃ *OHCM10* p.488), which are also characterized by dramatic cognitive decline. More recent studies, however, indicate that the cognitive symptoms of normal ageing are not associated with significant neuronal death. Rather, the decrease in brain volume is thought to be due to shrinkage of neuronal cell bodies and dendrites and a progressive loss of myelin integrity.

A reduction in synapse number, without accompanying neuronal death, is suspected of underlying impaired cognitive function during senescence. The loss of synapses is associated with a reduction in dendritic spine number and density (the sites of synapses) on cortical neurones that may be as much as 50% in individuals >50 years old.

Also accompanying synaptic loss, degeneration of myelinated axons in the cortex and white matter has been reported. As white matter represents the major connective pathways within the CNS, it is likely that the damage to this cortical connectivity underlies the synaptic loss. This myelin damage is manifest as white matter lesions (small scars in the brain's white matter) that can be detected using magnetic resonance brain imaging. They are found in healthy old people and accumulate with age.

Neuronal death with ageing may, however, be more significant in subcortical areas. Modulatory inputs to the hippocampus and cortex from cholinergic neurones in the basal forebrain and from brainstem monoaminergic neurones are reduced during senescence. Correlations exist between the degree of cholinergic cell loss and behavioural change, suggesting that the reduction of these neuromodulatory systems is likely to contribute to the impaired integrative function of the cortex with age.

Emotion

Emotions of anger, fear, pleasure, and contentment are expressed as behavioural patterns that include facial expressions, bodily demeanour, and autonomic arousal. Conscious recognition of the physiological reactions to emotion-inducing stimuli can be described as 'feelings'. In general, emotions influence all aspects of cognitive function (attention, memory, reasoning), yet we have restricted intentional control over them.

Key brain regions and their interactions that influence the processing of emotions include the prefrontal cortex, the amygdala, the hypothalamus, and the anterior cingulate cortex (Fig. 11.67). Connectivity between these and other brain areas enables the integration of emotional and cognitive processes.

The hypothalamus integrates and coordinates the motor and endocrine responses that constitute the behavioural expression of emotional states. Aggression is associated with neural activity in the medial hypothalamus. On the other hand, the lateral hypothalamus is active during expression of anger. Indeed, stimulating this region induces the behavioural profile associated with anger, though not the conscious experience of the emotion.

Expressions of anger are also associated with activity in other parts of the limbic system, particularly the cingulate cortex. Reciprocal connections between the cortex and the hypothalamus perform several functions:

- They allow the conscious experience of emotions in the form of feelings
- They activate appropriate bodily mechanisms in response to external stimuli
- They direct appropriate motor behaviour (e.g. avoidance) in response to the stimulus.

In addition, neocortical involvement may suppress emotional responses to minor stimuli.

The amygdala is critical for the processing of emotional signals that are expressed through changes in hypothalamic function. Fear, for example, is associated with autonomic (heart rate and blood pressure), endocrine (stress hormones), and motor behaviour changes, along with analgesia and potentiated somatic reflexes like the startle response. Fear conditioning describes the ability of a neutral stimulus to acquire fear-inducing properties when paired closely in time with a threatening event. Damage to the amygdala can impair the processing of signals of fear such as facial and vocal expressions. From a clinical point of view, such a mechanism may relate to the pathophysiology of phobic anxiety and post-traumatic stress disorders.

The amygdala also plays important roles in emotional conditioning, the consolidation of emotional memories, and associative learning involving reward and appetite. Modulation through β-adrenoceptors is essential for the enhancement of emotional memory.

The amygdala processes emotion-inducing stimuli through two pathways:

- Basic sensory input is received by the amygdala via a direct route from the thalamus and hippocampus. This elicits short-latency emotional responses, possibly without cortical (cognitive) involvement. This direct input may also prepare the amygdala for the reception of more complex information from cortical areas

● The prefrontal cortex receives processed sensory signals from the thalamus, as well as information concerning the output of the amygdala and homeostatic feedback concerning the rest of the body. A second pathway to the amygdala thus carries integrated cortical output, resulting in a slower response that also includes conscious awareness of the emotional experience.

The ventromedial prefrontal cortex is proposed to integrate signals originating within the body with other forms of cognitive information to modulate emotional intensity. This may enable decision-making in situations where the possible choices have subtly different emotional values. In this respect, earlier significant emotional events may be linked with patterns of physiological reactions. Alternatively, the role of the prefrontal cortex is suggested to modulate emotional decision-making in favour of longer-term adaptive rather than immediate gains.

Homeostatic inputs from autonomic, visceral, and musculoskeletal sources via the brainstem, hypothalamus, and somatosensory and cingulate cortices are integrated into the generation of feelings. Likewise, the recall of feelings associated with emotional experience recruits similar brain areas including brainstem nuclei, hypothalamus, and somatosensory and orbito-frontal cortices. The anterior cingulate cortex is implicated in processing of emotional information with respect to the regulation of mood.

The distinction between emotion and its mental representation as feelings suggests separate systems exist for the perception of emotion and feeling states. The involvement of the amygdala in emotional perception, but not in the pathways associated with the expression of feelings, supports at least partial distinction in the mechanisms underlying these two aspects of emotion.

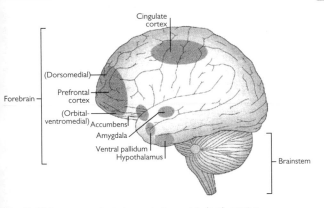

Fig. 11.67 Key structures in the human brain associated with emotion.

Reprinted by permission from Macmillan Publishers Ltd: *Nature Reviews: Neuroscience* 5:7, 583–9, Copyright (2004).

Depression

Depression and mania are disorders of affect, characterized by changes in mood ($\circleddash$ *OHCS11* pp.710–5). Two forms of depressive syndrome have been categorized:

- Bipolar depressive syndrome reflects oscillation between depression and mania, perhaps suggesting a biochemical imbalance. There is good evidence for a hereditary link to this form of depression
- Unipolar depressive syndrome or depression without associated mania. This is more common in older patients and can be associated with anxiety and agitation.

Symptoms associated with depression include:
- General feeling of apathy, misery, and pessimism
- Low self-esteem including guilt, inadequacy, and ugliness
- Indecisiveness and loss of motivation
- Retardation of thought and actions
- Sleep disturbance
- Loss of appetite.

Mania is associated with symptoms of excessive exuberance, self-confidence, enthusiasm, physical activity, irritability, impatience, and anger.

Abnormal function of brain regions associated with emotion is implicated in mood disorders. The cognitive aspects of depression such as memory impairment, feelings of worthlessness, guilt, and suicidality may be mediated by the neocortex and hippocampus. The involvement of the amygdala, striatum, and nucleus accumbens in emotional memory indicates these regions may be involved in decreased drive for pleasurable activity, reduced motivation, and anxiety. The hypothalamus may also be involved in the disruption of sleep (circadian rhythm) as well as drives for food or sex.

Box 11.2 summarizes the treatments for depression.

Box 11.2 Drug therapies for depression

(🔗 *OHPDT2* p.316.)

- Most drug treatments that alleviate the symptoms of depression have known effects on brain 5-HT and noradrenaline systems
- Tricyclic antidepressants (TCAs) (e.g. amitriptyline) are non-selective inhibitors of monoamine transporters that remove these transmitters from the extracellular space after synaptic release
- Monoamine oxidase inhibitors (MAOI) (e.g. phenelzine) inhibit the breakdown of monoamine neurotransmitters
- Selective serotonin re-uptake inhibitors (SSRIs) (e.g. fluoxetine) selectively inhibit 5-HT transporters
- Lithium is mainly used to control the manic phase of bipolar depression through mechanisms that may involve disruption of intracellular signalling pathways
- The actions of these drugs led to the monoamine theory of depression that suggests the condition is due to a functional deficit of monoamine (5-HT, noradrenaline, and dopamine) neurotransmission. Mania is thus regarded as being due to the opposite mechanism (i.e. a functional excess of these neurotransmitters)
- The monoamine hypothesis may have some basic validity in explaining the mechanism of action of antidepressant drugs. However, a unifying biochemical theory of affective disorders cannot be substantiated. The main problem is that the primary biochemical actions of drugs are very rapid but antidepressant effects usually take 2–4 weeks to develop. In addition, a proportion (~20%) of individuals do not respond to these treatments. This suggests that affective disorders are more than just a malfunction of monoamine neurotransmission. It is likely that antidepressant drugs promote secondary adaptive changes, probably through gene activation, that account for their therapeutic efficacy
- Dysregulation of the hippocampus and the hypothalamic–pituitary–adrenal (HPA) axis, → increased levels of circulating stress hormones (cortisol), can promote many symptoms of depression. The ability of antidepressant drugs to stimulate, through genomic mechanisms, the production of growth factors (e.g. brain-derived neurotrophic factor (BDNF)) that can promote synapse formation in the brain and normalize HPA axis function has been suggested to account for their delayed therapeutic effects.

Schizophrenia

Schizophrenia is a highly debilitating but common disorder with a lifetime risk in the general population of ~1%. A further 2–3% of the population has the related schizotypal personality disorder.

Schizophrenia is classified as a neurodevelopmental disorder that epidemiological studies suggest is predominantly, but not completely, attributable to genetic causes. Studies of identical (monozygotic) and non-identical (dizygotic) twins indicate ~50% and ~10% concordance, respectively—both considerably higher than the incidence in the general population.

Two types of symptoms characterize the disorder:

- Positive (type I) symptoms are manifest as psychotic episodes of the disorder and include auditory hallucinations, delusions, and incoherent or disordered cognitive processes
- Negative (type II or residual) symptoms characterize non-psychotic periods and represent the most unmanageable aspects of the disorder in that drug therapy is commonly ineffective. They include social isolation and withdrawal, decreased emotions, loss of drive, apathy, poverty of speech, and affective flattening.

The progression of the clinical signs of schizophrenia can be divided into four stages:

- During childhood, no obvious symptoms of the disorder are apparent (premorbid phase)
- During adolescence, non-specific behavioural changes appear, related to the development of negative symptoms (prodromal phase)
- The onset of the disorder, characterized by the first psychotic episode, occurs typically during the late teens to early twenties. Left untreated, the disorder will continue, with psychotic episodes interspersed by phases displaying negative symptoms, throughout adulthood (progressive phase). Also associated with disease progression are mood disorders and impaired cognitive function
- These symptoms persist during later life in the final residual stage of the disorder. Many sufferers never recover.

The dopamine theory of schizophrenia

Based on the strong correlation between the potency with which neuroleptic drugs antagonize dopamine D_2 receptors, and their effective therapeutic dose, the dominant hypothesis is that hyperactivity of brain dopaminergic pathways underlies the positive symptoms of schizophrenia.

Nevertheless, while neuroleptic drugs occupy D_2 receptors very quickly after oral administration, a delay of 1–2 weeks is needed before an overall decrease in dopaminergic activity is obtained and therapeutic effects are seen. This suggests that while the dopaminergic system may be integral to the mechanism of neuroleptic action, indirect actions of these drugs are also likely to be involved.

There are four major dopaminergic pathways in the brain (Fig. 11.68).

- Dopaminergic neurones originating in the ventral tegmental area (VTA), and projecting to nuclei of the limbic system, form the mesolimbic dopamine pathway. Projections to the nucleus accumbens, amygdala, hippocampus, anterior cingulate, and entorhinal cortex—areas

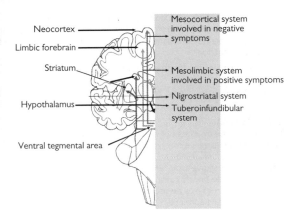

Fig. 11.68 Dopamine pathways and schizophrenia.

intimately involved in emotional function and memory—support the idea that hyperexcitability of this pathway mediates the positive symptoms of the disorder
- Dopaminergic neurones also originating in the VTA but projecting to the neocortex (in particular the prefrontal cortex) form the mesocortical dopamine pathway. These cortical areas are important in motivational planning, attention, and social behaviour. It is possible that abnormalities in these regions are important in the negative symptoms. Damage to cortical connections in these areas can result in similar symptoms. Thus, it is possible that reduced activity of this dopaminergic pathway is responsible for negative symptoms and this may explain their refractoriness to typical neuroleptic drugs
- Two additional dopaminergic pathways in the brain are of importance relative to unwanted effects of neuroleptic drugs:
  - The nigrostriatal dopamine system is found in the basal ganglia and projects from the substantia nigra compacta to the striatum (Parkinson's disease). Neuroleptic drugs acting on this pathway can induce unwanted extrapyramidal effects (movement disorders):
    - Acute dystonias are reversible effects associated with Parkinsonism-type symptoms of muscle rigidity, tremor, and loss of mobility. Onset is dose-dependent and the severity often declines with progressive treatment
    - Tardive dyskinesia is a chronic, irreversible effect involving involuntary movement of particularly the face and tongue, but also the trunk and limbs. These unwanted effects arise in 20–40% of cases and have a delayed appearance after prolonged neuroleptic treatment for months to years. Their underlying cause is unknown but changes in D_2 receptor sensitivity in the striatum and neuronal death in cortical motor areas may both contribute

- The tuberoinfundibular dopamine system projects from the hypothalamus to the pituitary (⊜ pp.789–90). Neuroleptic treatment can cause endocrine disturbances, notably lactation, due to increased prolactin production.

Box 11.3 summarizes the treatments for schizophrenia.

Box 11.3 Treatments and drug therapies for schizophrenia

(⊜ *OHPDT2* Ch. 4.)

- Drugs used in the treatment of schizophrenia are known as neuroleptics. Recurrence of psychotic episodes can be effectively controlled by chronic treatment. However, they are not useful for controlling the negative symptoms
- ~40% of cases are poorly controlled by classical neuroleptics, though newer atypical drugs may be useful in some of these refractory cases
- Classical or typical neuroleptics such as the phenothiazines (e.g. chlorpromazine), the thioxanthenes (e.g. flupentixol), and the butyrophenones (e.g. haloperidol) show a preference for antagonism at D_2 receptors but also block 5-HT$_2$, α-adrenergic, and muscarinic acetylcholine receptors
- Atypical neuroleptics show some effectiveness against negative as well as positive symptoms and, generally, have fewer 'extra-pyramidal' side effects. Examples include the dibenzazepine, clozapine, and the benzamide, sulpiride
- As well as being effective blockers of D_2 receptors (although D_4 and presynaptic D_1 receptors may also be blocked), their higher potency block of 5-HT$_{2A}$ receptors is regarded as an important factor in their different therapeutic and side effect profiles.

Epilepsy

- A common disorder that affects up to 1% of the population
- Of unknown origin, but can develop after brain trauma (e.g. infection, tumour growth, neurological disease); only one rare type of epilepsy has been traced to a single gene defect, but there is about 30% heritability of epilepsy
- Characterized by seizures that range in intensity (loss of concentration through to convulsive fit), frequency and impact (specific functions affected (partial seizure) through to global effects throughout the body (generalized seizure))
- Associated with high-frequency discharges from neurones in specific regions (partial seizure) or throughout the brain (generalized seizure)
- Generalized seizures are subcategorized into tonic–clonic type (grand mal) and absence seizure type (petit mal):
  - Tonic–clonic seizures involve an initial global spasm (which can involve biting, cessation of respiration, salivation, and defecation), followed by a series of convulsions for a period of several minutes. The patient often remains unconscious for several minutes after the seizure and usually feels dazed and confused after regaining consciousness
  - Absence seizures are less dramatic and occur in children: the child suddenly enters a trance-like state for several seconds, before returning to their activities, with no ill effects
- Seizures can be associated with enhanced release of excitatory neurotransmitters (most notably glutamate acting on NMDA receptors) and/or depressed release of inhibitory neurotransmitters (GABA), but there is no obvious reason for this in sufferers.

Box 11.4 summarizes the treatments for epilepsy.

Box 11.4 Treatments and drug therapies for epilepsy

(⊕ *OHPDT2* Ch. 4.)

The majority (~75%) of patients respond well to prophylactic drug therapies for epilepsy, although the need to take drugs continuously is associated with side effects. The main drug therapies are as follows:

- *Phenytoin* is used in most types of seizures (except absence seizures). It is related to the barbiturates and acts to stimulate inhibitory GABA receptors to depress neuronal activity. Though effective, this drug is associated with a wide range of side effects from confusion and headaches through to anaemia, foetal malformation (cleft palate), and even hepatitis in some patients.
- *Carbamazepine* is related to TCAs and is particularly effective in partial seizures. Its use is often associated with drowsiness, but more severe disturbances can also be experienced. Because of its effect on metabolism of other drugs, possible drug interaction(s) is an important consideration before prescription.
- *Valproate* is effective in both grand mal and petit mal. It is a simple carboxylic acid that increases GABA, possibly through postsynaptic effects and/or inhibition of GABA metabolism. Side effects are restricted to hair loss and mild hepatic problems in some cases, but it is implicated in spina bifida when used during pregnancy
- *Ethosuximide* is often used in absence seizures through its ability to inhibit T-type calcium channels.
- *Phenobarbital* (a barbiturate) is rarely used in epilepsy because of its sedative effect, and benzodiazepines are only used in cases where seizures are so frequent for the benefits to outweigh the sedative side effects.

Addiction

- Addiction or dependence is a compulsion to perform an action (e.g. take a drug) with a loss of control in limiting that behaviour. This compulsion or craving is a form of psychological dependence
- Intimately associated with the concept of addiction is tolerance, a reduction in effect with repeated execution of the action, and withdrawal (also known as physical dependence) or the appearance of symptoms associated with the termination of chronic expression of the behaviour
- Anxiety, depression, and dysphoria are symptoms common to withdrawal from a wide range of addictive behaviours. It is possible that these symptoms may also be motivating factors for perpetuating addictive behaviour
- The development of addiction is associated with positive reinforcement (i.e. an increased probability of a response to a given stimulus). In the context of drug use, reinforcement could be through direct effects on the probability of self-administration or indirect effects such as a drug's ability to enable reinforcing effects of associated but neutral stimuli
- While most frequently associated with drug use, other disorders such as impulse control disorders also display symptoms characteristic of addiction—namely, compulsion to perform the action, repeated performance, and withdrawal symptoms of dysphoria and depression. Such disorders are associated with a wide range of activities, including eating, exercise, gambling, sex, and shopping.

Brain pathways and addiction

See Fig. 11.69.

- The development of addiction has been associated with midbrain and forebrain systems that mediate motivated behaviour and natural reward mechanisms
- Ascending and descending pathways through the median forebrain bundle, particularly monoamine pathways, link several regions associated with motivated behaviour. Of particular importance are the VTA and the basal forebrain, including the nucleus accumbens, olfactory tubercle, frontal cortex, and amygdala
- Neurochemically, the mesolimbic dopamine pathway (➲ p.840–41), linking the VTA and the basal forebrain, is seen as an integral component in linking dependence and natural reward systems
- In addition, opioid, GABA, and 5-HT systems modulate the function of the VTA and basal forebrain during natural and drug reward-related behaviours indicating the convergence of several neurochemical systems in positive reinforcement mechanisms
- During withdrawal, monoamine levels, particularly dopamine in the VTA and 5-HT, decrease dramatically, suggesting that addiction involves neuroadaptive mechanisms that contribute to both dependence and withdrawal symptoms.

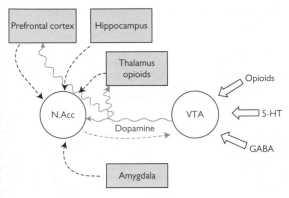

Fig. 11.69 Neural circuitry implicated in addictive mechanisms. Dashed lines indicate limbic input to the nucleus accumbens (N.Acc); the wavy connection represents the mesolimbic dopaminergic pathway believed to be important in reward mechanisms, as is the pathway from nucleus accumbens to VTA, represented by the broken grey connection. Various neurotransmitter systems indicated in white modulate VTA dopaminergic activity.

Drug addiction mechanisms

Cocaine and amphetamine

Cocaine and amphetamine are psychomotor stimulants that induce euphoria and act as reinforcers for drug self-administration. Although they increase levels of all monoamines the reinforcing effects crucially depend on dopamine release.

Impairing the mesolimbic dopamine pathway, particularly in the region of the nucleus accumbens and central nucleus of the amygdala, abolishes the reinforcing effects of cocaine and amphetamine.

Opiates

The reinforcing effects of opiates such as morphine and heroin involve actions at a specific receptor subtype, the μ-opioid receptor, found on neurones in the VTA and the nucleus accumbens. Blocking these receptors inhibit positive reinforcement by opiates. This does not, however, involve the mesolimbic dopamine pathway as opiate addiction is unaffected by its pharmacological or anatomical disruption.

Weak, long-acting μ-opioid receptor agonists, such as methadone, are used to alleviate physical dependence during drug withdrawal.

Nicotine

Nicotine has a direct reinforcing effect that involves activation of both the mesolimbic dopamine and opioid pathways through its agonistic actions at nicotinic acetylcholine receptors. Nicotine replacement, to alleviate withdrawal symptoms and psychological dependence, in conjunction with counselling, is the primary treatment used to help smokers give up. The α2-adreoceptor agonist, clonidine, may also be used. It acts presynaptically to reduce neurotransmitter release that may reduce the efficacy of the reward pathways.

Alcohol

The pathways associated with alcohol dependence are more complex. Alcohol shares sedative and anxiolytic properties with benzodiazepines and these may contribute to the reinforcing properties of both drugs. Alcohol and benzodiazepines potentiate GABA systems and this effect, in the central nucleus of the amygdala, is considered important for the development of dependence.

Activity in the mesolimbic dopamine pathway may also contribute to alcohol reinforcement, though this is not essential. Glutamate, 5-HT, and opioid systems have all been implicated in alcohol dependence.

Benzodiazepines are commonly used to alleviate withdrawal symptoms during the acute drying-out period, presumably because of their shared action on GABA systems. Disulfiram, an inhibitor of aldehyde dehydrogenase, promotes increased plasma acetaldehyde levels if alcohol is taken. This causes unpleasant symptoms (flushing, hyperventilation, tachycardia, panic) that can act as a form of aversion therapy. Acamprosate, a taurine analogue, reduces craving through an unknown mechanism, though it may be related to interactions with amino acid neurotransmission.

Degenerative diseases

Neurones in the CNS are post-mitotic cells and the potential for replacement is, in general, limited. As such, pathological neuronal loss is likely to have serious consequences for brain function. Neuronal death is a common feature of a group of disorders termed neurodegenerative disorders.

Neuronal death in neurodegenerative disease can occur by one of two processes:

- Necrosis (usually induced by acute injury) involves cell swelling, vacuolization, and lysis and is often associated with an inflammatory response
- Apoptosis (or programmed cell death) can be triggered by extracellular signals that occur normally during development but can also be triggered pathologically during neurodegenerative disease:
  - Apoptosis is characterized by cell shrinkage, nuclear chromatin and DNA fragmentation, and by the activation of caspases that break down certain intracellular proteins
  - Macrophages remove dead cells without inducing inflammatory responses.

Two mechanisms known to induce necrosis as well as to trigger apoptosis, and hence implicated in neuronal death associated with neurodegenerative disease, are oxidative stress and excitotoxicity.

- Oxidative stress refers to the excessive production of reactive oxygen species (ROS) including superoxide radicals (O_2^-), hydroxyl free radicals, and H_2O_2 when mitochondrial oxidative phosphorylation, and hence ATP production, is compromised. ROS damage important intracellular components including enzymes, DNA, and membrane lipids
- Glutamate-induced excitotoxicity involves excessive stimulation of calcium-permeable NMDA receptors and a catastrophic increase in intracellular calcium that further promotes glutamate release, activates membrane-damaging proteases and lipases, and also induces the production of ROS
- The importance of efficient mitochondrial ATP production and endogenous safety mechanisms to protect against ROS—e.g. enzymes such as SOD and catalase, and antioxidants such as glutathione, ascorbate, and α-tocopherol (vitamin E)—has led to the suggestion that compromising any of these mechanisms may underlie several neurodegenerative diseases.

In many neurodegenerative disorders, the causes of cell death can be attributed to inherited traits. However, many sporadic cases are idiopathic, suggesting non-genomic mechanisms may also be important. Thus, specific genetic mechanisms account for all cases of Huntington's disease ($\ominus$ *OHCM10* p.702). Genetic mutations also account for some cases of motor neurone disease ($\ominus$ *OHCM10* p.506), Parkinson's disease ($\ominus$ *OHCM10* p.494), and Alzheimer's disease ($\ominus$ *OHCM10* p.488). New variant CJD ($\ominus$ *OHCM10* pp.696, 697), on the other hand, is attributable to a transmissible infective agent.

Alzheimer's disease

(⊕ *OHCM10* p.488.)

- In Alzheimer's disease, neurones projecting from the entorhinal cortex to the dentate gyrus (hippocampal formation) and the principal neurones in the CA1 region of the hippocampus appear particularly vulnerable to death. This explains the usual course of the disease in which episodic memory is impaired early on (⊕ p.832)
- Pyramidal neurones in the cortical circuitry that link temporal, prefrontal, and parietal cortical areas also display vulnerability in Alzheimer's disease and the involvement of these areas in cognitive function may explain the later decline in cognition with disease progression
- A characteristic of Alzheimer's disease is a relatively selective loss of cholinergic neurones in the basal forebrain, → a reduction in the cholinergic innervation of the cortex and hippocampus. Attempts to alleviate this cholinergic loss using the cholinesterase inhibitors, tacrine or donepezil, has resulted in modest improvement in cognitive function in some patients
- Abnormal processing of amyloid precursor protein (APP) is regarded as a key step in pathogenesis of Alzheimer's disease, → extracellular accumulation of β-amyloid protein in the form of amyloid plaques
- Another characteristic of Alzheimer's disease is intraneuronal deposition of the phosphorylated microtubule associated protein, tau. When neurones die, phosphorylated tau filaments aggregate to form extracellular neurofibrillary tangles (NFT)
- Neurones die by apoptosis and necrosis in Alzheimer's disease. However, it remains to be established whether amyloid plaques and NFTs mediate neurotoxicity.

Parkinson's disease

(⊕ *OHCM10* p.494.)

- Parkinson's disease is manifest primarily as a disease of motor control resulting from the degeneration of nigrostriatal dopaminergic neurones in the basal ganglia
- Both apoptotic cell death and death induced by ROS and mitochondrial dysfunction have been implicated in Parkinson's disease
- Major treatment involves increasing brain dopamine levels by oral administration of the metabolic precursor, L-dopa. Dopamine itself cannot cross the blood–brain barrier. Brain levels of L-dopa are increased by the use of inhibitors of dopa decarboxylase (e.g. carbidopa) in the periphery
- Intraneuronal protein deposits, called Lewy bodies, are characteristic of Parkinson's disease
- A central role for Lewy bodies in Parkinson's disease neuropathology is suspected because mutations in genes encoding for their constituent proteins (e.g. α-synuclein) have been linked to some familial forms of the disease.

Motor neurone disease

(⮑ *OHCM10* p.506.)

Mutations in the *SOD-1* gene have been causally linked to amyotrophic lateral sclerosis, a form of motor neurone disease that results in the degeneration of motor neurones in the spinal cord, brainstem, and motor cortex.

Huntington's disease

(⮑ *OHCM10* p.702.)

- Huntington's disease is characterized by involuntary jerky movements that result from GABAergic cell death in the striatum. It is thought this leads to hyperactivity of the nigrostriatal dopamine pathway. Treatment of the symptoms is through dopamine D_2 antagonists (⮑ p.844) or GABA agonists such as baclofen
- Huntington's disease is inherited and attributable to the presence of high repeats of the DNA nucleotide sequence, CAG, in the huntingtin gene (*HTT*, that encodes for a protein implicated in regulating apoptosis and other cell death mechanisms).

New variant Creutzfeldt–Jakob disease

(⮑ *OHCM10* pp.696, 697.)

- New variant CJD is a type of spongiform encephalopathy characterized by a vacuolated appearance of the postmortem brain and associated with dementia and loss of motor coordination
- It is postulated that an infective agent—an abnormal form of a protein called a prion—triggers the disease. The abnormal prion is thought to lead to accumulation of insoluble prion protein in the brain, triggering neurotoxic processes.

Chapter 12

Infection and immunity

Bacterial structure

General features

Bacteria are prokaryotes, ~1–3μm in size, and lack a nuclear membrane and membrane-containing organelles. The cytosol contains ribosomes, inclusions such as glycogen or endospores, organic and inorganic molecules, plus a nucleoid (a nucleus that lacks a nuclear membrane).

Cell wall

Bacterial cell wall contains peptidoglycan, a sugar heteropolymer of cross-linked N-acetylglucosamine and N-acetylmuramic acid molecules. Inside this lies a lipid bilayer cell membrane containing proteins and carbohydrates (Fig. 12.1). Gram-positive bacteria have a thick cell wall containing lipoteichoic acid and an inner cell membrane. Gram-negative bacteria have a phospholipid outer cell envelope containing a high concentration of lipopolysaccharide and protein, a thin cell wall, and an inner cell membrane.

Additional features

Some bacteria have a polysaccharide capsule on the outside. Surface structures may include flagella aiding motility and pili or fimbriae contributing to adherence. Additional specific proteins and enzymes contribute to virulence or represent vaccine targets. Some bacteria contain spores and their position aids speciation.

Genomic organization

Bacterial genomes contain 1–6 million base pairs (Mb). DNA is linear and closed. In addition to genomic DNA, bacteria contain mobile genetic elements transferred between bacteria. Mobile DNA contains genes encoding resistance to antimicrobials. Examples of mobile DNA elements are:
- Plasmids: small circular DNA molecules containing up to 0.1Mb
- Transposons: contain insertion-sequence elements—enzymes that facilitate transposition of DNA between sites on chromosomes or plasmids
- Bacteriophages: viruses that infect bacteria and transfer bacterial DNA.

Protein synthesis

DNA-dependent RNA polymerase allows DNA to be copied into RNA. mRNA binds to an initiation site on the 30S ribosomal subunit. The 50S ribosomal subunit is bound to form a 70S complex to allow protein synthesis. tRNA is also present and used in protein synthesis.

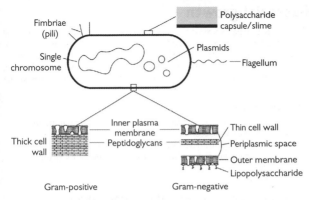

Fig. 12.1 Bacterial structure.

Bacterial growth

Minimal requirements are water, carbon, and inorganic salts. Many bacterial pathogens lack enzymes required for synthesis of key amino acids, nucleosides, or vitamins and these must be added. Synthesis of organic macromolecules allows growth and cell division. Assuming balanced replication, the daughter cell will contain the same amount of macromolecules as the parental bacterium.

Growth phases

- Lag phase: bacterium adapts to new environmental conditions and replication does not occur
- Exponential or logarithmic phase: bacteria replicate
- Stationary phase: limited availability of nutrients prevents further replication
- Death phase.

Regulation of gene transcription during growth

Changing environmental conditions require methods to adjust gene transcription:

- Nutrient sensing: gene transcription regulated by detection of specific nutrients
- Two-component signal transduction: sensor detects an environmental factor and phosphorylates a regulator that activates gene transcription
- Quorum sensing: changes in cell density alter levels of diffusible autoinducers that bind to transcriptional activators inducing gene transcription—in effect, quorum sensing is a mechanism of communication between bacteria.

Bacterial classification

- During Gram staining, the initial reagents, crystal violet and Gram's iodine, interact with bacterial ribonucleotides to give a purple colour. In Gram-positive bacteria, this remains, but in Gram-negative bacteria this colour is washed away by alcohol and acetone, and the bacteria stain red after counter-staining with safranin. This stain also allows separation into cocci (round) and bacilli (rod-shaped)
- Anaerobes require an environment lacking oxygen
- Metabolic tests, motility patterns, and spore formation present further features for classification
- Molecular techniques (sequencing genes and analysis of DNA) aid classification
- Rickettsiae and chlamydiae are obligate intracellular parasites that are Gram-negative coccobacilli. Most rickettsiae are unable to survive outside the cells they infect
- *Mycoplasma* spp. are unusual bacteria in that they lack a cell wall.

Bacterial pathogenesis

Bacterial pathogenesis

- Reflects the capacity of bacteria to cause disease
- Influenced by host factors and microbial factors
- Some bacteria replicate in the host without causing disease (commensals). Others penetrate tissues or elaborate toxins causing clinical disease (pathogens)
- Bacterial adaptations contribute to pathogenesis (virulence factors) but may only be expressed in a particular environment, e.g. inside macrophages
- Disease results from both microbial factors and the host response to bacteria
- A dysregulated inflammatory response leads to tissue damage.

Colonization

First stage in human diseases is often the ability to adhere to cutaneous or mucosal surfaces (colonization).
- Not unique to pathogens
- Pili, fimbriae, and polysaccharide capsules aid adherence
- Colonization may necessitate killing commensal flora by release of antibacterials (e.g. bacteriocin) or scavenging iron, hence limiting its availability for other bacteria.

Invasion

Tissue invasion is facilitated by enzymes that degrade matrix (e.g. hyaluronidase, elastase). Other bacteria may invade by instigating internalization by host cells and subsequently escaping these intracellular locations.

Avoidance of host defence

Host defence against bacteria includes opsonization of bacteria and phagocytosis. For intracellular bacteria, T-cells prime macrophage killing. Humoral immunity is also important. Capsule inhibits complement-mediated opsonization and phagocytosis. Fc-binding proteins or IgA proteases inhibit immunoglobulin. Some intracellular bacteria are efficiently phagocytosed but escape the phagolysosome or are able to resist antimicrobial molecules (e.g. denitrification inactivates phagocyte NO).

Toxins

Exotoxins are secreted virulence factors. Toxin-releasing bacteria need not invade tissue (e.g. intraluminal toxin release by colonizing bacteria or ingestion of preformed toxin in food). Two main groups of toxins are:
- Cytolytic: directly damage cell membranes and kill host cells
- Bipartite (A-B toxins): the B subunit binds to the host cell receptor and the A unit acts on the intracellular substrate. For example, cholera toxin B unit binds to GM_1, a sialoganglioside; the A unit induces ADP ribosylation and hence cAMP-mediated fluid secretion by enterocytes. Other toxins enter target cells by direct injection, as occurs with the type III secretion system of many Gram-negative bacteria. Some toxins act on extracellular targets (e.g. streptococcal pyrogenic toxins bind to the T-cell receptor $V\beta$ chain to induce cytokine release and toxic shock).

Endotoxin

Bacterial constituents, not secreted. Antimicrobials may enhance endotoxin release. The lipid A portion of lipopolysaccharide is responsible for many of the pathogenic features of Gram-negative bacteria including fever, shock, and activation of clotting. In this case, it is the host response to the bacterial component that results in dysregulated cytokine production (IL-1β, TNF-α, IL-6). Lipoteichoic acid in the Gram-positive cell wall induces similar effects.

Diagnosis

Commonly used techniques in the diagnosis of bacteria are summarized in ◆ Chapter 15. The most commonly employed diagnostic techniques are:
- Detection of bacteria in specimens by Gram staining (or other specialized stain) and visualization by microscopy
- Bacterial culture and detection of specific bacteria using diagnostic algorithms

In specific cases, other methods involving detection of bacterial antigens, serology, molecular techniques, and analysis of biopsy specimens are employed.

Antibiotics

These are antimicrobials produced by living organisms. Many antimicrobials are synthetic and the term antimicrobials is preferable.

Pharmacodynamics

Antimicrobials may be bactericidal (bacteria are killed) or bacteriostatic (growth is inhibited but bacteria are not killed). Bactericidal antimicrobials include β-lactams, aminoglycosides, and quinolones. Bacteriostatic antimicrobials include macrolides, tetracyclines, sulfonamides, and chloramphenicol. The lowest concentration of an antimicrobial agent that inhibits growth of bacteria after 18–24h of culture is called the minimal inhibitory concentration (MIC). For some antimicrobials, the time the antibiotic concentration is above the MIC determines efficacy (time-dependent killing) (e.g. penicillin and macrolides). For others, the peak dose achieved, C_{max}, is most important and the ratio of C_{max}/MIC predicts outcome (dose-dependent killing) (e.g. aminoglycosides and quinolones). Antimicrobials that demonstrate dose-dependent killing may also exhibit a post-antibiotic effect (ability to inhibit antibiotic growth after the concentration has fallen below the MIC).

Some antimicrobial combinations demonstrate synergy (combined effects greater than predicted from the sum of individual effects). Drug levels in specific sites (e.g. CSF) may be low—knowledge of antimicrobial penetration of selected locations predicts efficacy. Antimicrobials may fail to reach adequate concentrations in locations such as abscesses and a combination of surgery and antimicrobial therapy will be required. In the case of prosthetic materials (prosthetic joints or cannulae), resolution occurs in most cases by foreign body removal in addition to antimicrobials.

Pharmacokinetics

Some antimicrobials are well absorbed after oral administration, while others require intravenous administration. Antibiotics may need dose adjustment in individuals with liver or renal impairment, depending on the route of elimination. In addition, use in pregnancy or paediatrics may require dose adjustment because of altered metabolism or drug toxicity.

Allergy and drug toxicity

Antimicrobials are a frequent cause of allergy. This can range from mild skin reactions to life-threatening desquamation or anaphylactic shock. Adverse reactions vary but include fever, abnormal liver function tests, renal impairment, and decreased production of polymorphonuclear cells, lymphocytes, or other blood cells.

Mechanisms of action

Ideal antimicrobial targets are specific for prokaryotes. Table 12.1 summarizes common targets. A brief overview of the spectrum of action of these antimicrobials and of some of the bacteria inhibited is provided, but for more detail, the reader is referred to more specialized texts. Molecular targets for inhibition include:

• Cell wall synthesis: makes the bacteria susceptible to osmotic rupture, as bacteria are hyperosmolar compared to the host environment

- Bacterial protein synthesis: exploits selective differences between prokaryotic and eukaryotic ribosomes
- Folic acid synthesis: selectively inhibited in bacteria
- DNA synthesis: DNA gyrase is required for negative supercoiling of bacterial DNA, a prerequisite for DNA replication. Prokaryotic RNA polymerases may also be selectively inhibited.

Antibiotic resistance

A global problem that limits the efficacy of antimicrobial chemotherapy and adds to the cost. The major factor in the development of resistance is inappropriate use of antimicrobials in humans and in animals. Resistance may be intrinsic (e.g. Gram-negative bacteria are not susceptible to glycopeptides). Acquired resistance involves mutation of existing genes or acquisition of new genes. New genes are spread by mobile genetic elements (plasmids, transposons, bacteriophages) exchanged between bacteria. Antimicrobial use exerts a selective pressure that allows resistant strains to proliferate at the expense of susceptible strains. Some transposons contain both virulence factors and antimicrobial resistance genes in the same transposon-aiding propagation. Examples of resistance are summarized in Table 12.1. Mechanisms include:

- Production of enzymes that inactivate the antimicrobial agent
- Mutation in the molecular target for the antimicrobial agent
- Decreased penetration of the agent into the bacterium
- Active efflux of the drug out of the bacterium.

Increasingly, bacteria are resistant to multiple antimicrobials, e.g. meticillin-resistant *Staphylococcus aureus* (MRSA), vancomycin-resistant enterococci (VRE), extended-spectrum B-lactamase (ESBL)-producing Gram-negative bacteria, and multi-drug resistant strains of *Mycobacterium tuberculosis* (MDRTB).

Table 12.1 Antibacterial agents

Cellular target of inhibition	Specific targets	Class of antimicrobial	Examples	Spectrum	Mechanisms of resistance
Cell wall	Transpeptidases (penicillin-binding proteins) that mediate cross-linking of peptidoglycans	β-lactams	Penicillin cephalosporins, carbapenems	Broad, some more specific for GPC, others for GNB. Specific β-lactam antibiotics for *Staphylococcus aureus* and *Pseudomonas aeruginosa*	Enzymatic; β-lactamases Altered penicillin-binding proteins Decreased permeability of GNB efflux
	Addition of subunits to peptidoglycans	Glycopeptides	Vancomycin	GPC	Alterations of D-alanine-D-alanine component of stem peptide
Protein synthesis	Binds to ribosomal 50S subunit	Macrolides	Erythromycin	Mainly GPC; also *Mycoplasma* spp.. *Chlamydia* spp.. *Legionella* spp.. *Helicobacter* spp.	Ribosomal methylation to inhibit binding Efflux from bacteria
		Lincosamides	Clindamycin	GPC and anaerobes	Ribosomal methylation
		Chloramphenicol	Chloramphenicol	GPC and GNB but widespread resistance now limits use	Enzymatic; chloramphenicol acetyl transferase
		Oxazolidinones	Linezolid	GPC, mycobacteria spp.	Alteration of ribosomal target
		Streptogramins	Quinupristin/dalfopristin	GPC	Alteration of ribosomal target Efflux Drug inactivation
	Binds to ribosomal 30S subunit	Tetracyclines	Doxycycline	GPC and GNB including *Mycoplasma* spp., *Chlamydia* spp., *Rickettsia* spp., *Legionella* spp., *Borcella* spp. and spirochetes	Efflux Alteration of ribosomal target

		Aminoglycosides	Gentamicin	GPC and GNB. Particularly used for GNB in combination with β-lactam	Enzymatic; antimicrobial modification by amino-glycoside-modifying enzymes; decreased uptake into bacteria Efflux
Cell metabolism	Sulphonamides	Co-trimoxazole (trimethoprim-sulfamethoxazole)	Inhibition of folic acid synthesis	Mainly GNB. Some GPC. Also some parasites and *Pneumocystis jirovecii* (*carinii*)	Modification of targets to bypass metabolic block
DNA synthesis	Quinolones	Ciprofloxacin	Inhibit A subunit of DNA gyrase and topoisomerase IV	GNB, some activity against GPC	Altered DNA gyrase. Also altered permeability to drug or enhanced efflux
	Imidazoles	Metronidazole	Generation of DNA damage due to generation of reactive-intermediates	Anaerobes and some parasites	Can occur but precise mechanisms undefined
	Rifamycin	Rifampicin	DNA-dependent RNA polymerase	Mycobacteria, *Staphylococcus aureus*, *Legionella* spp., *Brucella* spp., and *Rhodococcus equi* Always in combinations	Mutations in RNA polymerase
Cell membrane	Polymyxins	Polymyxin B	Cationic polypeptides inserted into bacterial membranes and permeabilize bacteria	GNB	Unknown

GNB, Gram-negative bacilli; GPC, Gram-positive cocci.

Viral structure and classification

Viruses contain nucleic acid in the form of linear or circular RNA in a single strand (most RNA viruses) and DNA in a double strand (most DNA viruses) and in a (+) or (−) sense orientation.

- Viral genomes encode from three to >100 proteins
- The nucleic acid is surrounded by a protein coat (capsid), made of subunits that form a helical, icosahedral, or complex structure (Fig. 12.2)
- Nucleic acid and capsid form the nucleocapsid. Nucleic acid may complex with nucleoprotein
- Many viruses contain a lipid envelope. Unenveloped viruses are resistant to acid and desiccation and typically cause infection via the oro-faecal route
- Respiratory or parenterally acquired viruses typically contain an envelope. The envelope contains proteins that are glycosylated in the external portion.
- Classification is by presence of RNA or DNA, size of the genome, capsid configuration, or presence of envelope (Table 12.2). Grouped into families with individual members (e.g. the *Herpesviridae* include herpes simplex virus 1 and 2, varicella–zoster virus, Epstein–Barr virus, and cytomegalovirus).

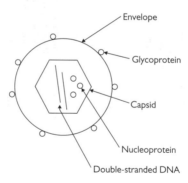

Fig. 12.2 Viral structure.

Table 12.2 Classification of viruses

Type of nucleic acid	Genome size (kilobases)	Virus	Family	Envelope	Capsid
RNA	7–8	Poliovirus	Picornaviridae	Absent	Icosahedral
	7–8	Norwalk virus	Calciviridae	Absent	Icosahedral
	7–11	HIV-1	Retroviridae	Present	Icosahedral
	10–12	Rubella virus	Togaviridae	Present	Icosahedral
	10–13	Yellow fever virus	Flaviviridae	Present	Unclear
	10–14	Influenza virus	Orthomyxoviridae	Present	Helical
	13–16	Rabies virus	Rhabdoviridae	Present	Helical
	16–20	Measles virus	Para-myxoviridae	Present	Helical
	19	Ebola virus	Filoviridae	Present	Helical
	16–27	Rota virus	Reoviridae	Absent	Icosahedral
	20–30	Corona virus	Coronaviridae	Present	Helical
DNA	3	Hepatitis B virus	Hepadnaviridae	Present	Icosahedral
	5	Parvovirus B-19	Parvoviridae	Absent	Icosahedral
	5–8	Human papillomavirus	Papillo-maviridae	Absent	Icosahedral
	36–38	Adenovirus	Adenoviridae	Absent	Icosahedral
	120–240	Herpes simplex virus	Herpesviridae	Present	Icosahedral
	130–380	Vaccinia virus	Poxviridae	Present	Complex

HIV-1, human immunodeficiency virus-1.

Viral life cycle

Viruses require a host cell in which to replicate by fission. Stages in the cycle include the following.

1. Attachment to host cell

- Capsid proteins (rotavirus), envelope glycoproteins (influenza virus) or so-called spike-proteins (surface protrusions; SARS-COV2) bind to host cell surface targets (receptors)
- The range of cells that express these viral receptors is a factor determining the range of cells capable of infection (tropism)
- Attachment is often a multi-step process involving several receptors (HIV-1 binds to host cells via the interaction of glycoprotein gp120 with CD4 and chemokine receptors).

2. Viral entry

Mechanisms of internalization include:
- Conformational changes: allows fusion with host membrane
- Receptor-mediated endocytosis involving clathrin-coated pits
- Direct penetration.

Internalized virus is uncoated and envelope and capsid are degraded by cellular proteases and/or acidification in cellular organelles to allow release of nucleic acid for viral replication.

3. Viral replication

- (+) sense RNA viruses: RNA functions as mRNA for translation by host ribosomes into protein or is copied by viral-encoded RNA polymerase into intermediate (−) sense RNA that acts as a template for (+) sense RNA generation during replication
- (−) sense RNA viruses: RNA is copied into (+) sense RNA by pre-formed RNA polymerase to allow transcription or function as an intermediate in transcription of (−) sense RNA for incorporation into progeny virions
- Double-stranded RNA viruses use the (−) sense RNA to produce (+) sense RNA
- Retroviruses contain RNA-dependent DNA polymerase; RNA is copied into DNA that integrates into the host genome to allow transcription (e.g. HIV-1). This strategy permits persistence of infection
- DNA-containing viruses transcribe DNA in the nucleus using host transcription machinery. Transcription follows a temporal pattern: early transcripts encode regulatory proteins and proteins involved in DNA replication while late transcripts encode structural proteins. Some DNA viruses also persist in the host cell but are present as small extra-chromosomal DNA-termed episomes rather than integrated into the host chromosomes (e.g. herpesviruses).

4. Viral assembly

Viral proteins are usually transcribed as a precursor that requires cleavage by specific proteases into individual proteins.
- Viral nucleic acid allows initiation of capsid nucleation
- Capsid subunits combine to form a complete capsid surrounding nucleic acid

- Enveloped viruses acquire envelope as the virus buds through nuclear, cytoplasmic, or ER membranes
- Surface proteins are incorporated and modified by processes such as glycosylation in the Golgi apparatus
- Matrix or tegument proteins are added
- Mature virions are released from the host cell
- Often, the maturation process is inefficient and the majority of viruses produced are incomplete and unable to infect further cells.

Viral pathogenesis

Cellular effects

Viral replication modifies host cell gene transcription with loss of normal cellular homeostasis. Expression of multiple host cell kinases is altered. Genes induced include host response genes, such as cytokines. Cell cycle may be modulated. Cell proliferation occurs with certain viruses. The immune response to viral proliferation contributes to cytopathology inducing cell death of directly infected and uninfected bystander cells. Cytopathic effects include induction of apoptosis. Viruses can modulate human leucocyte antigen (HLA), chemokine, or cytokine expression and induction of apoptosis enabling immune evasion. Possible outcomes of acute lytic infection include:

- Chronic infection: persistent replication continues and viruses are shed from infected cells
- Latency: virus persists in the host genome but is not transcribed until reactivated by specific stimuli
- Reactivation: specific stimuli induce transcription of latent virus. Immunosuppression may be an important trigger
- Transformation: ability to promote immortalization of cells → cancer formation. Chronic pro-inflammatory effects may contribute to malignant transformation by some viruses.

Systemic effects

Local replication may allow systemic spread. Detectable virus in the blood (viraemia) produces further replication in the reticulo-endothelial system resulting in increased levels of viraemia before seeding of target organs. Some viruses spread via other routes (e.g. peripheral nerves). Clinical symptoms result from the cytokines produced by viraemia or tissue damage locally or in target organs.

Host immunologic response

Includes humoral and cell-mediated immunity (cytotoxic T-lymphocytes (CTL)). Modified by age, nutritional status, immunosuppression, and vaccination status. Results in the production of cytokines that induce symptoms and facilitate viral clearance. Viruses have developed adaptations to allow immune evasion (e.g. alteration of surface proteins recognized by antibodies or CTLs).

Viral infection

Subclinical or results in clinical disease. The ability of a virus to cause human infection or disease is determined by viral and host factors.

Transmission

Various routes including oro-faecal, respiratory, touching an infected lesion, insect or animal bites, transfusion of blood, transplantation of organs, sexual exposure, or vertical transmission from mother to child. Some routes of infection (e.g. transfusion) are associated with a larger infecting dose and more severe disease.

Virulence

Determined experimentally by the quantity of virus required to cause cytopathic effects in cell culture or death to experimentally infected animals. Reflects host susceptibility and viral factors. Viral mutations alter virulence.

Host susceptibility

Individuals may be immune due to prior infection or immunization. This may prevent or modify infection so disease does not occur. Genetic make-up may modify susceptibility.

Viral therapy

Antiviral agents are described in Table 12.3. Viruses may develop resistance to antiviral agents by developing mutations in genes encoding the targets.

Table 12.3 Antiviral agents

Description	Agent	Spectrum	Target
Inhibitors of entry and disassembly	T-20 (enfuvirtide)	HIV-1	Blocks viral fusion with host cell membrane
	Rimantadine	Influenza A virus	Blocks viral uncoating after entry
	Pleconaril	Enteroviruses, rhinoviruses	Binds to capsid, blocks attachment and uncoating of the genome
Inhibitors of viral replication	Aciclovir, valaciclovir	HSV, VZ	Phosphorylated drug inhibits DNA polymerase
	Ganciclovir	CMV, HHV-6	As for aciclovir
	Foscarnet	HSV, VZ, CMV, HHV-6	Directly inhibits DNA polymerase
	Cidofovir	CMV, HSV, VZ, HHV-6, HHV-8	Phosphorylated drug inhibits DNA polymerase
	Zidovudine	HIV-1	Inhibits reverse transcriptase
	Lamivudine	HIV-1, HBV	Nucleoside reverse transcriptase inhibitor of HIV-1 and DNA polymerase inhibitor of HBV
	Tenofovir	HIV-1, HBV	As for lamivudine but nucleotide inhibitor
	Ribavirin	RSV, HCV, Lassa fever virus	Not fully elucidated; involves depletion of guanosine triphosphate and inhibition of nucleic acid synthesis
	Formivirsen	CMV	Antisense nucleotide; intravitreal injection inhibits ocular replication of CMV

Table 12.3 (Contd.)

Description	Agent	Spectrum	Target
Inhibitors of viral assembly and release	Zanamivir oseltamivir	Influenza A virus Influenza B virus	Inhibits viral neuraminidase and hence release of virus
	Protease inhibitors (e.g. lopinavir)	HIV-1	HIV protease inhibition blocks cleavage of polyproteins and inhibits production of mature virions
Immunomodulators	Interferon-α	HBV, HCV, HPV, HHV-8	Stimulates antiviral host responses that inhibit viral protein synthesis
	Imiquimod	HPV	Binds to toll-like receptor 7; stimulates interferon-α expression

CMV, cytomegalovirus; HBV, hepatitis B virus; HCV, hepatitis C virus; HHV, human herpesvirus; HIV, human immunodeficiency virus; HPV, human papillomavirus; HSV, herpes simplex virus; RSV, respiratory syncytial virus; VZ, varicella–zoster virus.

Prions

Proteinaceous infectious particles that lack nucleic acids. Resistant to procedures that hydrolyse nucleic acids. Cause chronic, progressive neurological diseases characterized by reactive astrocytosis and often spongiform cellular changes. Result in accumulation of an abnormal protein—prion protein (PrP). Examples of prion diseases include Kuru, Creutzfeldt–Jacob disease (CJD), and new variant CJD. Diagnosis may involve identification of PrP in biopsy material. Treatments are investigational.

Fungi

Classification

- Yeasts (e.g. *Candida* or *Cryptococcus* spp.)
- Moulds—filamentous fungi (e.g. *Aspergillus* spp. and *Trichophyton* spp.)
- Dimorphic fungi—yeasts in tissue but grow *in vitro* as moulds and include the North American dimorphic fungi (e.g. Histoplasmosis).

Dermatophyte infections are superficial fungal infections of the skin and related structures caused by moulds. Yeasts are ovoid structures that replicate by budding. Moulds are tubular structures—hyphae that grow by longitudinal extension and branching. Fungal replication involves the production of spores that may be either asexual or sexual.

Structure and pathogenesis

Fungal cell walls contain chitin and polysaccharides and fungal endotoxin stimulates pro-inflammatory cytokines. Cryptococci contain a polysaccharide capsule, but most fungi do not. Inside the cell wall is the sterol-containing cytoplasmic membrane. Enzymes involved in sterol (in particular, ergosterol) synthesis are targets of antifungal drugs. Fungi produce exoenzymes; proteases digest tissue components such as keratin to facilitate invasion, and phospholipases also contribute to pathogenicity. Immunologic defects predisposing to fungal infection include decreased polymorphonuclear phagocyte function, defects in physical barriers (often associated with prosthetic devices such as cannulae), and impaired cell-mediated immunity.

Diagnosis and therapy

Diagnosis includes microscopy after special staining, culture, serology, antigen detection, molecular techniques, and analysis of pathologic specimens. The principal antifungal agents and their targets are summarized in Table 12.4. New antifungal targets are being identified but resistance is also emerging.

Table 12.4 Antifungal agents

Class	Agent	Spectrum	Target
Echinocandins	Caspofungin	Yeasts and moulds	Inhibition of $\beta1,3$-glucans in the cell wall
Polyenes	Nystatin, amphotericin B	Yeasts and moulds	Bind to membrane sterols such as ergosterol
Azoles	Fluconazole	Yeasts	Blocks C-14α methylase and inhibits ergosterol synthesis
	Itraconazole, voriconazole	Yeasts and moulds	Same as fluconazole but spectrum extended to moulds
Allylamines	Terbinafine	Moulds (dermatophytes)	Inhibits squalene epoxidase required for ergosterol synthesis; concentrated in nails and hair
Other	Griseofulvin	Dermatophyte infections	Concentrated in nails and hair; inhibits microtubule polymerization
Cytosine analogue	Flucytosine	Yeasts	Inhibits thymidylate synthetase and so DNA synthesis
Immuno-modulators	GM-CSF	Moulds and yeasts	Stimulates macrophage microbicidal killing of fungi

Parasites

Spectrum

Parasites include unicellular organisms, protozoans (e.g. *Plasmodium falciparum*), and multicellular organisms (e.g. helminths—worms and flukes) and ectoparasites (including insects whose life cycle only involves interaction with the external surface of the host). Many parasitic infections are geographically restricted to tropical counties and occur in temperate climates due to travel or immigration. Parasites cause major global illness (e.g. malaria).

Life cycles

Unique—often with stages of development in hosts other than humans. Knowledge of intermediate hosts (helminths, insects, and other mammals) is important in prevention. *P. falciparum* causes malaria and is spread by infected female *Anopheles* spp. mosquitoes that release sporozoites when biting. These pass to the liver hepatocytes and mature into schizonts and are released from the liver as merozoites that infect erythrocytes. Within erythrocytes, maturation to ring trophozoites and schizonts occurs. Asexual erythrocytic schizonts mature and are released as merozoites that infect other erythrocytes. Alternatively, sexual development in the erythrocyte allows formation of male and female gametocytes. Ingestion of these by a mosquito results in gamete fusion, formation of a diploid zygote, oocyst formation, and release of sporozoites to the salivary gland for infection of further humans.

Pathogenesis

Genomics and proteomics provide molecular insights. For example, in malaria, red cells express *P. falciparum*-infected erythrocyte membrane protein-1 (*PfEMP1*) inducing cytoadherence to endothelial cells, obstruction of small capillaries, and → complications such as cerebral malaria. Host responses to parasites involve cell-mediated immunity. Many protozoan infections are particular problems in individuals with HIV infection.

Intracellular protozoans such as *Leishmania* spp. require Th1 responses with TNF-α, IFN-γ, and IL-12 production to prime macrophage killing of intracellular parasites. For other parasites, a Th2 cytokine bias with release of IL-4, IL-5, and IL-10 predominates.

Increased levels of eosinophils are observed with many helminthic infections. Humoral immunity is important for some enteric protozoan infections such as *Giardia lamblia*. Intact splenic function is required in protection against some parasites, including erythrocytic parasites.

Diagnosis and treatment

Diagnosis involves microscopy, antigen detection, serology, molecular techniques, and analysis of pathology specimens. A selection of antimicrobial agents and the targets they act upon are summarized in Table 12.5.

Table 12.5 Antimicrobials active against parasites

Agent	Parasite	Target
Quinine, chloroquine, mefloquine	*Plasmodium* spp.	Haem polymerase inhibitor, blocks conversion of free haemoglobin into malarial pigment. Full mechanism of quinine unclear
Atovaquone	*Plasmodium* spp.	Inhibits co-enzyme Q and hence cellular respiration and pyrimidine synthesis
Sodium stibogluconate	*Leishmania* spp.	Dysregulation of parasitic metabolism, but exact mechanism unclear
Suramin	*Trypanosoma brucei*	Inhibits glycerol-3-phosphate oxidase and dehydrogenase and hence energy metabolism
Nifurtimox	*Trypanosoma cruzi*	Generation of reactive oxygen species that damage the trypanosome
Sulfadiazine plus pyrimethamine	*Toxoplasma gondii*	Inhibits dihydropteroate synthetase and dihydrofolate reductase to block parasitic folic acid synthesis
Metronidazole	*Giardia lamblia*, *Entamoeba histolytica*	DNA damage
Praziquantel	*Schistosoma* spp. and tapeworms	Stimulates calcium influx, leads to paralysis
Albendazole	Multiple worms	Numerous targets including inhibition of microtubule assembly and impaired uptake of glucose
Ivermectin	Some nematodes, e.g. *Strongyloides stercoralis* and *Onchocerca volvulus*	Opens glutamate-gated chloride channels, paralysing pharyngeal pumping activity
Diethylcarbamazine	Tissue nematodes	Multiple activities, hyperpolarization → paralysis and membrane damage
Niclosamide	Tapeworms	Uncouples oxidative phosphorylation → break-up of the scolex
Permethrin, malathion	Ectoparasites including *Sarcoptes scabiei*	Kill adults and ovicidal

Innate immunity

Evolutionarily conserved; present in invertebrates and vertebrates. Allows differentiation of microbial non-self from self. Rapid response. Germ-line encoded receptors recognize a limited number of highly conserved microbial products.

Physical barriers

The skin and mucosa block invading microorganisms. Breached by prosthetic appliances (e.g. catheters) or by tissue damage (e.g. burns). Epithelial cells can internalize microorganisms and secrete cytokines. Commensal flora block colonization by pathogenic microorganisms. Secretions contain antimicrobial factors such as lysozyme, defensins, collectins, iron-binding proteins, natural antibodies (antibodies against colonizing microorganisms that may cross-react with a pathogenic strain) and have a low pH.

Complement pathways

Comprising >30 proteins that contribute to innate host defence, link with adaptive immunity, and aid clearance of immune complexes. Cascade of sequentially activated serine proteases (Fig. 12.3).
- Classical pathway: C1 complex binds to antibody on the bacterial surface
- Lectin pathway: mannose-binding lectin (MBL) binds to bacterial mannose
- Alternative pathway: C3b binds to surface hydroxyl groups on bacteria.

These pathways activate cleavage of other components, in particular C3 and C5. Complement enhances phagocytosis of opsonized bacteria (C3b fragments), chemotaxis (migration of leucocytes to sites of infection; C5a), polymorphonuclear leucocyte (PMN) activation (C5a and C3a), and bacterial lysis (C5b-9 membrane-attack complex). Complement also enhances antibody response and immunologic memory, hence linking innate and adaptive responses.

Collectins

Contain a collagenous domain linked to a calcium-dependent lectin; include C1q, MBL, and surfactant proteins –A and –D. MBL binds to microbial carbohydrates and this activates two associated serine proteases that cleave C3.

Pathogen-associated molecular patterns (PAMPs)

PAMPs are conserved patterns of molecules on microbial structures recognized by the immune system via pattern-recognition receptors (PRRs). PAMPs include lipopolysaccharide (LPS), lipoteichoic acid (LTA), peptidoglycan, flagellin, mannan, glucan, bacterial DNA, and double-stranded RNA. PRRs are secreted, endocytic, or signalling and widely expressed in a nonclonal pattern.

Toll-like receptors (TLRs)

Activate transcription factors of the nuclear factor-κB (NF-κB) family and induce pro-inflammatory and antimicrobial gene transcription in response to PAMPs. Signal via multicomponent molecular complexes that may

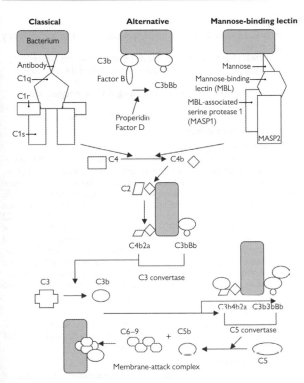

Fig. 12.3 Complement pathways.

contain other molecules functioning as the PRR (e.g. LPS binds to CD14 and this complex binds to MD-2 and the TLR4 molecule to induce signalling). For other TLRs, such as TLR2, heterodimers of different TLR are required for signalling. TLRs contain a leucine-rich extracellular domain and a cytoplasmic domain similar to the IL-1 receptor. TLRs and their ligands include TLR2 (LTA and lipopeptides), TLR3 (double-stranded RNA), TLR4 (LPS), TLR5 (flagellin), and TLR9 (CpG motifs in bacterial DNA). Signal transduction involves IL-1 receptor associated kinases and mitogen-activated protein kinase kinases.

Nucleotide-binding oligomerization domain (Nod) proteins

Nods are intracellular receptors that respond to bacterial components such as components of peptidoglycans. Nod signalling results in NF-κB induction.

Phagocytes

In addition to humoral factors, the principal effectors of innate immunity are phagocytes.
- Mononuclear phagocytes are monocytes in the blood and differentiate into macrophages in tissue. The resident phagocyte is the long-lived tissue macrophage
- PMNs are short-lived phagocytes and, along with monocytes, are recruited to sites of infection
- Eosinophils contribute to host defence against parasites.

Phagocyte functions include:

Internalization of microorganisms

Macrophages internalize non-opsonized microorganisms using PRR (mannose receptor, macrophage scavenger receptors, etc.). Viruses may be internalized by receptor-independent endocytosis involving direct membrane invagination. Phagocytes internalize opsonized microorganisms using Fc receptors that bind immunoglobulin (FcγRs also bind C-reactive protein or serum amyloid protein) and complement receptors (CR) that bind C3 complement cleavage products (CR1 also binds C1q and MBL). Integrins also contribute to phagocytosis. Internalized microorganisms are contained in phagolysosomes, but some microorganisms have developed adaptations to escape the phagolysosome and avoid intracellular killing. Other organisms are resistant to phagocytosis due to factors such as a polysaccharide capsule. Opsonization may reverse a microorganism's resistance to phagocytosis.

Cytokine production

Phagocytes produce cytokines to activate (or downregulate when appropriate) innate and adaptive immune responses. Primes phagocyte killing of microorganisms and regulates the immune response.

Killing of microorganisms

Microorganisms are killed intracellularly and extracellularly. Antimicrobial mechanisms include:
- ROS: superoxide generated by the NAD(P)H oxidase system, undergoes further reactions to generate alternate ROS (e.g. hydroxyl radical, hydrogen peroxide). Myeloperoxidase (MPO) catalyses the reaction between hydrogen peroxide and halide ions to produce potent antimicrobial molecules. PMN are major producers of ROS. Some of ROS effects may be via activation of proteases
- Reactive nitrogen species (RNS): NO, produced by inducible NO synthase (iNOS), reacts with ROS to produce cytotoxic peroxynitrite. Important for killing of intracellular pathogens in macrophages, although the importance of NO in human defence mechanisms is questionable
- Proteases: acid or neutral. Include cathepsins such as cathepsin G
- Cationic proteins: proteins of different size, positively charged to interact with the negatively charged surface of microorganisms (e.g. bactericidal/permeability-increasing protein (BPI); a component of PMN granules, active against Gram-negative bacteria), defensins (small arginine-rich peptides), and cathelicidins
- Lactoferrin: bacteriostatic; deprives bacteria of iron required for growth
- Low pH: contributes to microbial killing, but many bacteria can withstand low pH.

Antigen presentation

Although not as effective at antigen presentation as dendritic cells (DCs), macrophages also function as antigen-presenting cells (APCs). TLRs play a role in upregulating the expression of co-stimulatory molecules (CD80, CD86 on APCs) and help link innate to adaptive immune responses. TLRs also aid DC maturation.

Natural killer (NK) cells

Effectors of cellular innate response, especially antiviral responses. Large granular lymphocytes that do not contain T-cell receptors (TCR) and do not require prior stimulation. Possess surface receptors—killer-cell inhibitory receptors (KIR)—that bind to major histocompatibility complex (MHC) class I antigens expressed on almost all normal nucleated cells. This prevents target killing. Viral-infected (and tumour) cells may down-regulate MHC class I antigen and, in the absence of KIR engagement, NK cells kill targets by lysis or apoptosis. NK cells can also kill viral-infected cells coated with antibody.

Adaptive (acquired) immunity

Present only in vertebrates. Development is unique to every individual and not passed on to offspring. Specific recognition of small details of molecules (termed antigens) results in immunologic memory. Allows response to foreign antigens but also response to self-antigens (controlled by deleting auto-reactive somatic cells). Requires stimulation and resultant clonal expansion before maximal immune response occurs—hence delayed. Diversity generated by somatic rearrangement of genes encoding antibody and the TCR.

Antigens

Description

An antigen is a specific compound that is bound to by a specific antibody or TCR. If it induces an immune response, it is termed an immunogen. Characteristic features of immunogens include:

- Non-self, recognized as foreign antigens
- Chemical complexity
- High molecular weight (small-molecular-weight compounds that are immunogenic—haptens—need to be complexed to a carrier protein)
- Susceptibility to antigen processing by degradation of parent molecule (proteins undergo proteolysis).

Epitopes

A portion of the antigen (the epitope) binds non-covalently to a unique antigen binding site on antibody or TCR. The epitope is usually 5–15 amino acids long (or equivalent size for non-peptides). Peptides are strongly immunogenic. Carbohydrates are immunogenic, particularly when present in structures such as glycoproteins. Lipids are only immunogenic when combined with a lipid carrier. Nucleic acids may function as immunogens in auto-immunity but rarely in infection. B-cells bind soluble, non-processed antigens present in microbial proteins, carbohydrates, or lipids, while T-cells recognize processed antigen derived from peptides presented by APC after proteolytic degradation. Super-antigens activate many T-cells with different antigen specificities but common characteristic TCR Vβ segments, resulting in high level cytokine production. *Streptococcus pyogenes* and *Staphylococcus aureus* can express superantigen toxins and induce toxic shock.

Major histocompatibility complex

MHC genes and gene products

T-cells respond to antigen bound to a MHC molecule on the APC (MHC restriction of T-cell responses). MHC molecules bind to only selected antigens, so contributing to the selection of antigen for presentation.

MHC class I molecules

HLA A, B, and C. Consist of an α chain bound to β_2 microglobulin. Expressed on all nucleated cells.

MHC class II molecules

HLA DR, DQ, DP. Consist of an α and a β chain. Expressed constitutively on B-cells, DC, and thymic epithelium and inducible on T-cells, macrophages, and endothelial cells.

Expression

Each individual expresses both parental genes for each HLA antigen (co-dominant expression). There are seven HLA DR genes, so an individual expresses six class I and up to 20 class II molecules on each cell expressing class I and II molecules. HLA genes differ between individuals due to genetic polymorphism, and these variants are termed alleles. This has consequences in transplantation, where it is highly unlikely two individuals will have identical HLA haplotypes. HLA typing describes each allele by locus, allele type (which corresponds largely to serologic group), and subtype (determined by PCR) (e.g. DRB1*0702). Certain HLA haplotypes are associated with increased risk of certain diseases (e.g. tuberculosis).

MHC presentation of peptide

MHC molecules bind peptide fragments in the polymorphic region (peptide groove). This region also binds the TCR. MHC class I molecules bind peptides derived from endogenous antigen (e.g. from intracellular viruses or parasites) and present antigen to CD8 T-cells (MHC class I binds CD8). MHC class II molecules bind peptides derived from degradation of exogenous peptides (e.g. phagocytosed bacteria or viruses) for presentation to CD4 T-cells (MHC class II binds CD4). Most large antigens, as found in microorganisms, have multiple different epitopes, at least one of which will bind successfully to a HLA allele ensuring an immune response.

Antigen presenting cells

DCs, macrophages, and B-cells perform antigen presentation. DCs are the only APCs capable of inducing primary immune responses and, hence, immunological memory. Immature DCs capture antigen in peripheral tissue. Subsets of immature DCs contribute to innate immunity by producing interferon (IFN) α.

After antigen capture, DCs migrate to regional lymphoid tissue, where DC maturation is associated with upregulation of MHC molecules, and antigen presentation occurs. DC antigen presentation to naive T- and B-cells results in activation and the subset of DCs involved in antigen presentation can determine whether a type 1 (Th1) or type 2 (Th2) immune response is generated.

DCs are separated by surface markers in humans; myeloid (CD11c+) and plasmacytoid (CD11c−) DC exist. DCs include interstitial DCs of the dermis and Langerhans cells of the epidermis.

Macrophages presentation is less efficient but occurs during secondary immune responses. B-cell antigen presentation occurs when B- and T-cells have both already been primed by antigen.

T-cells

Thymic-derived lymphocytes that induce cell-mediated immunity. Split into subsets (e.g. CD4+ and CD8+). Directly interact with antigen via TCR.

TCR complex structure

Transmembrane protein with one α and one β chain, each containing constant and variable regions. The variable regions interact to form three hypervariable regions (complementarity determining regions—CDRs 1, 2, and 3). The TCR complex is formed by two separate heterodimers of the $\alpha\beta$ chains associated with one $\gamma\varepsilon$ and one $\delta\varepsilon$ heterodimer and one $\zeta\zeta$ homodimer. All three components of CD3 ($\gamma\varepsilon$, $\delta\varepsilon$, and $\zeta\zeta$) contribute to TCR signal transduction via immunoreceptor tyrosine-based activation motifs (ITAM). On antigen engagement, TCR ITAMs bind cellular kinases that activate cell signalling. A minor subset of T-cells, $\gamma\delta$ T-cells, have TCR containing $\gamma\delta$ heterodimers (distinct from the $\gamma\delta$ chains of CD3).

TCR diversity

Generated by having multiple alternative genes encoding separate segments of each chain of the TCR present in the germline DNA (the inherited chromosome)—same principle as immunoglobulin diversity. As lymphocytes differentiate, these genes are rearranged and an individual clone of lymphocytes contains DNA that encodes just one gene for each segment (Fig. 12.4). Each chain is made up of one copy of the possible V and J genes combined with a C gene. The β chain is made up of one copy of the possible V, D, and J genes combined with a C gene. This gene rearrangement requires the products of the recombination activation genes RAG1 and RAG2 to allow enzymatic cleavage during recombination. TCR show allelic exclusion (only one parental chromosome is used), so all TCR in a clone have identical antigen specificity. TCR gene rearrangement allows for the generation of ~10^{15} separate TCR.

T-cell selection

T-cell maturation occurs in the thymus. The TCR expressed ($\alpha\beta$ or $\gamma\delta$) is determined and TCR gene rearrangement occurs. T-cells at this stage are CD4+/CD8+ (dual positive). If TCR binds antigen, the cell survives (positive selection); if not, dies by apoptosis. The cells become educated to recognize the antigen only if presented by MHC (MHC restriction). Interdigitating DC present antigen to MHC-restricted T-cells and, if the reaction is of too high affinity, the cells die by apoptosis (negative selection or central tolerance). This prevents autoreactive T-cells. After negative selection, T-cells downregulate either CD4 or CD8. T-cells are now MHC-restricted and self-tolerant. Mature T-cells traffic to regional lymphoid organs or peripheral tissues.

T-cells that have not previously encountered antigen are termed naïve and selectively home to regional lymphoid tissue, while activated or memory T-cells selectively migrate to peripheral tissues. Some autoreactive T-cells escape to the periphery but undergo a form of apoptosis mediated by Fas ligand and the pro-apoptotic Bcl-2 family member Bim (peripheral tolerance).

TCRA locus

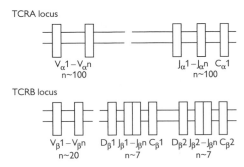

Fig. 12.4 Organization of the genes encoding the α and β chains of the human T-cell receptor.

Antigen presentation to T-cells

MHC-bound peptide interacts with the most variable region of the TCR (CDR3), while the CDR1 and CDR2 regions interact with conserved regions of the MHC. The interaction between the MHC-peptide complex and the TCR is low affinity and insufficient to activate the T-cell. In addition, CD4 or CD8 bind to conserved regions of MHC class II or I respectively, allowing adhesion and signal transduction via specific cellular kinases. Co-stimulatory pairs must also bind before T-cell activation (Fig. 12.5): CD40 on the APC with CD40 ligand (CD154) and B7.1/B7.2 (CD80/CD86) on the APC with CD28. Later in the activation process, B7.1/B7.2 may bind with CTLA-4 and inhibit further activation. The co-stimulatory molecule interactions help localize molecules involved in the signal transduction induced by TCR engagement. Adhesion molecules help stabilize the APC–T-cell interaction: CD58 with CD2 and CD54 with CD11a/CD18. TCR engagement results in signal transduction involving Src family tyrosine kinases that phosphorylate ITAMs on the intracytoplasmic tails of CD3. ZAP-70, another tyrosine kinase, is then activated, as are, ultimately, transcription factors, including NF-κB and nuclear factor of activated T-cells (NF-AT), resulting in gene transcription. Initially, cells are activated and proliferate. This is associated with cytokine production.

T-cell effector functions

Activation of T-cells results in the development of characteristic phenotypes:
- Th cells: the effector function is influenced by the nature of antigen and APC. Th1 CD4 T-cells produce characteristic cytokines (IL-2, IFN-γ, TNF-β). Th1 responses are evoked by bacterial and viral antigens and activate CD8 T-cells, NK cells, and macrophages to kill cells containing intra-cellular microorganisms. Th2 cells produce IL-4, IL-5, IL-10, and IL-13. This triggers IgE production by B-cells and activation of eosinophils in response to parasites and allergens
- CTLs: CTLs are CD8 T-cells adapted to kill cells expressing specific antigen bound to MHC class I. Activation induces target (microorganism infected, tumour or transplanted cells) killing. CTLs can be activated

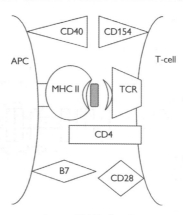

Fig. 12.5 Antigen presentation to a CD4-T cell involves presentation of the peptide bound to the MHC class II molecule of the antigen presenting cell (APC) to the T-cell receptor (TCR). The CD4 molecule and the TCR both bind to the MHC class II molecule. T-cell activation also requires co-stimulation via CD40 (binding to CD154) and B7 (binding CD28).

by Th1 cells or by DC presentation of antigen. CTLs kill using granules containing perforin and granzyme and, to a lesser extent, using Fas ligand. This results in lysis or apoptosis of the target, depending on the pathway. Some CD4 T-cells also function as CTLs
• Memory T-cells: clonal downsizing follows initial proliferation of an antigen-specific clone. Activated cells are killed by apoptosis (activation-induced cell death). A proportion of the clone survives as memory T-cells. Memory T-cells (CD45RO+) express different antigens from naïve T-cells (CD45RA+, CD62L+). Memory T-cells are more easily activated by antigen and may not require co-stimulation by CD28. Professional suppressor T-cells are T-regulatory cells, which retain a unique ability to suppress other immune responses
• Suppressor T-cells: T-regulatory cells express immunomodulatory cytokines such as IL-10 and TGF-β CD4+, CD25+ regulatory T-cells prevent activation of autoreactive T-cells that have not been removed by other mechanisms of tolerance
• γδ T-cells: found in the skin, respiratory tract, and intestine and increase during infection, especially tuberculosis.

B-cells

Produce antibody and can present antigen.

Antibody structure

Antibodies are proteins (immunoglobulins) that are secreted or membrane-bound. The basic units of immunoglobulin are two light chains and two heavy chains as shown in Fig. 12.6. Disulfide bonds hold these together.

The light chains are identical and contain a variable domain and a constant domain. Each domain is stabilized by an intra-chain disulfide bond that creates a loop in the peptide chain that forms a compact globular structure. The heavy chains are also identical and contain a variable region and three constant domains.

Each light chain associates with one heavy chain. A disulfide bond at the hinge region holds the two heavy chains together. The light chains of any given immunoglobulin are of one of two classes, κ or λ. The heavy chains are made of one of five classes (isotypes) separated into subclasses (both heavy chains are identical in a given molecule).

The region containing the variable and first constant domains is termed the Fab fragment and the region containing the second and third constant domains is the Fc fragment. The heavy chain determines biologic activity: half-life in the circulation, Fc receptor binding, etc.

Within each variable domain lie three hypervariable regions (complementarity determining region (CDR)). The three adjacent CDR form the specific antigen binding site. Antigen–antibody binding is non-covalent and requires a close fit. The strength of this association is the affinity. Flexibility at the hinge regions allows the two antigen binding sites to interact with antigens on a large structure (e.g. a bacterium). Multivalent antigens have numerous interactions of different affinity and the overall binding energy is the avidity.

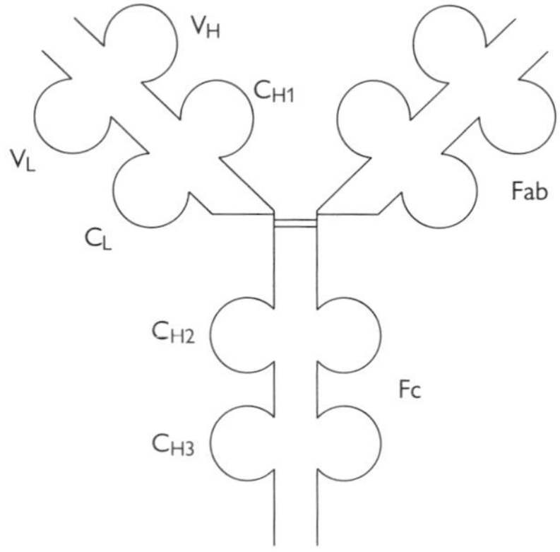

Fig. 12.6 Antibody structure. Immunoglobulin consists of a light chain containing a variable domain V_L and a constant domain C_L. The heavy chain consists of one variable V_H and three constant domains C_{H1-3}. The segment containing the C_{H2-3} of both heavy chains is the Fc portion and the two segments containing V_L, V_H, C_L, and C_{H1} are each termed Fab fragments.

Genetics of antibody diversity

Germline DNA encodes multiple genes for each segment of antibody light or heavy chain. During early B-cell maturation gene rearrangement occurs; selected genes for each variable segment are aligned and transcribed into RNA. Recombination requires RAG1 and -2.

For light chains, there are ~40 V genes and four or five J genes encoding the variable domain. For κ, there is one C gene, but for light chains there are four alternative C genes, each associated with one of the four J genes. For heavy chains, there are ~50 V genes, 20 D (diversity) genes, and six J genes that encode the variable domain. The D and J genes together encode the CDR 3 region.

Any B-cell clone makes antibody containing only one class of light chain of fixed antigen specificity. Genes from only one parental chromosome are used (allelic exclusion). There are multiple C genes, one for each isotype and subtype of immunoglobulin. Those closest to the J genes are Cμ and Cδ.

During the early stages of development, B-cells undergo VDJ recombination and alternative splicing allows translation of protein with either a M or D immunoglobulin isotype heavy chain. At later stages of differentiation (requires cytokines and specific antigen), alternative isotypes such as IgG, E, or IgA are produced (class switching).

Random association of V(D)J genes and heavy/light chains creates diversity. Junctional diversity (deletion or alterations in nucleotide sequence at the sites of gene rearrangement), insertional diversity (addition of nucleotides at V–D or D–J junctions), and somatic hypermutation (point mutations in the V(D)J genes occurring during secondary immune responses create increased antibody affinity) all enhance diversity.

Biological functions of antibody

Agglutination (clumping of insoluble antigen) and precipitation (clumping of soluble antigen) by IgG, IgM

Opsonization. Antibody is bound to microorganisms and the Fc portion binds to the Fc receptors of phagocytes

Complement activation. Complement cleavage products enhance opsonization, induce chemotaxis of leucocytes to sites of infection, or directly lyse bacteria

Antibody-dependent cell-mediated cytotoxicity (ADCC). Natural killer cells also express Fc receptors but the engagement of these leads to killing of the target cell not phagocytosis

Neutralization of toxins or viruses

Immobilization of motile bacteria. Antibodies may bind to structures such as flagella

Protection of the foetus and neonate. Maternal IgG crosses the placenta and protects the foetus and neonate, which is incapable of producing its own antibody.

IgM is the initial antibody produced during infection or immunization. IgG is produced later but constitutes the majority of serum immunoglobulin and has the longest half-life. IgA is the major immunoglobulin in secretions in the gut or respiratory tract and in combination with lysozyme is bactericidal although it is unable to activate complement.

IgM exists as a pentameric structure with five immunoglobulin units bound by a J chain, while secreted IgA consists of two units bound by a J chain. IgA also contains a secretory component when present in secretions. IgE is important in host defence against parasites such as worms and can also lead to allergic manifestations such as anaphylaxis and atopy (Table 12.6).

B-cell differentiation

Early differentiation in the bone marrow. Cells successfully rearranging antibody genes express pre-B-cell receptor (BCR) and undergo positive selection; the remainder die by apoptosis. Immature B-cells express monomeric IgM as a BCR. If these cells encounter self-antigen, they are deleted by apoptosis or become inactivated (anergy). This results in negative selection and self-tolerance. Mature B-cells express surface IgM and IgD and subsequent foreign antigen binding results in activation. This occurs in the germinal centres of secondary lymphoid organs. B-cells proliferate and differentiate. Mature antibody-producing B-cells are termed plasma cells and can demonstrate antibody class switching.

B-cell effector populations and functions

- Plasma cells. Differentiated B-cells; secrete high levels of antibody. Lack cell surface immunoglobulin
- Memory B-cells. Long-lived non-proliferating B-cells available for antigen rechallenge. Express surface immunoglobulins other than IgM or IgD. Somatic hypermutation allows affinity maturation
- B-1 B-cells. Minor subpopulation found in body cavities such as pleura and peritoneum. Produce low-affinity IgM in response to bacterial polysaccharide.

Table 12.6 Characteristic features of immunoglobulin isotypes

Characteristic	IgG	IgM	IgA	IgD	IgE
% total Ig in plasma	80	6	13	<1	<1
Location	Intra/extravascular	Intravascular	Secretions/intravascular	B-cell surface	Mast cells, basophils, secretions
Structure	Monomeric	Pentameric	Dimeric	Monomeric	Monomeric
Protein subunits	None	J	J and S	None	None
Half-life (days)	23	5	6	3	2
Special features	Placental passage	Primary antibody produced	Present in milk and secretions	B-cell receptor	Mediates atopy

B-cell antigen presentation. Not a specific subpopulation. B-cells constitutively express MHC class II and possess co-stimulatory molecules such as B7, upregulated with activation and CD40 to allow antigen presentation to CD4 T-cells. Plasma cells do not express surface antibody and cannot present antigen. Adhesion molecules also aid interaction.

B-cell activation

Antigens with multiple repeating epitopes (e.g. polysaccharides) directly activate B-cells (thymus-independent antigens). These antigens do not induce class switching or memory cells. Protein antigens have single epitopes and require additional signals from CD4+ Th cells to activate B-cell (thymus-dependent antigens). The nature of the cytokines produced by Th determines the class of antibody produced. CD40/CD40 ligand interaction is critical for B-cell proliferation and antibody class switching. B-cell activation results in signal transduction. Immunoglobulins activate Src family kinases to phosphorylate ITAMs on signalling molecules associated with immunoglobulin in the BCR complex (Igα and Igβ). Other molecules, such as CD19 and CD21, also enhance BCR complex signal transduction. When phosphorylated, ITAMs activate Syk (a kinase) to activate pathways that allow transcription factors such as NF-κB and NF-AT to translocate to the nucleus and stimulate gene transcription (immunoglobulins and cytokine receptors). Th cytokines enhance B-cell proliferation and differentiation.

Lymphatic organs

Lymphocyte maturation occurs in the primary lymphoid organs: bone marrow for B-cells, thymus for T-cells. The thymus has an outer cortex with epithelial cells and thymocytes, where T-cells initially mature, and an inner medulla, where T-cell selection occurs in association with DCs and macrophages. B-cell maturation occurs in the foetal liver but, after birth, occurs in the bone marrow. Mature lymphocytes are released into the circulation. Differentiation and proliferation occur in secondary lymphoid organs: the lymph nodes, spleen, and MALT. The spleen contains red pulp and white pulp. Antigens in the blood are concentrated in the spleen and antigen presentation leads to lymphocyte proliferation and differentiation in the white pulp. The germinal centres are B-cell-containing regions, and T-cell-rich areas are peripheral in the white pulp. Lymph nodes receive antigen from peripheral tissues via afferent lymphatics. The cortex contains lymphoid follicles with germinal centres (B-cell-rich) and paracortical (T-cell-rich) areas. The medulla contains sinuses, and plasma cells migrate to the medulla to secrete antibodies. The efferent lymphatics carry antibody and primed T-cells from the lymph nodes.

Immunization

This may be passive (administering immunoglobulin) or active (vaccination). Vaccines may be administered to the whole population or specific groups at high risk of an infection. Types of vaccines include:

Toxoids. Produced by inactivation of toxins (e.g. tetanus). Highly immunogenic

Polysaccharide. Polysaccharide antigens are T-cell independent and fail to produce immunologic memory. Protein conjugate vaccines link polysaccharide to an immunogenic protein and allow immunologic memory and immune response in children <2 years old

Recombinant antigens. Recombinant DNA technology allows the expression of virus surface antigen in yeast cells and its purification for vaccination. Used in many viral vaccines, including hepatitis B

Live attenuated organisms. Administration of microorganisms that express common antigens to the pathogen being immunized against but of attenuated virulence due to laboratory manipulation. Examples include the oral polio vaccine and bacille Calmette–Guérin (BCG) used to vaccinate against *Mycobacterium tuberculosis*. Live attenuated vaccines should not be used in immunosuppressed individuals in whom disseminated infections may result

Whole inactivated microorganisms. Some vaccines use whole microorganisms that have been inactivated (e.g. influenza virus vaccine). Many of the older heat-killed vaccines are relatively ineffective and are rarely used (e.g. older cholera vaccines)

Newer strategies. These include using virus-carrier vaccines with a harmless virus (such as vaccinia virus or adenovirus) genetically engineered to express a gene encoding the immunizing antigen, synthetic peptide vaccines, or DNA vaccines in which DNA plasmids are injected. These strategies may be combined with DNA vaccination followed by virus carrier or peptide (prime-boost regimens) and are under investigation in infections for which immunization has been unsuccessful (e.g. HIV).

Immunity following vaccination wanes with time and revaccination (a booster) may be required after a set period of time.

Cytokines

Small-molecular-weight proteins that regulate cellular effector functions ﬁ the immune system (Table 12.7). Approximately 20 cytokines identified ﬞ date. Role in immune homeostasis but also contribute to disease patholog Some have direct effector functions on microorganisms (e.g. antiviral e fects of IFN-α).

Cells usually respond to a particular pattern of multiple cytokines syr ergistically (the observed effect is greater than the sum of the individu effects) or antagonistically. Function is regulated by expression of the cytc kine and its receptor.

Chemokines are related chemotactic cytokines. They can be groupe into families depending on the arrangement of amino terminal cysteine re idues (e.g. CC and CXC chemokines where C is a cysteine residue and ﬞ a non-conserved amino acid). IL-8 is a CXC chemokine. Chemokines er hance adherence of leucocytes to the endothelium, transmigration throug blood vessels, and activation of the leucocyte in inflamed tissue.

Cytokine effects include systemic effects (e.g. fever) and release of acut phase protein reactants including CRP and MBL. Cytokine responses t antimicrobial products (such as bacterial LPS and superantigens) mediat signs of sepsis (fever or low body temperature, increased heart rate, ir creased respiratory rate, increased or decreased numbers of PMNs) and c septic shock (signs of sepsis with a low blood pressure that fails to correc despite fluid resuscitation).

Table 12.7 Cytokines and their functions

Cytokine	Cell of origin	Function
IL-1(IL-1α, IL-1β)	M/M, EN, B, o	Lymphocyte activation, macrophage stimulation, fever and acute phase response
IL-2	T	Activation and proliferation of T-cells, co-factor in B-cell proliferation
IL-4	T, MA, BA	Growth factor for B-cells and Th2 T-cells; inhibits Th1 T-cells; stimulates IgG and IgE production
IL-5	T, MA	B-cell growth and antibody production; eosinophil differentiation
IL-6	T, M/M, EN, o	T-cell activation; B-cell antibody production; haematopoietic progenitor cell growth; acute phase response
IL-8	M/M, EN, EP, o	PMN chemoattraction
IL-10	M/M, T, o	Inhibits macrophage cytokine production; inhibits Th1 response, stimulates antibody production
IL-12	M/M, B	Activates Th1 T-cells and NK cells
IFN-α	M/M, T, o	Inhibits RNA viruses; downregulates IL-12; expands memory T-cells; upregulates MHC class I molecules
IFN-β	EP, F	Inhibits RNA viruses; upregulates MHC class I molecules
IFN-γ	T, NK	Activates macrophages; activates Th1; inhibits Th2 responses
TNF-α	M/M, T, NK, o	Activation of macrophages, PMN, endothelial cells; co-factor for B-cell and T-cell proliferation; fever and septic shock
TGF-β	M/M, T, o	Downregulates pro-inflammatory cytokines; wound healing
Granulocyte-monocyte-stimulating factor	T, M/M, EN, F, o	Growth of PMN and macrophages; enhances function

B, B-cell; BA, basophil; EN, endothelial cell; EP, epithelial cell; F, fibroblast; MA, mast cell; M/M, monocyte, macrophage; NK, natural killer cell; o, others; T, T-cell.

Treatment

The immune response to infections can be enhanced or, during transplant-ation/autoimmune disease, suppressed. Therapeutic interventions include:

- Active immunization. Used for selected microorganisms and being investigated in the therapy of other diseases such as cancer
- Passive immunization. Transfer of specific factors such as immunoglobulin
- Adoptive transfer. Cells can be transfused into individuals. Examples are transfusion of bone marrow stem cells or PMNs to individuals after bone marrow transplantation or CTLs, removed from individuals and primed in the laboratory by DCs to respond to a specific virus (e.g. Epstein–Barr virus infection)
- Cytokines. Specific cytokines (IFN-α, TNF-α, IL-2) are used therapeutically to enhance immune responses or may be blocked using inhibitors or soluble receptors (anti-TNF antibodies, soluble IL-2 receptors). Colony-stimulating factors such as GM-CSF or G-CSF may be used in neutropenic hosts
- Steroids. Used to suppress inflammation (◑ p.904) and lymphocyte activation. May modify inflammation in certain infections such as meningitis or be used to prevent autoimmune diseases or transplant rejection
- Cytotoxic drugs. Azathioprine is a purine antagonist that blocks RNA and DNA synthesis. Mycophenolate mofetil selectively inhibits purine synthesis in lymphocytes and blocks T- and B-cell proliferation. Cyclophosphamide inhibits DNA metabolism by alkylating DNA. Methotrexate inhibits folic acid-dependent DNA biosynthesis. Both agents are cytotoxic for lymphocytes and used in immunosuppression
- Ciclosporin and tacrolimus. Inhibit *IL-2* gene transcription by binding to a cytoplasmic receptor and inhibiting calcineurin—a cell signalling phosphatase. Sirolimus, a similar drug, inhibits IL-2 function by blocking IL-2 receptor signal transduction but has effects on the signalling of many other cytokines
- Antibody therapy and inhibition of co-stimulation. Used in transplantation. Anti-lymphocyte globulin or anti-CD3 antibody (OKT3) inhibit T-cell activation by transplanted antigens. Antibodies to IL-2R, e.g. daclizumab, may represent a more selective approach to decreasing T-cell activation. Experimental approaches involve using CTLA-4 to bind B7 and block co-stimulation
- Miscellaneous therapies. Recombinant host defence molecules are being studied in investigational protocols (e.g. bactericidal/permeability-increasing protein). Abnormalities in clotting cascades are a feature of sepsis in association with certain patterns of cytokine production. Activated protein C is being used to treat certain kinds of bacterial sepsis.

Inflammation: overview

If the human body was not exposed to infective agents, such as viruses, bacteria, and protozoa, or to the effects of trauma, then inflammation would be an unnecessary process. As it is, inflammation is a vital process which repels infections and initiates repair and regeneration after trauma. This should be borne in mind when studying the process of inflammation, since it is often seen as an entirely adverse process: patients regard inflammation as manifest by abscesses, redness, or pain—as the disease, rather than a reaction to it. Doctors aim many of their treatments at reducing the effects of inflammation.

Acute and chronic inflammation

Inflammation is traditionally classified into acute and chronic forms which would appear to relate to the time course of each type. This is not the case for all inflammatory processes. For example, infectious mononucleosis (glandular fever) is an acute illness caused by infection with the Epstein–Barr virus but is characterized by a purely 'chronic' inflammatory process from its onset.

The classification into acute and chronic types does describe the predominant inflammatory cell type present in the process—neutrophils in acute inflammation, and macrophages and lymphocytes in chronic inflammation—which is a useful distinction. However, it might be better to label these types 'neutrophil-predominant' and 'macrophage/lymphocytes-predominant' rather than acute and chronic.

Cellular mediators of inflammation

Neutrophils (PMLs)

(⊃ pp.436.)
- Short-lived cells, 10–20h in blood, which are produced in the bone marrow
- Can be thought of as the first-wave, front-line rapid deployment defenders who appear rapidly at a site of inflammation and die for the overall 'good' of the body
- Very effective killers of bacteria
- Cell surface receptors for C3b component of complement, Fc portion of antibody tails—these facilitate phagocytosis of microorganisms
- Contain vacuoles with lysosomal enzymes and oxidating protein complexes which kill ingested bacteria
- Release some chemical mediators of inflammation.

Eosinophils

(⊃ pp.436.)
- Can be thought of as variants of neutrophils but with design favouring the killing of multicellular parasites, such as enteric worms, rather than bacteria
- Longer-lived than neutrophils— 4 days in blood, weeks in tissue
- Found in relatively large numbers in certain tissues, especially the lamina propria in the GI system, even in the absence of disease
- Cytoplasmic granules containing cationic proteins which can kill worms (and other cells) by binding to the cell surface
- Have an apparently misdirected action in asthma where they accumulate in the bronchial mucosa and the cationic proteins damage the ciliated epithelium.

Mast cells

- Have cytoplasmic granules containing sulphated glycosaminoglycan to which chemical mediators of inflammation, including histamine, are reversibly bound
- Main reaction is degranulation which involves release of granule contents very rapidly into surrounding tissue
- Key components of type I hypersensitivity (⊃ p.910).

Macrophages

- Derived from blood-borne monocytes
- Long-lived cells; months in tissues
- Found in second wave of cells in acute inflammation, after neutrophils
- Have phagocytic functions and vacuoles with digestive and free radical generating systems; less effective at killing common bacteria than neutrophils but more effective at killing atypical bacteria such as *Mycobacteria*
- Some specialist macrophages, e.g. dendritic macrophages in skin, present antigens to T-cells
- Activated macrophages produce many chemical mediators of inflammation and immune response including lymphokines, tumour necrosis factor

- Can fuse together to form multinucleated giant cells, often in response to material that is too large for a single cell to phagocytose, e.g. exogenous material such as silica particles
- When activated, may take on an epithelioid appearance with abundant eosinophilic cytoplasm. A cluster of epithelioid macrophages surrounded by a rim of lymphocytes is called a granuloma and is characteristic of certain causes of inflammation, especially infective mycobacteria such as tuberculosis and leprosy.

Chemical mediators of inflammation

There are a vast number of chemical mediators of inflammation, as the list indicates, and the actions of these, and the cellular responses to them, regulate the inflammatory response.

Because there are so many factors, it is very difficult to dissect out the role of a single mediator in any particular inflammatory situation. However, many therapeutic agents have been developed against specific mediators, so it is important to know something about their actions so that response to these agents can be predicted. With the range of mediators described, it is easy to lose sight of the basic functions of them:

• To bring other cellular mediators of inflammation to the site in an appropriate number and at an appropriate time
• To increase blood flow and blood vessel permeability so that cellular mediators can get to the site of inflammation and so material can be carried away.

Complement system

This is a cascade system of proteins which may be activated in a number of different ways in acute inflammation:

• By the Fc portion of antibodies which have combined with specific antigen—the classical pathway of activation
• By the endotoxins of Gram-negative bacteria—the alternative pathway
• By enzymes releases by dying cells in tissue necrosis
• By some products of the kinin and fibrinolytic systems.

Once the complement system has been activated, it produces a number of proteins which are active in inflammation:

• C3a and C5a—chemotactic for neutrophils, increases vascular permeability, causes release of histamine from mast cells
• C567 complex—chemotactic for neutrophils
• C56789 complex—the membrane-attack complex which punches holes in cell membranes → cell lysis
• C4b, C2a, C3b—all opsonize bacteria by binding to them; macrophages have specific receptors for these proteins.

Vasoactive amines

Histamine—causes vasodilatation and increased permeability of blood vessels.

Nitric oxide

Potent vasodilator, inhibitor of platelet and monocyte adhesion and, at high concentrations, a powerful cytotoxic agent.

Kinin system products

Bradykinin—vasoactive and pain-stimulating functions.

Clotting system products

Factor XII (Hageman factor)—activates the coagulation, fibrinolytic, and kinin systems.

Arachidonic acid metabolites

- Prostaglandins—potentiate increased vascular permeability, inhibit or stimulate platelet aggregation
- Leukotrienes—vasoactive properties.

Cytokines

Interleukins.

Patterns of inflammation

Acute neutrophilic

This is the prototypic pattern of acute inflammation which may be found in an abscess of a localized bacterial infection. It is characterized by masses of neutrophils which phagocytose and kill the bacteria. Many of these neutrophils will die during this process producing an amorphous mass of lysed nuclei and cytoplasm.

Acute progressing to chronic

This occurs in the later stages of neutrophil-rich inflammation. Macrophages arrive in great numbers and phagocytose the debris. Some fibroblasts may appear and produce fibrous scar tissue to ablate the abscess cavity. Lymphocytes and plasma cells will be less common, unless there is a persisting infection.

Eosinophilic

Acute inflammation in which eosinophils, rather than neutrophils, predominate occurs with parasitic, rather than bacterial, infections.

Chronic granulomatous

Granulomas occur in special types of chronic inflammation and this may give clues as to the cause of the inflammation:
- Mycobacterial infections:
  - Tuberculosis
  - Leprosy
- Other specific infections:
  - Syphilis
  - Invasive parasitic infections
- Idiopathic inflammatory diseases:
  - Sarcoidosis
  - Crohn's disease
  - Granulomatosis with polyangiitis (formerly known as Wegener's granulomatosis)
- Hepatic reaction to some drugs:
  - Allopurinol
  - Phenylbutazone
  - Sulphonamides.

Systemic effects of inflammation

Raised temperature

Any substantial focus of acute inflammation causes a pyrexia due to the release of endogenous pyrogens (e.g. IL-2) from inflammatory cells which set the thermoregulatory area of the hypothalamus to a higher temperature.

Cachexia

There is often a negative nitrogen balance in substantial chronic inflammation with considerable weight loss, hence the historical term for tuberculosis—'consumption'.

Treatments

The wide range of mediators involved in the inflammatory response represent the various targets of anti-inflammatory therapy. However, the highly complex and interactive nature of the inflammatory response limits the effectiveness of highly specific drugs that only target isolated inflammatory pathways.

Steroidal anti-inflammatory drugs

(→ *OHPDT2* Ch. 6)

Glucocorticoids have an anti-inflammatory effect, primarily through inhibition of transcription of the gene for IL-2 required for cloning of Th2 cells crucial to the inflammatory response. In addition, they inhibit transcription of other inflammatory cytokines, including TNF-α, IL-1, and IFN-γ, as well as the expression of inducible inflammatory enzymes, like phospholipase A_2 (and, consequently, PAF and prostanoids), COX-2, and inducible NO synthase. The broad impact that glucocorticoids have on different inflammatory pathways is instrumental in their effectiveness.

Non-steroidal anti-inflammatory drugs

Perhaps the best-known anti-inflammatories are the NSAIDs (→ *OHPDT2* Ch. 15), most of which do not require a prescription. These drugs are usually taken to reduce the effects of minor aches or pains and for headaches, but they are also sometimes recommended for rheumatic pain.

NSAIDs act by irreversibly inhibiting COX enzymes (COX-1 and COX-2), → reduced prostanoid synthesis. This inhibition has a number of anti-inflammatory effects, including:

- Reduced vasodilatation in response to prostaglandins PGE_2 and PGI_2 (prostacyclin). As a result, there is less oedema and swelling. This is probably the main effect in headache
- Reduced sensitization of nociceptive nerve endings to 5-HT and bradykinin
- Reduced fever.

Aspirin is the best-characterized NSAID and its action is known to be primarily mediated by acetylation of a specific serine residue in the COX enzyme to irreversibly block its action. Ibuprofen binds to a different site in the enzyme, to the same effect. Some of the actions of paracetamol might also be mediated via inhibition of COX, but it is not generally recognized to be an anti-inflammatory drug.

The major side effect of NSAIDs is gastric bleeding due to a loss of the protective effects of prostaglandins in the stomach, resulting in increased gastric secretion and reduced gastric blood flow.

Antihistamines

Calcium-mediated release of histamine from mast cells is partly responsible for the first two elements of the so-called triple response to a mild insult to the skin:

- The reddening due to arteriolar vasodilatation
- The wheal due to increased permeability of venules.

(The 'flare' in surrounding tissue is due to release of vasodilators (e.g. CGRP) from nerve endings of sensory nerves in the vicinity of the insult.)

Histamine also stimulates sensory nerve endings to cause itching and, at a systemic level, stimulates bronchoconstriction associated with the immediate phase of asthma (⮕ pp.422–3).

These actions of histamine are mediated via stimulation of H_1 histamine receptors, and antagonists of this specific subclass of receptors are commonly referred to as antihistamines (not to be confused with H_2 receptor antagonists that are used to treat and prevent peptic ulcers).

Although H_1 receptor antagonists should theoretically have an anti-inflammatory effect through their inhibition of the inflammatory responses, in reality, their use is limited to treatment of allergy (rhinitis), insect bites, and drug hypersensitivities. Many of the H_1 receptor antagonists characterized to date have non-specific effects, through inhibition of muscarinic, 5-HT, and α-adrenoceptors. These non-specific actions contribute to the drowsiness caused by some antihistamines and mediate some of the effects of antihistamines that are prescribed for motion sickness or even sedation.

Bradykinin

Bradykinin is an endothelium-dependent vasodilator that acts to relax vascular smooth muscle, primarily through G-protein-coupled B_2 receptors on endothelial cells, stimulating the release of PGI_2 and NO. However, it can also cause spasm in the bronchial tree and is associated with many of the symptoms experienced in allergic reactions (including vasodilatation, increased vascular permeability, and pain).

Icatibant is a peptide antagonist for B_2 receptors, but is not used therapeutically. A number of non-peptide antagonists are under investigation, which may be more suitable as orally active agents for use in allergy.

Cell involvement: regulation

An infection causes some damage to tissues in the body; the amount of damage and the type of tissue determine what happens after the infection has been eradicated:

- If the damage does not cause the death of cells, then the cells will recover and the tissue will return to normal
- If there is death of cells in a tissue that is capable of regeneration (e.g. the liver), then the regenerative cells will divide to replace those that died and, after some remodelling, the tissue will return to normal
- If there is death of cells in a tissue that is incapable of regeneration, then the tissue will heal by repair, which will lead to formation of scar tissue and some loss of function.

Cell involvement

Fibroblasts

- Primary cell involved in repair
- Recruited by multiple growth factors (e.g. platelet-derived growth factor, IL-1) from platelets, endothelial cells, macrophages, and neutrophils
- Produce collagen and fibronectin to constitute fibrous connective tissue which changes with age to relatively acellular bands of collagen.

Macrophages and neutrophils

- Present from the initiating infection
- Not directly involved in repair but secrete chemical mediators which attract fibroblasts
- Macrophages may continue with debris-removing activities.

Endothelial cells

- Present in granulation tissue at the start of the repair process
- Growth is stimulated by angiogenic factors such as vascular endothelial growth factor.

Regulation

Repair, like many processes in the body that occur in response to injury, is a double-edged sword:

- Production of fibrous tissue to replace cells that have died allows retention of the structural integrity of the organ (e.g. a fibrous scar at the site of a myocardial infarct prevents the heart wall from rupturing at that point)
- Production of fibrous tissue can cause loss of function in the tissue surrounding it (e.g. fibrosis in the lungs can cause physical constriction of blood vessels → pulmonary hypertension and right-sided heart failure)
- Overproduction of fibrous tissue can cause a much larger fibrous scar than was necessary to repair the immediate damage with distortion and loss of function in the surrounding tissue (e.g. a keloid scar on the skin).

Regulatory processes
- Positive factors for formation of fibrous tissue:
  - Production of growth factors that recruit fibroblasts
  - Production of factors that increase vascular permeability (e.g. vascular endothelial growth factor) and so lead to increased deposition of plasma proteins such as fibrinogen and fibronectin
- Factors which inhibit formation of fibrous tissue:
  - Matrix metalloproteinases which break down proteins in the extracellular matrix, such as collagen, laminin, and fibronectin.

Examples of repair and its consequences

Lung—after slowly resolving bacterial pneumonia (➲ p.425):
- If the elastic walls of the alveoli are damaged, then the lung cannot regenerate
- Fibrin and inflammatory cells fill the alveoli
- Fibroblasts are recruited
- Fibroblasts synthesize collagen
- Fibrin and inflammatory debris are removed by macrophages
- Final state = fibrous scarring in lung with reduced expansion of lungs (thus reduced vital capacity on respiratory function tests) and reduced area for gas exchange (so reduced VO_2max).

Liver—after years of hepatitis C infection:
- Liver cells have the ability to regenerate so a single episode of damage does not usually initiate the repair process; however, continuing damage while the cells are attempting to regenerate does lead to repair
- Fibrous tissue forms between liver cells
- Bands of fibrous tissue form between portal tracts with intervening regenerative nodules (cirrhosis)
- The fibrous tissue obstructs flow through the portal venous system → portal hypertension
- Replacement of most hepatic tissue by fibrous tissue leads to liver failure.

Skin—after recurrent bacterial abscesses:
- Skin is relatively resistant to damage and will heal without much repair
- However, an abscess destroys many cells and leaves a cavity
- The cavity is lined by a wall of fibrous tissue after repair
- The fibrous tissue may contract to obliterate the cavity which will just leave a scar
- If the cavity is not obliterated, then recurrent bacterial infections will occur, → greater scarring (e.g. hidradenitis suppurativa = recurrent bacterial infections in the apocrine skin glands in the axilla and groin).

Hypersensitivity

Overview

The body has a number of immune systems which function to prevent or limit infections from viral, bacterial, or parasitic organisms. These systems may produce some adverse effects within the body, but this is more than balanced by the prevention of damage that would occur if the infective organism established an infection within the body. However, sometimes these immune systems react against objects that are not infective organisms (e.g. pollen grains in hayfever) and then the damage caused by the immune system itself is disadvantageous and inappropriate. This is termed a hypersensitivity reaction.

It should always be borne in mind that hypersensitivity reactions are caused by systems that normally prevent infections, since any therapy that is used to reduce the effects of a hypersensitivity reaction may lead to a vulnerability to infection.

Hypersensitivity reactions are classified into different types according to the specific system which is mediating the reaction.

Type I hypersensitivity

See Fig. 12.7 (➔ *OHCM10* pp.794–5).

This is mediated by mast cells and is typified by hayfever. Mast cells have IgE antibodies on their surface which are directed against a specific antigen, particular types of pollen in the case of hayfever. When an antigen binds to the antigen-specific site on these antibodies, a signal is transmitted through the mast cell which causes it to degranulate, releasing the contents of granules in its cytoplasm into the immediate surrounding environment. These granules contain many chemical mediators of acute inflammation, such as histamine, which lead to vasodilation and oedema in the surrounding tissues—manifest as nasal obstruction and watery secretion in hayfever. Severe type I hypersensitivity reactions manifest as anaphylaxis (➔ *OHCM10* p.794).

There are a number of steps in this pathway which can be blocked to prevent this hypersensitivity reaction:

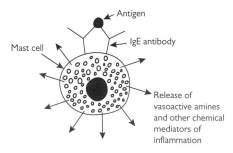

Fig. 12.7 Type I hypersensitivity. The binding of the antigen to IgE molecules on the surface of mast cells leads to degranulation of those cells with release of chemical mediators of acute inflammation.

- Affected individuals can avoid the precipitating antigen (but this can severely limit their lifestyle in the case of ubiquitous environmental antigens such as pollen)
- The number of mast cells in the affected organ can be reduced by long-term administration of immunosuppressants such as steroids
- The cell membrane of the mast cells can be stabilized, so it is less likely to degranulate, by administration of disodium cromoglicate
- If all these fail, then the effects of the chemical mediators of acute inflammation can be blocked using drugs such as antihistamines.

Type II hypersensitivity

See Figs 12.8 and 12.9.

This is mediated by specific antibodies directed against and by both cells and activation of the complement system. An example is the breakdown of red blood cells in a Rh blood antigen-positive foetus of a mother who is Rh antigen negative and has been exposed to Rh antigen-positive blood in the past (usually by a previous pregnancy) (➲ p.432). The maternal antibodies which are directed against the Rh antigen cross the placenta into the

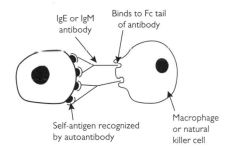

Fig. 12.8 Type II hypersensitivity. Mediated by macrophages binding to specific antibodies which have bound to antigens on a host cell. The macrophage could phagocytose the antigen-covered cell or a natural killer cell could induce its death.

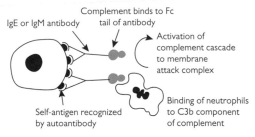

Fig. 12.9 Type II hypersensitivity. Mediated by complement which binds to the tails of the specific antibody and either leads to formation of the membrane attack complex and lysis of the cell or attraction of neutrophils by the C3b component.

foetal circulation, where they bind to the Rh antigens on the surface of the foetal red blood cells. The binding of the antibodies to the antigen causes a conformational change in the structure of the antibody tail region which then binds the first component (Clq) of the complement protein cascade. This leads to production of activated complement components (e.g. C3b), which form a 'membrane attack complex' that attaches to the membrane of the red blood cells. This forms a hole in the membrane, which causes cell lysis by uncontrolled influx of water into the cell. This process leads to a haemolytic anaemia in the foetus, which may be so severe that it requires intrauterine blood transfusions. In this instance, the best method of treatment is prevention by ensuring that all Rh-negative mothers receive injections of antibodies directed against the Rh antigen during pregnancy and immediately after birth. Thus, any foetal red blood cells that do get into the maternal circulation are immediately lysed by these antibodies before the maternal immune system has time to react against them.

Type II hypersensitivity can also be mediated by cells, such as macrophages or NK cells, which bind to the tails of antibodies that have bound antigen and then phagocytose the cell or release factors which kill it.

Type III hypersensitivity

See Fig. 12.10.

If the ratio of antigen to antibody occurs within a certain range, then the antigens will crosslink antibodies to form large complexes which will lodge in membranous filtration systems in the body (e.g. the glomerulus or synovial membrane). Components of the complement system will bind to the antibody tails and damage will occur from the activated complement components themselves and from the inflammatory cells which they attract. Examples of immune complex disease include the glomerulonephritis, arthritis, and endocarditis that occur in rheumatic fever after a streptococcal throat infection.

Type IV hypersensitivity

See Fig. 12.11.

This is mediated by cells rather than antibodies. A common example is nickel allergy which may occur from exposure to this metal in wrist watches or metal components of clothing. Nickel, by itself, is too small to induce an immune response (hypersensitivity would ensure extinction of the species if reactions were elicited to pure elements), but if it binds to proteins in the body, such as keratin in the skin, then it can produce an allergenic complex.

The antigenic complex is presented to T-cells by specific types of macrophages. The T-cells which are specifically reactive with the antigen divide to produce an expanded clone of cells. When the antigen is encountered again, it is recognized by these T-cells (again, usually when presented on the surface of specific types of macrophages) and they secrete a number of cytokines, including IL-12 and IFN-γ), which produce an inflammatory reaction. These cytokines cause an accumulation of activated macrophages at the site of inflammation. Neutrophils are not usually involved in this type of hypersensitivity reaction.

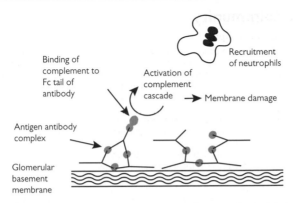

Fig. 12.10 Type III hypersensitivity. Large complexes of antigen and antibody lodge in membranous filtration systems in the body (e.g. the glomerulus in the kidney) where they activate the complement cascade.

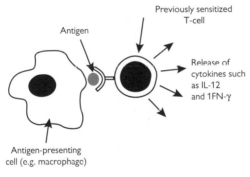

Fig. 12.11 Type IV hypersensitivity. Cell-mediated hypersensitivity caused by sensitized T-cells releasing cytokines after binding to the antigen.

Autoimmunity

General principles

An autoimmune reaction occurs when the body mounts an immune re-action to some intrinsic component within it (e.g. thyroid epithelial cells) rather than against some extrinsic object (e.g. bacteria, pollen, nickel). There are many regulatory processes in the body which are designed to prevent autoimmune reactions but there are some mechanisms which can breach these:

- Exposure of a 'hidden' antigen to the immune system:
  - Some tissues in the body are not patrolled by cells from the immune system
  - Some tissues in the body are in sealed compartments which are not exposed to cells from the immune system (e.g. proteins within myocardial cells)
  - If a tissue is damaged and proteins that immune cells have not been exposed to during the period of immune tolerance *in utero* and early childhood are released, then an immune reaction will be mounted against them (e.g. Dressler's syndrome (➜ *OHCM10* p.698) following myocardial infarction)
- Cross-reaction between an extrinsic and intrinsic antigen:
  - Some infective organisms possess antigens that stimulate an immune reaction, but these antigens are similar to antigens in intrinsic body components so the antibodies or activated lymphocytes that are formed against the infective organism cross-react with these intrinsic components to produce an autoimmune reaction
  - An example is rheumatic fever (➜ *OHCM10* p.142), where the immune reaction directed against the cell wall of streptococcal bacteria cross-reacts with heart valve tissue.

Autoimmune reactions may be mediated by any of the body's immunological mechanisms (➜ pp.880–91). Autoimmune disease is often classified into:

- Single organ—where a single area in the body is affected, presumably due to the antigen only being present in that organ. Examples include:
  - Hashimoto's thyroiditis (➜ *OHCM10* p.220)—where the epithelial cells in the thyroid are destroyed by activated lymphocytes → hypothyroidism, when there are insufficient numbers of cells left to produce normal amounts of thyroxine
  - Grave's disease (➜ *OHCM10* p.218)—where, unusually, a stimulatory autoantibody is produced which binds to the thyroid-stimulating hormone receptor on thyroid epithelial cells, causing excess production of thyroxine and, thus, hyperthyroidism
- Systemic—where the whole body is affected due to widespread presence of the antigen. Examples include:
  - Systemic lupus erythematosus (➜ *OHCM10* p.554)—where there are many autoantibodies present which are mainly directed against targets in the cell nucleus. The skin, joints, and kidney are the most commonly affected sites
  - Sjögren's syndrome (➜ *OHCM10* p.710)—with autoimmunity directed against the salivary and lacrimal glands, with activated T-lymphocytes the effector cells.

Transplantation

Transplantation of organs is now a common treatment (at least as far as the supply of donor organs allows) for failure of individual organs such as kidney, liver, heart, and lungs. The major difficulty with transplantation is preventing the new host from rejecting the organ by an immune process.

Every individual has a large number of different antigens expressed on their cells which are highly variable. These include blood group antigens and the HLA (major histocompatibility loci antigens). The only organs that would not be rejected without immunosuppression would be those in a site in the body that does not have immune surveillance (e.g. the cornea) or those from an identical twin. For all other transplants, a number of methods of preventing rejection have to be employed:

- Matching of donor and recipient:
  - Match blood group type
  - Match HLA types as far as is possible
- Immunosuppressive therapy:
  - Corticosteroids
  - Azathioprine
  - Ciclosporin
  - Anti-lymphocyte antibodies
- Monitoring of graft function and rejection:
  - Functional measurements (e.g. glomerular filtration rate in kidneys)
  - Biopsy and histological examination
  - Adjustment of immunosuppressive therapy on the basis of these results.

Transplantation of bone marrow (commonly used to increase the dosage of chemotherapy that can be given in the treatment of malignancy) is an interesting reversal of the usual problems of transplant rejection since, in this case, it is the transplanted cells that are the immune system and can thus mount an immune attack on the host—graft-versus host-disease.

Deficiency of humoral immunity

Most common inherited immunodeficiency. Results in frequent infections—
otitis media, sinusitis, pneumonia due to *Streptococcus pneumoniae* or
Haemophilus influenzae, and recurrent gastroenteritis. Treated by immunoglobulin infusion.

- X-linked agammaglobulinaemia:
  - Mutation in a B-cell cytoplasmic tyrosine kinase; maturation failure
  - Absence or profound decrease in B-cells and antibody
- Common variable immunodeficiency:
  - Not hereditary and exact mechanism unclear
  - B-cell maturation abnormal and numbers normal or low
  - IgA and IgG (±IgM) low
  - May not present until third decade
  - Autoimmunity and cancers associated
- Transient hypogammaglobulinaemia:
  - Low levels of IgG after 6 months of age but normal B-cell numbers
  - Self-resolving
- Selective immunoglobulin deficiency:
  - Normal B-cell numbers
  - IgA deficiency most common
  - Autoimmune manifestations
  - If associated IgG2 deficiency, frequent infections
- Hyper-IgM syndrome:
  - Majority X-linked
  - Lack CD40 ligand
  - IgM high; IgG, IgA, and IgE low
  - B-cell numbers normal
- Secondary defects in humoral immunity: occur with malignancy
  (lymphoreticular malignancies, e.g. multiple myeloma; ➔ *OHCM10*
  p.368), protein malnutrition, splenectomy (➔ *OHCM10* p.373), sickle
  cell disease (➔ *OHCM10* p.340), bone marrow transplantation (➔
  OHCM10 p.364), and AIDS (➔ *OHCM10* pp.398–403) (especially in
  children).

Deficiency of cell-mediated immunity

T-cell defects. Susceptible to opportunistic infections: cytomegalovirus, *Candida* spp., *Pneumocystis jirovecii (carinii)* pneumonia (PCP), mycobacterial infections, and *Toxoplasma gondii*.

- Congenital thymic aplasia (DiGeorge's syndrome):
  - Autosomal dominant
  - Thymic aplasia
  - Absent or few mature T-cells
  - B-cell numbers normal but defective responses to T-cell-dependent antigens
  - Absence of hypoparathyroid glands causes low serum calcium
- Functional T-cell defects: normal T-cell numbers but mutations involving the TCR complex or ZAP 70 tyrosine kinase
- Chronic mucocutaneous candidiasis:
  - Selective defect involving response to *Candida* spp.
  - Chronic skin and mucocutaneous candidiasis only manifestation
- Secondary defects in T-cell function:
  - HIV (→ *OHCM10* pp.398–403) infection most common acquired cause
  - Also, transplantation, immunosuppressant drugs, malignancies (e.g. Hodgkin's lymphoma; → *OHCM10* p.360)
  - Pregnancy and advanced age decrease T-cell function.

Combined humoral and cell-mediated deficiency

Severe combined immunodeficiency disease (SCID)

- Most patients die within 2 years unless receive successful bone marrow transplantation
- Heterogenous disorders
- X-linked SCID:
  - Associated with mutations in the γ-chain of the IL-2 receptor
  - T-cells low, B-cells present but cannot produce antibody
- Other forms autosomal recessive:
  - SCID associated with mutation in JAK3 tyrosine kinase, similar to X-linked form
- Adenosine deaminase deficiency (ADA):
  - Leads to a defect in purine metabolism of lymphocytes
  - Low T- and B-cells
  - Treatment options include gene therapy
- Recombinase deficiency:
  - Mutations in the *RAG* genes
  - Absent T- and B-cells.

Bare lymphocyte syndrome

- T- and B-cell numbers normal but APCs lack MHC class II molecules, so fail to present antigen.

Wiskott–Aldrich syndrome

- X-linked mutations in a specific protein cause combined immunodeficiency
- Normal T- and B-cell numbers
- Functional defects; low IgM and IgG, and high IgA and IgE levels
- Low platelets; allergic reactions, including severe eczema; and increased incidence of cancer.

Ataxia–telangiectasia

- Autosomal recessive defect in a tyrosine kinase required for DNA repair
- Neurologic (unsteady gait) and ocular features associated with lymphopenia and low IgA and IgE ($\pm$low IgG)
- Increased malignancies.

Defects in phagocyte function

Susceptibility to pyogenic bacteria and fungal infections.

Chronic granulomatous disease (CGD)

- X-linked and autosomal recessive defects in components of NAD(P)H oxidase which catalyses: $NAD(P)H + 2O_2 \rightarrow NADP^+ + 2 \cdot O_2^- + H^+$
- Generation of $2 \cdot O_2^-$ (superoxide) is defective
- Infections with catalase-positive organisms; *Staphylococcus aureus* and the fungus *Aspergillus* spp. most frequent
- Recurrent pulmonary infections and abscesses of skin, bone, and liver
- Obstructive complications can involve gut or genitourinary tracts due to formation of granulomatous lesions
- Diagnosis by detecting normal PMN numbers but decreased superoxide production (nitroblue tetrazolium test or measurement of dihydrorhodamine 123 fluorescence by flow cytometry)
- Treatment of infections: prophylactic antibiotics and IFN-γ injections.

Leucocyte adhesion deficiency

- Autosomal recessive defects in leucocyte adherence of variable severity
- Most common form due to failure to make certain integrins that share a common polypeptide
- High PMN numbers but no abscesses and defective complement-dependent phagocytosis (lack CR3)
- *Diagnosis:* absence of specific integrins on leucocytes
- Recurrent respiratory, GI, and skin infections due to *Staphylococcus aureus* and Gram-negative bacteria.

Myeloperoxidase deficiency

- Autosomal recessive—most common defect of PMNs, but usually clinically silent
- Fail to convert hydrogen peroxide to the more potent microbial molecule, hypochlorous acid.

Neutropenia

(➔ *OHCM10* p.330.)
- Congenital forms rare
- Cyclic neutropenia characterized by decreased counts and infections every 3 weeks
- Acquired due to chemotherapy or, occasionally, other drugs
- Bacterial and fungal infections
- Granulocyte colony stimulating factor (G-CSF) used to boost counts
- Prophylactic antimicrobials when pyrexial (➔ *OHCM10* p.352) (neutropenic regimen).

Chédiak–Higashi syndrome

- Autosomal recessive condition characterized by large lysosomes in PMN
- Recurrent bacterial infections and ophthalmological and neurological complications.

Job's syndrome

- Autosomal dominant, recurrent infections, and severe eczema
- High levels of IgE and chemotactic defects.

Other defects

Mononuclear cell defects

- Defective production of IFN-γ/IL-12 or receptor mutations effect monocyte/macrophage function
- Infections include atypical mycobacterial infections such as *Mycobacterium avium* complex and *Salmonella* spp.
- Treatment includes IFN-γ.

Anti-TNF-α therapy

- Mycobacterial infections or other infections with intracellular pathogens.

TNF-α receptor-associated periodic syndromes (TRAPS)

- Gain-of-function mutations in TNF-α receptor
- Recurrent fever in absence of infections.

Complement deficiency

- Most are autosomal recessive defects in individual complement components
- Deficiency of early classic or alternative pathway components leads to recurrent bacterial pneumonia and autoimmune conditions (e.g. systemic lupus erythematosus; ➜ *OHCM10* p.554)
- Late complement deficiencies associated with recurrent and atypical *Neisseria* spp. infections, including meningitis
- Diagnosed by complement component assays
- MBL deficiency associated with recurrent respiratory and GI infections.

Splenic dysfunction

- Usually acquired; occasionally congenital
- Sickle cell disease leads to autosplenectomy
- Increased frequency of infections with encapsulated bacteria; *Streptococcus pneumonia*, *Haemophilus influenzae*, *Neisseria meningitides*, and parasites; *Babesia microti*, *Plasmodium* spp.
- Immunization against encapsulated bacteria is essential.

Chapter 13

Growth of tissues and organs

Atrophy

Definition
Decrease in the size of an organ or tissue as a result of a decrease in size of the constituent cells and/or their number.

Physiological atrophy
- Remnant structures—during development (e.g. thyroglossal duct)
- Organs—after a physiological stimulus to hyperplasia/hypertrophy has been removed (e.g. uterus after birth, skeletal muscles after retirement from weight training).

Pathological atrophy
- Ischaemia (e.g. myocardium in chronic ischaemic heart disease)—the cells appear to decrease in size to reduce their metabolic needs in the face of ischaemia, to maintain survival, even if this is at the expense of some loss of function. Therapeutically, this can be important, since if the blood supply can be increased (e.g. by stenting or coronary artery bypass grafting), then the cells can increase in size again with a concomitant increase in function. (In the heart this is sometimes known as 'hibernating' myocardium)
- Immobility—skeletal muscle rapidly atrophies if immobilized
- Denervation—denervated tissues undergo general atrophy which is most marked in muscle
- General inadequate nutrition—if the body is starved of calories and protein, then protein is taken from skeletal muscle with consequent atrophy. This is starkly illustrated by pictures of malnourished people in famine conditions, but it should be remembered that this can occur in the immediate postoperative period in patients who do not have an adequate food intake but do have increasing nutrient needs
- Removal of endocrine stimulus—tissues that respond to hormones, and endocrine glands themselves, undergo rapid atrophy if the trophic hormone specific to them is removed. This is a very important clinical consideration in patients receiving long-term systemic corticosteroid therapy. Such patients will have atrophic adrenal glands because the exogenous steroids will stop secretion of ACTH. Thus, if the body has a sudden need for additional corticosteroids (e.g. during the stress of a major operation), the atrophic adrenal glands are unable to respond. In these conditions, medical staff have to play the role of the pituitary and prescribe increased doses of corticosteroids to cover the episode
- Ageing—atrophy certainly occurs in many tissues with ageing, but it is not clear whether this is due to factors associated with ageing, such as decreased mobility, or an intrinsic ageing process.

Atrophy and apoptosis
The pathological causes of atrophy can also stimulate apoptosis of cells if the stimulus is prolonged. This has the important implication that the tissue cannot return to its state prior to the stimulus, even if this is completely reversed.

Hypertrophy

Definition

The increase in size of an organ or tissue due to an increase in the size of cells in that tissue.

Requirements

- Organ or tissue where cells cannot divide
- Stimulus to cell division—usually increase in 'work', for example:
  - Skeletal muscle: biceps in weightlifters, quadriceps in sprinters
  - Cardiac muscle: marathon runners, patients with aortic valve stenosis.

Complications of hypertrophy

- Obstruction of adjacent tissue
- Infarction of tissue if it outgrows its blood supply.

Pathological hypertrophy

In some instances, hypertrophy occurs when no appropriate stimulus is present. This results in an organ which is oversized for the amount of 'work' it is required to do. The most illustrative example of this is hypertrophic obstructive cardiomyopathy ($\bigodot$ *OHCM10* p.152) where the heart muscle, especially in the septum, hypertrophies to a massive extent despite no extra work being demanded of the heart. This leads to two potentially fatal complications (arrhythmia main cause of sudden cardiac death):

- Myocardial ischaemia and infarction—as the mass of muscle outgrows its blood supply
- Obstruction to blood flow out of the left ventricle—with reduced blood pressure and syncopal acts (which may be relieved by the apparently paradoxical treatment with β-adrenergic receptor blocking drugs to reduce the strength of the cardiac contractions).

Hyperplasia

Definition

The increase in size of an organ or tissue due to an increase in the number of cells in that tissue.

Requirements

- Organ or tissue where cells can divide
- Stimulus to cell division—usually increase in 'work', for example:
  - Thyroid: thyroid-stimulating antibody in Graves' disease (◑ *OHCM10* p.218)
  - Lymph node: reaction to viral infection, e.g. infectious mononucleosis (◑ *OHCM10* p.405)
  - Prostate: unknown—?lifelong hormonal stimulus (benign prostatic hyperplasia) (◑ *OHCM10* p.642)
  - Skin (callous): repeated removal of upper layers of epidermis by abrasion/wear.

Complications of hyperplasia

- Obstruction of adjacent tissue (e.g. urethra in prostate, trachea behind thyroid)
- Infarction of tissue if it outgrows its blood supply.

Nomenclature of neoplasms

General

The naming of tumours is frustratingly inconsistent and, in the end, the only way to be sure about the names and significance of tumour names is to memorize them.

As a start, the word 'tumour' simply means a swelling (which could include swelling after trauma such as burns or contusions), so 'tumour' is not the best word to use as a generic term for growths such as cancers. A better term for these is neoplasms (literally new growth).

One also needs to bear in mind the nomenclature that the general public use—usually 'cancer' to mean any malignant growth. If a doctor uses the word 'cancer' or 'carcinoma', then the average patient tends to assume that this will be the most virulent form and will kill them in a few months—at least until you have explained the detail. It is better to start with some less emotive term so that all the details about treatment and prognosis can be explained without the patient being in a blind panic about imminent mortality, especially when the tumour is relatively innocuous, such as cutaneous basal cell carcinoma.

General naming conventions

See Table 13.1.

There are some 'rules' which make some of the names of neoplasms easier to decipher:

- All tumours tend to be denoted by the suffix -oma (e.g. carcinoma, lipoma). However, some reactive, non-neoplastic conditions also have this suffix (e.g. granuloma)
- Benign tumours of mesenchymal (connective tissue) origin tend to end in -oma (e.g. lipoma = a benign tumour of fat, rhabdomyoma = a benign tumour of striated muscle)
- Malignant tumours of mesenchymal origin tend to end in 'sarcoma' (e.g. sarcoma = generically a malignant tumour of mesenchymal origin, liposarcoma = a malignant tumour of fat, rhabdomyosarcoma = a malignant tumour of striated muscle)
- All malignant tumours of epithelial origin are denoted carcinoma. These are further subdivided into the specific pattern of differentiation, which usually represents the epithelial cell type where the tumour arose (e.g. adenocarcinoma in the glandular-lined part of the GI tract (stomach and more distally), squamous cell carcinoma on the skin and in the oesophagus, transitional cell carcinoma in the urinary tract)
- Benign tumours of epithelial origin may have names specific to their site but, generically, have names again related to the epithelial cell type at that site (e.g. adenoma = benign epithelial tumour in a glandular lined organ (such as the colorectum), papilloma = polypoid benign epithelial tumour of either squamous or transitional cell origin).

Eponymous terms for neoplasms abound and are very frustrating, e.g. renal cell carcinoma (which is a sensible descriptive name) is also known as Gravitz's tumour, which gives no information about the origin or behaviour of the tumour. To complicate matters, this particular tumour may also be known as hypernephroma because, histologically, its appearance is similar to the adrenal gland (which lies above the kidney).

Table 13.1 Tumour naming conventions

Tissue	Benign tumour	Malignant tumour
Glandular epithelium	Adenoma	Adenocarcinoma
Squamous epithelium	Squamous papilloma	Squamous cell carcinoma
Transitional epithelium	Transitional papilloma	Transitional cell carcinoma
Melanocytes	Naevus/lentigo	Malignant melanoma
Striated muscle	Rhabdomyoma	Rhabdomyosarcoma
Smooth muscle	Leiomyoma	Leiomyosarcoma
Fat	Lipoma	Liposarcoma
Nerve sheath	Neurofibroma	Neurofibrosarcoma
Lymphocytes	–	Lymphoma*
Glial cells	–	Glioma*

*These two tumours are examples of the inconsistencies of neoplasm nomenclature, where the name seems to indicate a benign tumour but where the tumour is actually malignant. Melanoma would be another example, but it is usually prefaced by 'malignant' in most medical literature.

The morphology of neoplasia

The pathway of progression from a normal tissue through to an invasive and metastasizing tumour is extremely complex and will be delineated by many changes, most of which will only be detectable at molecular level. However, there is a reasonably well-defined set of morphological changes that can be identified in sections under the light microscope, and which help our understanding of the pathways of carcinogenesis. These morphological changes are most easily seen in epithelial surfaces, in small tissue biopsies or in individual cells that have been scraped from or fallen from (exfoliated) a surface.

Metaplasia

The change in a cell type from one fully differentiated pattern to another fully differentiated pattern (e.g. bronchial epithelium from ciliated columnar to squamous epithelium in cigarette smokers).
• This is not necessarily a premalignant change but often represents a response to a deleterious environmental factor, such as cigarette smoke, which may itself be carcinogenic.

Dysplasia

See Fig. 13.1.
A term applied to epithelial surfaces when there is disorder of maturation of the cells. This is most clearly seen in an epithelial surface such as the squamous epithelium which covers the uterine cervix. The changes include:
• Division of cells above the normally proliferating basal cell layer, as evidenced by higher mitotic figures in higher levels of the epithelium
• Loss of polarity of the nuclei of the epithelial cells
• Lack of differentiation of the epithelial cells (e.g. squamous cells failing to produce keratin).

Carcinoma *in situ*

See Fig.13.2.
A term reserved for a severe degree of dysplasia in an epithelial surface where there is no discernible differentiation in cells between the base and top of the epithelium, but also no evidence of stromal invasion. As soon as cells breach the basement membrane at the base of the epithelium, then this term no longer applies and the lesion is an invasive carcinoma.

Microinvasive carcinoma

A pragmatic term which is used to describe tumours which have invaded through the basement membrane but only into the surrounding stroma to a small degree. This does not represent a distinct biological category but is used in selection of therapy for patients in relation to certain organ systems (e.g. microinvasive carcinoma of the uterine cervix can be treated by local resection rather than a formal hysterectomy and lymph node clearance).

Invasive carcinoma

Any tumour which has substantially invaded through the basement membrane of the epithelium. The important implication is that it can gain access to lymphovascular channels and metastasize to distant sites.

Fig. 13.1 Dysplasia. Fully differentiated squamous epithelium on the left; dysplastic epithelium on the right, showing loss of nuclear polarity and no differentiation towards flattened keratinocytes at the top of the epithelium.

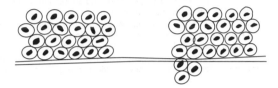

Fig. 13.2 Carcinoma *in situ* on the left, showing no difference in differentiation between the bottom and top of the epithelium. On the right, cells have invaded through the basement membrane so this is now invasive, rather than *in situ*, carcinoma.

Metastasis

Definition

Metastasis of a tumour is the spread of that tumour to a distant site in the body and its subsequent growth at that site. This is distinguished from extension into adjacent parts of the body by direct spread. To metastasize, tumour cells must travel to the distant site in either blood vessels or lymphatics, or by moving across a body cavity such as the pleura, peritoneum, or meninges. Most people who develop cancer, and subsequently die of it, do not die from the effects of the primary tumour (e.g. the breast cancer) because that has been completely removed by surgery when it was first diagnosed. They die of the effects of metastases in other parts of their bodies (e.g. replacement of the liver by metastatic tumour and death from liver failure). It is very important to know the mechanisms of metastasis, because the most effective cancer therapies will be those that are directed against these.

Processes in metastasis

See Fig. 13.3.
1. Detachment from adjacent epithelial cells—epithelial cells are normally attached tightly to adjacent epithelial cells by cell adhesion molecules. Cancer cells lose these attachments and can detach from adjacent cells.
2. Invasion through the basement membrane—until cancer cells have invaded through the basement membrane, a cancer is said to be *in situ*, with the implication that it does not have access to veins or lymphatics and so cannot metastasize. The basement membrane is a relatively impassable barrier made predominantly of collagen, so tumour cells must produce suitable enzymes (e.g. collagenases) to digest this. They also need some motility to pass through the damaged membrane.
3. Invasion through connective tissue—once through the basement membrane, the cancer cells must move through connective tissue, again digesting structural fibres that hinder their progress using appropriate enzymes such as collagenases.
4. Invasion into blood vessels/lymphatics—this again requires cell motility and enzymes to digest structural components in the vessel wall, although lymphatics have a very thin wall that is easily penetrated.
5. Survival in the blood vessel or lymphatic channel—once in the vessel, cancer cells are exposed to the host immune system including lymphocytes, which may recognize them as 'foreign' and destroy them through responses that include activation of natural killer cells. To evade such defences, tumour cells may aggregate tightly together so that the central tumour cells are protected, or shed proteins that are recognized as 'foreign' from their surface so that immune cells bind to these rather than the cell surface of the tumour cells.
6. Extravasation from the blood vessel or lymphatic channel—using the same mechanisms as intravasation.
7. Growth at the distant site—most cells in the body require some growth factors to induce growth and then prevent apoptosis. Such growth factors are often derived from surrounding stromal cells, but cancer cells often develop the ability to produce their own growth factors (autocrine production). A group of tumour cells can grow to

a diameter of 1mm, but to increase in size beyond that, new blood vessels need to grow into the tumour because the central cells can no longer be supplied simply by diffusion from the outside of the tumour. To do this, cancer cells produce angiogenic factors which induce the growth of capillaries from the surrounding stromal tissue into the tumour.

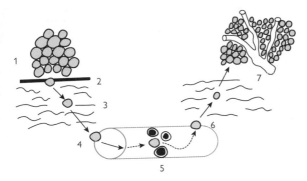

Fig. 13.3 The processes involved in metastasis. The numbers refer to the stages described in the text.

With all these processes required for a cancer cell to successfully metastasize, it can be imagined that few cancer cells achieve this. However, a primary tumour may contain millions of cells, so the probability that a few will successfully metastasize is quite high.

Therapeutic targets in the metastatic process

Any of the steps described can be targeted to try to block the metastatic process. At its earliest stage, it could be possible to cause cancer cells to produce the cell adhesion molecules that they have stopped producing and thus prevent detachment of some tumour cells. If the actions of tumour cell enzymes, such as collagenase, could be blocked at a local level, then this might prevent invasion through the basement membrane and connective tissue. At the moment, the most promising targets lie at the other end of the metastatic process, where drugs that block the development of blood vessels have shown much promise in experimental models.

Molecular mechanisms of neoplasia

There is no simple molecular mechanism for the development and progression of neoplasia—it is not a simple mutation of one gene that produces cells which grow too fast and metastasize. Instead, cancer arises through multiple gene mutations and/or silencing mechanisms (such as hypermethylation) in a population of cells and may be modified by the surrounding environment of those cells. The genes which are mutated will vary between different types of cancer, but will also show considerable variation between individual tumours of the same type. There are some general principles that apply to this process:

* Most mutations are not in the germ line but are acquired during an individual's life—hence the steady increase in cancer incidence with increasing age
* Although there may be a similarity in the pattern of gene mutations in the same type of cancer (e.g. mutation of the *HRAS* gene is common in colorectal cancer), each individual tumour has a unique pattern of genetic abnormalities
* In the later stages of tumour progression (which may be the time when the tumour presents clinically), there is widespread abnormality of the genome which may include substantial aneuploidy due to breakdown of chromosomal structure. Thus, there will be many genomic abnormalities in samples from these tumours and this may not be representative of the important early pathogenetic genomic changes.

The molecular mechanisms of neoplasia will become much clearer during the current decade, and the overall picture will become more integrated and cohesive. Until this knowledge is available, the description of these mechanisms is necessarily less comprehensive.

Proto-oncogenes

Genes in which overactivity (e.g. by a gain-of-function mutation) leads to the development of cancer are called proto-oncogenes. This name gives the impression that they are sitting in the genome waiting to mutate and cause cancer rather than fulfil a useful function in a cell. This is unfortunate, since they are genes which, generally, have a critical role in the normal control of cell growth and differentiation.

Ras genes as examples of proto-oncogenes

* *RAS* genes are mutated in about one in four human cancers
* *RAS* genes code for GTPase proteins which transmit signals from cell surface growth factor receptors
* Mutation in *RAS* genes can produce a protein which is always active, even when there is no signal from growth factor receptors, which induces uncontrolled cell proliferation
* Only one copy of a *RAS* gene needs to be mutated to have this effect. Thus, the mutation has a dominant effect.

Tumour suppressor genes

Genes in which underactivity (e.g. by a deletion mutation) leads to the development of cancer are called tumour suppressor genes. Again, analogous with proto-oncogenes, this name tends to imply that the main role of these genes in cell function is to actively suppress tumours. However, they have important functions in normal cell biology.

The retinoblastoma gene as an example of a tumour suppressor gene

- Retinoblastoma is a rare tumour of the eye which can be inherited or sporadic; the gene was discovered from studies of inherited cases
- The retinoblastoma gene codes for a protein which is involved in the control of the cell cycle
- Mutation in both copies of the gene to produce non-functional or absent proteins will lead to lack of regulation of the cell cycle and uncontrolled proliferation
- Individuals who inherit a mutant retinoblastoma gene only need to mutate the single normal copy of the gene to produce a tumour-producing phenotype
- The retinoblastoma gene is said to have a recessive pattern of action with respect to tumour formation because both copies of the gene have to be mutated before there is an effect.

The 'mutator' phenotype

Since the body is continually exposed to environmental carcinogens, and DNA replication during mitosis and meiosis produces many errors, it is actually surprising that cancer is not much more common and that individuals rarely develop more than one or two tumours in a lifetime. That this is so is testament to the efficiency of the cellular mechanisms which repair damage to DNA or prevent cells with DNA damage from replicating. If a cell acquires deficiencies in these protective mechanisms, it will gain more and more damage to its DNA, which may cause loss of even more protective mechanisms.

Cells which have an early loss of DNA damage detection and repair mechanisms are said to have a mutator (or replication error) phenotype because of their predisposition to acquire mutations.

HNPCC as an example of a mutator phenotype

- HNPCC = hereditary non-polyposis colorectal cancer
- Due to a deficiency of one of a group of DNA mismatch repair proteins (e.g. MSH1, MLH1)
- Individuals with HNPCC are born with one defective allele and the other allele is silenced by either mutation or hypermethylation
- Colorectal cancers in HNPCC occur at an earlier age, have a mucinous pattern of differentiation, and occur more commonly in the right side of the colon than sporadic colorectal cancer
- Tumours also occur at other sites including endometrium, ureter.

Commonly mutated genes in human cancers

p53

- Over 50% of human cancers have mutated *p53* genes
- Both copies of the gene contain deletion mutations in most breast, lung, and colon cancers
- Most mutations are acquired during life
- Individuals with a mutated *p53* gene in their germ line have a 25× risk of developing cancer (Li–Fraumeni syndrome)

- p53 protein is a nuclear protein that controls the transcription of other genes which mediate apoptosis and arrest of the cell cycle
- Levels of p53 increase with DNA damage, thus preventing cells from replicating damaged DNA
- If DNA damage is repaired, then cell allowed to continue in cell cycle
- If DNA damage not repaired, then cell induced to apoptose
- Therefore *p53* = tumour suppressor gene
- Non-functional p53 proteins lead to progressive accumulation of DNA damage in successive generations of cells.

APC

- *APC* = adenomatous polyposis coli gene
- About 80% of sporadic colorectal cancers show loss of both *APC* genes
- Individuals born with one deleted *APC* gene develop hundreds of adenomas in the colorectum by the age of 20 years, and inevitably develop colorectal cancer
- APC protein is located in the cytoplasm where, among other functions, it marks β-catenin for degradation ubiquitination
- Deleted APC → increased cytoplasmic β-catenin → increased nuclear β-catenin → increased transcription of genes which cause cellular proliferation
- Therefore *APC* = tumour suppressor gene.

Cancer chemotherapy

Definition

The treatment of malignant disease with antiproliferative agents (aka cyto-toxic chemotherapy). Until recently, there were no 'magic bullets' for cancer therapy. Most of these drugs cause non-specific DNA damage, ei-ther leading to cell death by apoptosis or preventing cell division. The ma-jority of cellular biochemical processes are identical in normal and malignant cells, but malignant cells are characterized by uncontrolled proliferation, and may fail to recognize and repair DNA damage. In contrast, normal tissues exposed to chemotherapy demonstrate temporary loss of proliferating cells, with a variety of side effects, but then recover through damage repair or replacement of cells from normal precursor or stem cells.

Classes of chemotherapy agents by mechanism of action

- DNA binding: direct alteration of DNA by alkylating agents (e.g. nitrogen mustard, cyclophosphamide) or platinum complexes (cisplatin, carboplatin)
- Antimetabolites: block synthesis of purines and pyrimidines, essential for DNA synthesis (e.g. 5-fluorouracil, methotrexate)
- Antimicrotubule: interferes with mitosis (e.g. vinca alkaloids, taxanes)
- Topoisomerase: inhibition leads to DNA damage (anthracyclines, e.g. doxorubicin; topo-I, e.g. irinotecan; topo-II, e.g. etoposide).

Chemotherapy side effects

Many such drugs affect organs which are dependent on cell renewal to maintain normal tissue integrity:
- Bone marrow: neutropenia, infection, thrombocytopenia, anaemia
- GI tract: nausea, vomiting, diarrhoea, mouth ulcers
- Skin: hair loss
- Gonads: infertility.

Others have more specific normal tissue effects:
- Antimicrotubule: peripheral nerve damage
- Anthracyclines: cardiomyopathy.

Chemotherapy dose and scheduling

Generally, increasing doses result in increasing cell kill, both tumour and normal tissue. Chemotherapy is commonly delivered at the highest safe dose once every 3–4 weeks, in order to allow normal tissue recovery, in particular bone marrow. However, some drugs (e.g. 5-fluorouracil) are ad-ministered continuously, at low dose, with relatively reduced side effects but increased tumour cell kill.

Combination chemotherapy

Successful eradication of some childhood leukaemias and adult lymphomas was achieved in the 1960s by combining three or more chemotherapy drugs. The principles behind combining these drugs are to use agents which are known to be active against the cancer, have different mechanisms of action, and different toxicity profiles, allowing safe delivery of each drug at full dose. However, even with such combined regimens, the majority of common cancers are not curable with chemotherapy alone.

Role of chemotherapy in different cancers

- Cure of advanced disease (e.g. lymphoma, leukaemia, testicular cancer)
- Cure of microscopic residual disease after surgery (e.g. breast and colorectal cancer)
- Palliative (non-curative) treatment of advanced disease (e.g. lung cancer).

Molecular targeted chemotherapy

Recent advances in understanding of the molecular changes in cancer cells have led to the development of tumour-specific treatments (e.g. imatinib for chronic myeloid leukaemia). This drug was designed to target the mutated protein (BCR-Abl tyrosine kinase) which drives leukaemic cell growth. Such treatments are highly effective in controlling malignant disease but also have little impact on normal tissues which lack the mutated protein.

Radiobiology

Definition

Study of the effects of ionizing radiation on normal and malignant cells. This is particularly important in the treatment of cancer with high-energy X-rays (radiotherapy). When an X-ray beam passes through living tissue, energy is absorbed, resulting in free radical generation and a variety of cellular effects, mainly due to DNA damage. This damage may be repaired, but if not repaired, may result in cell death or subsequent failure of cell division or cell survival with altered (mutated) DNA.

Normal tissues

The severity of damage to normal organs depends on the dose of radiation, measured in Gray (1Gy =1J kg^{-1}), and the volume of tissue treated. For each organ, a tolerance dose can be defined, below which full recovery will follow irradiation. Many tissues require continual cell renewal to maintain their integrity, and radiation exposure of these early responding tissues produces biological effects within 1–4 weeks. For example:

- GI tract: mucosal ulceration (e.g. sore mouth, diarrhoea)
- Skin: erythema and desquamation (similar to sunburn)
- Bone marrow: myelosuppression (white cell count, platelets, then red blood cells)
- Gonads: fall in sperm count, ovarian failure.

These acute effects recover within 2–6 weeks through normal cell proliferation—except loss of fertility, which may be irreversible even after low-dose radiotherapy.

Other tissues exhibit late responses, expressing damage months or even years after radiation, and for these, damage may be irreversible, e.g. lung fibrosis, spinal cord damage (myelitis).

Fractionation

The biological effects of a given dose of radiotherapy are markedly altered when the dose is administered in divided doses (fractions), with relatively greater sparing of normal tissue damage compared with cancer cells. Most curative radiotherapy uses multiple, daily, small fractions (around 2Gy/fraction) to a total dose of 60–70Gy. Small fraction size is particularly important in minimizing late radiotherapy damage to normal tissue. The overall treatment time for a course of radiotherapy may be adjusted by changing the dose per fraction and the number of fractions per day. For a given total dose of radiotherapy, shortening the treatment time increases the early effects on normal tissues but increases tumour cell kill. A short treatment time may be particularly important for fast-growing cancers.

Oxygen effect

Hypoxia causes relative resistance to radiation, and the abnormal vasculature supplying cancers can produce hypoxic areas within the cancer—a potential cause of failure to eradicate the cancer with radiotherapy.

Radiosensitivity of cancers

Malignant cells vary in their response to radiotherapy (Table 13.2).

Table 13.2 Radiosensitivity of cancers

Sensitive	Intermediate	Resistant
Hodgkin's	Breast	Melanoma
Seminoma	Lung	Glioma

Carcinogenesis

Normal somatic cells which survive radiation but sustain unrepaired DNA damage may, by chance, have DNA mutations which will eventually result in a malignant phenotype, producing a radiation-induced cancer. This process may take many years.

Clinical use of radiotherapy

- Curative treatment of macroscopic cancer (e.g. carcinoma larynx, cervix, prostate)
- Curative treatment of microscopic cancer after surgery (e.g. carcinoma breast)
- Palliative treatment of advanced cancer (e.g. carcinoma bronchus).

Monoclonal antibody therapy

Antibodies are raised against target cell-specific antigens, often using 'humanized' mice, generated by transferring human immunoglobulin genes into mice (transgenic mice).

Antibodies are usually then bound to a therapeutic agent (drug, enzyme or radioactive agent) for targeted delivery to the cells in question. Alternatively, the monoclonal antibodies increase the visibility of the neoplastic cells to the immune system, prompting antibody-dependent cell-mediated cytotoxicity.

Neoplastic cells have often developed complex defence mechanisms to suppress the innate immune system. Compound immunotherapy can be used to overcome the multiple levels of tumour defence against the immune system – sometimes known as 'checkpoint therapy'—each defence is referred to as a checkpoint.

Radioimmunotherapy is an alternative often used for blood-borne cancers such as lymphomas. In this case, the targeted antibody is linked to a radioactive moiety to improve selectivity of the radiotherapy for diseased cells.

Table 13.3 Some common and emerging immunotherapy drugs used in cancer

Mab	Alternative name	Target(s)	Mode of action
Bevacizumab	Avastin®	• Colorectal cancer • Non-small cell lung cancer • Metastatic renal cell cancer • Glioblastoma multiforme	• Inhibition of vascular endothelial growth factor (VEGF) important in tumour angiogenesis
Pertuzumab	Perjeta®	• HER-2 positive breast cancer	• Inhibits dimerization of HER2–HER3 and subsequent signalling via PI3K/Akt
Trastuzumab	Herceptin®	• HER-2 positive breast cancer	• Antibody-dependent cellular cytotoxicity after binding to HER2+ cells
Rituximab	Rituxan®, MabThera®	• Non-Hodgkin's lymphoma • Chronic lymphocytic leukaemia	• Antibody-dependent cellular cytotoxicity • Complement-mediated cytotoxicity • Apoptosis
Yttrium-90 Ibritumomab Tiuxetan	Zevalin®	• Non-Hodgkin's lymphoma • Other lymphomas	• Radioimmunotherapy
Sipuleucel-T	Provenge®	• Prostate cancer	• Antibody-dependent cellular toxicity

Chapter 14

Medicine and society

Introduction

Medicine cannot be practised without considering the societal contexts within which it exists: a new 'wonder drug' for hypertension will be ineffective if a patient does not take it; an expensive, complicated treatment for an advanced stage of a disease would be unnecessary if that disease could be prevented by more simple means; a patient who is efficiently diagnosed with gluten-sensitive enteropathy (coeliac disease) but who does not understand a doctor's description of sources of gluten in the diet will continue to suffer from the effects of that disease.

There also needs to be an overall view or 'tally' of disease in society to look out for the rise of new diseases (e.g. *Helicobacter*, HIV, severe acute respiratory syndrome (SARS), swine flu and COVID-19 in the past three decades) and to balance the need and availability of resources for particular diseases. These areas are covered by specialisms such as psychology, sociology, epidemiology, behavioural sciences, and health economics, but are considered here under the generic title of medicine and society.

Measuring disease in a society

An individual doctor working in a hospital is unlikely to spot significant changes in the rate of a specific disease in the locality because the rate at which that doctor sees patients is determined by the availability of resources such as clinic appointments. General practitioners are better placed to detect short-term rises in diseases such as influenza, but they will not be able to detect changes in less common diseases because they are likely to see only one or two cases per year. The specialism of epidemiology (and the related specialism of public health medicine) provide techniques for measuring the overall burden of disease in a society. There are a few definitions which are important:

- *Prevalence*—the number of people in a defined population who have a disease at any one time point, e.g. all those with rheumatoid arthritis
- *Incidence*—the number of people in a defined population who are diagnosed with a disease in a defined time period, e.g. the number of women diagnosed with breast cancer in a year.

It can be seen that the prevalence and incidence of a disease can vary considerably depending on the biological course of that disease. If a disease is chronic, e.g. persisting for years—such as rheumatoid arthritis, then there may be many people in a population with that disease, i.e. a high prevalence, but the incidence could be low. If the time course of a disease is rapid then there can be a high incidence but a low prevalence, e.g. an influenza epidemic will have a high incidence in 1 year but a very low prevalence because people recover from flu within 10 days.

Epidemiological methods can detect changes in the incidence/prevalence of a disease in a particular population and this may lead to further investigations into the pathogenetic mechanisms of that disease. It is important here to recognize the differences between causes of disease and risk factors for a disease:

- *Cause*—is a factor which has been scientifically proven to be an integral part of the mechanism of causing a disease, e.g. cystic fibrosis is caused by a mutation in the cystic fibrosis transmembrane conductance regulator gene.
- *Risk factor*—is a factor which is identified as being statistically significantly associated with a disease but has not yet been proven to be an integral part of the disease mechanism. Risk factors are often identified in epidemiological studies, and widely reported in the popular media (e.g. 'a daily glass of red wine protects against prostate cancer'), but their association may only occur because they themselves are closely associated with another factor that is a real cause of the disease. Thus, risk factors identified by epidemiological studies then need laboratory studies, using the classical scientific method of proving or disproving the null hypothesis, to show whether they are real causes of the disease in question.

Detection of disease

In order to calculate the prevalence or incidence of a disease there have to be methods of detecting that disease in a population. Methods include:

* *Specific surveys*—e.g. healthcare workers going out into a population and performing tests that detect a disease, such as sputum testing for tuberculosis
* *Notifiable diseases*—in many countries there are specific disease which must be notified to authorities by healthcare workers whenever the disease is diagnosed. In the UK, such diseases include poliomyelitis, anthrax, cholera, tuberculosis, whooping cough (all infectious diseases)
* *Cancer registries*—in many countries there is a flow of information from hospitals to registries whose aim is to record all cases of cancer and follow-up of those cases. There are obvious problems in establishing reliable flows of information so the figures are not wholly accurate
* *Death certificates*—death certificates include a doctor's opinion of the cause of death of the person and this might be regarded as a useful source of information on diseases that cause death. However, many studies have shown that unless an autopsy has been performed, this information can be inaccurate, often misidentifying the pathology within a body system or even incorrectly implicating failure of a particular system as the cause of death
* *Industrial diseases*—there are some diseases which are specifically caused by work in industry which are recorded in government statistics, e.g. asbestos-related lung disease, including mesothelioma and asbestosis.

Assessing the effectiveness of treatment of a disease

Another important aspect of public health medicine/epidemiology is measuring and assessing the effectiveness of a treatment of a disease. This can only be done by taking a society-wide view of the benefits and costs of a treatment. Many aspects of this are controversial and institutions that have been set up to advise on or regulate new treatments, e.g. the National Institute for Health and Care Excellence (NICE) in the UK, are often subject to much pressure from patient groups and specialist doctors. In assessing treatments, a number of factors have to be taken into account:

* *The effectiveness of the treatment*—if a treatment is prescribed to the correct type of patient for the correct type of disease how effective is the treatment? Some treatments are almost self-evidently highly effective, e.g. administration of thyroxine for hypothyroidism and other hormone-replacement therapies, an antibiotic to which a bacterial organism is sensitive in a patient with septicaemia (large randomized controlled trials were not required to demonstrate the effectiveness of penicillin). Other treatments are much more difficult to assess and do require large randomized controlled trials with long-term follow-up, e.g. antihypertensive drugs, cancer chemotherapy
* *The cost of a therapy*—this is important because extremely expensive treatments that only produce a small benefit may be too costly to be accepted as routine treatment in a state-funded health service. The whole cost of the treatment needs to be calculated and not just the simple cost of the drug, e.g. an anti-cancer drug (such as the new monoclonal antibody therapies) may cost £10,000 per patient per year but this cost may increase to £30,000 a year if the additional costs of administering the drug (e.g. inpatient admission for intravenous infusion) and monitoring the patient (e.g. blood tests to detect neutropenia) are included
* *The resources required to administer the treatment*—even if the financial cost of a treatment is thought to be good value when the benefit of the treatment is assessed it may be that there are insufficient resources available at that point in time to allow the treatment to implemented across a whole healthcare system, e.g. insufficient nurses trained to give chemotherapy.

Some of the measures used to measure the benefit of a therapy include:
* *Life expectancy*—the average number of years of life predicted to be left for a person of a particular age. Thus, treatments can be evaluated by the number of years that they increase life expectancy in patients. However, this does not take into account the quality of those 'extra' years
* *Quality-adjusted life years (QALYs)*— a widely used measure that takes account of the amount of extra years of life a treatment will on average give a patient and the quality of life of those extra years. A treatment that gave 10 years extra life with no reduction in the quality of life in those years could be described as giving 10 QALYs. A treatment that gave 10 years of extra life but with reduced quality of life (e.g. amputation of a leg for a malignant tumour) would be described as

having <10 QALYs but >0 QALYs and the point which it lies between the two points would depend on a fairly subjective assessment of the reduction is quality of life, e.g. if it was thought to reduce the quality of life by 20% it would produce $10 \times 0.8 = 8$ QALYs. QALYs are a more useful measure of benefit than raw life years but the quality adjustment can be difficult to assess

- *Disability-adjusted life years (DALYs)*—if a disease does not cause premature death, then QALYs will not be a useful measure of benefit (it will always be zero), so the concept has been extended to DALYs, where the number of years with/without a disability is included. This is very useful for chronic diseases that cause a significant reduction in the quality of life of a person but do not usually cause premature death; such diseases include osteoarthritis, rheumatoid arthritis, and dental caries.
- *Healthy life years (HLY)*—a variant of QALYs and DALYs, describing the number of disease-free years of life. It is often used in calculating composite 'health' scores for comparisons between countries, especially in the European Union.

The sociocultural context of medicine

The doctor–patient relationship

A hypothetically 'technically perfect' doctor who could diagnose any disease with which a patient presented and prescribe optimal treatment would be very ineffective if he or she did not establish a satisfactory working relationship with his or her patients. It is no use being able to diagnose diseases if you cannot explain that diagnosis in a way that the patient can understand and act upon. Treatment will be ineffective if a doctor does not empathize with a patient sufficiently well to understand the context of the illness within the patient's overall life situation. In fact, a doctor is unlikely to make the correct diagnosis if he/she does not establish a satisfactory working relationship because he/she will not elicit all the symptoms and signs that will point to the correct diagnosis. Many aspects of the doctor–patient relationship are best learnt in simulated and real-life interviews with careful guidance, but there are some ground rules that can be learnt before this happens. A very useful formulation is that given by the General Medical Council in the *Good Medical Practice* guidelines:

'To fulfil your role in the doctor–patient partnership you must:
a. Be polite, considerate and honest
b. Treat patients with dignity
c. Treat each patient as an individual
d. Respect patients' privacy and right to confidentiality
e. support patients in caring for themselves to improve and maintain their health
f. Encourage patients who have knowledge about their condition to use this when they are making decisions about their care.'[1]

One interesting area of the doctor–patient relationship is determining the reason that the patient has come to consult the doctor, especially in general practice, which is usually the first port of call for any patient. While a doctor could assume that the patient has come to consult them about the illness which the patient is describing to them, this would often be incorrect as demonstrated by sociological research, such as that of Irving Zola, who identified five triggers to consultation[2]:

1. The occurrence of an interpersonal crisis.
2. The perceived interference with social or personal relations.
3. 'Sanctioning' (pressure) from others.
4. The perceived interference with vocational or physical activity.
5. The 'temporalizing of symptomatology'—setting a deadline for action (often manifest as a rush of consultations just before Christmas in the UK).

While these are at least tangentially related to the 'illness' that the patient is describing, the doctor needs to find out which are these are the most relevant to this particular patient, because this will influence the advice and treatment that is recommended. For example, a man may present with the

---

1 General Medical Council (2006). *Good Medical Practice*. London: General Medical Council, pp. 20–1.

2 Zola IK (1966). Culture and symptoms—an analysis of patient's presenting complaints. *Am Sociolog Rev* **31**, 615–30.

symptoms of urinary outflow obstruction due to benign prostatic hypertrophy which is causing little current interference with his life but he has come at this point in time because his wife's brother has just been diagnosed with prostate cancer. In such circumstances, this man may not need immediate therapy for his prostatic hypertrophy but may need the 'reassurance' of a prostate-specific antigen blood test to reduce the probability of a diagnosis of prostate cancer.

The 'clinical iceberg' is a related concept which points out that there is often little difference between the symptoms experienced by people in the general population who do not go to consult their doctor and those who do. For example, levels of pain from osteoarthritis are similar, so there must be some other precipitating factor for the consultation.

Illness behaviour

It is easy to observe different patterns of behaviour that people exhibit when they have an illness, even from friends and associates when they contract a viral respiratory infection ('man flu' etc.). These patterns of behaviour will be influenced by a person's upbringing and current life situation, including employment and religious beliefs. However, there are some generic patterns of behaviour that can be usefully identified because they provide some information which may be useful when treating a patient with that pattern of behaviour:

- *Denial*—a pattern of behaviour where the patient does not consciously acknowledge that they have the disease from which they are clearly suffering or by their behaviour, do not accept the presence of the disease. This may be seen as a beneficial adaptive behaviour when faced with an imminent terminal illness, although acceptance of such a diagnosis may be more beneficial for the person and their family. May occur with other illnesses at different times in a person's life. For example, 'denial' of type 1 diabetes mellitus in teenage life with refusal to take insulin and repeated hospital admissions in diabetic ketoacidosis
- *Medicalization*—a behaviour pattern where all symptoms and behaviours are seen as having a strong relationship to defined medical illness rather than more 'rational' explanations such as significant life events and so on. Doctors and other healthcare professionals may well induce this behaviour in patients by overemphasis on the medical aspects of a patient's overall life situation. It differs from hypochondriasis in that it does relate to a real, rather than imagined, illness or state. There is long-standing debate over whether the medical profession has turned normal life events, such as childbirth, into 'diseases' by the process of medicalization
- *Risk taking*—this is behaviour which threatens a person's health because they are not making rational decisions about the risks attached to a certain behaviour. Of course, this varies from the very obvious, e.g. unprotected sexual intercourse with multiple partners in a population with a high prevalence of HIV infection, to more widespread and subtle behaviour such as being overweight/obese over a long period of time despite knowledge of the health risks attached to this state (which many doctors themselves exhibit)

- *Hypochondriasis*—a strongly held belief by a person that they have a particular disease in the face of all evidence to the contrary. Can be mild and as such be adaptive behaviour to an unsatisfactory life situation, but can be so strongly held as to constitute a monoideistic delusion.

Further information

Farmer R, Lawrenson RE, Miller D (2004). *Epidemiology and Public Health Medicine*. Oxford: Blackwell Publishing.

Morrison V, Bennett P (2006). *An Introduction to Health Psychology*. Edinburgh: Pearson Education.

Scrambler G (2008). *Sociology as Applied to Medicine*, 6th edn. London: Elsevier.

Techniques
of medical sciences

Molecular genetics: introduction

Molecular genetics is the study of the structure and function of genes at a molecular level. This includes both the use of genetics to understand inheritance patterns, and also molecular biology to investigate why certain mutations cause particular traits or diseases.

Genomics is the application of automated oligonucleotide sequencing and computerized data retrieval and analysis to sequence an organism's entire genetic complement.

- Of obvious interest to the medical sciences field is the Human Genome Project which had sequenced the entire human genome by the early 2000s, and subsequent whole genome sequencing projects such as the 1,000 and 100,000 Genome Projects
  - Readily searchable database of gene sequences, annotated with biological information (and protein function if known), are freely available

Genome-wide association studies (GWA studies or GWAS), also known as whole genome association studies (WGA studies or WGAS), are the comparison of the genomes of individuals of the same species to identify inter-individual variations in the form of single nucleotide polymorphisms (SNPs; ➋ p.222). These differences can then be associated with certain traits, such as diseases, when the genetic variations are statistically more frequent in individuals with the trait/disease. This technique has followed genes to be 'associated' with particular diseases, such as diabetes, helping to characterizing the molecular pathway of the diseases.

One recent approach has been to combine GWAS with metabolomics, the systematic study of small-molecule metabolite levels in samples such as plasma and urine, allowing SNPs to be associated with particular changes in the metabolome (i.e. the entire set of small metabolic within a biological sample). This can suggest potential functions for uncharacterized genes (functional genomics) based on the phenotype of metabolite changes.

Electrophoresis

Electrophoresis is the separation of complex mixtures of solutes (e.g. peptides, proteins, DNA, RNA) by virtue of their electrical charge and size, with separation, usually enhanced by using a solid matrix with sieving properties.

- Agarose gel electrophoresis is commonly used to separate fragments of DNA or RNA:
  - The solid matrix used is agarose (a polysaccharide extracted from algae (seaweed)):
    - The degree of size separation depends on the percentage of agarose, but usually 1–2% (weight/volume of buffer) is appropriate
  - The agarose gel is submersed in a tank of the same buffer, usually a solution of Tris-acetic acid-EDTA (TAE) or Tris-boric acid-EDTA (TBE)
  - Samples are loaded into wells pre-formed in the gel, with loading dye to increase the density of the solution so it stays in the wells, and to allow visualization of how far migration has progressed
  - An electrical field is applied across the gel, and the samples move towards the positive electrode due to the overall negative charge of the DNA/RNA molecule—the smaller the component, the further it will travel in a given period of time
  - The DNA/RNA fragments can be visualized by staining with ethidium bromide (traditionally, but more recently alternative less toxic DNA stains), either added to the gel or post running, which binds to DNA/RNA and fluoresces under ultraviolet light
  - Fragment size can be estimated by running standards (a DNA or RNA ladder) on the gel
- Electrophoresis can also be used to separate proteins, a process known as SDS-polyacrylamide gel electrophoresis or SDS-PAGE:
  - The solid matrix consists of linear chains (multimers of polyacrylamide) cross-linked with bis-polyacrylamide molecules, formed in reactions catalysed by ammonium persulphate and TEMED (N, N, N´, N´-tetramethylethylenediamine)
  - The percentage of polyacrylamide will affect the separation (around 10% is suitable for protein up to around 100kDa), or gradient gels (e.g. 10–20%) can be used to increase the range of separation
  - Sodium dodecyl sulphate (SDS) is included to denature the protein and to coat it with negative charges (approximately two SDS molecules per amino acid), so that the proteins will separate purely on molecular mass
  - Protein sizes can be estimated by running standards on the gel.

DNA cloning

The technique of gene cloning has allowed individual gene products to be studied in detail, both at a functional and a structural level.

- DNA cloning involves the amplification of a particular DNA fragment:
  - If the sequence is known, it can be amplified using gene-specific primers by the polymerase chain reaction (PCR) described on **➔** p.960
  - If part of the sequence is known, a labelled DNA probe to this fragment can be used to identify the gene
  - If there is an antibody to the protein, cells transfected with the gene can be identified by screening
  - If the function of the gene is known, it can be identified by testing cells transfected with the gene for that particular response
- The usual starting material for these types of studies is a cDNA library:
  - This is constructed by isolating all the mRNA from a particular cell that is known to express the protein of interest
  - This is converted into cDNA using the enzyme reverse transcriptase (RT):
    ○ The primer for the RT reaction is often a 15–18mer of poly(T) that hybridizes to the poly(A) tail of the mRNA
    ○ The cDNA is then cloned into a vector using restriction enzymes (REs) and amplified in *Escherichia coli*
    ○ These clones can then be screened to identify which of them contains the gene of interest (by the methods previously mentioned)
    ○ To represent all of the mRNA that was present in the cell, the cDNA library needs to have many thousands of clones.

REs and vectors are essential tools for DNA cloning.

- Isolated from bacteria, REs cut DNA at specific sequence of four or more bases:
  - Can produce either 3´ or 5´ overhangs ('sticky ends') which can be joined by DNA ligase to a compatible DNA fragment that has a complementary overhang, or 'blunt ends' which have no overhang and can be ligated to any other blunt end
  - The joining of two unrelated DNA fragments results in a recombinant DNA molecule
- Vectors (or replicons) are carrier DNA that can independently replicate in a host organism to produce multiple copies of itself:
  - Commonly used vectors are plasmids (replicate in *E. coli*). Plasmids need to have certain features: an origin of replication, gene coding for antibiotic resistance for selection of transformed cells, multiple cloning site (MCS: a number of unique RE sites close together that allow the insertion of the gene of interest), promoter(s) flanking the MCS to allow the cloned gene to be expressed
  - Other types of vector include bacteriophages, cosmids, bacterial artificial chromosomes (BACs), and yeast artificial chromosomes (YACs):
    ○ Cosmids, BACs, and YACs have the advantage of being able to take larger inserts than plasmids and bacteriophages
    ○ YACs have been useful in mapping eukaryotic genes which can be up to several million base pairs in length.

If the sequence of a gene is known, it can be amplified using PCR (Fig. 15.4; ➋ p.960).
- The PCR product is visualized by electrophoresis on an agarose gel (➋ p.953)
- PCR is often used to look for gene expression by first generating cDNA from mRNA isolated from a particular tissue—a technique known as RT-PCR (➋ p.964):
  - In combination with sequencing, RT-PCR can be used as a diagnostic tool for diseases involving known point mutations
  - These molecular techniques are also highly valuable in a research environment.

There are a number of techniques that allow the recognition of either nucleic acid or protein from samples:
- Northern blotting (➋ p.990) uses a labelled antisense DNA probe to look for copies of mRNA which have been separated by size on a gel and blotted onto a filter. The probe will hybridize to its complementary message allowing determination of the size and quantity of the mRNA transcript in the sample
- Southern blotting (➋ p.961, Fig. 15.3) is similar to northern blotting except that the template is RE-treated DNA rather than mRNA
- Western blotting involves separating proteins in a sample by size on a denaturing polyacrylamide gel and transferring them across onto a membrane (usually made of PVDF) by electrophoresis:
  - The membrane is then probed using an antibody to the protein(s) of interest
  - The antibody binding can be quantified using an enzyme-conjugated secondary antibody (e.g. anti IgG-horseradish peroxidase) that can be assayed colourimetrically, in a process similar to that for an ELISA (Fig 15.12, p.984).
  - This technique allows the identification/quantification of individual proteins from a mixture such as a cell or tissue extract.

DNA sequencing

Two different techniques of DNA sequencing were developed in the late 1970s—the Maxim–Gilbert and the Sanger methods. The Sanger method is the one commonly used today, and has been automated for high through-put sequencing projects such as the Human Genome Project.

Maxim–Gilbert method of chemical cleavage

- Double-stranded DNA is cut into fragments and the end of one strand labelled with a ^{32}P-dNTP using part of *E. coli* DNA polymerase I (the Klenow fragment)
- The DNA is chemically treated in separate reactions so as to induce it to break into fragments at each of one of the bases, using conditions that only cause a couple of breaks per molecule
- Statistically, therefore one should get all the size fragments that end with such a base
- Running these on a polyacrylamide gel will separate these fragments and allow the sequence to be read from an autoradiograph (photograph on X-ray film).

Sanger method of interrupted enzymatic cleavage

- Based on the random interruption of the synthesis of a DNA strand in a reaction analogous to the PCR
- Four separate reactions are run, each with dNTPs and a single ddNTP species:
  - ddNTP is a dideoxynucleoside triphosphate, with an H on C3 of the sugar moiety (rather than the OH in a dNTP)
  - These can be incorporated into DNA by a polymerase but not extended due to the lack of a 3′OH for the next nucleotide
  - The dNTPs and ddNTP are at such concentrations that the dNTP is incorporated randomly at every position
  - The ddNTP is labelled so that the differently sized fragments produced can be visualized
  - Traditionally, ^{32}P-ddNTPs were used, but now it is more common to use ddNTPs that are labelled with fluorescent molecules, a different colour for each base
- The reactions are run on polyacrylamide gels to separate the fragments. Either an autoradiograph is taken (for ^{32}P-ddNTPs) of the gel (Fig. 15.1), or it is read with a laser in an automated sequencer for the more commonly used fluorescent ddNTP method (Fig. 15.2)
- Direct sequencing of PCR products from patient tissue has allowed the relatively easy diagnosis of genetic diseases:
  - One or more small regions of genomic DNA can be amplified by PCR
  - These fragment(s) can be sequenced directly to determine if a mutation exists in the region(s) of the gene amplified
  - Has the advantage of being quick, accurate, and relatively cheap
  - Has been used for diagnosis of, e.g. haemophilia, cystic fibrosis.

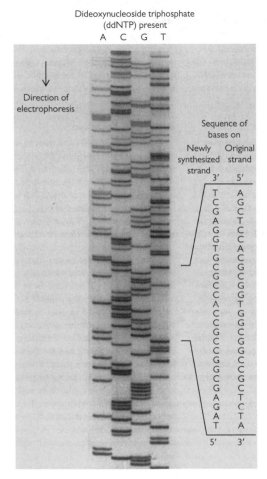

Dideoxynucleoside triphosphate
(ddNTP) present
A C G T

Direction of
electrophoresis

Sequence of
bases on

Newly   Original
synthesized   strand
strand
3'   5'

Newly synthesized strand (3')	Original strand (5')
T	A
C	G
G	C
A	T
G	C
G	C
T	A
G	C
C	G
C	G
C	G
A	T
C	G
C	G
G	C
C	G
C	G
G	C
G	C
C	G
G	C
A	T
G	C
A	T
T	A

5'   3'

Fig. 15.1 Photograph of autoradiograph of polyacrylamide sequencing gel.

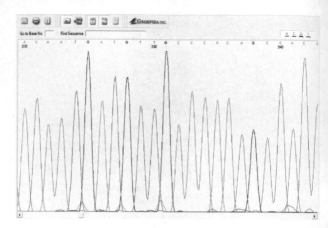

Fig. 15.2 Example of DNA sequencing trace using fluorescent ddNTP technique.
Image courtesy of David Meredith.

Techniques involving DNA

- Southern blotting:
  - Southern blot analysis involves cutting DNA with RE(s), running the fragments on an agarose gel, transferring onto a membrane, and probing with a labelled complementary DNA probe to the gene of interest (Fig. 15.3)
  - Useful in genotyping, e.g. gene knock-out animals
- PCR: allows detection of a specific DNA by amplifying it using primers (short sequences of synthetic DNA (oligos) that bind to a specific sequence of target DNA) and thermostable DNA polymerase in a thermocycler machine (Fig. 15.4). PCR products are usually analysed by agarose gel electrophoresis (➲ p.953), providing a non-quantitative estimate of DNA content. There are, however, quantitative PCR methods such as real-time PCR (qPCR, ➲ p.964):
  - Two developments made PCR a much more feasible and routine technique:
    - ○ The isolation and purification of thermostable DNA polymerase from bacteria that have evolved to live in hot springs and geothermal vents in the ocean. The DNA polymerase is able to withstand repeated periods of temperature of 95°C and has maximal processivity at 72°C (extension rates of ~1000bp/min)
    - ○ The manufacture of thermal cyclers ('PCR machines'): programmable heating blocks which can rapidly change temperature, thus automating PCR
- DNA fingerprinting:
  - DNA fingerprinting is a technique used in forensic science to identify whether a sample found at a crime scene matches that from a particular individual:
    - ○ DNA is isolated and digested with REs
    - ○ The fragments are separated by electrophoresis on an agarose gel and visualized by ethidium bromide staining
    - ○ Due to the sequence variability (polymorphisms) and differences in the length of multiple repeat sequences in introns, the band pattern is unique to an individual
  - Closely related individuals will have less polymorphisms in their genomes
  - This allows a similar approach to be used for paternal testing
  - A PCR-based technique can also be used with primers that hybridize with strongly conserved sequences but spanning variable number repeat regions
- PCR screening for disease:
  - Where a disease is known to be caused by a specific mutation in a gene, it is possible to use PCR to screen for carriers and even to detect mutations in cells taken from a foetus *in utero*:
    - ○ A good example of this is the most common mutation causing cystic fibrosis, the loss of three bases resulting in an amino acid deletion (ΔF_{508})
    - ○ A PCR product made from primers spanning this region of the gene will be three bases shorter in cystic fibrosis than wild-type
    - ○ This difference can be seen when the PCR products are run on a suitable gel.

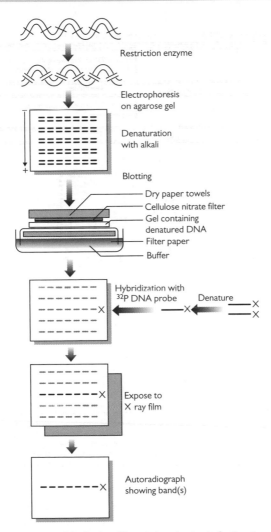

Fig. 15.3 Diagram of the Southern blot technique showing site fractionation of the DNA fragments by gel electrophoresis, denaturation of the double-stranded DNA to become single-stranded, and transfer to a nitrocellulose filter.

Reproduced with permission from Mueller RF, Young ID (2003). *Emery's Elements of Medical Genetics*, 11th edn. Churchill Livingstone.

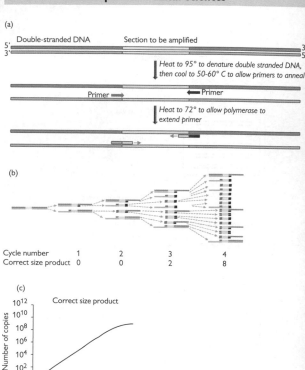

Fig. 15.4 Principles of PCR. (a) The three stages of the PCR cycle. (b) There is exponential amplification of the region of interest, whereas longer PCR products undergo linear amplification. Thus, after several cycles the correct sized product predominates. (c) There is a linear region of amplification, followed by non-linear region as reagents are exhausted. Classical PCR is non-quantitative and usually analysed by electrophoresis on an agarose gel and visualization by ethidium bromide staining.

Techniques involving RNA

- Northern blotting is essentially the same in principle as Southern blotting (Fig. 15.3; ➔ p.961), except that the starting material is isolated mRNA and the probe is usually single-stranded DNA:
  - mRNA can be isolated as its poly(A) tail will bind to oligo-d(T) based resins
- RNA techniques involving reverse transcription:
  - Most other techniques involving RNA require conversion of mRNA into DNA (termed cloned or cDNA). This can be done using an oligo-d(T) primer or random hexamers and the enzyme reverse transcriptase (RT)
  - Once mRNA has been converted into cDNA, it is possible to amplify specific sequences using PCR. The whole process is often called RT-PCR. This can be used to see if a particular gene is expressed in a cell type or tissue
  - Classic PCR is not quantitative
  - However, PCR can be used to quantify the level of mRNA expression using variations of the technique:
    - Semi-quantitative PCR: known amounts of an engineered DNA that binds the same primers but gives a different sized product to the wild-type are added to the sample
    - Comparing the intensity of the reaction bands on an ethidium bromide agarose gel allows estimation of the original mRNA concentration
    - Real-time PCR (qPCR or RT-qPCR when coupled to reverse transcription for RNA quantification): the products of the PCR reaction (amplicons) are detected quantitatively as they are made (i.e. in 'real time') during the exponential phase of amplification, rather than at the end of the reaction, This requires fluorescent detection reagents, which can be either dyes binding to double-stranded DNA (e.g. SYBR green) or additional fluorescent probes (e.g. TaqMan probes). RT-qPCR allows a very accurate estimation of the number of copies of mRNA in the starting sample[1]
- The genes expressed by a particular cell or tissue can be seen using an expressed sequence tag (EST) library:
  - RT-PCR is carried out to make a cDNA copy of all the genes expressed
  - The resulting cDNAs are sequenced and the results deposited in a database
  - This allows the genes to be identified, and the number of clones of a gene indicate its relative frequency of expression
- DNA microarrays:
  - A technique to detect changes in gene expression at the mRNA level in cells involves the use of DNA microarrays, also known as 'gene chips'
  - DNA microarrays have oligonucleotides representing up to the entire gene complement spotted on them in an ordered array

1 Bonetta L (2005). Prime time for real time PCR. *Nat Meth* **2**, 305–12.

- cDNA is prepared (e.g. from two sets of the same cells grown under different conditions), with one set labelled with red dye (condition 1) and the other green (condition 2)
- Both cDNA samples are hybridized to the gene chip
- Scanning with lasers reveals which genes are expressed under which condition:
  - Red spot signifies only condition 1
  - Green spot, only condition 2
  - Yellow spot, both conditions
  - Variations in between, e.g. orange, would indicate expression in condition 1 > in condition 2
- Computer analysis allows identification of genes up- and down-regulated
- Care must be used in interpreting microarray data in isolation: this technique does not provide any information as to whether changes in transcription detected are translated into functionally relevant changes in protein expression
- RNAase protection assay:
  - The RNAase protection assay (or nuclease protection assay) is used to identify the presence of and to quantify individual species of RNA in a heterogenous sample of RNA isolated from a cell
  - The extracted RNA is hybridized with an antisense RNA or DNA probe complementary to the sequence(s) of interest, forming a double-stranded RNA or RNA–DNA complex
  - The probe is generated by cloning a fragment of the gene of interest into a plasmid and using an RNA polymerase to synthesize an antisense RNA transcript:
    - It is labelled by including radioactive (usually ^{32}P) uridine 5′-triphosphates in the reaction mixture
  - After hybridization, ribonuclease enzymes that non-specifically cleave only single-stranded RNA (or S1 ribonuclease that cleaves only single-stranded DNA if a DNA probe is used) is added, which digests any free probe, ultimately to free nucleotides
  - The reaction mixture is separated by electrophoresis on a polyacrylamide gel, and the presence of intact probe is seen as a distinct band on a radiograph
  - This technique is very sensitive but does not reveal the size of the native RNA, as the size of the band represents the size of the probe, and not of the cellular RNA (northern blotting allows size determination, ⊃ p.964)
- *In situ* hybridization (Fig. 15.5):
  - This technique can be used to localize a particular mRNA species in:
    - Fixed tissue, either sectioned or a whole sample (e.g. an embryo)
    - Unfixed tissue (e.g. cryosections of fresh, frozen brain tissue)
  - It reveals which cells are expressing the mRNA (unlike PCR where a tissue/cell sample is homogenized)
  - The tissue sections are mounted onto microscope slides
  - They are then permeabilized either by detergent (e.g. Tween) or enzyme (proteinase K) treatment to allow the labelled probe access to the cellular mRNA:
    - The probe is an antisense RNA made by *in vitro* transcription

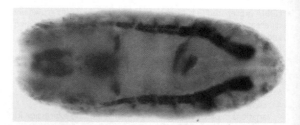

Fig. 15.5 Expression of the amino acid transporter gene, *path*, in a late-stage whole-mount *Drosophila melanogaster* embryo (dorsal view), analysed by *in situ* hybridization. *Path* is expressed ubiquitously, but at higher levels in specific tissues (as highlighted here in the midgut, brain, and sensory nervous system).

Image courtesy of Deborah Goberdhan, Department of Physiology, Anatomy and Genetics, University of Oxford.

- ○ Labelling of the probe used to be either with ^{32}P or ^{35}S (with radiography used to detect signal). It is now nearly always non-radioactive, such as with digoxigenin, biotin, or fluorescein
- ○ Detection of the probe is with an antibody (immunohistochemistry; ➲ p.984) visualized either by an enzymatic colorimetric reaction or fluorescence, or by direct fluorescence of fluorescein-labelled probes
- Fluorescent *in situ* hybridization (FISH) is an important subtype of *in situ* hybridization used for chromosome mapping:
  - A specific fluorescently labelled DNA probe is used to identify specific chromosomes or chromosomal regions (e.g. can check the number of copies of chromosome 21 as a test for Down syndrome).

Proteomics: introduction

The sequencing of the human genome by the year 2000 was a major scientific achievement and has provided a huge amount of invaluable information. We now know approximately how many functional genes there are in the human genome (~25,000), their location, and, in some cases, their function. However, this knowledge now needs to be integrated with knowledge about the proteins which these genes produce, how these proteins interact, and how their expression varies in different types of cells and in disease.

This has led to a need for techniques that can identify and quantitate proteins in cellular samples—the science of proteomics. The number and scope of these techniques is increasing all the time. The following includes a short description of the most popular current techniques—protein extraction and mass spectrometry.

Protein extraction

To analyse the proteins in a sample, they must first be extracted from the cells within which they are contained.

Two-dimensional gel electrophoresis

This is a simple and popular technique for separating a mixture of proteins into its constituents (Fig. 15.6):

- The protein extract is mixed with a non-charged detergent, β-mercaptoethanol and urea to solubilize, denature, and disassociate all the proteins
- The resulting mixture is put in a tube of polyacrylamide gel (PAGE) with a gradient of pH across it, and an electric current is applied
- Each protein migrates to its isoelectric point, where the pH of the surrounding gel is such that the protein has no overall positive or negative charge and so does not move
- This can produce resolution into about 50 protein bands
- The tube of gel is soaked in a solution containing the detergent SDS, which is highly negatively charged and so masks the native charge of a protein, making the overall charge now proportional to molecular weight
- The tube of gel is attached to a slab of polyacrylamide gel and an electric current is applied across the gel in a direction at 90° to the isoelectric focusing
- The proteins now migrate according to molecular weight
- This can produce resolution into >1000 proteins
- Proteins may be visualized by stains (such as Coomassie or silver) or by autoradiography (if the proteins have been radioactively labelled *in vivo*)
- A sample will typically produce thousands of individual spots and it would be difficult to identify any difference in pattern between samples by simple inspection. Images of these gels are scanned into a computer system which then compares gels from two different samples and identifies those spots which are different in the two

• In this way, for example, the differences between proteins in normal tissue and a cancer arising from that tissue could be identified.

When these differential spots have been located, the specific proteins within these spots need to be identified. This is usually done by cutting the spot out of the gel and subjecting it to a further analytical step such as mass spectrometry.

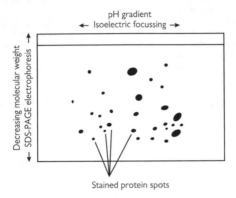

Fig. 15.6 Two-dimensional gel electrophoresis.

Mass spectrometry

A protein has a typical precise mass (if allowance for processes such as glycosylation is made), but it is possible that more than one protein has the same mass (or the same mass within the resolution of the measuring system). However, if a protein is cleaved using a specific enzyme, such as trypsin, then the resultant peptide fragments will have a specific range of molecular weights that can be used to precisely identify the whole protein by mass spectrometry—a peptide 'fingerprint'.

The rather lengthily named 'matrix-assisted laser desorption ionization-time-of-flight spectrometry' (MALDI-TOF) is the most common method:

- The enzyme-digested peptide mix is dried onto a slide (ceramic or metal)
- A laser heats the mixture causing formation of an ionized gas
- The ionized particles are accelerated in an electric field towards a detector
- The time taken for a peptide to reach the detector is a function of its mass and charge
- The data from the detector is matched with information on protein databases to give definitive identification of the protein (a task requiring considerable computing resources)

Recently, liquid chromatography with tandem mass spectrometry (LC-MS/MS) has become the method of choice for proteomic analysis. Furthermore, other '-omic' specialties have evolved:

- Metabolomics: attempts to measure all metabolites
- Lipidomics: attempts to measure all lipids (subdiscipline of metabolomics).

While measurement of every one of a particular species in a sample is un-realistic, these techniques are gaining prominence for establishing molecular 'fingerprints' in complex disease states.

Mass spectrometry is a mainstay of these disciplines, but Fourier transform infrared and nuclear magnetic resonance spectrometry are also commonly used.

Structural proteomics: analysing protein structure and function

Protein structure

X-ray crystallography

- 'Gold standard' for protein structure
- X-rays are diffracted by atoms in the protein:
  - In a well-ordered crystal, the scattered waves reinforce each other at certain points to give spots
  - This can be converted into an electron density map and then a 3D atomic structure by computer (e.g. Fig. 1.5; → p.13)
- At highest resolution, can identify the position of every atom in a molecule, except hydrogen
- The main problem is getting good-quality crystals of pure protein:
  - Membrane proteins have proved especially difficult.

Nuclear magnetic resonance

- Does not require large amounts of protein to be crystallized:
  - Useful for hard-to-crystallize proteins
  - As protein is in solution, can visualize structure changes (e.g. on binding substrate)
- A small volume of solution is placed in a strong magnetic field:
  - This causes nuclei with magnetic moment (spin) to line up with the magnetic field (e.g. protons)
  - Pulses of radiofrequency electromagnetic radiation excite the nuclei and misalign them
  - As they realign, they release radiofrequency electromagnetic radiation. This radiation release is affected by adjacent atoms
  - By knowing the protein sequence, it is possible to compute a 3D structure from these data of which nuclei are close to each other.

Protein function

Fluorescence resonance energy transfer (FRET)

- This is a specialist application of confocal microscopy (→ p.976)
- Proteins believed to be interacting are labelled with different colour fluorescent proteins (e.g. GFP (green) and CFP (cyan))
- The proteins are co-expressed in the same cell
- Excitation of the CFP fluorochrome usually results in cyan light out; however, if the proteins are close enough (1–10nm), then the GFP will be excited by the CFP emission and green light will be seen instead (Fig. 15.7). This energy transfer is called FRET.

Affinity chromatography/co-immunoprecipitation

- The protein of interest is covalently bound to a column matrix
- Cell extract is run through this column and interacting protein(s) non-covalently stick to the immobilized protein
- Any proteins adhering to the protein of interest can be eluted and identified using mass spectrometry
- Co-immunoprecipitation is similar, except that an antibody to one of the proteins is used to precipitate out the protein complex

• This can then be probed with antibodies to candidate associating protein(s):
  • Method requires proteins to be associated together tightly enough for the complexes to survive co-immunoprecipitation.

Yeast two-hybrid assay

• 'Bait' protein DNA is fused to DNA-binding domain of a gene activator protein
• 'Prey' protein DNA is fused to transcriptional activation domain— usually make from a cDNA library
• If bait and prey associate, then the gene activator and transcription activation domains will interact with each other and turn on a reporter gene (Fig. 15.8)
• From this, the positive prey protein DNA can be selected and sequenced
• Reverse two-hybrid assay can also be used to look for mutations in DNA or chemicals that disrupt a proven association through the two-hybrid system.

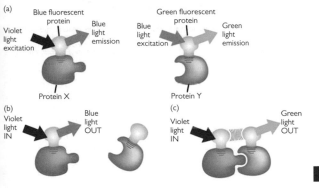

Fig. 15.7 Fluorescence resonance energy transfer (FRET). To determine whether (and when) two proteins interact inside the cell, the proteins are first produced as fusion proteins attached to different variants of GFP: (a) protein X is coupled to a blue fluorescent protein; (b) if protein X and Y do not interact, illuminating the sample with violet light yields fluorescence from the blue fluorescent protein only; (c) when protein X and Y interact, FRET can now occur.

Reproduced with permission from Alberts B et al. (2002). Molecular Biology of the Cell, 4th edn. Garland Science, Taylor & Francis LLC.

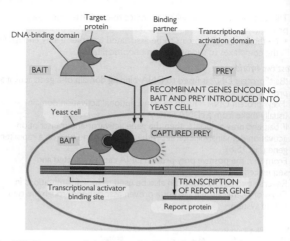

Fig. 15.8 The yeast two-hybrid system for detecting protein–protein interactions (see text).

Reproduced with permission from Alberts B et al. (2002). *Molecular Biology of the Cell*, 4th edn. Garland Science, Taylor & Francis LLC.

Microscopy

Light microscopy

- Involves either passing visible light through, or reflecting visible light off, a sample and collecting it with one or more lenses to give a magnified image:
  - The image can be seen by eye, or recorded photographically (conventionally or digitally)
- Light microscopy is limited to a resolution of about $0.2\mu M$, and the sample needs to be dark or strongly refracting to get a good image
- Light from out of the focal plane also limits resolution
- Native cells and tissues often lack the contrast to be studied by light microscopy without staining (➔ pp.982–83):
  - Staining is with dyes on fixed (i.e. dead) tissue
  - The tissue processing and staining can produce artefacts
- At least some of these limitations can be overcome by more advanced optical techniques which non-invasively increase the contrast of the image, e.g. phase contrast microscopy.

Fluorescence microscopy

- A variant on light microscopy involves using high-energy light (often lasers) to illuminate the sample
- Fluorescent compounds absorb energy at one wavelength and emit it at another (lower) frequency
- Optical filters allow imaging of just the emitted fluorescent signal
- Fluorescence can come from within the native sample (autofluorescence) or more normally from:
  - Labelling with a chromophore-conjugated antibody, e.g. with FTIC or fluorescein (green) and TRITC or rhodamine (red)
  - Genetically fusing a fluorescent protein to the protein of interest:
    - The best known of these proteins is GFP
    - Small genetic variations in GFP had produced other colours, e.g. blue (BFP), cyan (CFP), yellow (YFP).

Confocal microscopy

- An enhancement on fluorescent microscopy is confocal microscopy
- Traditional light and fluorescent microscopy resolution are limited by light from out of the focal plane
- Confocal microscopy allows the production of an optical section through a sample by using a pinhole to exclude light that is out of focus (Fig. 15.9):
  - A two-dimensional picture can be built up by scanning the sample, producing an 'optical section'
  - A 3D image can also be produced by focusing and scanning at different levels in the sample and using computerized image processing to 'stack' the optical sections.

Electron microscopy

- Uses a beam of electrons rather than light to illuminate the specimen—the shorter wavelength allows higher resolution (at least 100-fold higher magnification than for light microscopes)

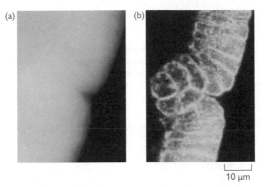

10 μm

Fig. 15.9 Two images of the same intact gastrula-stage *Drosophila* embryo, stained with a fluorescent probe for actin filaments: (a) conventional, unprocessed image is blurred by presence of fluorescent structures above and below the plane of focus; (b) this out-of-focus information is removed in confocal image.

Reproduced with permission from Alberts B et al. (2002). *Molecular Biology of the Cell*, 4th edn. Garland Science, Taylor & Francis LLC.

- Electrostatic and electromagnetic 'lenses' control and focus the electron beam to produce a focused image:
  - Images can be made on film or by specialized digital cameras
  - As specimen is in a vacuum, can only use fixed tissue
- There are different kinds of electron microscopy, the most commonly used in biology are:
  - Transmission electron microscopy (TEM)—electrons pass though ultra-thin sections of sample, and the more electron dense the structure, the darker it appears in the image (Fig. 15.10)
  - Scanning electron microscopy (SEM)—the sample is scanned by a focused electron beam and the reflected energy is measured, allowing large specimens to be imaged with such a large depth of field that the 3D shape is well represented
  - Samples need to be stained to increase electron absorbance (contrast):
    - Heavy metals are used, such as osmium, lead, palladium, tungsten, and uranium
    - For SEM, the sample is coated with stain, such as gold, palladium, or tungsten
- TEM can be combined with immunohistochemical techniques (➔ p.984) with antibodies conjugated to small (e.g. 10nm) gold particles (Fig. 15.10).
- Some biomolecules are not compatible with the harsh conditions of TEM (high vacuum conditions, intense electron beams) and are destroyed. Cryo-electron microscopy (cryo-EM) uses frozen samples and gentler electron beams, followed by sophisticated image processing, to allow high-resolution images to be generated:
  - This technique won the Nobel Prize for Chemistry in 2017

- Has been useful for protein structures, especially for those which do not form ordered crystals so cannot be studied by X-ray crystallography (➔ p.972)
- For serial block face SEM, an integral microtome cuts through the sample block while taking sequential SEM images. These images can then be vertically stacked and analysed by computer to give a 3D image of the structure:
  - False colour can be added to allow clear visualization of particular structures within the cell (Fig. 15.11).

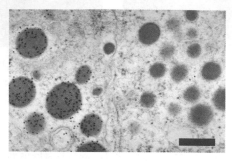

Fig. 15.10 Immunogold labelling for growth hormone in secretory granules in the anterior pituitary. The cell on the right does not contain growth hormone. (Scale bar = 400nm.)

Image courtesy of Helen Christian.

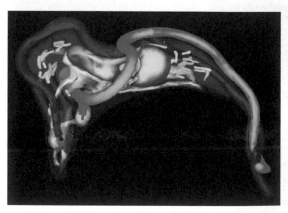

Fig. 15.11 A 3D image of a post-mitotic trypanosome about to undergo cytokinesis (cell division), reconstructed from sequential images taken on a serial block face SEM. False colour has been added to highlight particular structures of interest, including the original and new flagellum (green and purple, respectively), two nuclei (darker blue), and two kinetoplasts, one of each of which will go to each daughter cell on division.

Image courtesy of Laura Smithson and Sue Vaughan, Department of Biological and Medical Sciences, Oxford Brookes University.

Cytology

Definition

The examination of isolated cells rather than cohesive tissue histology (➔ p.982).

Methods of cytological sampling

- Exfoliative cytology—uses cells that are shed or scraped from body surfaces (e.g. bronchial epithelial cells that are coughed up in sputum, epithelial cells from the uterine cervix removed using a plastic brush)
- Aspiration cytology—uses cells which have been aspirated from a solid organ using a fine gauge needle and syringe (e.g. cells from a breast lump).

Uses

- Diagnosis of malignancy—e.g. detection of bronchial carcinoma by exfoliated cancer cells in sputum
- Screening for malignancy—e.g. detection of cervical intra-epithelial neoplasia (CIN) in cells scraped from the cervix in the UK national cervical screening programme
- Detection of microorganisms—e.g. *Trichomonas vaginalis* in cervical smears, *Pneumocystis jirovecii (carinii)* in bronchoalveolar lavage from immunocompromised patients.

Technique

- Smearing of cells directly onto glass slides or cytocentrifuging from a transport medium
- Brief fixation of cells using an alcohol
- Staining of cells by a histochemical method such as Papanicolaou
- Interpretation by a cytopathologist or cervical screener
- Issuing of written report.

Histology

Definition

Literally, examination of tissue (from the Greek *histos* = tissue). However, in current usage, it refers to the examination of tissue by light microscopy after it has been processed into thin sections. Although the histological appearances of tissue are altered by many artefact-inducing processes, these artefacts are reproducible, and there is a huge body of knowledge about the histological appearances of disease, which has built up over the past 150 years.

Uses

- Tumour diagnosis—histology is the primary method of definitive tumour diagnosis in current medical practice. Although imaging (such as computed tomography or magnetic resonance imaging) may provide very clear views of definite masses which are presumed to be tumour (e.g. multiple masses in the liver which are most likely to be metastases), there are some non-neoplastic processes which can produce masses on imaging (e.g. abscesses), so a definite tissue diagnosis by histology is still required
- Tumour staging—the extent of tumour spread (e.g. metastasis to local lymph nodes) is an important determinant of an individual patient's prognosis. Although modern imaging is, again, very valuable in this process, histology is still the main method of assessing tumour stage in surgically resected tumours (e.g. colorectal cancer)
- Non-neoplastic diagnosis—the histological appearances of many non-neoplastic diseases are reproducible and specific to a particular diagnosis. Inflammatory conditions of the skin are a good example of a spectrum of diseases which can be diagnosed by histology
- Assessment of the body's response to disease—inflammatory and fibrotic processes are easily assessed by histology, and so histology may play an important role in determining the extent of a non-neoplastic disease. An excellent example is hepatitis C—histological examination of a liver biopsy can determine the amount of inflammatory activity (and therefore the predicted response to treatment such as interferon therapy) and also the amount of damage that has already been caused to the liver (by assessing the amount of fibrosis from none through to frank cirrhosis)
- Detection of microorganisms—although microbiological culture is generally more sensitive than histology, there are some circumstances where histology is an effective technique for detecting microorganisms (e.g. *Helicobacter pylori* in endoscopic gastric biopsies).

Technique

Histology is a relatively simple technology that has changed little over the decades. The most significant advance has been the use of specific antibody stains (➔ p.984). The basic steps are:

- Fixation of the fresh tissue by a chemical solution (usually formaldehyde) which cross-links proteins
- Selection of a sample of the tissue if it large; small biopsies are examined in their entirety

- Processing of the tissue from the formaldehyde solution through progressive dehydration by alcohol into 100% liquid alcohol
- Embedding of the tissue into paraffin wax
- Cutting of thin (5–10μm thick) sections from the wax block
- Mounting of the wax sections onto a glass slide
- Staining of the section on the glass slide, usually by haematoxylin and eosin dyes which stain nuclei, dark purple, and cytoplasm, pink
- Interpretation of the sections by a histopathologist (a medically qualified doctor with postgraduate training in histology)
- Issue of a written report of the histological findings.

Although relatively simple, the tissue processing and staining is labour intensive and takes 1–2 days to complete.

Frozen sections

If an immediate histological diagnosis is required (e.g. an unexpected finding of disseminated tumour at an exploratory laparotomy), then a small tissue sample can be frozen in liquid nitrogen and sections cut and stained from this within a few minutes.

Immunohistochemistry

Immunohistochemistry allows the identification of proteins in cells and tissue samples (or thin sections of tissue), using specific antibodies (Figs 15.9–15.12). Once the tissue has been incubated with the primary antibody (i.e. the one raised to the protein of interest), the antibody binding pattern can be visualized in a number of ways, usually involving the use of a secondary antibody (2°Ab) that recognizes the primary one:

- The 2°Ab can be conjugated to an enzyme such as horseradish peroxidase, which catalyses a colorimetric reaction to stain the area(s) of the cell/tissue where the protein is present (Fig. 15.12). This is a relatively old-fashioned approach, largely replaced by fluorescent staining (Fig. 15.9; p.977), as described further on ➔ p.976:
  - The staining is visualized by microscopy (either light or fluorescent/confocal)
- Alternatively, the 2°Ab can be conjugated to small (e.g. 10nm) particles of gold, which can be localized by electron microscopy (Fig. 15.10, ➔ p.978).

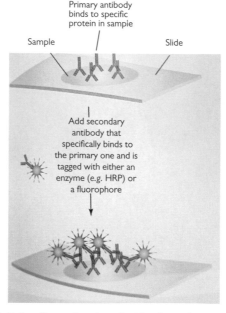

Fig. 15.12 Outline of immunohistochemical staining of a sample, using either horseradish peroxidase (HRP)-labelled or fluorescently labelled secondary antibodies.

Adapted from fig 13.19 Ahmed et al Biomedical Science Practice 2ed OUP, p313.

Fluorescent immunohistochemistry uses a similar approach, except instead of being conjugated to an enzyme, the 2°Ab is conjugated to a fluorescent chromophore.

- Usual chromophores include FITC (fluorescein) or Cy3 (green), and TRITC (rhodamine) or Cy5 (red)
- Dual labelling of two proteins allows identification of areas of co-expression (red + green = yellow staining).

Much higher levels of detail can be imaged using an electron microscope (at least 100-fold higher resolution than with a light microscope).

- In this case, the 2°Ab can be conjugated to a small particle of gold, which is electron dense and therefore will appear black
- Use of different size particles (e.g. 5nm and 10nm) allows more than one protein to be localized.

Although not strictly immunohistochemistry, one modern approach has been to genetically engineer proteins that are endogenously fluorescent. This can be done by adding the coding sequence for the ~30kDa protein, GFP, originally isolated from the jellyfish *Aequorea Victoria*.

- Experimental mutations in the gene for GFP has allowed the development of an enhanced GFP (EGFP) and other colours, e.g. cyan (CFP) and yellow (YFP)
- The use of these GFP-fusion proteins has allowed imaging of live samples.

Diagnostic uses

Immunohistochemistry has a very wide application in diagnostic histopathology to determine the precise nature of a tumour. This, in turn, allows the optimal therapy to be planned for the patient. Increasingly, individualized tumour therapies are being developed which rely on a specific protein being present in a tumour—the therapy itself is often an antibody, so the tumour of an individual patient will require immunohistochemical staining to identify the presence or absence of this protein.

Common clinical uses of immunohistochemistry include:

- Oestrogen receptor assay in breast cancer: all breast cancers are immunohistochemically stained for the oestrogen receptor protein and if a tumour does contain oestrogen receptor ('ER positive'), then the patient will be treated with the anti-oestrogen drug, tamoxifen, or one of its derivatives
- HER2 receptor assay in breast cancer:
  - If a breast cancer overexpresses the epidermal growth factor receptor 2 (HER2) protein, then it may respond to the drug trastuzumab (Herceptin®), which is a monoclonal antibody directed against this receptor. Again, immunohistochemistry is required to determine the presence of this protein
  - In the UK, at present, Herceptin® is used in the palliative treatment of metastatic breast cancer
- Typing of lymphomas: lymphomas (malignant tumours of lymphoid cells) are a heterogeneous group of tumours with widely differing prognoses. Most lymphocytes have a relatively similar appearance by conventional light microscopy, so immunohistochemistry is widely used to detect specific cell surface proteins which allows classification of the tumour into a specific lymphoma type.

Microbiological stains

Microorganisms can be viewed in samples by microscopy after treatment with special stains.

- Gram stain. Allows detection of Gram-positive and -negative bacteria, cocci, and rods. Also stains yeast. Some bacteria stain poorly
- Ziehl–Neelsen stain. Used to detect *Mycobacterium tuberculosis* and related mycobacterial species that are acid-fast and stain poorly with the Gram stain. Can be modified to detect *Nocardia* spp. Other stains such as fluorochromes (e.g. auramine-rhodamine) are more sensitive for screening specimens for *Mycobacteria* spp.
- Potassium hydroxide (KOH). Used on wet mounts to detect fungi. Calcofluor white staining and fluorescence microscopy is more sensitive. In biopsy specimens, silver stains such as periodic acid–Schiff are used to detect fungi
- Direct fluorescent antibodies (DFA). A monoclonal antibody specific to a microorganism may be tagged with a fluorochrome and used to detect the microorganism in a specimen
- Giemsa stain. Used on blood films to detect parasites (malaria in particular) and, occasionally, other intracellular pathogens
- Electron microscopy. Detection of viruses in certain samples
- Direct microscopy on unstained samples. Used on unicellular parasites and the larger ectoparasites
- Histologic specimens. Special stains and pathological features (e.g. granulomata) aid diagnosis.

Antigen detection

Available for specific microorganisms and samples (e.g. CSF). Detect antigen by latex agglutination (aggregation of particles) or enzyme immuno-assay (release of light). DFA and fluorescence microscopy is also a rapid means of antigen detection.

Culture

Bacteria

- *Media.* Cultured on solid or liquid media that can be enriched to optimize growth of specific bacteria, selective to allow only growth of certain bacteria or to contain an indicator so that bacteria with a certain characteristic (e.g. the ability to ferment a specific sugar) are detected:
  - Blood agar is an enriched all-purpose media
  - MacConkey agar is selective for common Gram-negative bacteria and contains an indicator so that lactose fermenters appear red
  - Broth cultures allow inoculation of greater volumes and are more sensitive
- *Conditions.* Automation allows early detection of bacterial growth by a characteristic such as microbial production of carbon dioxide. Growth conditions (temperature and atmospheric conditions) are adjusted depending on the requirements of specific microorganisms. Anaerobes require culture without oxygen
- *Identification.* Bacteria are speciated by colony morphology on solid media (including haemolysis pattern on blood agar), Gram stain, biochemical tests, serology, motility pattern, flagella, and spores. Multiple tests require computer-based programs for precise identification.

Mycobacteria

Grow on specific solid (e.g. Lowenstein–Jensen) and liquid media. Identification is by growth rate, colony morphology, biochemistry, and the use of molecular techniques.

Fungi

Grow on specific agar and liquid media (e.g. brain–heart infusion). Identification includes detecting yeasts or hyphal elements, the pattern of sporing bodies, and biochemical tests. Culture of mycobacteria and fungi is slow (often requiring weeks) and dangerous (necessitating a biosafety level 3 laboratory).

Viruses

Cultured in cell lines and identified by the pattern of cytopathic effect or reaction with specific fluorescent antibody.

Serology

Used for detection of viruses, parasites, fungi, and fastidious (difficult to culture) bacteria.

Requires identification of specific IgM or a fourfold increase in IgG antibody titre. Antibody is detected by using a secondary antibody with a tag that allows detection of bound antibody–antigen complexes. Detection may be by agglutination, precipitation, complement fixation, or radioactivity, but the majority of commercial assays now use indirect fluorescence or enzyme-linked immunosorbent assay (ELISA) (Fig. 15.13; ➲ p.989). ELISA can be adapted to detect IgG or IgM antibody or antigen.

Western blotting allows microbial proteins to be separated by SDS-PAGE (➲ p.953) and transferred (usually using an electrical current) to a nitrocellulose membrane. Reaction of patient sera with the proteins is detected by a specific pattern of bands when treated with a secondary antibody.

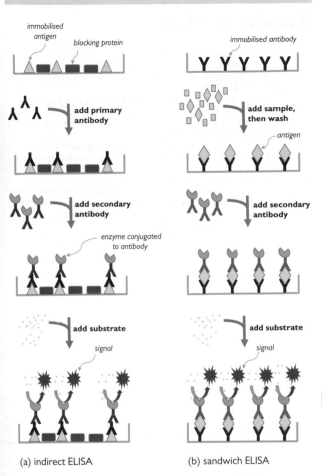

(a) indirect ELISA   (b) sandwich ELISA

Fig. 15.13 Principles of enzyme-linked immunosorbent assay (ELISA).
a) indirect ELISA: the antigen is bound to the wells of the test plate and blocking
protein is added to stop non-specific antibody binding to the well. Antibody is added,
and binds to the antigen. A secondary antibody, with an attached enzyme, is added
which binds to the primary antibody. Addition of enzyme substrate results in the
formation of a signal, e.g. a coloured product or light emission. b) sandwich ELISA:
the primary antibody is bound to the wells of the test plate, and sample added. The
antigen binds and the non-binding sample is washed off. A second antibody to the
antigen, with an attached enzyme, is added and a signal is generated on addition of
substrate (as above).

Molecular techniques

- PCR (➔ p.954). Allows detection of a variety of micro-organisms. Microbial DNA is amplified using primers (short sequences of DNA) that bind to a specific sequence of microbial DNA. Analysis can be qualitative or quantitative (qPCR; ➔ p.964). RNA can also be amplified using RT-PCR (➔ p.964)
- Molecular probes. Chemiluminescent DNA probes target sequences of ribosomal RNA and aid speciation of *Mycobacterial* spp. in culture. *In situ* hybridization allows the detection of nucleic acid derived from microorganisms in biopsy specimens
- Restriction patterns. Microbial DNA is cleaved using endonucleases (cleave at specific sites). The pattern of fragments on a gel provides a molecular fingerprint, useful in epidemiological studies
- 16S ribosomal RNA (16S rRNA) sequencing. Useful when bacteria cannot be cultured or when identification is equivocal with phenotypic tests such as biochemical testing. Hypervariable regions of 16S rRNA allow speciation. Can be combined with PCR when bacteria not cultured as in culture-negative bacteria
- Proteomics. Identification of the entire proteome expressed by a microorganism can enable speciation.

Resistance testing

- Disc diffusion susceptibility testing. An antimicrobial agent diffuses from an impregnated disc onto an agar plate containing bacteria or fungi (Fig. 15.14a)
- MIC testing. Microorganisms are grown in a liquid media with a known concentration of antimicrobial agent. Can be adapted to mycobacteria, fungi, and viruses, to solid media, or to use with a strip impregnated with drug (E-test; Fig. 15.14b)
- Molecular techniques. Detection of genes associated with resistance (e.g. the *mecA* gene for meticillin resistance in *Staphylococcus aureus*). Resistance mutations in mycobacteria or viruses are detected by PCR (genotypic assays) and are compared with MIC estimates (phenotypic assays).

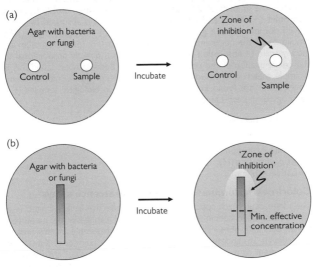

Fig. 15.14 (a) Disc diffusion susceptibility test. An impregnated disc is placed on an agar plate containing bacteria or fungi. When there is antimicrobial (fungicide) agent impregnated in the disc (sample) it clears a 'zone of inhibition' in the bacteria (fungi), whereas the control does not. (b) E-test form of minimum inhibitory concentration (MIC) testing. A strip impregnated with a gradient of antimicrobial (fungicide) agent (highest concentration at top) clears a 'zone of inhibition' in the bacteria (fungi), with it possible to estimate the minimum effective concentration.

Photometry and spectrophotometry

Flame photometry

Flame photometry has long been used to detect metal ions (Na^+, K^+, Mg^{2+}, and Ca^{2+}). The principle underlying the technique is that the different metal ions burn with different coloured flames. Thus, a solution containing such ions can be sprayed through a flame, and the change in flame colour detected by a spectrophotometer. This technique used to be routinely used to establish Na^+ and K^+ levels in the blood, although nowadays, ion-selective electrodes are often used for this purpose.

Spectrophotometry

The colour that a substance appears to our eye is determined by those wavelengths of white light that are reflected by it—all the other wavelengths are absorbed. A spectrophotometer is an instrument that determines the amount of light of different wavelengths—usually between ultraviolet (~200nm) and infrared (~600nm)—that is absorbed by a dissolved compound of a given depth (usually 1cm—the pathlength). The amplitude of the peaks in the 'absorption spectrum' (measured in an arbitrary scale of absorption units from 0.0–3.0) is directly proportional to the concentration of the compound present (Beer–Lambert law). This is a simple means of measuring the concentration of highly coloured compounds that are found in high concentrations in biological samples (e.g. haemoglobin). Beer–Lambert law:

$$\text{Absorbance} = \varepsilon cl$$

where:
ε = molar extinction coefficient ($M^{-1}\,cm^{-1}$)
c = concentration (M)
l = pathlength (cm).

Colorimetric, luminescence, and fluorescence assays

The range over which spectrophotometry is effective is very limited (e.g. for haemoglobin, concentrations of 0.5–5μM can easily be measured; higher concentrations require dilution to be measurable by this means; lower levels are undetectable). Unfortunately, the majority of substances that we want to measure are not coloured (although most absorb in the ultraviolet spectrum) or do not lend themselves to direct spectrophotometric analysis. However, a number of colorimetric, luminescence, and fluorescence assays have been developed for specific compounds. These assays rely on highly specific reactions of the substance of interest with an added reagent. The reaction gives rise to a highly coloured, luminescent, or fluorescent compound that can be detected in a spectrophotometer, luminometer, or fluorometer. Many of these assays are now available in kit form for use with 96-well plates that can be read in a plate-reader for very rapid throughput of large numbers of samples.

Radioimmunoassay, ELISA, chromatography

Radioimmunoassay (RIA)

RIA is often the technique of choice for determining the concentration of specific peptides and proteins in blood or urine samples. Kits are now available for a huge range of peptides and proteins. The principle involves introducing the sample into a tube that is coated with a known amount of the specific antibody for the antigen of interest.

- An excess of radio-labelled antigen is then added to the tube, which will bind to any antibody that is still available
- The tube is then washed out, leaving the antibody bound to a mixture of unlabelled ligand from the sample and radio-labelled ligand
- The amount of the antigen in the sample is inversely proportional to the amount of radioisotope present, as determined by measuring the radioactivity in a γ-counter and comparing the reading to those of standards of known amount. It is important to note that the standard curve for RIAs is not linear—it is usually fitted with a cubic spline curve.

Enzyme-linked immunosorbent assay

ELISA (⊃ p.989, Fig. 15.13) is a popular means of detecting peptides and proteins that does not require the use of radioisotopes. ELISAs are generally purchased in kits consisting of pre-formed 96-well plates coated with an antibody specific to the antigen of interest.

- After addition of sample to the wells, a secondary enzyme-linked antibody is added to the wells, which binds to the antibody–antigen complex
- Addition of the colourless substrate for the enzyme that is linked to the antibody results in the formation of a coloured product, the amount of which is measured spectrophotometrically in a plate-reader and is proportional to the amount of enzyme bound and, consequently, the amount of antigen present.

Chromatography

Chromatography works on the principle that substances can be separated according to their specific chemical characteristics (e.g. charge, molecular size, and/or partition in aqueous/organic solvents). In this way, we can separate and measure the concentrations of the components of very complex mixtures, including blood plasma, tissue, and cell extracts.

- Chromatography usually utilizes a 'mobile phase'—a solvent in the case of liquid chromatography (high-performance liquid chromatography; HPLC) and an inert gas for gas chromatography (GC)—that passes through or over a 'stationary phase' in a column
- The greater the interaction of the substance of interest with the stationary phase, the longer it will take to pass through the column, giving it a longer 'retention time'
- The effluent from the column is passed through a detector, which can either measure the absorbance in the ultraviolet region of the spectrum or by electrochemical means. Compounds appear as 'peaks' on a chart—the higher the peak, the more of the compound

is present. The precise amount present can be analysed by comparing the peak height or area to those of standards of known concentration. Sometimes, the specificity of HPLC can be improved by exposing the components of the mixture to a reagent that will fluoresce when it reacts with a particular target molecule. The amount of the fluorescent product can be detected downstream of the column using a fluorescence detector

• A recent refinement to chromatography techniques is their combination with mass spectrometers that can be used to determine the molecular mass of the substance after separation by gas chromatography (GC/MS) or HPLC (LC/MS). Identification of unknown substances requires tandem mass spectrometry, whereby the mass of the intact molecule of interest is detected in the first mass spectrometer prior to shattering of the molecule and identification of the daughter ions generated. Comparison of these features with online databases usually narrows the field to one or two possible candidates.

Flow cytometry and fluorescence-activated cell sorting (FACS™)

- Flow cytometry is a technique for analysing and quantifying small particles, such as cells, which are suspended in a stream of liquid
- A laser is shone through the stream of liquid, and detectors measure the forward and side scattered light and fluorescence (if required):
  - The forward scattered light signal allows determination of the cell volume
  - The side scattered light allows interpretation of intracellular features (e.g. shape of the nucleus, amount of intracellular organelles)
  - Cells with a fluorescent tag (e.g. labelled with a chromophore-conjugated antibody or expressing a GFP-tagged protein) can be counted (quantified)
- Fluorescence-activated cell sorting (often known as FACS™, an acronym trademarked by Becton, Dickinson, and Company, USA) is a specialization of flow cytometry:
  - The parameters of the liquid stream are adjusted so that the cells are relatively widely spaced within the stream
  - A vibrating mechanism breaks the fluid stream into droplets that (statistically) contain only a single cell
  - A fluorescent detector identifies fluorescent-positive cells just before the stream breaks into droplets, and these droplets are given an electrical charge
  - Charged droplets are diverted by an electrostatic deflection system, thus separating fluorescent cells from non-fluorescent ones.

In vitro, ex vivo, in vivo

In order to examine a physiological or pathological process, or to determine the characteristics of drug action, it is necessary to work with live cell cultures or tissue samples. A range of different techniques is available to help elucidate biological processes.

In vitro

Literally, this term means 'in glass' and can equally apply to experiments with cell cultures, tissue homogenates, or functionally intact pieces of tissue. These are very useful techniques for providing clues as to the cellular mechanisms involved in physiological, pathological, or pharmacological processes, without the complications of the highly complex physiology of a whole organism. This simplicity is, however, also the major limitation of *in vitro* investigation: extrapolation of *in vitro* results to the *in vivo* situation is inherently dangerous on the basis that the complex integrated systems and metabolic processes of the whole body are likely to have a bearing on the processes involved *in vivo*. Table 15.1 summarizes the benefits and drawbacks of *in vitro* experiments.

Ex vivo

This term is often confused with *in vitro*. *Ex vivo* should only be applied to experiments that are conducted on cells or tissue that have been removed from an animal or human, after having been subjected to a drug treatment *in vivo*. Under these conditions, the therapeutic intervention is subject to all the metabolic processes that occur in a fully functional animal or human, but the end-point of the experiment is a specific test that is carried out on excised tissue.

In humans, these experiments are only possible when a realistic end-point can be achieved from easily obtainable tissue (e.g. skin biopsy, blood samples), but in animals, these experiments might be terminal (i.e. the animal is killed prior to tissue removal), allowing any tissue sample to be used to determine the drug effect (e.g. blood vessel, brain tissue, liver homogenate). The *in vivo* element of these experiments (human or animal) means that they require permission from the respective authorities.

In vivo: animal experiments

Clearly, where possible, the most accurate reflection of a drug effect on a physiological or pathological process will be obtained from experiments *in vivo*. However, this is a highly emotive issue because it usually involves vivisection in laboratory animals. In almost all cases of drug development, this is an essential step, prior to ethical approval for clinical studies, to highlight any unforeseen side effects or problems.

As with any experiments, *in vivo* studies should be designed to test specific hypotheses, with clear, achievable end-points. Ethical review panels have to be assured that any animal suffering will be minimized, or preferably avoided altogether, before permission will be granted; in UK, there are stringent regulations associated with the Scientific Procedures Act (1986), requiring licensing of both individuals involved in the research, as well as projects. Consideration must be given to the best species to use as a model for the human condition in question, and power calculations should be conducted to determine how many animals are likely to be needed in each group to ensure a reasonable chance of observing a statistically significant difference between placebo and treatment groups. Where possible, randomized, blinded, crossover studies should be considered, to improve the robustness of the results obtained.

- Mice are widely thought to be the best model organism for studying human disease due to their (patho)physiology being so similar to ours[2]:
  - Mouse genes can be mutated randomly (by exposure to radiation or chemicals, e.g. ENU)
    o This approach has been aided by advances in high throughput sequencing
- More sophisticated genetic approaches can be used to target particular genes, e.g. gene trapping and gene targeting (Fig. 15.15):
  o Gene trapping is random whereas gene targeting is specific for a gene of interest
  o This technology won its inventors the Nobel Prize in Physiology or Medicine in 2007[3]
- There are disease model mouse strains that have genetically engineered susceptibility to particular diseases such as diabetes, hypertension, and Huntington's disease
- One of the biggest challenges is to identify the phenotype of the mutant mice:
  o As there are a number of worldwide consortia undertaking large-scale studies, there needs to be consistency in phenotyping.

2 Rosenthal N, Brown S (2007). The mouse ascending: perspectives for human-disease models. *Nat Cell Biol* **9**, 993–9.

3 Available at: http://nobelprize.org/nobel_prizes/medicine/laureates/2007/

Table 15.1 Advantages and disadvantages of various *in vitro* techniques

	Advantages	Disadvantages
Cultured 'immortalized' cell lines	• Relatively cheap (purchase of only one sample of seed cells required for an indefinite number of culture generations)	• Immortalization will change the phenotype—results may not reflect the response of the 'mortal' cells *in vivo*
	• Rapid results	• Measures tend only to be of biochemical markers rather than the real physiological response
	• Human cell lines available	
	• Real-time changes in ion flux or morphology can be observed using fluorescent markers	
Primary cultures	• Human cells can be harvested (ethical approval required) or bought commercially	• Requires harvesting of cells from tissue
		• Bought cells are expensive
	• In some cases, functionality can be observed (e.g. isolated cardiac myocyte contraction)	• Cells may progressively change phenotype through progressive generations
	• Real-time changes in ion flux (e.g. Ca^{2+}) or morphology (e.g. cell cytoskeleton) can be observed using fluorescent markers	• A cell of a particular type might require interaction with different cell types to produce a normal physiological effect
Isolated tissues or organs	• Real-time functional responses can be observed easily (e.g. muscle contraction, blood vessel function) and correlated with changes in biochemical mediators	• Fluorescence imaging more difficult than with cell cultures
		• Human specimens are difficult to obtain, except skin biopsies or 'surgical debris' (e.g. discarded organs after transplant, excess tissue that has been biopsied for diagnostic purposes, tissues from amputated limbs)—experiments with human tissue require ethical approval
	• Whole organ function can be investigated (e.g. heart, kidney)	
	• Impact of drugs on function can be easily measured under equilibrium conditions (useful for determining efficacy and affinity)	

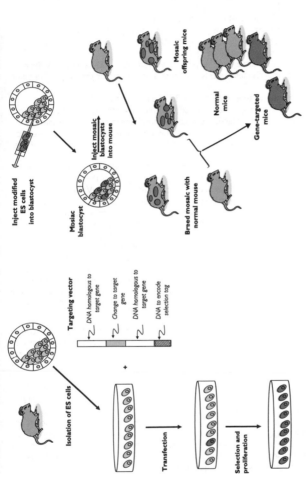

Fig. 15.15 Method of gene targeting in mice.

Clinical studies

Ethical approval should be sought at the earliest opportunity—evidence of *in vitro* and *in vivo* studies in animals is usually required, together with toxicology data. Clinical trials proceed in a standardized fashion:

• Phase I: a small study (20–80 subjects) to help evaluate the correct dosing range, the safety of the drug, and any side-effects. First-in-man studies are often carried out in healthy volunteers before trying them in patients from the target group

• Phase II: a larger study group is involved (100–300 patients). A clear end-point is identified, which must be achieved for the drug to be taken further

• Phase III: a large study (1000–3000 patients) to confirm the drug efficacy, perhaps in comparison to an existing therapy and usually also compared to a placebo group ('dummy' treatment with the same appearance as the study drug but without the active agent). Side-effects are closely monitored and information collected as to the safety of the drug prior to marketing

• Phase IV: post-marketing studies that help to determine the optimal use of the drug and to clarify the risks and benefits of the drug

• Expanded access protocols: clinical trials (phases I–IV) are necessarily conducted on a very restricted group of patients, minimizing any confounding factors that might compromise the findings of the studies. As a result, a large number of potential patients who might benefit from the drug on trial are excluded. In some cases, the pharmaceutical company might apply for 'expanded access' to allow patients who do not meet the strict criteria for the trials (e.g. with regard to age and gender, complex medical history) to gain access to the trial drug. This is usually only granted in cases where there is no credible alternative therapy, particularly if the disease is life-threatening. Furthermore, there should be no evidence from the trials that precede the application to suggest that there might be detrimental effects.

As with *in vivo* animal studies, great consideration must be given to minimize any potential pain or suffering of the subject in clinical trials, with the added caveat that the subject must be fully informed of the procedure and possible implications. Patients must sign a consent form to that effect. End-points must be clearly stated prior to the experiment and, unless the nature of the experiment dictates otherwise, the trial should be of a double-blind (both patient and researcher are unaware of whether the administered agent is placebo or test drug), randomized, crossover (some patients receive drug first, followed at a later date by placebo or vice versa) style. This is not always possible—e.g. it is sometimes necessary to run parallel groups, one of which receives placebo while the other receives drug treatment. In these cases, it is essential that the groups be matched as closely as possible (e.g. medical history, age, gender).

In very large trials, an interim analysis is usually carried out: if the drug is found to be detrimental, or indeed highly beneficial, the trial may be stopped on the grounds that it is unethical to continue.

Index

Note: Tables, figures, and boxes are indicated by an italic *t*, *f*, and *b* following the page number.

Important numbers (1)

Body fluid compartments (70kg person)

total body fluid	42L
extracellular	14L (plasma 3L; interstitial fluid 10L; transcellular fluid 1L)
intracellular	28L

Composition

	Extracellular	Intracellular
Na^+	140mM	15mM
K^+	4mM	140mM
Ca^{2+}	2.4mM	$0.1\mu M$
Cl^-	110mM	less than extracellular, but variable with tissue
HCO_3^-	25mM	10–20mM (according to intracellular pH)

Blood

plasma glucose (fasting)	4mM
plasma osmolarity	300mOsm
P_aO_2	100mm Hg (13.3kPa)
P_vO_2	40mm Hg (5.3kPa)
P_aCO_2	40mm Hg (5.3kPa)
P_vCO_2	46mm Hg (6.1kPa)
blood pH (arterial)	7.4
systemic arterial blood pressure	120/80mm Hg
pulmonary arterial blood pressure	15/5mm Hg
resting cardiac output	5L min^{-1}

Blood cells: normal counts

(cells L^{-1} differential count)

erythrocytes	5×10^{12} (haematocrit 45%)
leucocytes	$\sim 7 \times 10^9$ of which:
neutrophils	40–70%
lymphocytes	20–40%
monocytes	~6%
eosinophils	~3%
basophils	<1%